A FIELD
EAST
REPTILES

BY STEPHEN SPAWLS, KIM HOWELL,
HARALD HINKEL AND MICHELE MENEGON

Second Edition

BLOOMSBURY WILDLIFE
LONDON · OXFORD · NEW YORK · NEW DELHI · SYDNEY

Bloomsbury Wildlife
An imprint of Bloomsbury Publishing Plc
50 Bedford Square, London, WC1B 3DP
29 Earlsfort Terrace, Dublin 2, Ireland

BLOOMSBURY, BLOOMSBURY WILDLIFE and the Diana logo are trademarks of
Bloomsbury Publishing Plc

First published in the United Kingdom 2018

Text © 2018 by Stephen Spawls, Kim Howell, Harald Hinkel and Michele Menegon
Photos © 2018 by Stephen Spawls, Kim Howell, Harald Hinkel and Michele Menegon, and by photographers as credited in the photo captions.

Stephen Spawls, Kim Howell, Harald Hinkel and Michele Menegon have asserted their right under the Copyright, Designs and Patents Act, 1988, to be identified as Authors of this work.

All rights reserved. No part of this publication may be reproduced or transmitted in any form or by any means, electronic or mechanical, including photocopying, recording, or any information storage or retrieval system, without prior permission in writing from the publishers.

No responsibility for loss caused to any individual or organisation acting on or refraining from action as a result of the material in this publication can be accepted by Bloomsbury or the author.

A catalogue record for this book is available from the British Library.

Library of Congress Cataloguing-in-Publication data has been applied for.

ISBN: PB: 978-1-3994-0481-5; ePub: 978-1-4729-3562-5;
ePDF: 978-1-4729-4309-5

2 4 6 8 10 9 7 5 3 1

Design by Jocelyn Lucas

Printed and bound in India by Replika Press Pvt. Ltd.

To find out more about our authors and books visit www.bloomsbury.com and sign up for our newsletters.

Frontispiece: Micrelaps vaillanti Guineafowl Snake. Anthony Childs, Meru National Park

Pelusios broadleyi
Lake Turkana Hinged Terrapin
Roberto Sindaco

CONTENTS

INTRODUCTORY ESSAYS
INTRODUCTION 5
How to Use this book: The Text and the Maps 7
East African Reptiles and Their Zoogeography 9
Observing, Gathering Data and Collecting Reptiles 10
Reptile Conservation 12
Identifying Reptiles 14
Some Brief Notes on Reptile Taxonomy 16
What are Reptiles? 18
Safety and Reptiles 18
Acknowledgements 20
Further Reading, Literature and Resources 23

SPECIES ACCOUNTS
CHELONIANS 25
Family Testudinidae (Land Tortoises) 28
Family Cheloniidae (Sea Turtles) 40
Family Dermochelyidae (Leatherback Turtles) 45
Family Trionychidae (Soft-shelled Turtles) 47
Family Pelomedusidae (Side-necked Terrapins) 50

LIZARDS 65
Family Eublepharidae (Eublepharid Geckoes) 69
Family Sphaerodactylidae (Clawed Geckoes) 71
Family Gekkonidae (Typical Geckoes) 73
Family Scincidae (Skinks) 127

Family Lacertidae (Lacertids or 'Typical Lizards') 179
Family Cordylidae (Girdled Lizards and Their Relatives) 205
Family Gerrhosauridae (Plated Lizards and Their Relatives) 213
Family Agamidae (Agamas) 220
Family Chamaeleonidae (Chameleons) 242
Family Varanidae (Monitor Lizards) 309

WORM LIZARDS **315**
Family Amphisbaenidae (Worm Lizards) 315

CROCODILES **325**
Family Crocodylidae (Crocodiles) 327

SNAKES **335**
Family Typhlopidae (Blind Snakes) 341
Family Leptotyphlopidae (Worm or Thread Snakes) 363
Family Boidae (Boas) 375
Family Pythonidae (Pythons) 378
Family Lamprophiidae (African Snakes) 387
Family Colubridae (Colubrid Snakes) 478
Family Elapidae (Cobras, Mambas and Relatives) 542
Family Viperidae (Vipers) 567

APPENDICES
Glossary 598
Line Drawings 600
Index 613

Philothamnus punctatus Speckled Green Snake Michele Menegon

'The scientist does not study nature because it is useful, he studies it because he delights in it, and he delights in it because it is beautiful. If nature were not beautiful, it would not be worth knowing, and if nature were not worth knowing, life would not be worth living'
Henri Poincaré, scientist and philosopher, 1854–1912

'Good natural history is a source of timeless, priceless information for the biological sciences'
Harry W. Greene, American herpetologist, 1945–

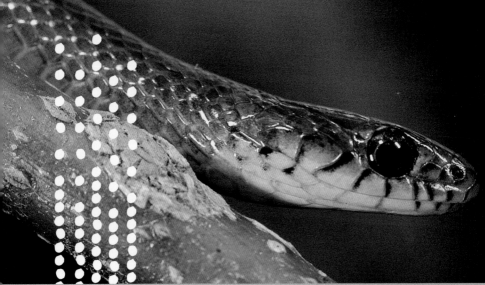

Natriciteres sylvatica
Forest Marsh Snake
Michele Menegon

INTRODUCTION

In 2002 we published a field guide to the reptiles of East Africa. Since that date there have been many changes in the East African herpetological field. New species have been found in East Africa and described. Existing species have been split, animals have been recorded in East Africa for the first time; range extensions for existing species have been recorded. As a result of systematic studies, particularly those based on DNA and other biological molecules, many species have been reclassified, regrouped and their scientific names changed; new relationships have been discovered.

Our original guide described 432 species of reptile from East Africa; the new list contains 492 species. Well over 100 species have had their generic names changed. It is time for an updating, to document that progress, and we hope that this book serves that purpose. In order to produce a portable soft-backed book we have reduced the size of some of the original introductory material, and also abbreviated many species accounts without (we hope) losing too much essential material. Those who have a copy of our original field guide will find that the running order has changed slightly, particularly with the snakes, to reflect the latest evolutionary relationships, although where there is no obvious hierarchy we have tried to retain something of the original order. There is some debate over the order of appearance of the various groups and taxa in the evolutionary record. The general consensus is that, in order of major splits, the turtles and tortoises are the oldest group, then crocodiles, then lizards, including amphisbaenians, and finally snakes. However, in this book we have followed what might be called a 'traditional' running order; with the turtles and tortoises first, then lizards, amphisbaenians, crocodiles, and finally snakes. Within the generic accounts, if species are unequivocally closely related we have grouped them in order, if the situation is not clear cut then they are alphabetical.

As we said in the first edition, this book is not the 'last word' in East African herpetolo-

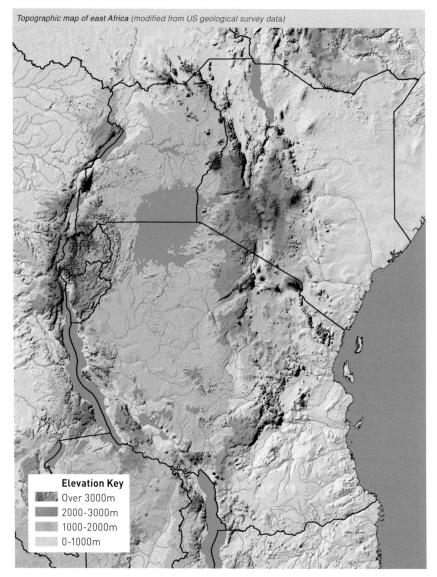

Topographic map of east Africa (modified from US geological survey data)

Elevation Key
Over 3000m
2000-3000m
1000-2000m
0-1000m

gy. There will never be a last word. There is a lot to be discovered. New discoveries, range extensions and taxonomic changes are always occurring. Biological data on the animals themselves is continuously being updated. Huge areas of East Africa have yet to be visited by herpetologists. And the field of African herpetology is desperately short of workers. Throughout this book, you will encounter such expressions as 'never knowingly seen alive', or 'never photographed alive', or 'details unknown'. There is much research to be done. Be part of it. The active enthusiast, be they scientist or amateur, can make a big difference. So get into the field. Observe, document, photograph and collect; source what is appropriately called the 'ground truth'. The distribution records of even the most common of our reptiles are patchy. Many of the range extensions in this new edition result from a single observation by an enthusiast in a herpetologically unexplored area. Even if you object to collecting live animals you may find specimens dead; preserve them or freeze them and take them to a museum. Get in contact with zoologists at local institutions; a list of these institutions is given at the end of this chapter in the section entitled 'Further reading, literature, resources and institutions'. The section also includes details of journals and magazines that publish both amateur and professional material, and the authors' contact details. Publish your data. As Jim Al-Khalili has said, "what is the point of learning new things about how the world works if you don't tell anyone about them?'

HOW TO USE THIS BOOK, THE TEXT AND THE MAPS

This book is meant for use in the laboratory, in the home and in the field. Put it in your field bag or your car. When you see a reptile, leaf through the book and try to match it with the picture. If it matches, then check the distribution map; is it in the right area? Then look at the text; we have tried to emphasis key points of identification. Check the maximum size. If you don't have this book, or the animal doesn't appear to be shown, take a photograph and make some notes. Particularly useful identification pointers are the colours and patterns of the animal, its location, its size and any behaviour it displays. With turtles and tortoises, the size, colour and shape of the shell is useful; as is the locality (in or near water, or not). With lizards the head and body shape

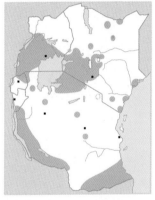

and any head ornamentation (spikes, horns, ear holes visible). With snakes, the colour, pattern, size, shape of the head, the type of pupil and the behaviour may give clues to its identity. However, remember to be cautious when approaching certain reptiles to check identification, particularly snakes; see the notes entitled 'Safety and reptiles' towards the end of this introduction.

The maps show the distribution of each species, as known at present; the yellow shading indicates where it is known to occur. A question mark indicates an unconfirmed record, for reasons usually explained in the text. The black squares are the capital cities of each

Lygodactylus picturatus Yellow-headed Dwarf gecko. Michele Menegan, Zanzibar.

country (not records), plus Dar es Salaam, and Tabora in Tanzania for orientation purposes. If you have identified a species but it seems to be well outside its known range, check the habitat; is it suitable? A forest chameleon will not be in semi-desert. However, reptile

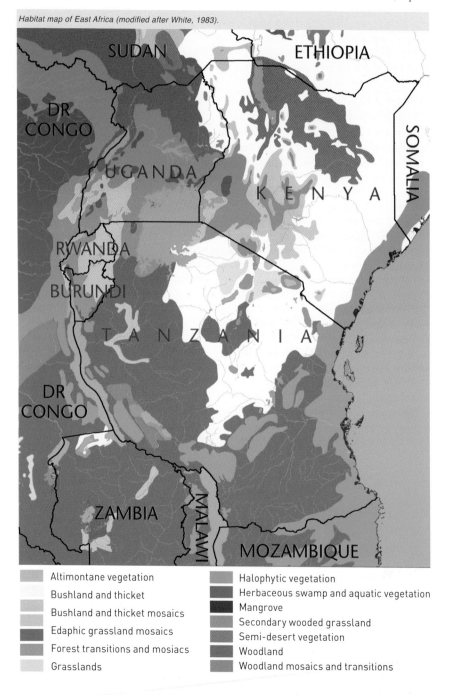

Habitat map of East Africa (modified after White, 1983).

- Altimontane vegetation
- Bushland and thicket
- Bushland and thicket mosaics
- Edaphic grassland mosaics
- Forest transitions and mosiacs
- Grasslands
- Halophytic vegetation
- Herbaceous swamp and aquatic vegetation
- Mangrove
- Secondary wooded grassland
- Semi-desert vegetation
- Woodland
- Woodland mosaics and transitions

distributions are poorly known in East Africa, and big range extensions can occur; if your record represents a range extension and you have a dead specimen or a good photograph, forward it to an expert or a museum (details at the end of this chapter) noting exactly where it was found.

EAST AFRICAN REPTILES AND THEIR ZOOGEOGRAPHY

The five countries of East Africa (Tanzania, Kenya, Uganda, Rwanda and Burundi) have an area of over 1.8 million square kilometres. It is an astonishingly varied environment. The altitude varies from sea level to snow-capped mountains just under 6000m; the vegetation includes true desert, semi-desert, savanna, woodland, alpine moorlands and a wide range of forests, from hot coastal forest through true rain forest to cold montane forest. East Africa is also riven by two gigantic, north-south rift valleys, which have created a landscape of block-faulted mountains, valley floors, escarpments, lakes and volcanic mountains. On the eastern side of our region there is a chain of ancient crystalline mountains, the Eastern Arc. The maps on p6 and p8 show variation in altitude and vegetation. Reptiles and amphibians are present throughout the East African landscape, being absent only from areas above the snowline. About 230 species of lizard and a similar number of snake species occur in East Africa, there are also 25 species of turtle and tortoise, three crocodiles and ten worm lizards, a diverse tropical fauna. In general, species numbers decrease with altitude; the higher the fewer, because all reptiles need external heat. At Watamu on the Kenyan coast, just above sea level, about 48 species of snake have been recorded, at Nairobi at 1,600m altitude, some 26 snake species are known, at Limuru (2,300m) only nine types of snake occur and at Kiandongoro on the Aberdare mountains at 2,900m only two species of snake are recorded. Rainfall is also significant, more rainfall equals more reptiles; at low altitude in dry areas there will be fewer species than in areas with higher rainfall.

Most of our East African reptiles are savanna animals and fall into three main groups: animals of the relatively moist savannas of south-eastern Africa (Zambezian savanna); animals of the drier savannas of west and central Africa (Sudan savanna); and species

Afrotyphlops lineolatus Lineolate Blind snake, Michele Menegon.

Python regius Royal Python, Barry Hughes.

from the dry country of north-east Africa, the Somali-Maasai fauna. There are also a few Palaearctic species, originating in the northern half of Africa, that reach East Africa. Our coastline is inhabited by some reptiles and amphibians typical of the East African coastal mosaic fauna, dwellers in the moist thicket and woodland that may once have extended almost continuously from the shores of north-eastern South Africa and Mozambique to southern Somalia. The animals of the great Guinea-Congolian forests of west and central Africa are represented by a number of forest species on the western side of East Africa, many of which reach their easternmost limits in the Albertine Rift forests of Uganda, Rwanda and Burundi, others extend to the forests of western Kenya. There are also a number of endemic reptile species found only within East Africa. Most of these unique animals are associated with hills, mountains and forests, in particular the volcanic hill forests of Kenya and the ancient East Arc mountains that extend from southern Kenya through Tanzania to Mlanje mountain in Malawi. An equally important endemic fauna occurs on the watershed mountains on both sides of the Albertine Rift on the western border of our region. This is why it is important that these regions are protected.

The present structure of the East African reptile and amphibian fauna has been largely shaped by three main factors. First, the geological events (formation of rift valleys, volcanic activity, watershed movement, river capture, flooding and erosion); second, the various climatic changes over the last 30,000 years and beyond (variation in rainfall, changes in lake levels, shrinkage and expansion of forests and deserts); and third, human activity.

OBSERVING, GATHERING DATA AND COLLECTING REPTILES

In general, reptiles are best left in the wild, that is where they live and they are an important part of the ecosystem. In addition, many areas in East Africa are protected, and animals there may not be collected. It is our hope that the users of this book enjoy watching reptiles in wild places, and their successful identification adds to the enjoyment. However, we are also aware that in certain circumstances collecting reptiles is justified. Consequently, we give here some advice on observing, recording, finding and collecting reptiles. The notes on collecting are aimed at those with appropriate permission from the relevant authorities.

A camera is very useful. The rise of the small versatile digital camera and the camera phone has made a huge difference to the gathering of field data. Many of us now take such handy portable devices into the field. With them, if you see a reptile, you can photograph it, and the image can be sent worldwide. More often than not, the species can be identified from the picture. A list of suitable fora is given at the end of this section, and as authors we are always pleased to receive such data; many range extensions and new records that appear in this book are the result of an enthusiast photographing a wild reptile and sending us the image. We don't intend to give detailed photographic advice here, but in photographing reptiles, the following pointers may be useful. First, get as close as safely possible to your subject (but be careful with snakes; unless you know your subject is harmless then stay at least twice the snake's length away from it). If light is low, use flash. Take a few shots, and keep the camera as steady as possible (important with small cameras and phone cameras). Make sure the head is clearly visible and in sharp focus. Activate the GPS function if you have one.

Weather is significant for fieldwork. In East Africa the broad climatic picture is of dry and wet seasons. The wet season is the best time for observing reptile (and amphibian) activity. The activity of these two vertebrate groups is greatly reduced during the dry season, although some lizards and diurnal snakes like sand snakes may still be active.

Most reptiles are secretive; they spend much of their time hidden. Those in the open, if approached, tend to freeze or move away unobtrusively. The casual observer in the wild may see a few, but a little expert hunting will increase your chances of seeing these creatures. There are various useful techniques you can use. Turning over ground cover such as rocks, logs and vegetation heaps will expose hiding animals, but take care where you put your fingers, and replace the cover afterwards. Lifting up rock flakes, breaking up fallen logs, and removing dead bark from trees may reveal animals, but bear in mind that this destroys hiding places (and may also expose you to a big dangerous snake!). Areas around rural housing or abandoned buildings are often prime reptile habitat. In forest areas, raking through leaf litter will often expose lizards and snakes. When walking in the bush, keep your eyes open. Move slowly, look around. You can also look for tracks in sandy areas and follow them back to where the animal is hiding; it may be buried, or in a hole. If you sit and watch a termite mound on a warm day animals may emerge to bask. In open areas lizards are often active. Rocky hills, especially those with sloping sheet rock areas and cracks, are usually excellent places to see lizards. In open country, isolated big trees are important refuges.

Water sources are good places to see reptiles. Frog-eating snakes may be in waterside vegetation. Birds and some mammals, especially squirrels, may give away the presence of a snake. If you hear birds or mammals mobbing something, it is worth investigating.

Looking for reptiles at night can be very rewarding, although the risks of any sort of night activity must always be carefully assessed in East Africa. You will need a good lamp or torch. At the beginning of the rainy season, amphibian activity increases dramatically, and this may attract snakes. Many types of snakes are active at night, as are most geckoes. Some reptiles, especially chameleons and green snakes show up clearly at night; they look pale against dark vegetation. However, apart from crocodiles, reptile eyes do not reflect light well, thus it is no good looking for reptilian eyeshine. In urban areas, walls and drains act as traps and snakes may be found crawling along them. Storm drains often trap animals. Geckoes may be hunting around lamps at night. Diurnal lizards may be sleeping on vertical rock faces. Reptiles moving at night often make a surprising amount of noise, especially in well-vegetated areas; once in a while turn off your lamp and listen. Bear in mind that if you want to look for reptiles at night, before dark you should check if there is a watchman or security personnel present. Never walk at night in areas you haven't surveyed during the day; you might get lost, or wander into danger (a cliff, a swamp, or into the presence of big game). Wear stout footwear at night and take care not to tread on snakes. Be aware also of the possible presence of suspicious security personnel.

Driving slowly at night (20–30km/h) can also be rewarding, especially on quiet tarred roads, with your headlights dipped. You may encounter geckoes and snakes either lying on the surface to absorb heat or crossing the road especially if the road is wet. However, night driving in East Africa can be extremely dangerous. Hazards include unlit and unroadworthy vehicles, roadblocks, people and livestock on the road, and the generally low standards of driving; you may encounter drivers with no concept of roadcraft, defensive driving, speed awareness or consideration for others. And at night you cannot see hazardous situations developing. Never be tempted to road cruise on a busy East African road at night.

Useful equipment for the reptile collector may include cloth bags with sewn-in drawstrings, and screw-topped plastic jars for little animals. Remember to keep containers with live animals in them away from sunlight or excessive heat, and take care not to squash them. A hooked stick is useful for turning ground cover, removing tree bark, poking about in holes and pressing down snakes. For dangerous snakes a grab stick, pair of tongs and a pair of goggles (for spitting cobras) will be needed. A pair of industrial leather gloves is also useful for protecting the hands when digging. Other useful equipment might include a small shovel or trowel for digging out holes, and some plastic tubing for sticking down holes so you don't lose your way when digging them out. Lizards may permit close approach and can be noosed using fishing line on a light rod (e.g. an old radio aerial or a long thin stick). Animals can sometimes be flushed out of holes with chloroform, petrol, mothballs or smoke, but such techniques may permanently damage the refuge, kill the animal in the hole or start a bush fire, so their use should be carefully considered. Tortoise enthusiasts have developed a useful technique for locating a tortoise in a hole, using a snooker or pool cue; a tortoise poked with a pool cue makes a most distinctive thwock.

A small pair of binoculars is useful. You don't always have to walk in East Africa to observe the smaller wildlife; the quiet watcher with binoculars in an elevated spot can often see a surprising amount.

Local help is often available in East Africa. People will come and offer their services and they often have a lot of local knowledge. If you have a book with you and speak the local language, you can show pictures of what you want. But never ask or allow people to catch snakes for you; if someone gets bitten there will be major problems.

Serious scientific collectors will need preservative and containers. Valuable specimens are often found dead. A 10% formalin solution or a 70% alcohol solution can be used to preserve specimens. In general, formalin is a more efficient preservative, but pure alcohol preserves long strands of DNA; necessary for molecular taxonomy, so if you have both, preserve some bits of the animal (a bit of skin or muscle will do) in a little container of alcohol and put the rest in formalin. All specimens should be labelled; mark the label with soft pencil. Locality is the most important piece of data; a specimen without a location is

Rhampholeon spinosus Rosette-nosed Chameleon, Stephen Spawls.

virtually useless. Larger animals should be cut open along the abdomen, and/or preservative injected into the soft body parts. In an emergency you can preserve specimens with methylated spirit, alcohol (neat vodka, gin or whisky), petrol or concentrated salt solution. Specimens should be deposited in a public-access collection, such as a museum.

Always consider safety in the field. Anyone intending to collect dangerous snakes should be thoroughly experienced, and be part of a team with adequate medical support. Snakebite treatment may involve blood transfusions. The field worker should also have suitable protective clothing, food, drink and medical supplies; consider taking a supply of antivenom.

REPTILE CONSERVATION

Reptiles have an image problem. A lot of people dislike them, and snakes and certain lizards are actively feared and killed if encountered. Environmental education may help with this problem, but even in highly developed countries many people do not like reptiles and amphibians. It is for this reason that places such as reptile parks serve such an important function, in bringing the public face to face with reptiles and amphibians and thereby giving them a chance to appreciate their beauty and usefulness. For the same reason, any enthusiast who keeps local reptiles (with appropriate permission, if needed) and shows them to their friends is benefiting conservation. In addition, reptiles are an important part of ecosystems. They eat pests, and they possess both known and untapped potential in culinary, economic and medical terms. The benefits of reptiles and biodiversity need to be emphasised in schools and colleges.

In many ways the future is bright for East Africa's reptiles. Largely as a result of the electronic age there has been a huge rise in awareness of the beauty and charms of our region's fauna, particularly among young people. There has also been a significant increase in interest in all aspects of East Africa's natural history and biodiversity, and social media has enabled heightened awareness and a rapid response to events of environmental significance; consider the fascination with (and widespread condemnation of) recent cases of the wanton killing of African lions, or the public anger over proposed excisions of national parks. All five East African countries have active tourist industries that employ many people and bring in outside money. Consequently their governments have a financial interest in protecting wild places and their inhabitants. There is a large, reasonably well-financed, well-developed and protected national park and reserve system, helped to some extent by the forest reserves, protecting biodiversity. In addition, many reptiles are relatively small and secretive, they are not as vulnerable to habitat loss and exploitation as larger wild animals and can often survive on farmland and in suburban areas; Nairobi's hill suburbs are still full of Jackson's chameleons, and Blanding's tree snakes survive in gardens and ornamental woodland around Kampala.

But at the same time, threats exist. Few species have been thoroughly assessed in terms of current distribution and required habitat. In a country like Kenya, with expanding population and rapid development, habitats are severely under threat, from farming, logging, urbanisation and mineral exploitation. Even national parks can be at risk from government action if they are deemed to stand in the way of progress. Some reptile species are in demand for the pet trade. Others are unable to exist in harmony with humanity; large reptiles (especially snakes) tend not to be tolerated by people, no matter how innocuous they are. At one time, large Mole Snakes could be regularly found on the shores of Lake Naivasha, they are no longer there, even though they are harmless. In the species accounts we have given details if a taxon has been assessed by the International Union for the Conservation of Nature (IUCN). They publish the Red Lists of Threatened Species and definitions of the threat levels can be found online. We have added our own comments where pertinent. If there is no comment on conservation status it means the species has not been assessed and no major threats have been identified.

Without a doubt, some species are at risk. East Africa's vulnerable reptiles are the endemic species, especially those with limited ranges that live in areas suitable for agriculture or exploitation. Many live in tiny habitats, which are often forested. With population increases, these forests and wild lands are vulnerable, liable to be exploited for their timber and potential use for farming. For example, many East African endemic chameleons live in tiny forested areas; they are not only at risk from harvesting for the pet trade, but also from deforestation, and possibly climatic change as well. Consider the Kenya Horned Viper, *Bitis worthingtoni*, endemic to the high country of the central rift valley in Kenya; this is prime farming land. If you look at Google maps you will see that a huge proportion of the high grassy savannas it favours have disappeared, given over to smallholdings and big mechanised farms. We have literally no idea how many are left, to what extent the population is fragmented, and how viable are the remaining populations. At present, our conservation priorities should be to identify vulnerable habitats, especially those with no protection, document their fauna and flora and develop and implement protective strategies. Often a single small well-protected reserve will suffice to conserve a rare reptile and its habitat.

We would like to draw attention to the importance of museums in conservation activities. The range of almost all species described in this book is based upon museum specimens. Without those specimens, ranges would be unknown, taxonomic status impossible to clarify, and no conservation strategy would be possible. There is a modern school of thought that decries collecting, that states that rare creatures should be always left alive; that range mapping should be done using field observations. This may be possible with larger mammal and bird species, but it is not possible with amphibians and reptiles. There are few observers and few specimens; reptiles and amphibians are hard to see and find, the difference between many similar species is not obvious in the field and field identification is difficult. Museum specimens are necessary, and it is most important that specimens are deposited in local scientific collections, to assist researchers; details of suitable institutions are given in the section at the end of this chapter, entitled 'Further reading, literature and resources'.

IDENTIFYING REPTILES

Some species are easy to identify, others are not. We have provided dichotomous keys and these will be useful to distinguish similar species, although they are not infallible (and, as you will find, there are some species that can only be distinguished by characters such as differences in DNA; which rules out any possibility of amateur identification). But many reptiles *are* easily identifiable in the wild. If you are going into the field it is worth having a good look at the pictures in this book before you go, to get an idea of what you might encounter.

There are a number of common species that a naturalist in Kenya may come across, and it may be worth having a look at our pictures of those species and familiarising yourself with them. The two most commonly encountered tortoises in East Africa are the Leopard Tortoise *Stigmochelys pardalis* or one of the hinged tortoises *Kinixys*. After a rainstorm, especially at low altitude, you might well come across a Helmeted Terrapin *Pelomedusa subrufa*, looking for a temporary pool. Common lizards that you are likely to see include the Striped Skink *Trachylepis striata*, widespread in most towns and cities, the Variable Skink *Trachylepis varia*, often in gardens, and various species of dwarf gecko *Lygodactylus*, often on trees, fences and walls during the day. Around lamps at night you are almost certain to see a Tropical House Gecko *Hemidactylus mabouia*. In highland towns, you might well

Psammophis mossambicus Olive Sand Snake, Michele Menegon.

see a Blue-headed Tree Agama *Acanthocercus atricollis*, and on rocks at lower altitudes in Kenya, the Kenya Red-headed Rock Agama *Agama lionotus* may be seen; in Tanzania it is largely replaced by the Dodoma Agama *Agama dodomae*. Common chameleons of low altitude include the Flap-necked Chameleon *Chamaeleo dilepis* and the Slender Chameleon *Chamaeleo gracilis*; in the rainy season you might find one on a road. Along a river, especially at altitudes below 1,500m, you may be lucky enough to see a Nile Monitor *Varanus niloticus*, and its close relative, the White-throated Savanna Monitor *Varanus albigularis*, may be spotted out in the savanna.

Few snakes are common, in any sense, in East Africa; you are sure to see some lizards if you walk in East Africa's wild places, but are unlikely to see a snake. However, the two most common snakes in East Africa are probably the Brown House Snake *Boaedon fuliginosus* and the White-lip *Crotaphopeltis hotamboeia*, both nocturnal. By day, a fast moving snake that crosses your path may well be a sand snake *Psammophis*. In some areas, there is a reasonable chance of seeing a Boomslang *Dispholidus typus*, and along rivers in the National Parks careful observation might reveal one of the rock pythons.

We give below some general rules to help with field identification of our region's reptiles. They are applicable to East African animals only and you should use them with caution, particularly where snakes are concerned; and do not attempt to apply rules from elsewhere, which could lead to fatal mistakes. Our diagrams (Figures 1 to 31, pages 600-611) show how to identify crucial features and count important scales.

Most chameleons can be readily identified by their swivelling eyes set in turrets, their grasping feet and prehensile tail. Monitor lizards are large, most are over 70 cm long, and no other East African lizard grows that big. Worm lizards look like worms. Large male agamas have broad, bright heads, usually green, blue, red or pink. The smaller agamas often have blue on the throat. Agamas can also be identified by their broad heads and very thin necks.

Geckoes are the only lizards active at night. Most have vertical eye pupils and soft skin. Skinks have shiny bodies, are often striped and/or brightly coloured, with little limbs and no distinct narrowing of the neck. Lacertid lizards are small, fast-moving, often striped and most common in arid country. Plated lizards have an obvious skin fold along the flanks, and are relatively large.

In East Africa there is no single way to tell a harmless snake from a dangerous one. Any fat snake over 4m long must be a python. Any snake with more than one conspicuous stripe running along the body is probably back-fanged and not dangerous. Any snake over 2m long is probably dangerous. Any grey, green or greenish tree snake over 1.3m long is almost certainly dangerous, it will probably be a mamba, boomslang or vine snake. If it is over 1.3m long and inflates the front half of its body in anger, it will probably be a dangerous back-fanged tree snake (boomslang or vine snake).

Any snake that spreads a hood, flattens its neck or raises the forepart of the body off the ground when threatened is almost certainly dangerous, and will probably be an elapid (other possibilities include night adders and the rufous beaked-snake). Any snake with conspicuous bars, cross-bands, rings or V-shapes, especially on the neck or the front half of its body, on the back or on the belly is probably dangerous. Most cobras have dark bars or blotches on the underside of their necks. A fat-bodied snake, with a sub-triangular head, that lies quietly when approached, is probably a viper or adder. Small black, dark grey or brown snakes with tiny eyes, no obvious necks and a short fat tail that ends in a spike will almost certainly be burrowing asps.

Most bush vipers are a mixture of greens, blacks and yellows, with broad heads and thin necks, and are usually in a bush or low tree. A little snake with conspicuous pale bands on a dark body is probably a garter snake. Any snake with rectangular, sub-rectangular or triangular markings on its back or sides, or rows of semi-circular markings along the flanks, is probably a viper. A snake that forms C-shaped coils and rubs them together making a noise like water falling on a hot plate, will be either a carpet viper (dangerous) or an egg-eater (harmless). Any small snake with a blunt rounded head *and* a blunt rounded tail, with eyes either invisible or visible as little dark dots under the skin, with tiny scales that are the same size all the way around the body, will be either a blind snake or a worm-snake, and harmless.

SOME BRIEF NOTES ON REPTILE TAXONOMY

Taxonomy is that branch of science concerned with classification; working out the relationships between organisms, and providing a way of distinguishing between them. The most useful taxonomic unit is the species; it is an essential concept. It is important that we know what species are, and that they are robustly defined. No faunal list can be drawn up unless we can say what species occur there; no researcher can say what organism they worked on or with unless the species can be unequivocally identified, no new locality records can be added if the specimen cannot be identified to species.

Species have been defined in many ways and the debate continues, but essentially a species is a group of organisms that are each other's closest relatives, can interbreed and produce fertile offspring. Us humans are a single species. The species is a dynamic unit, for obvious reasons. Life on earth started once and diversified; species change over time, new species are constantly being formed as organisms, and groups of organisms, evolve. All species are given unique two-part scientific names: the first or generic name applies to a genus, or group of very similar organisms; the second to that actual group of inter-breeding organisms. For example the Black Mamba has the scientific name *Dendroaspis polylepis* (which is Latin for "tree snake with many scales"). No other species is allowed this name. There are three other species of mamba, all have the generic name *Dendroaspis*, and all have different specific names; for example the Green Mamba has the scientific name *Dendroaspis angusticeps*. For some species of organism a third name, the subspecies, may be added where there are two (or more) sub-groups within the population that somehow differ; for example there are two subspecies of the Speckled Sand Snake *Psammophis punctulatus*; one has a single dorsal stripe, *Psammophis punctulatus punctulatus*, the other has three dorsal stripes, *Psammophis punctulatus trivirgatus*. In this book, we have occasionally mentioned subspecies in the identification section, where relevant.

Reptile taxonomy is made easier by the diversity of reptiles themselves, and the fact that most are covered with scales, which can be counted. Tortoises and turtles are identified by their size, shell shape and scale counts. Lizards are often identified by a mixture of scale counts, body shape and colour, snakes by scale counts, colour, teeth and body shape. These characters have proved useful for over 250 years of herpetological taxonomy. Recently, a new set of tools have become available to laboratory-based taxonomists; the techniques that enable the isolation of standardised bits of DNA, and the computing power that allows the analysis of that DNA. With these tools, new forms can be defined. New evolutionary lineages and relationships can be discovered and old ones modified through the differences in matching sections of DNA. Although traditional reptile taxonomy involving the detailed descriptions of animals in terms of morphology, appearance and scale counts is still useful, especially with new forms that differ clearly from other members of the genus, the analysis of DNA has enabled researchers to fine-tune their identification, and often split what seemed in the past to be a single species into several species, evolving separately. Even the approximate date at which the two forms began to diverge, and accumulate differences, can be ascertained. This is a very useful development. But it has certain disadvantages. Often, those separate lineages cannot be told apart in the hand, or even under the microscope, and this can lead to taxonomic confusion. For example, the Helmeted Terrapin *Pelomedusa subrufa*, a single water tortoise widespread throughout sub-Saharan Africa and not particularly variable in appearance, has recently been split into 12 full species. Useful though this is in a broad sense, the authors of this work have made no attempt to construct a key that would enable any researcher to distinguish between species in the hand or even in the laboratory; individual animals can only be assigned to species by analysis of their DNA. This creates difficulties, particularly in Africa where few scientists have access to a

Mecistops cataphractus Slender-snouted Crocodile, Stephen Spawls.

molecular lab. But, like its subjects, taxonomy is evolving. A visual key may be possible. And in Africa, there is a lot to be discovered.

We should briefly like to say something about English or common names. We have tried to use names by which the species are known to English-speaking naturalists in East Africa. Other names may be used elsewhere. For example, the little dark frog-eating snake *Crotaphopeltis hotamboeia* has been always known to naturalists in East Africa as a White-lip and we call it that; in southern Africa it is often called a Herald Snake. There have been recent attempts to call the big tree cobra *Pseudohaje goldii* 'Goldie's Tree Cobra'. This is technically correct since it is named after Sir George Goldie, but Charles Pitman, in his classic work *A Guide to the Snakes of Uganda*, in 1938 called it Gold's Cobra, and it has been known to East African naturalists as Gold's Tree Cobra ever since; we retain that name.

WHAT ARE REPTILES?

Reptiles are a varied class of tetrapod ('four-footed') vertebrates, although some have lost their limbs. The group includes living snakes, lizards, turtles and tortoises, crocodiles and the tuatara. There are also many extinct forms in the class, including dinosaurs (some scientists include birds in the group, as they are closely related to modern crocodiles, but the following brief discussion of reptilian characters does not apply to birds). Although diverse, reptiles share many features. They have a dry waterproof skin and most reproduce by internal fertilisation, followed by the laying of amniotic, self-contained eggs, which develop without parental involvement. There is no larval stage, the offspring look like their parents. Most reptiles are ectotherms; meaning they get their body heat from outside sources and this explains why there are no nocturnal lizards at altitudes over 2,200m in East Africa; the night-time temperatures are too cold. The lack of internal heating means that reptile energy budgets are low, and thus they eat relatively less food than mammals; some snakes survive on six or eight meals a year. Although snakes are largely solitary, social behaviour is shown by some lizards, which live in structured colonies. Many reptiles show ritual courtship behaviour and some males fight over females. Reptiles are an old group; the oldest clearly reptilian fossil is over 300 million years old. There was a tremendous radiation of reptiles between 250 and 65 million years ago, including many species of dinosaur, followed by a mass extinction. But reptilian evolution is occurring rapidly today with a radiation of smaller forms. Modern reptiles are well represented in East Africa today, with over 490 species known. These numbers will rise as the East African reptile fauna becomes better known. It is to be hoped that at the same time the exploitative effects of humanity upon the environment will not plunge some into extinction.

SAFETY AND REPTILES

Reptiles in East Africa can usually be safely observed with enjoyment, sometimes from quite close quarters, provided a little care is taken to avoid disturbing the animal and to minimise the risk to yourself. There is a pleasure in watching the activities of a little agama colony, the slow progress of a tortoise through the savanna, or the probing search by a sand snake of holes that may conceal prey, that too few people pause to enjoy. Take time to observe reptiles. However, a few important points should be noted.

No East African turtle, tortoise or lizard is venomous, despite local legends that indicate some lizards (especially agamas and chameleons) have a venomous bite. In addition, no species has poisonous excreta. Thus chelonians, lizards and harmless snakes may be safely approached. But do not go too close to an unidentified snake, as it may be dangerous. A distance of twice the snake's length is as close as you should get, unless you suspect the snake is a spitting cobra, in which case do not go closer than five metres. Beware of approaching

Kinixys erosa Forest hinged tortoise, Kate Jackson.

any snake, particularly a large one, that appears to be agitated (spread hood, raised head or neck, hissing loudly, opening its mouth) as it may strike or rush forward if cornered.

Beware of crocodiles, they are dangerous. The Nile Crocodile *Crocodylus niloticus* is the only East African reptile that may attack humans in order to eat them. Even quite small crocodiles may attack people. You should never wade or swim anywhere that Nile Crocodiles might live (basically, any water body, up to about 1,800m altitude), and this includes lakes, rivers and even quite small dams. It is advisable to stay at least 8 or 10m from the water's edge.

Care must be taken if you wish to catch or handle a reptile. Most will bite if cornered or seized. Some chelonians can deliver a powerful bite and handlers have lost fingers. A bite from a smaller lizard can hurt, a big lizard like a large agama or a monitor can give a severe, skin-breaking bite, and may scratch with their claws or lash with their tails. We don't advise handling big lizards unless you are experienced.

Unless you are an expert, you should never handle snakes. This includes apparently dead ones; they might not be dead. Be prudent if offered the opportunity to handle a snake at a snake park or reptile exhibition; do the handlers seem professional? Remember also that reptiles usually have bacteria on their skin and in their mouths. These are unlikely to cause any problems, provided you are not bitten, that you thoroughly wash your hands after holding amphibians and reptiles, and avoid touching your eyes and mouth while handling. If you demonstrate reptiles to people, ensure that they wash their hands after handling an animal.

In other publications (our original field guide, and *The Dangerous Snakes of Africa* by Stephen Spawls and Bill Branch) we have dealt at length with avoidance of snakebite, and how to treat it, we included details of how to deal with venom in the eye after an encounter with a spitting cobra. We refer interested persons to those publications. We have given a few brief comments on venoms and symptomology for all dangerous snake accounts. A few rules for avoiding snakebite are given here.

A number of East African snakes are dangerously venomous. To avoid being bitten always treat snakes (and other reptiles) with respect. Do not try to catch or kill them unless you are an expert. Snakes do not make unprovoked attacks. If confronted with a snake, move backwards slowly. To avoid being bitten by a snake, always look where you are going. Wear adequate shoes; don't put your hands or feet in places you cannot see. Use a lamp or torch at night. Don't sleep on the ground or gather firewood after dark. Clear suitable snake hiding places from near your home, especially if you have small children. If someone gets venom in their eyes after an encounter with a spitting cobra, the eyes should be gently washed out with large quantities of water (or other bland fluids such as milk), and the victim taken for a medical check-up. If someone has been bitten by a snake, treat it as a medical emergency, abandon other plans and organise rapid safe transport to a large hospital (clinics and health posts are unlikely to be able to adequately treat a major snakebite case). Do not try to catch or kill the snake. Leave the wound alone, don't start cutting it, giving shocks, using ice etc. Immobilisation therapy (keeping the bitten area, especially limbs, as still as possible, often by splinting) may be useful in some cases, as detailed in our other publications. But don't let fears about snakebite put you off going into wild places; the risks are very small.

ACKNOWLEDGEMENTS

The original edition of this book was prepared more than 16 years ago. The team were largely based in Africa. Digital photography was in its infancy. Electronic communications were often slow and erratic. We relied on snail mail and personal contact to source literature and illustrations. There was no centralised database of herpetological material, and often crucial revisions were not widely disseminated.

Naja nigricollis Black-necked Spitting Cobra, Daniel Hollands.

Things have improved dramatically, particularly in three areas: the speed of availability of research material; the widespread use and reduced cost of digital and phone cameras; and the availability of centralised sources of information. We have made constant use of a crucial online resource, the Reptile Database, available at reptile-database.org. It is run by Peter Uetz and Jiri Hosek; we salute them. Their efforts, aided by an army of volunteers, have made our work far easier. On the database all species are listed, with pertinent literature. Of equal importance, the team there scan the journals, critically read the revisionary material and accord it the appropriate significance. We struggled to obtain taxonomic material in the previous edition, sometimes we were not aware of revisions. Now that information is available instantly through the database. We offer also our thanks to the museums and institutions with East African collections and their curators; a detailed list was given in the previous edition but we re-mention here the National Museum in Nairobi, the University of Dar es Salaam, and Dr Charles Msuya, the Natural History Museum in London, the California Academy of Sciences and the Museum Alexander Koenig in Bonn.

We would also like to thank the field herpetologists who have helped us. Four African institutions and their curators stand out: Sanda Ashe, Anthony Childs, Royjan Taylor, Charles Wright and Kyle Ray at Bio-Ken Snake Farm, at Watamu on the Kenyan coast; Joe Beraducci at MBT's Snake Farm and Reptile Centre at Arusha, Tanzania; Patrick Malonza, Victor Wasonga, Beryl Bwong and Vincent Muchai at the Herpetology Section, the National Museum in Nairobi; and Deon Naude and Lynn and Barry Bale at Meserani Snake Park, near Arusha. They have made a major difference. All have live collections and allowed us to freely photograph their specimens; those collections so often included rare animals that a field worker might struggle to see, even in years of field work. In addition, the Bio-Ken team generously allowed us free access to their photographic database; and many of the fine pictures here came from that, while in Arusha Joe Beraducci frequently put aside his work in order to ferret out reptilian lifestyle information for us. Anthony Childs kindly offered us hospitality, and expertise, at the Emakoko, and allowed us to photograph his personal collection. Harry Greene kindly allowed us to use his apposite quotation on natural history. Harald thanks the Kigali Museum of Natural History. Terry Stevenson and John Fanshawe kindly let us use their habitat maps from 'Birds of East Africa'.

Two old friends who worked with us on the last edition are no longer with us; James Ashe and Alex Duff-MacKay. Safari *njema*, Jim and Alex!

Our thanks are also due to the photographers who have allowed us the use of their superb images; they have greatly strengthened this book, both in terms of photographic coverage and distributional data. We thank all those who have sent us images, and apologise for any omissions; we had so many fine pictures, we couldn't use them all. We thank Bryan Adkins, Tony and Ali Allport, Susie Allan, Jonas Arvidson, Kenny Babilon, Wolfgang Boehme, Andrew Botta, Bill Branch, Gary Brown, David Bygott, James Christian, Zarek Cockar, Rolf Davey, Bob Drewes, Anthony Childs, Davide Dal Largo, Tim Davenport, Job de Graaf, Richard de Jong, Max Dehling, Craig Doria, Leejiah Dorward, Gerald Dunger, Dietmar Emmrich, Squack Evans, Vincenzo Ferri, Brian Finch, Florian Finke, Sean Flatt, Paul Freed, Toby Gardner, Frank Glaw, David Gower, Eli Greenbaum, Thomas Hakannson, Olivier Hamerlynck, Oliver Hawelitschek, Claudia Hemp, Daniel Hollands, Barry Hughes, Caspian Johnson, Kate Jackson, Miroslav Jirku, Walter Jubber, Chege wa Kariuki, Odile Keane, Rungwe Kingdon, Johan Kloppers, Will Knocker, Patrick Krause, Dirk Kreulen, Matthijs Kuijpers, Malcolm Largen, Daniel Liepack, David Lloyd-Jones, Dudley Lucas, Quentin Luke, Maggie Mahan, Patrick Malonza, Johan Marais, Andrew R Marshall, Dino Martins, Justin Kenya Mathews, Michelle Dawn Mathews, Tomas Mazuch, Mike McLaren, Konrad Mebert, David Moyer, Martha Mutiso, Kim Nel, Petr Necas, Darcy Ogada, Olivier

Pauwels, Thomas Price, Fabio Pupin, Darrell Raw, Eduardo Razzetti, Dave Richards, Dennis Rodder, Beate Roll, Norbert Rottcher, Jose Rosado and the President and Fellows of Harvard College, Mike Roberts, Steve Russell, Ignas Safari, Edwin Selempo, Philip Shirk, Terrill Shrock, Robert Sindaco, Tim Spawls, Ton Steehouder, Jan Stipala, Lulu Sturdy, Nelly Swila, the Frontier-Tanzania team, Colin Tilbury, Jean-Francois Trape, Casper Van de Geer and the Watamu Turtle Watch, Luke Verburgt (Enviro-Insight), Lorenzo Vinciguerra, Jens Vindum, James Vonesh, Washington Wachira, Philip Wagner, Victor Wasonga, Frank Willems, Elvira Wolfer, Wolfgang Wuster and Andre Zoersel.

We thank also those who critically read sections of the manuscript and added their input; in particular Tomas Mazuch, for his expertise on the *Hemidactylus* and the herpetofauna of north-east Africa, Wolfgang Wuster for his expertise with African elapids and a strengthened cobra key, and Philipp Wagner for his kind input into the Agama section. We also offer here our gratitude to Dr Don Broadley, sadly no longer with us. Don's assiduous revisionary work on African herpetology continues to smooth our paths. Barry Hughes helped us constantly, making many visits to the library in the Natural History Museum to scan and mail us literature; our thanks are also due to David Gower and Patrick Campbell at the Museum. Steve would also like to thank his old friend Jonathan Leakey who kindly made available the field notebooks of C.J.P. Ionides, Tanzanian herpetologist; the natural history information here has proved invaluable; thanks to Julia Leakey and Dena Crain for getting the material to us. Steve also thanks his many field companions and those who gave hospitality in East Africa, especially his sons Jonathan and Tim, Daniel Hollands, Conrad and Linda Thorpe, Jonathan Leakey and Dena Crain, Glenn Mathews, Tom Butynski and Yvonne De Jong, Joy and Ian MacKay and Rungwe Kingdon. For daily stimulating discussion Steve is grateful to his fellow science lecturers at City College Norwich Janet Cross and Ian Cummings, and also his Heads of School, Clare Turner and Jerry White. Steve also thanks his wife, Laura, for her endless and unstinting support. Kim wishes to

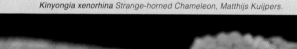

Kinyongia xenorhina Strange-horned Chameleon, Matthijs Kuijpers.

thank his wife, Prof. Imani Swilla, and daughter Nelly Swilla for patience, understanding and support during this task. Michele Menegon wishes to thank the director of the Science Museum of Trento, Michele Lanzinger, for his 'wild' vision of the role of science museums, the Lipparini family for their continued support and Anna for being the wonderful person she is. And Finally, we thank Jim Martin, at Bloomsbury for his kindly unstinting support and Jocelyn Lucas for her beautiful and elegant design.

FURTHER READING, LITERATURE AND RESOURCES

The internet has smoothed our path regarding herpetological resources. For any interested East African herpetologist, the reptile database mentioned above should be the first port of call. Also useful is the Kenya Reptile Atlas (www.kenyareptileatlas.com) and a Facebook forum East African Snakes and other reptiles, at https://www.facebook.com/groups/662521540444058/. An excellent resource is the recently established iNaturalist project on the Herpetofauna of the Eastern Afromontane at http://www.inaturalist.org/projects/eastern-afromontane-herpetofauna. This is linked with the database of the herpetological collection of the MUSE, the Science Museum of Trento, where is possible to upload photos of reptiles and amphibians taken across the Eastern Afromontane and get them identified by a community of scientists and passionate herpetologists.

Through the internet you will also find references to the 'classic' works of East African herpetological literature, including older material by authors such as Charles Pitman, Don Broadley, Arthur Loveridge, Raymond Laurent, Gaston de Witte and H. W. Parker, and newer material by authors like Bill Branch, Tomas Mazuch and Malcolm Largen. Two fine books with pan-African treatment include Colin Tilbury's *Chameleons of Africa*, and Bill Branch's *Tortoises, Terrapins and Turtles of Africa*. Bill has also published a handy photographic guide to the snakes and other reptiles of East Africa. For a visually stunning overview of Kenya's highland chameleons read *Mountain Dragons* by Jan Stipala. Also useful, although hard to obtain, is *The Dangerous Snakes of Africa* by Stephen Spawls and Bill Branch, and *Venomous Snakes of Africa* by Maik Dobiey and Gernot Vogel. Of relevance to East Africa is the superb, beautifully illustrated herpetological literature of Southern Africa; especially books by Johan Marais, Bill Branch and Don Broadley; for West and Central Africa there are two fine books by Jean-Francois Trape and co-authors. Some excellent work on the African herpetofauna has been done by Luca Luiselli, Godfrey Akani, Eli Greenbaum, Kate Jackson, Jens Rasmussen (sadly missed) and Stephen Goldberg.

We mention also a most useful revisionary paper by Alex Pyron and co-authors 'A phylogeny and revised classification of Squamata, including 4,161 species of lizards and snakes'; this can easily be found online. Likewise, for any health professional involved or interested in snakebite treatment, searching for 'Guidelines for the prevention and clinical management of snakebite in Africa' will retrieve for free a superb book by Dr David Warrell, a pioneer of tropical snakebite treatment. Two other websites have proved endlessly useful, the Biodiversity Heritage Library and the IUCN Red List. A number of scientific journals regularly publish African herpetological material, including *Zootaxa*, and the *African Journal of Ecology*. The magazine of the East African Wildlife Society, *Swara*, often has herpetological articles. The only organisation solely devoted to African herpetology is the Herpetological Association of Africa, which produces a scholarly journal and a regular newsletter. Details of all these can be sourced from the internet; and most (especially *Swara* and the HAA) welcome contributions. Institutions involved with East African herpetology include the Herpetology Section at the National Museum in Nairobi, and the associated Nairobi Snake Park and Nature Kenya, Bio-Ken Snake Farm at Watamu, Kenya Wildlife

Service, the Department of Zoology and Wildlife Conservation at the University of Dar es Salaam, Meserani Snake Park on the eastern outskirts of Arusha, MBT's Snake Park between Arusha and Moshi, Tanzanian National Parks, the Zoology department at Makerere University, Kampala, Uganda Reptile Village outside Entebbe. Details of how to contact all of these can be found via the Internet. Bio-Ken at Watamu have call out teams in strategic places around Kenya who will respond to sort out a problem snake. Living, preserved and dead material can be donated to the National Museum in Nairobi, which has the most comprehensive preserved collection of East African reptiles. Finally, we authors are also always pleased to receive information and pictures relating to any aspect of East African herpetology; our contacts are as follows: Stephen Spawls at stevespawls@hotmail.com, Kim Howell at kimhowellkazi@gmail.com, Harald Hinkel at haraldhinkel@hotmail.com and Michele Menegon at mmenegon@gmail.com.

Trioceros hoehnelii Von Hoehnel's Chameleon, Stephen Spawls, Nyahururu.

ORDER TESTUDINES
CHELONIANS (TURTLES AND TORTOISES)

Pelomedusa subrufa
Helmeted Terrapin
Michele Menegon

ORDER TESTUDINES

Instantly identifiable and well-known in folklore and fables, turtles and tortoises are a distinctive and ancient order of reptiles; fossil chelonians over 200 million years old have been found. The chelonians are identified by their protective shells; with hard outer scales covering an inner, bony layer. Inside, the ribs are fused to the shell. The names used for the different groups in the order vary; we have used the useful collective term chelonian, not all will agree with this. In East Africa a tortoise is a land dwelling chelonian, terrapins and turtles are respectively small and big water-dwelling animals. Worldwide, around 330 species are known; in East Africa there are seven land tortoises, five sea turtles, two big freshwater turtles and 13 terrapins. Many species can totally or partially withdraw their head and limbs inside the shell. They are long-lived; an Aldabran Tortoise lived over 150 years in captivity. They may take several years to reach maturity. In many species adult males have a concave plastron (the underside of the shell), the male goes on top while mating and the concavity stops him slipping off. Males often have thicker tails than females. All chelonians lay eggs, the females excavates a flask-shaped pit in soil or sand. In some species the sex of the young is temperature-dependent, with cooler nests producing mostly male offspring. Land tortoises largely eat plants (some will take invertebrates and carrion); most water-dwelling species are carnivorous. Due to their size and shape they take a long time to gain heat, so are rarely found above 1,500m altitude in East Africa. Chelonians have a sense of direction; sea turtles migrate across thousands of miles of ocean to lay eggs on their home beach, translocated land tortoises have found their way home over many kilometres.

Small land chelonians are fairly defenceless and have many enemies, including predatory birds, snakes, monitor lizards and small carnivores, little sea turtles may be taken by birds, crabs and big fish. Land tortoises are at risk from bush fires; it is estimated that a single major bush fire in western South Africa killed over 100,000 tortoises. Lions and leopards may occasionally take a tortoise, but large chelonians have one main enemy; humans. And what an enemy! Some are collected for the pet trade (the East African endemic Pancake Tortoise is particularly at risk). Their habitat is exploited and disappearing. Large turtles are trapped and killed in fish nets. In the past, sea turtles were killed for their shells ('tortoiseshell') and meat; on many tropical beaches the eggs are still harvested in huge numbers. Nesting beaches are being exploited for development; and while turtles and beach tourism can co-exist, the night lights and the scavengers associated with humanity often create problems for sea turtles. In parts of southeast Asia, many species of chelonians are threatened by their collection in huge numbers for the food trade, due to a superstitious (and totally irrational) belief that eating the meat of a long-lived and hardy animal will make the consumer equally hardy and long-lived. Education, aided to a large extent by social media, is reducing the rate at which these charming animals are ruthlessly exploited and killed, but there is a long way to go. And yet, if properly kept, tortoises make engaging long-lived pets if they have access to sunlight and green plants.

Family Testudinidae: Tortoises

The order Testudines is split into two sub-orders, Cryptodira and Pleurodira. In East Africa all land tortoises, sea turtles and the two large soft-shells are in the Cryptodira. The name means 'hidden neck', they all withdraw their head into the shell by flexing the neck into an S-shape, in a vertical plane, and at the same time, in terrestrial species, the legs are withdrawn, and the front legs protect the head. The freshwater terrapins of the genera *Pelomedusa* and *Pelusios* are in the sub-order Pleurodira; the name means 'side-neck' and these terrapins withdraw the head sideways in defence, leaving one eye exposed. They have webbed hind feet.

KEY TO THE EAST AFRICAN CHELONIAN GROUPS AND SOME SPECIES

1a   Neck exposed (bent sideways) when head is withdrawn into shell. Sub-order Pleurodira......2
1b   Neck hidden when head is withdrawn into shell, or head not withdrawable into shell. Sub-order Cryptodira......3

2a   Plastron has no hinge, head never reticulated or spotted......*Pelomedusa*, helmeted terrapins [p. 51]
2b   Plastron with hinge at front, head often shows reticulations or spotting...... *Pelusios*, hinged terrapins [p. 52]

3a   Limbs modified as flippers, with 0–2 claws, in the sea or on beaches......4
3b   Limbs not modified as flippers, with 3–5 claws, not in the sea......8

4a   Carapace leathery, with seven longitudinal ridges, flippers clawless..... *Dermochelys coriacea*, Leatherback Turtle [p. 45]
4b   Carapace not leathery, with a symmetrical arrangement of horny plates, flippers with 1–2 claws......5

5a   Carapace with four pairs of costals......6
5b   Carapace with five or more pairs of costals......7

6a   Hooked beak present, four prefrontal head shields, usually two claws on each limb.....*Eretmochelys imbricata*, Hawksbill Turtle [p. 42]
6b   Beak not hooked, two prefrontal head shields, usually with a single claw on each limb.....*Chelonia mydas*, Green Turtle [p. 43]

7a   Carapace with five pairs of costals, head large, colour red-brown.... *Caretta caretta*, Loggerhead Turtle [p. 41]
7b   Carapace usually with 6–9 pairs of costals, adults olive in colour, young olivaceous black.... *Lepidochelys olivacea*, Olive Ridley Turtle [p. 40]

8a   Shell with scutes or scales......9
8b   Shell covered with skin......12

9a   Shell flattened, soft and flexible.... *Malacochersus tornieri*, Pancake Tortoise [p. 35]
9b   Shell not flattened, tough and inflexible......10

10a   Adult shell has hinge at rear of carapace.... *Kinixys*, hinged tortoises [p. 30]
10b   Adult shell without hinge......11

11a   Shell yellow and black, or uniform brown in very large animals.... *Stigmochelys pardalis*, Leopard Tortoise [p. 28]
11b   Shell black or grey.... *Aldabrachelys gigantea*, Aldabran Giant Tortoise [p. 36]

12a   Plastron with skin flaps that cover withdrawn rear limbs, range southern Tanzania.... *Cycloderma frenatum*, Zambezi Flap-shelled Turtle [p. 48]
12b   Plastron without skin flaps over rear limbs, range northern Kenya and Uganda. *Trionyx triunguis*, Nile Soft-shelled Turtle [p. 47]

Aldabrachelys gigantea
Aldabran Tortoise
Michele Menegon

SUB-ORDER **CRYPTODIRA**
FAMILY **TESTUDINIDAE**
TORTOISES

A family of land-dwelling tortoises, with about 60 species in the group, in 16 or so genera, eight of which occur in Africa. Recent taxonomic work has caused several name changes among East African members. The Leopard Tortoise, originally in the genus *Geochelone*, now has its own genus, *Stigmochelys*; a new species of hinged tortoise, *Kinixys zombensis*, has been resurrected out of *Kinixys belliana*, and after much debate the Aldabran Giant Tortoise is placed in the genus *Aldabrachelys*.

Dependent upon green plants, land tortoises in East Africa are best seen in the rainy season in the savanna and semi-desert national parks. All land tortoises are on CITES Appendix II, meaning they can only be legally exported under permit.

LEOPARD TORTOISE *Stigmochelys pardalis*

IDENTIFICATION: A big, yellow and black land tortoise with a domed, steep-sided shell, a blunt bullet-shaped head, and a hooked beak. The costal and vertebral scales sometimes form blunt, tent-like peaks. The limbs are stout, five claws present on the front feet, four on the back. The anterior surface of the front legs is covered with 3–4 rows of big, non-overlapping scales. Buttock tubercles are present and well-developed. Males have a longer thicker tail than females, with no terminal spike. Adult shells usually 30–50cm long, but may be up to 72cm in dry areas of northern Kenya, hatchlings 5–7cm long and 23–50g; big adults are usually female. Mass usually 7–14kg but up to 60kg. As illustrated, the colour changes during life, the often dotted, black-edged scales of hatchlings changing to yellow or tan, with light or heavy black speckling, dots or radial patterns; in age the markings often fade to a dull uniform brown or grey.

HABITAT AND DISTRIBUTION: Semi-desert, dry and moist savanna, rocky country, coastal woodland and thicket, from sea level to about 1,500m altitude, sometimes slightly higher. Within East Africa, known from the savanna and grassland of north and north-east Tanzania, south-east Kenya to the coast, and skirts the edge of the central Kenyan highlands to north-west Kenya and low eastern Uganda, also known from Kidepo in north-eastern Uganda. Records sporadic from northern Kenya (Sololo, Tarbaj, Wajir Bor, Malka Murri, near Mt Kulal). Apparently unrecorded from most of Uganda, southern and western Tanzania (save between Lake Rukwa and northern Lake Malawi/Lake Nyasa). Elsewhere, west to the Sudan, south to eastern South Africa. Conservation Status: Least Concern on the IUCN Red List. Leopard Tortoises occur and are thus protected in many large East African national parks, including Tsavo, Mara-Serengeti and Mikumi.

Stigmochelys pardalis

NATURAL HISTORY: Diurnal but shelters during the heat of the day. Active in the rainy season, especially mornings, when mating activity occurs. Males fight during the breeding season, butting and pushing their rivals. The males are equally rough with the females during mating, this is accompanied by asthmatic wheezing and hissing. The female digs a deep pit in an open area at dusk for her eggs, softening the soil with cloacal water or urine. On average 4–21 eggs (maximum 30) are laid in darkness, the female then depresses the soil over the pit, using her plastron. Several clutches may be laid each year. Incubation times 4–12 months, depending upon on the temperature. The nests may be excavated by small mammalian carnivores, and the hatchlings are eaten by snakes, birds, monitor lizards and carnivores (even lions will take them). Thus the young ones feed nervously from

Stigmochelys pardalis Leopard tortoises. Top left: Maasai Mara. Top right: Watamu hatchlings. Bottom left: Tsavo River sub-adult. Bottom right: Arusha albino, Stephen Spawls.

cover, dashing out for a bite, but once they reach 20cm or so, they are safe from almost everything (except humans, occasionally lions and hyenas) and can feed and move about more safely. They eat a wide variety of plants, and are recorded eating carrion, and chewing bones and eating hyena faeces to obtain calcium. They have territories, home ranges of 20–350 hectares recorded. Translocated individuals are known to have returned home over distances up to 50km. They are at risk from fire, particularly in savanna areas where the vegetation has become very thick. In parts of East Africa they are respected and known as Mzee Kope ('old man tortoise'); Maasai warriors may decorate them with ochre. Often kept as pets, they are hardy and tame well, but clutches buried in the ground rarely hatch at altitudes above 1,500m as it is too cold.

HINGED TORTOISES *KINIXYS*

A genus of attractive, omnivorous, small (maximum 40cm, usually 25cm or less) yellow, tan or brown tortoises, confined to sub-Saharan Africa and Madagascar. Eight species are known; four in our area. They are most active during the rains, activity is reduced in the dry season; some aestivate. Adults usually have a fibrous hinge on both sides near the back of the carapace (upper shell), enabling the rear shell to shut, protecting the hindquarters and back legs (they tend to hide head first in holes). Their taxonomy is complicated due to the variability of diagnostic external characters (patterns, shell shapes, etc.). Recent molecular taxonomic analysis indicates two forest and six savanna lineages, but the small number of specimens used does not clarify their distribution. However, it appears that in addition to the two known savanna species in East Africa, a third, re-elevated species, the Eastern Hinged Tortoise *Kinixys zombensis* probably occurs on the East African coast and south-eastern Tanzania.

KEY TO THE EAST AFRICAN MEMBERS OF THE GENUS *KINIXYS*

1a	Rear of carapace strongly serrated, in forest......*Kinixys erosa*, Forest Hinged Tortoise [p. 30]	3a	Adults with a variable carapace pattern, either radial or uniform, looks rounded, adult carapace not so wide, ratio of width to plastron length usually less than 0.71 ...*Kinixys belliana*, Bell's Hinged Tortoise [p. 32]
1b	Rear of carapace not sharply serrated, in savanna......2		
2a	Carapace depressed or flattened, shell with a zonary (ringed) pattern......*Kinixys spekii*, Speke's Hinged Tortoise [p. 33]	3b	Most scales on upper carapace with a radial pattern, looks squat, adult carapace relatively wide, ratio of width to plastron length usually 0.72 or more.......*Kinixys zombensis*, Eastern Hinged Tortoise [p. 34]
2b	Carapace convex, shell with a radial (spoked) pattern or no pattern......3		

FOREST HINGED TORTOISE *Kinixys erosa*

IDENTIFICATION: A big hinged tortoise, in forest. Has a slightly concave shell, the scales are black, bright yellow and warm brown, the head is rounded with an elevated snout and the beak has a single cusp. The marginal scales are often upturned at the edges, especially to the rear, giving a bizarre serrated appearance to the back end. The front feet have five claws, the rear feet four. The anterior of the front foot is covered by 4–5 rows of large, overlapping scales. No buttock tubercles, the tail has a claw-like tubercle at the tip, males have longer and thicker tails than females. Average size 15–25cm but may reach 37cm or more, males larger than females. Hatchlings 4–6cm. Adult shell is a vivid mix of bright yellow, brown and black, often the lower half of the pleural scales is bright yellow, giving the impression of a yellow side-stripe. The individual plastron scutes have black radial patches, the seams cream or yellow.

Kinixys erosa

HABITAT AND DISTRIBUTION: Low to mid-altitude forest, said to prefer moist areas; at altitudes between 700 and 1,500m in our area. Two Ugandan records: Nabea in the Budongo Forest and the Mabira Forest, known from just west of Lake Kivu in the DR Congo, so might be in Rwanda. Elsewhere, west to the Gambia, south-west to the Congo River mouth. A fossil record from Songhor in western Kenya indicates that area was forested. Conservation Status: Very widespread in the great central African forests, in undeveloped areas. However, listed as Data Deficient on the IUCN Red List, and is on CITES Appendix II. It is collected both for food, local medicine (said to treat infertility in women) and for the pet trade in West Africa, and thus may be under threat in populated areas; in some bushmeat markets in West Africa the numbers traded have collapsed, indicating overexploitation.

NATURAL HISTORY: Sometimes diurnal but unusually for a tortoise it may be active at night, and is known to travel long distances, up to 275m. Most active in the rainy season in West Africa. The home range is several hectares and usually includes a water body. Shelters under logs, in holes, leaf litter and vegetable debris. Presumably defends itself like other hinged tortoises, i.e. withdraws the limbs and closes the shell. The males fight, but rarely. Mating occurs throughout the year in central Africa, both during the day and at night; the male ramming the female and even turning her over. Between 1–10 large eggs (average 5–8) are laid, either buried or cached under vegetation, and more than one clutch may be laid in a season; the females are sexually mature at 20cm length. The eggs weigh 17–28g, diameter roughly 4cm and can take from 5–10 months to hatch. It is omnivorous, eating mostly fungi (husbandrists fatten their captives on mushrooms), also eats plants, invertebrates and carrion. Prized as a food by some forest peoples, it is hunted with dogs and located by its distinctive smell. They swim well, and in swamp forest in West Africa may actually disperse by swimming with rising flood waters, although a captive Ugandan specimen sank in a bucket.

Kinixys erosa Forest hinged tortoise. Left: Kate Jackson, Eastern Congo. Right: Jean-Francois Trape, West Africa.

BELL'S HINGED TORTOISE Kinixys belliana

IDENTIFICATION: A small brown, or yellow and brown hinged tortoise, in savanna, with a slightly domed shell (especially at the back), a pointed head and a beak with a single cusp. A hinge is usually present in adults at the rear of the shell. The back end of the shell is steep and flattened in adults. The front feet have five claws, the rear feet four. The anterior of the front foot is covered by several rows of large, overlapping scales. No buttock tubercles, the tail has a terminal spine, males have longer and thicker tails than females. Average size 14–18 cm, maximum about 22cm, hatchlings 3.5–4cm. The shell is usually yellow or tan, with black, radial (spoke-shaped) patterns, which may be heavy or light, on each scale. In large males the pattern usually fades to brown or grey. The plastron is yellow.

Kinixys belliana

HABITAT AND DISTRIBUTION: Dry to moist savanna, woodland and coastal thicket, from sea level to nearly 2,000m altitude. Absent from dry eastern and northern Kenya (save the northern border, centred on Moyale). Occurs from western Kenya, through eastern to most of northern Uganda. An isolated record from Kamuhanda, Kigali, Rwanda. Present on Zanzibar, widespread in northeast Tanzania, into Kenya and up the coast to the lower Tana River. Isolated Tanzanian records from Tatanda and Lindi. However, the elevation of the Eastern Hinged Tortoise *Kinixys zombensis* to a full species has created problems with coastal records. Elsewhere, south to South Africa, west to Senegal.

Kinixys belliana Bell's Hinged Tortoise, Stephen Spawls, Entebbe.

NATURAL HISTORY: Diurnal, usually moving during the early to mid-morning and late afternoon, most active during the rainy season. May aestivate during the dry season in holes, under boulders or in rock cracks, where they also shelter at night. If approached, it responds by withdrawing the limbs and closing the shell; if picked up it may urinate, defecate and hiss. They have large home ranges but they occur at low densities, 2–3 per hectare. The males fight in the breeding season. Breeding details poorly known, but females in South Africa lay 2–7 elongate eggs, roughly 3×4cm, in a pit in the shade of a rock or bush. Several clutches may be laid in a season. Omnivorous, eating fungi, plants, small invertebrates (especially millipedes) and carrion. There is a seasonal shift in diet; they eat mostly plants in the west season and animals in the dry. Captive specimens fond of red fruits and known to live up to 22 years.

SPEKE'S HINGED TORTOISE Kinixys spekii

IDENTIFICATION: A small hinged tortoise, in savanna, with a flattened shell, the scales are yellow or tan with darker margins, the head is rounded and the beak has a single cusp. A hinge is usually present in adults at the rear of the shell. The front feet have five claws, the rear feet four. The anterior of the front foot is covered by several rows of large, overlapping scales. No buttock tubercles, the tail has a terminal spine, males have longer and thicker tails than females. Average size 15–19cm, maximum about 22cm, hatchlings 3–6cm. Hatchlings have a fairly rounded shell. Adult shell is usually tan, yellow or light brown, with dark brown (sometimes black) margins, the marginal scales often have black or brown sutures (edges), giving the impression of short black vertical bars on the sides. The plastron is yellow or tan, sometimes with black speckling or edging. Note: theoretically this species may be distinguished from the Eastern and Bell's hinged tortoises by its zonary pattern and more flattened shell, but intermediate specimens are not uncommon, especially on the Kenyan coast where all three species occur.

Kinixys spekii

HABITAT AND DISTRIBUTION: Dry and moist savanna and coastal thicket, from sea level to 1,600m altitude, often in rocky areas, around inselbergs and crystalline hills. Occurs westward from the Kenyan coast through Tsavo and Ukambani to the Kapiti and Athi Plains, north to the Tana River, west across northern Tanzania to Ujiji, Kigoma and southern Burundi. Isolated records from the southern Kerio Valley, Eastern Rwanda and the Ruzizi Plain. Records lacking from most of south-central Tanzania, no definite Uganda records. Elsewhere, west into Shaba (DR Congo), south to northern Botswana and northeast South Africa.

Kinixys spekii Speke's Hinged Tortoise. Left: Stephen Spawls, Watamu. Right: Caspian Johnson, Ugalla.

NATURAL HISTORY: Diurnal, usually moving during the early to mid-morning and late afternoon, most active during the rainy season. May aestivate during the dry season in holes, under boulders or in rock cracks, where it also shelters at night, often uses exfoliation cracks and rock fissures for shelter, in eastern Kenya may be found in the same sort of places as the Pancake Tortoise *Malacochersus tornieri*. If approached, withdraws the limbs and closes the shell, if picked up it may urinate, defecate and hiss. In southern Africa they are known to have large home ranges and species density is low, 2–3 per hectare. The males fight in the breeding season. Breeding details poorly known, 2–6 elongate eggs recorded, roughly 3×4cm, mass 20–30g. Nest sites are invariably in the shade of a rock or a bush. On the Kapiti Plains, hatchlings appear in April and May. Omnivorous, like Bell's Hinged Tortoise, eating a range of fungi, insects, millipedes and snails, also carrion; mostly fungi and plants in the west season and animal food in the dry. A South African animal was eaten by a young lion.

EASTERN HINGED TORTOISE   Kinixys zombensis

IDENTIFICATION: A small hinged tortoise. Closely resembles Bell's Hinged Tortoise. The shell is slightly domed and steep at the back, the head is rounded and the beak has a single cusp. The front feet have five claws, the rear feet four. The anterior of the front foot is covered by several rows of large, overlapping scales. Average size is 17–20cm, maximum about 22cm, hatchlings 3–6cm. Hatchlings have a fairly rounded shell and their individual carapace scutes have a yellow ring, with a darker centre and margin. In adults the markings appear more radial, but gradually fade. The plastron is yellow or tan, sometimes with black speckling. Really big adults fade to a uniform dull yellow-brown or grey.

Kinixys zombensis

HABITAT AND DISTRIBUTION: Dry and moist savanna, woodland, coastal thicket and forest; in our area at altitudes from sea level to around 400m. In Kenya, known only from the coastal strip, widespread across south-eastern Tanzania, thence southwards along the east coast through Mozambique to eastern South Africa.

NATURAL HISTORY: As for other hinged tortoises; diurnal, most active during the mornings and evenings in the rainy season; aestivates during the dry season. Eats fungi, plants, small invertebrates and carrion; in the Shimba Hills often eats millipedes. Usually lays from 2–7 eggs, occasionally up to ten. Interestingly, many of the Shimba Hills specimens have cracked shells; it is speculated they have been trodden upon by elephants.

Kinixys zombensis Eastern hinged tortoise. Left: Bio-Ken archive, Kilifi. Right: Daniel Liepack, Shimba Hills.

PANCAKE TORTOISES / FLAT TORTOISES
MALACOCHERSUS

A monotypic genus containing a single species, the Pancake Tortoise. They have a curiously flat, flexible shell, an early museum curator asked the collector how he had managed to soften the shell. Their other adaptation to life in rock crevices (apart from flexibility) is their extraordinary climbing ability.

PANCAKE TORTOISE / FLAT TORTOISE *Malacochersus tornieri*

IDENTIFICATION: A small, broad, flat tortoise with a strange, soft flexible shell. The snout is rounded, the beak has two or three cusps. The body is broad, flat and flexible, due to the thin, weak carapace bones. The front feet have five strong digging claws, the rear feet four. The anterior of the front foot is covered by big overlapping scales. Males have longer and thicker tails than females. Average size 13–16 cm, maximum size about 18 cm, hatchlings 3–5 cm, 10–12g. Colour brown; the shells of juveniles are vivid with orange and black festoon markings, in adults the centre of each carapace scale is usually brown or tan, with radial markings, which may be fine or broad, big adults often become uniform brown, or brown with faded black speckling; the plastron also has radial markings.

Malacochersus tornieri

HABITAT AND DISTRIBUTION: Found at low altitude, rarely above 1,200m, in and around small rocky hills of the ancient crystalline basement, in dry savanna, always near rocks. The Kenya population occurs south from Mt Kulal and Mt Nyiro through the Ndoto Mountains and Mathews Range, along the rocky escarpments of the western Uaso Nyiro River, in Samburu NR and Shaba (although not known from Meru National Park), south through the inselbergs of Tharaka and the low eastern foothills of Mt Kenya, south-east into Ukambani, to as far south as Ithumba Hill in Tsavo East. Also in north-central Tanzania, south from Lakes Eyasi and Manyara through Tarangire and the dry central plateau to Ruaha National Park. Isolated records from Mkomazi National Park, the Serengeti kopjes and west of Smith Sound in northern Tanzania. A small population was recently located in Zambia, at Nakonde, just across the border from Tunduma in Tanzania. Conservation Status: IUCN Vulnerable A1bd; exploited in Tanzania as it is easily found in its rocky hiding places.

NATURAL HISTORY: Diurnal, usually active during the morning and later afternoon. May occasionally bask. Active about one day in four. When inactive, shelter in rock cracks and under large boulders; up to 11 have been found in one crevice; in eastern Kenya they are often found in true pairs in such places. Pancake tortoises have extraordinary climbing ability, they can ascend steep rocks and will climb out of wire netting enclosures. If threatened, they wedge themselves in their cracks with their claws and carapace and are almost impossible to dislodge. They will slide off rocks if located in the open. The males fight in the breeding season, they then pursue a female, biting at her legs and neck. Mating recorded in several months: January, February, March, April and May. One large elongate egg (rarely two), 5×3 cm is laid, often in July or August, so as to hatch at the beginning of the rainy sea-

son, which in eastern Kenya is November and December. More than one clutch may be laid per year. Incubation time in captive animals varies from 3–6 months. Diet: herbs, succulents, flowers, fruit, beetles and other arthropods.

Malacochersus tornieri Pancake Tortoise, Left: Stephen Spawls, Dodoma. Inset: Anthony Childs, Tsavo East Plastron Colours.

ALDABRAN AND SEYCHELLOIS TORTOISES
ALDABRACHELYS

A genus with three species, two of which are extinct. Gigantic land tortoises, the two extinct forms lived on Madagascar, the only living species originates from Aldabra atoll in the far western Indian Ocean. There has been vigorous and acrimonious debate on the true name, origins and status of the single surviving species.

ALDABRAN TORTOISE *Aldabrachelys gigantea*

IDENTIFICATION: A huge, black, grey or brown land tortoise, the second largest in the world, with a domed, steep-sided shell, a blunt bullet-shaped head, and a beak that is weakly hooked. Introduced on Zanzibar. The limbs are very stout, five big blunt claws present on the front feet, four on the back. The anterior surface of the front legs is covered with big, non-overlapping scales. Buttock tubercles are not present. Males have a longer thicker tail than females, with no terminal spike. Maximum size 304kg (a huge male in the Seychelles), adult shells usually 70–90 cm long, mass 40–100 kg, hatchlings 8–9 cm long. The shell is grey, brown or black, hatchlings and juveniles black. Males have a concave plastron.

Aldabrachelys gigantea

HABITAT AND DISTRIBUTION: On Aldabra, they live on the coral plateau, in enclosures on Zanzibar. Conservation Status: The population on Changuu (Prison) Island, off Zanzibar, was seriously threatened by poaching, many were stolen, consequently most of the babies were relocated in the grounds of Livingstone House on the main Zanzibar Island. A few captive specimens are maintained along the Kenyan coast. Strictly protected on Aldabra, a world heritage site, with a population of 15,000 or more.

NATURAL HISTORY: An island tortoise, active by day, in any suitable habitat including beaches, mangroves, coral plateau and grassland, but avoids the heat of the midday by sheltering, in thickets, under shady trees, tree trunks, may partially submerge in mud wallows or pools. They float without difficulty and can survive for a long time at sea; in 2013 one specimen washed up on the beach at Kiunga, Kenya, alive and well (although crusted with barnacles), having presumably drifted from Aldabra. The female digs a deep pit at dusk for her eggs, softening the soil with cloacal water or urine. On average 4–16 spherical (5cm diameter) eggs are laid in darkness, in a grassy or scrubby area. Babies appear between October and December on Aldabra, 3–6 months after laying. Aldabra tortoises eat a variety of plants and also carrion. Known to live to 152 years, but 65–90 years more usual.

Aldabrachelys gigantea Aldabran Tortoise. Stephen Spawls, Seychelles hatchling.

Eretmochelys imbricata
Hawksbill Turtle
Watamu Turtle Watch

SUB-ORDER **CRYPTODIRA**
SUPERFAMILY **CHELONIOIDEA**
SEA TURTLES

The marine turtles are the remnants of a huge, ancient group, that entered the sea about 100 million years ago. Some were huge; **Archelon** was over 3m long. Seven species of sea turtle, in six genera now remain. They are adapted to a life in the sea; they cannot withdraw their limbs, which are modified into flippers, and they can excrete excess salt through the tear ducts. They come ashore only to lay eggs; up to 1,000 in a season, and the sex of the clutch is determined by incubation temperature; higher values produce all females. Officially, all are protected, but in reality they, and their eggs, are often harvested (they are fairly helpless on shore) and their nesting beaches and feeding areas threatened by development, artificial lighting, coastal armouring, pollution and beach erosion. On land, they look dull-coloured, clumsy and ridiculous, but a free turtle in the sea is a creature of colour, grace and beauty, it is important that they are not allowed to become extinct.

Caretta caretta
Loggerhead Turtle
Watamu Turtle Watch

SUB-ORDER **CRYPTODIRA**
SUPERFAMILY **CHELONIOIDEA**
FAMILY **CHELONIIDAE**
MODERN SEA TURTLES

Advanced sea turtles that have strengthened limbs to increase their swimming efficiency, they have a hard shell. Fossils are known from the Cretaceous period, there are 33 genera, five of which are still living; four of these are represented in East Africa.

RIDLEY SEA TURTLES *LEPIDOCHELYS*

Ridleys are the smallest sea turtles, with broad shells. They are known for their *arribadas*, huge congregations (often many thousands) of Ridley turtles coming ashore on a single beach on one night to lay eggs simultaneously. Theoretically, there are so many eggs that the predators cannot cope, and no matter how many they eat, many remain to hatch safely. Sadly, humanity has found ways of dealing with the lot; Ridleys are now endangered, and few huge *arribadas* remain.

Two species of Ridley turtle are known. Kemp's Ridley *Lepidochelys kempi* is a critically endangered species found largely in the western Atlantic, it nests only on Mexican beaches. The Olive Ridley is found in East Africa.

OLIVE RIDLEY TURTLE Lepidochelys olivacea

IDENTIFICATION: The smallest sea turtle, with a smooth, round carapace, slightly wider than long. The general impression is of a deep-bodied, dark, plate- or saucer-shaped turtle. The head is sub-triangular, the snout is short and not compressed, the beak is slightly hooked and not bicuspid; the edges of the jaws are smooth. Two claws are distinguishable by the time the animal reaches a carapace length of 28.5cm. The tail is short in hatchlings and juveniles, longer in adult males. Maximum shell length about 70cm, and mass up to 45kg, average 50–60 cm, juveniles carapace length 4.3–4.5 cm. They reach sexual maturity in 7–9 years. The carapace is olive-grey, top of head brown with the scales outlined in yellow-white, dorsal skin surfaces and flippers dark grey, lower jaw yellow-brown, leading edges of the flippers light yellow to brown. Hatchlings are olivaceous-black, with plastral keels, lower jaw and cutting edges of upper jaw with white patches, throat grey. The trailing edges of the flippers have white flecks. The skin of the ventrum is greyish-white.

Lepidochelys olivacea

HABITAT AND DISTRIBUTION: Circumtropical, in the warm waters of the Atlantic, Pacific and Indian oceans. Usually coastal, protected shallow waters, sometimes in the open sea. In East Africa, sporadic records along the coast include the Tana delta, Watamu, Ungwana Bay, Mombasa, Pemba, Unguja (Zanzibar) island, Kunduchi near Dar es Salaam and Lindi, two nested on Maziwi Island in February 1974 and there are a few nesting records from Mombasa and Kilifi areas in Kenya (although many nest on the islands off Mozambique). Conservation Status: Vulnerable A2bd. Probably the most common sea turtle worldwide, but endangered due to trade in its limb and neck leather, also often accidentally taken in shrimp trawls.

NATURAL HISTORY: Olive Ridleys favour shallow water but will forage in deep water, they have been taken in prawn trawls at 110m depth. Sometimes in large 'herds', moving together between nesting beaches and foraging areas and sleeping communally on the surface. In some places (not in East Africa, where there are only a few nesting records), hundreds of females may come ashore simultaneously to nest (an *arribada*), usually on a rising afternoon tide; in the past *arribadas* of over 46,000 turtles have been recorded. Mating occurs offshore from the nesting beaches at the time of nesting. The female lays two clutches (occasionally three), 17–60 days apart, the interval controlled by tides and weather. About 110 (range 30–168) white, soft-shelled round eggs are laid, averaging 4×3.5cm at a depth of about 50cm. She returns to nest at one- to two-year intervals. The nest construction and egg-laying process is rapid, usually taking less than an hour, the female finishing off by banging the sand down over the nest with the thickened sides of her plastron. Incubation is short, 40–60 days. Olive

Olive Ridley Turtle, Paul Freed, Atlantic specimen.

Ridleys are carnivorous, feeding on marine invertebrates, squid and fish, they will take crabs in mangrove areas.

LOGGERHEAD TURTLE *CARETTA*

A genus with a single living species, although there are a number of fossil forms.

LOGGERHEAD TURTLE *Caretta caretta*

IDENTIFICATION: A big reddish-looking turtle with a huge, broad head. The snout is relatively short, the beak is slightly distinctly hooked and unicuspid; the edges of the jaws are smooth but have broad surfaces which allow crushing of hard-shelled prey. There are two claws on all four flippers; over time the outer claw usually becomes recessed, at maturity it is barely visible. The tail is longer in adult males than females. Maximum size in African adults 98.5cm, slightly larger elsewhere, maximum mass 140kg, average size is around 70–90 cm. Hatchlings are 4–5 cm long. The carapace of hatchlings is grey-brown when dry, pale red-brown when wet, becoming vivid streaked red-brown in adults, each scale with a dark edging. Occasionally very dark brown individuals occur, and rarely albino hatchlings are found. The underparts and skin are dark brown to black in juveniles but the centre of the plastral shields are light, the beak and eyelids are black or dark brown, becoming lighter in adults.

Caretta caretta

HABITAT AND DISTRIBUTION: World-wide, in tropical and temperate waters, usually fairly near coastlines and in the vicinity of reefs, but has been seen 240km from the land. Sometimes enters river estuaries. Fairly rare on the East African coast, but known from Watamu, Mombasa, Zanzibar, Kilwa Masoko, Mafia and Songo Songo Island. The Local Ocean Trust at Watamu estimates out of every 1,500 turtles recorded per year, about five are loggerheads. No breeding records for East Africa, but they bred on the north-eastern coast of South Africa and southern Mozambique. Conservation Status: Vulnerable A2b.

Caretta caretta Loggerhead Turtle Watamu Turtle Watch.

NATURAL HISTORY: In southern Africa juveniles spend the first three years of life drifting in offshore waters before returning to the coastline. They are omnivores, eating invertebrates (crabs, clams, jellyfish) and seagrass. They become sexually mature when the shell is about 60–70cm long, at an age of 10–30 years. Females may breed only once in 2–3 years. They mate in the summer in southern Africa, by day and night. Up to five clutches, and a total of 500 eggs may be laid each season, the female emerging at night to dig a body pit and then a flask-shaped egg pit. Incubation time is from 45–70 days. Individuals tagged in South Africa have been recaptured in Tanzania, having travelled over 2,500km.

HAWKSBILL TURTLE ERETMOCHELYS

A genus with a single living species.

HAWKSBILL TURTLE Eretmochelys imbricata

IDENTIFICATION: A small sea turtle, with bird-like beaked jaws (hence the common name) and an oval carapace. The beak is not hooked and no cusps are present, its edge is smooth. The forelimbs have very long digits, each flipper has two claws. Males are distinguished from females by having a longer, narrower shell, longer tail and a small plastron concavity. Maximum shell length about 90cm (occasionally larger), average 50–80 cm, the hatchlings have a carapace 4–5 cm long. The juvenile carapace is uniform brown, the plastron is dark, each scale has a large dark spot. The adults are beautifully marked, the carapace scales are warm translucent amber, beautifully marked with radiating red, honey, yellow, black and brown streaks. The plastron is yellow or tan, with brown markings, the flippers are black, the scales yellow-edged, the head is yellow with black centred scales.

Eretmochelys imbricata

HABITAT AND DISTRIBUTION: Circumtropical, in the warm waters of the Atlantic, Pacific and Indian oceans, sometimes in warmer temperate waters. Usually in shallow coastal waters, bays, areas with mangroves and estuaries, coral reefs and rocky areas with coral. One of the two most commonly-seen turtles on the East African coast (the other is the Green Turtle *Chelonia mydas*). Recorded from most coastal localities including Lamu, Malindi, Watamu, Kilifi, Mombasa, the Tanzanian islands, Dar es Salaam, Tanga and Lindi. Conservation Status: Critically Endangered A2bc; heavily persecuted both in the past and present for its shell ('tortoiseshell').

Eretmochelys imbricata Hawksbill Turtle Left: Watamu Turtle Watch. Right: Stephen Spawls, Hatchling, Seychelles.

NATURAL HISTORY: Forages in shallow water, especially around reefs, often nervous but can become habituated to humans. The juveniles feed mostly on floating vegetation, and will scavenge around weed slicks, but the adults are carnivorous, favouring invertebrate prey such as sponges, jellyfish, molluscs and fish. They become sexually mature at 8–10 years. Mating takes place in shallow water off the nesting beaches. Nesting occurs both by day and night. Two to four clutches are laid, at intervals of 15–19 days, between December and February in East Africa. Fifty to 200 eggs are laid each time, roughly 4cm in diameter, and take about two months to hatch. The flesh of some hawksbills turtles may be toxic, as a result of feeding on poisonous corals and algae.

GREEN TURTLE *CHELONIA*

A genus with a single living species.

GREEN TURTLE *Chelonia mydas*

IDENTIFICATION: A fairly big, hard-shelled, deep-bodied sea turtle with a small blunt head, the beak is not hooked and has no cusps. Younger individuals are flattish, big adults can develop huge domed shells; they look sub-circular in profile. The cutting edge of the horny jaw covering is serrated, this may be an adaptation to grazing. The shell is ovate and smooth, the thin scales do not overlap. There is a single claw on each of the flippers, young ones have two claws on the front flippers. The males have longer tails than females. Maximum carapace size about 1.4m, maximum mass in the Indian Ocean 200kg (Atlantic specimens larger), average 90cm to 1.2m, hatchlings 4–5 cm. The colour is quite variable, olive-green, yellow-green, greenish-brown or black; the vertebrals may be streaked or uniform. Hatchlings are dark, with bronzing on the vertebrals and a white border and plastron, the juvenile carapace is dark grey at 20cm length.

Chelonia mydas

HABITAT AND DISTRIBUTION: Tropical and warm temperate seas. It migrates across open sea, but likes to forage in warm, shallow areas with abundant vegetation. One of the two most common turtles on the East African coast, and recorded from most coastal localities including Kilwa, Dar es Salaam, the big Tanzanian islands (Pemba, Zanzibar, Mafia), Tanga,

Chelonia mydas Green Turtle Left: Watamu Turtle Watch. Middle: Watamu Turtle Watch. Right: Michele Menegon hatchling Tanzania.

Mombasa, Kilifi, Watamu, Malindi and Lamu. The most common nester on the East African coast. Most often observed where there are seagrass beds. Conservation Status: Endangered A2bd. Despite nominal total protection, often harvested on the East African coast, mostly for meat and oil, but the decorated shells are also sold, and under threat from coastal development, with altered beaches, night lights and associated scavengers.

NATURAL HISTORY: Often foraging in relatively shallow water, in East Africa enters the landward lagoons inside the reef. Sometimes rests, basking, on the surface and may bask on the shore, possibly to gain vitamin D. Can become habituated to humans where not persecuted. Green turtles become sexually mature in 10–15 years. Mating occurs in the water off the laying beaches, the males hook the enlarged claws on their front flippers over the leading edge of the female's shell. Nesting occurs throughout the year, in small numbers in a number of areas, with a peak between December and May and a low between September and November. The females come ashore at night, they move up the beach by a curious 'humping' movement, moving both forelegs together (all other turtles crawl up the beach by using their flippers alternately). Between 12 and 240 circular eggs (usually 100–150) are laid; the female returns to the beach several times, at intervals of 10–20 days, to lay further clutches. Incubation takes usually 45–60 days, and all the hatchlings emerge at once. Juveniles feed on small invertebrates, adults largely herbivorous, grazing on algae, seaweed and seagrass, but may also take marine invertebrates.

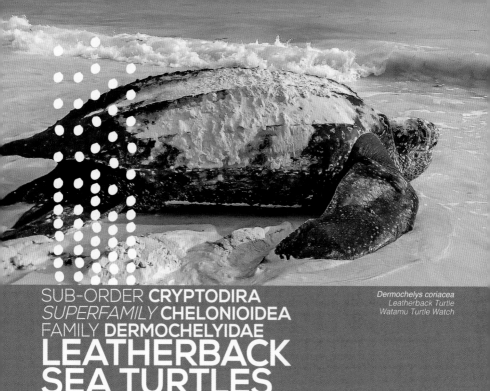

SUB-ORDER **CRYPTODIRA**
SUPERFAMILY **CHELONIOIDEA**
FAMILY **DERMOCHELYIDAE**
LEATHERBACK SEA TURTLES

Dermochelys coriacea
Leatherback Turtle
Watamu Turtle Watch

A family of giant sea turtles, the leatherbacks, characterised by the lack of nasal bones, extreme reduction of the carapace bones and a unique internal shell of small polygonal bones. There are four fossil genera but only a single living species.

LEATHERBACK TURTLE *Dermochelys coriacea*

IDENTIFICATION: Unmistakeable; a huge, black or dark brown turtle without horny plates, it is covered with a smooth leathery skin that resembles vulcanised rubber. On its back there are five long ridges, one on each side and five on the plastron. The snout is blunt, the head is huge and rounded. The beak is bicuspid, short and hooked, with sharp edges. The neck is short and thick. The flippers are huge and have no claws. The males have concave plastra and are rather flattened in profile, their tails are longer than the hind flippers, the female tail is half as long. Maximum carapace length is about 1.8m, maximum mass about 850kg; average size 1.3–1.7m, hatchlings are about 6cm long. Adults are black or deep brown, spotted with white, the skin ridges are paler, the plastron and underparts of the head are whitish, with some pink and grey-black coloration usually present. There is a curious distinct pink spot on the dorsal surface of the head of an adult, just above the pineal gland. In juveniles the vertebral ridges resemble lines of pale beads and the white edging to the flippers is conspicuous.

Dermochelys coriacea

HABITAT AND DISTRIBUTION: Found throughout the waters of the Atlantic, Pacific and Indian oceans, ranging to high latitudes, it is tolerant of cold conditions. Essentially an animal of the deep sea, but there are a fair number of East African records, including strandings at Kilifi, Lamu and Pemba, and offshore sightings off Pemba, Zanzibar, Mtwara, Watamu and Shimoni; Watamu Turtle watch report regular sightings of 'very large black turtles'. In 2014 a Leatherback nested at Watamu, this is a first for the East African coast, they don't like beaches with coral, although they nest in numbers in Mozambique and South Africa. Conservation Status: Vulnerable A2bd. At risk from coastal development, loss of nesting beaches, etc., and they sometimes die after eating floating bits of plastic, mistaking these for jellyfish.

NATURAL HISTORY: Highly adapted to life at sea, they can dive deep (up to 1,200m, although 50–60m is more usual) and stay down for an hour or more. The circulatory systems in their flippers and the oil content of their flesh are indicative of a warm-blooded animal, they can metabolise fat to generate internal heat and their huge size aids heat retention; one specimen had a body temperature 18°C higher than the water, such regulation is not possible by behavioural methods. They have extremely rapid skeletal and body growth, it is suggested that they are sexually mature at 3–5 years. They undertake long journeys, and enter cold currents to feed. Courtship and mating have never been described. In Maputaland, South Africa, gravid females come ashore on moonless nights between December and January, preferably those with high tide around midnight. They lay 50–170 (average about 100), billiard-ball-sized spherical or ellipsoidal eggs. The female comes back to the beach every 9–11 days, until she has deposited about 1,000 eggs. The eggs take 53–75 days to hatch. The hatchlings emerge at night to avoid predation by ghost crabs, gulls and fish; the nests themselves are also at risk from jackals. Leatherbacks have weak jaws and adults feed largely on jellyfish; moving into high latitudes in summer to follow the jellyfish flotilla.

Trionyx triunguis
Nile Soft-shelled Turtle
Stephen Spawls

SUB-ORDER **CRYPTODIRA**
SUPERFAMILY **TRIONYCHOIDEA**
FAMILY **TRIONYCHIDAE**
SOFT-SHELLED TURTLES

The softshells are a family of freshwater turtles without hard scales, their big disc-shaped shells are covered with skin; their familiar name referring to the three claws on their forelimbs. Over 30 species are known; two species occur in East Africa: the Nile Soft-shelled Turtle *Trionyx triunguis*, the only species in that genus, and the Zambezi Flap-shelled Turtle *Cycloderma frenatum*; two species of *Cycloderma* occur in Africa. Softshells have long necks, long tubular snouts and prominent eyes on top of the head, enabling them to see above the surface and breathe easily in water without exposing much of the head. They are edible and seriously exploited for food in parts of their range, as they have an oily but palatable flesh.

NILE SOFT-SHELLED TURTLE *Trionyx triunguis*

IDENTIFICATION: A large, flat, soft-shelled water turtle, without scales; body and shell covered with skin. It has a long, pointed tubular nose, prominent yellow eyes on top of the head and a long neck, which can be completely withdrawn under the shell. Carapace oval. The limbs are broad and webbed, with conspicuous nail-like claws on three of the four toes. The males have longer thicker tails than the females. Shells of large females up to 1.2m long, average 50–80 cm, hatchling size 3–4cm. Males are usually smaller than females. Mass up to 25kg. Shell brown, olive or dull green, spotted to a greater or lesser extent and edged yel-

Trionyx triunguis

low-white, limbs brown with yellow spots, whitish-yellow below.

HABITAT AND DISTRIBUTION: Permanent lakes, dams, large and small rivers, in deep and shallow water, known to enter the sea. From sea level to 1,500m altitude. In East Africa known from Lake Turkana, Lake Albert and the White Nile downstream from Murchison Falls. Probably occurs on the Daua River on Kenya's north-eastern border, known from the Juba River in Somalia. Elsewhere: west to Senegal, south-west to northern Namibia, north along the Nile to Egypt and the eastern Mediterranean coast. Not found in the Tana or any Kenyan rift valley lakes other than Turkana, as they were never connected to the Nile system. Conservation Status: Not assessed. On CITES Appendix III; may be traded with permits. Widespread in west and central African rivers and lakes, and thus probably under no threat, but eaten in parts of its range; extinct in Egypt.

Trionyx triunguis Nile Soft-shelled Turtle, Stephen Spawls, Lake Turkana.

NATURAL HISTORY: Long-lived, adults have survived 50 years in captivity. Spends most of its life in the water, active by day, possibly by night. Shy, but may bask. Shuffles down in shallow water, partially concealing itself with sand or mud. Courtship details poorly known, but a captive male was observed on top of a female, biting her neck vigorously. Nests in the dry season to avoid flooding, several clutches, of between 25–60 eggs, may be laid in a 25cm deep pit in the sand. In West Africa incubation time is 70–80 days. Mostly carnivorous, eating fish which it snatches from ambush by 'gulp feeding'; the neck is shot forwards, the mouth opens and the throat expands, sucking water and prey into the vacuum. Other prey items include amphibians, reptiles, crabs, insects and floating plant material. Capable of aquatic respiration, absorbing up to 70% of oxygen through the skin. Bites fiercely if incautiously handled. Eaten by the people living around Lake Turkana.

AFRICAN FLAP-SHELLED TURTLES *CYCLODERMA*

A genus containing two African species, one in our area. Like other softshells, their big disc-shaped shells are covered with skin. They have paddle-like limbs, and there is a curious big flexible skin flap on each side of the rear plastron, which conceals the back legs when they are withdrawn.

ZAMBEZI FLAP-SHELLED TURTLE/ZAMBEZI SOFT-SHELLED TURTLE *Cycloderma frenatum*

IDENTIFICATION: A big, flat, soft-shelled water turtle, without scales; body and shell covered with skin. It has a fairly long, pointed tubular nose, bright yellow eyes and a long striped neck, which can be completely withdrawn when threatened, which partially closes the front of the shell. Carapace oval. The limbs are broad and webbed, with conspicuous long, sharp black claws. The males have longer thicker tails than the females. Shells of large adults up to 55cm long, average 30–50cm, hatchling size 4–5cm. Adult mass up to 18kg. The shell is brownish with irregular, black, sub-rectangular mottling, that may form irregular dark crossbars, but may be dull brown or greeny-brown with faint marking in large adults. The shell margin is speckled yellow or white and has grey-white edges, the plastron is pale (pink, ivory, cream or dead white), sometimes with black speckles or streaks. The head is striped cream and black in juveniles, brown and black in adults, chin and throat paler, limbs mottled grey and white. A xanthic (yellow, without pigment) specimen was collected in the Selous Game Reserve.

Cycloderma frenatum

HABITAT AND DISTRIBUTION: In clear streams, rivers, lakes (and presumably swamps) in southern Tanzania; at altitudes between sea level and around 1,000m. Known from Lake Malawi/Lake Nyasa, the Rovuma, Kilombero and Rufiji Rivers; might be in the Ruaha River. Elsewhere, south to Mozambique and south-eastern Zimbabwe. Conservation Status: Lower Risk/Near Threatened; it is fairly widespread, shy and hard to catch, so probably not under any threat as yet.

NATURAL HISTORY: Poorly known. Spends most of its life in the water, active by day, possibly by night. Shy, but known to bask on the water surface and on shorelines. May shuffle down in shallow water, partially concealing itself with sand or mud. Courtship details not known. Comes ashore at night to lay eggs and migrates across flood plains. Clutches average 15–22, maximum 25 spherical eggs, roughly 3cm in diameter, laid between January and April. A Lake Malawi/Lake Nyasa female laid her eggs at night in January, in a 40cm pit, 12m from the shore. Incubation period long, 8–11 months, hatchlings have been collected between December and February. It is carnivorous, eating fish (which it snatches from ambush with a rapid neck movement), aquatic insects and crabs, possibly amphibians. It digs for snails and mussels. When handled, flapshells will try to scratch and bite. Preyed on by crocodiles and otters, and this turtle and its eggs are eaten by people living around Lake Malawi/Lake Nyasa.

Cycloderma frenatum Zambezi Flap-shelled Turtle Left: Rufiji River. Middle: Ruvuma River, Right: Walter Jubber, Selous Game Reserve (Xanthic).

Pelusios broadleyi
Lake Turkana Hinged Terrapin
Roberto Sindaco

SUB-ORDER **PLEURODIRA**
SUPERFAMILY **PELOMEDUSOIDEA**
FAMILY **PELOMEDUSIDAE**
AFRICAN SIDE-NECKED TERRAPINS

A sub-order of three families and just under 100 tropical species; one family occurs in our area.

The sub-order Pleurodira are an ancient group, and occur only in South America, Africa, Madagascar and Australia; their ancestors probably lived on the supercontinent Gondwana. There are two genera in the primitive aquatic family Pelomedusidae; the hinged terrapins *Pelusios* and the helmeted terrapins, *Pelomedusa*.

HELMETED OR MARSH TERRAPINS *PELOMEDUSA*

Originally, this African genus contained a single species, the Helmeted Terrapin *Pelomedusa subrufa*, widely distributed throughout sub-Saharan Africa, living mostly in temporary water bodies. A recent review based on DNA indicates that *Pelomedusa* is a species complex with at least twelve evolutionary lineages or clades; names have now been given to most of these lineages. At least three such species occur in East Africa but they cannot (as yet) be distinguished by any external characters, only by differences in DNA sequences. Thus the notes below apply to all three East African forms, although we have recorded what is presently known of their individual distributions.

HELMETED TERRAPIN *Pelomedusa species*
HELMETED TERRAPIN *Pelomedusa subrufa*
NEUMANN'S MARSH TERRAPIN *Pelomedusa neumanni*
TANZANIAN MARSH TERRAPIN *Pelomedusa kobe*

IDENTIFICATION: A flat, smelly terrapin with a big broad head and a protruding snout, with two (sometimes three) little tentacles (barbels) under the chin, which are used during mating. The neck is long and muscular, for seizing prey and righting themselves if upside down. The shell is hard, fairly flat and thin. There is no plastron hinge. The limbs are broad and webbed, with sharp claws. The males have longer thicker tails than the females, they also have narrower, flatter shells and grow larger than females. Shells of large adults up to 32cm long, average 15–20cm, hatchling size 2.5–3.5cm. Adult mass up to 2.5kg. The carapace may be black, grey or brown, often coated with green algae. Plastron yellow, brown or black. In juveniles the bridge (where plastron joins the carapace) may be vertically barred black and white/cream.

Pelomedusa species

HABITAT AND DISTRIBUTION: Dry and moist savanna and semi-desert, from sea level to about 1,600m altitude. Usually in temporary or small water bodies; even small rain pools. Rarely found in larger rivers and lakes, where it is at risk from crocodiles. Known localities for the three species are as follows: Helmeted Terrapin: north slope of Mt Kilimanjaro; Neumann's Marsh Terrapin: Kakamega, Lake Victoria, Mangara River (Lake Manyara); Tanzanian Marsh Terrapin: Naberera (Maasai Steppe, Tanzania), Tabora, just south of Morogoro. All other records from East Africa cannot be assigned with any certainty to any of the three known species, but the following notes apply to the complex. Widespread within East Africa, but records lacking from many areas. Not found in the high central Kenyan rift valley. A handful of records from northern Kenya (Laisamis, Buna, Lokomarinyang, the entire length of the Kerio River), few records from northern or south-western Uganda or southern Tanzania, where it almost certainly occurs, probably in suitable areas of Burundi, (it is known from the Ruzizi Plain in the DR Congo and in northeastern Rwanda), but no definite records.

NATURAL HISTORY: Aquatic, basks in the water, floating on the surface, or lying on the bank. They bury themselves in mud when water bodies dry up, and may stay there for several years. They emerge after rain (this sudden appearance of many individuals gives rise to the legend that they have dropped from the sky). They walk long distances across country looking for pools, they can smell water and may climb into tanks. They can run relatively fast. If picked up they will scratch, bite and produce a foul-smelling fluid from glands on the flanks. Mating takes place in the water, the male rubs his snout against the female's hind quarters, if she is receptive he grasps her shell with his feet and rubs his barbels onto her head, he may expel a stream of water from his nostrils over her head. Eggs are deposited in a pit dug by the female near the water. Clutch size usually 10–30 eggs, maximum 42, roughly 3.5×2.5cm, mass 10–15g. Hatchlings were collected in April and May in the Athi River area. Helmeted terrapins are opportunistic carnivores, eating any small animal they can seize, depending upon how big they are. Even drinking birds may be taken, and carrion; occasionally plant material is eaten; in Nigeria bigger specimens took rodents and birds occasionally. Captive specimens will take food from the hand and have lived up to 16 years.

Top left: *Pelomedusa subrufa* (Helmeted Terrapin) Stephen Spawls, Botswana. Lower left: *Pelomedusa kobe* (Tanzanian Marsh Terrapin) Michelel Menegon, Selous Game Reserve. Right: *Pelomedusa neumanni* (Neumann's Marsh Terrapin) Stephen Spawls. Top: Nairobi. Bottom: Watamu, hatchling.

HINGED TERRAPINS *PELUSIOS*

An genus of medium to large hinged water terrapins with dark shells, living mostly in permanent water bodies, although a few inhabit temporary pools; often seen basking. At present, 17 species are recognised, mostly in sub-Saharan Africa, a couple reach the islands of the western Indian Ocean and Madagascar. Ten species are known in East Africa. Adults can be identified to the genus in the hand by the presence of a hinge on the plastron, enabling them to withdraw the head and front limbs and shut the front of the shell, if a hand-held specimen is touched on the nose or forelimbs it will shut up as described. The shell is thick and domed, often with a vertebral ridge or keel; it may change colour during life, and in big adults is often extensively cracked and chipped.

KEY TO THE EAST AFRICAN MEMBERS OF THE GENUS PELUSIOS
This is a difficult genus to key out, but the locality will give clues; only a handful of species have a wide distribution in East Africa. The key refers mostly to shell features, look at the accompanying diagrams.

1a Plastron orange, neck and shoulder skin bright yellow, adults less than 12cm.....*Pelusios nanus*, Dwarf Hinged Terrapin [p. 58]
1b Plastron not orange, neck and shoulder skin not bright yellow, adults more than 12cm.....2

2a Plastral forlobe at least twice as long as length of interabdominal seam......3
2b Plastral forlobe about 1.5 times as long as length of interabdominal seam......5

3a Head buff, a broad black Y-shape on top of the head, extending to the neck, carapace with a distinctive black cen-

tral longitudinal stripe, in western Uganda......*Pelusios gabonensis*, Forest Hinged Terrapin [p. 57]
3b Head dark with light vermiculations, carapace without a black central longitudinal stripe......4
4a Plastron centrally yellow, in Albert Nile......*Pelusios adansoni*, Adanson's Hinged Terrapin [p. 53]
4b Plastron and bridge dark brown, streaked and spotted, around Lake Turkana......*Pelusios broadleyi*, Lake Turkana Hinged Terrapin [p. 54]
5a Posterior rim of carapace usually strongly serrated, sometimes sinuate, black border on both plastron lobes...... *Pelusios sinuatus*, Serrated Hinged Terrapin [p. 60]
5b Posterior rim of carapace smooth or slightly serrated......6
6a Plastron with a well-developed constriction at the level of the abdominal-femoral seam......*Pelusios subniger*, Pan Hinged Terrapin [p. 61]
6b Plastron without a well-developed constriction at the level of the abdominal-femoral seam......7
7a Plastron usually black in East Africa......*Pelusios rhodesianus*, Zambian Hinged Terrapin [p. 59]
7b Plastron not usually black in East Africa......8
8a Only in eastern Kenya and south-east Tanzania......*Pelusios castanoides*, Yellow-bellied Hinged Terrapin [p. 55]
8b In Uganda or around the Lake Victoria basin......9
9a Intergular longer than broad, only in Lake Albert......*Pelusios chapini*, Congo Hinged Terrapin [p. 56]
9b Intergular broader than long, in Lake Victoria basin as well as in north and west Uganda......*Pelusios williamsi*, William's Hinged Terrapin [p. 63]

ADANSON'S HINGED TERRAPIN *Pelusios adansoni*

IDENTIFICATION: A medium-sized hinged terrapin with a yellow-brown, brown or black shell, with a hinged plastron, a broad head and a slightly projecting snout, with two little tentacles (barbels) under the chin. The shell is hard, fairly deep and rounded. The limbs are broad and webbed, with sharp claws. The males have longer thicker tails than the females. Shells of large adults up to 22cm long, average 13–17cm, hatchling size unknown but probably 3–5 cm. Adult mass unknown. The shell carapace is usually blackish-brown, lightening to brown laterally. The plastron is dull yellow centrally, black at the sides. The head is grey-brown dorsally, with yellow vermiculations, yellow at the back; there is sometimes a yellow stripe from the orbit to the tympanum.

Pelusios adansoni

Pelusios adansoni (Adanson's Hinged Terrapin) Left: Roger Barbour, captive.

HABITAT AND DISTRIBUTION: Rivers, streams and backwaters in the savanna of central and west Africa, north of the forest and south of the Sahara. No definite records from East Africa, but probably in the Nile north of Lake Albert in north-west Uganda, as it is known from just downstream at Gondokoro in the Sudan, thence westward to Senegal and the Cape Verde Islands.

NATURAL HISTORY: Little known. Presumably similar to other hinged terrapins, i.e. diurnal, basks in exposed sites, lays hard-shelled eggs in a pit dug by the female near the water's edge. Several clutches, of up to 15 eggs, roughly 3×2cm, may be laid in a season. Feeds mostly on aquatic invertebrates, especially molluscs. If their pools dry up they bury themselves in the mud and aestivate.

LAKE TURKANA HINGED TERRAPIN *Pelusios broadleyi*

IDENTIFICATION: A small hinged terrapin, known from Lake Turkana and associated streams. The shell has a knobbly central keel, most pronounced in juveniles. Unusually, the plastron is small and somewhat rigid in this species, and is rarely closed. The limbs are broad and webbed, with sharp claws; there are big transverse scales on the anterior surface of the front legs. The males have longer thicker tails than the females. Shells of large adults up to 15.5cm long, average size unknown, hatchlings 2.5–4cm. The shell carapace is grey or red-brown, each scale finely speckled, the lower third of the costals is sometimes very light, giving the impression of a pale ring around the lower carapace. The plastron is brown to black, sometimes with yellow blotches at the centre, in juveniles the plastron and bridge may be extensively blotched yellow, orange or pink. The head is dark, with light vermiculations, the chin and neck lighter, the eyes yellow or green.

Pelusios broadleyi

Pelusios broadleyi *(Lake Turkana Hinged Terrapin) Upper left: Victor Wasonga, Loyengalani Upper right: Eduardo Razzetti, Hatchling. Bottom left: Tomas Mazuch. Bottom right: Robert Sindaco.*

HABITAT AND DISTRIBUTION: A Kenyan endemic species, originally described from seasonal streams at Loyengalani, south-east shore of Lake Turkana, also recorded at Koobi Fora and in the Omo Delta; at 300–450m altitude. Probably occurs throughout Lake Turkana and in the lower Kerio River. Conservation Status: Vulnerable D2; due to the small size of its habitat, as known so far.

NATURAL HISTORY: Presumably similar to other hinged terrapins, i.e. diurnal, basks in exposed sites (although Loyengalani and Koobi Fora are hot and arid spots, so might be nocturnal, the types were collected at night), lays hard-shelled eggs in a pit dug by the female near the water's edge, carnivorous, presumably aestivates if the stream dries up in the dry season.

YELLOW-BELLIED HINGED TERRAPIN *Pelusios castanoides*

IDENTIFICATION: A medium-sized (up to 23cm) hinged terrapin with an olive-brown shell, with a hinged plastron, a long neck (it can right itself using the neck if inverted), a small flattened head and a pointed snout, with two little tentacles (barbels) under the chin. The beak is strongly bicuspid. The shell is smooth, domed and elongate. The limbs are broad and webbed, with sharp claws. The males have longer thicker tails than the females. Shells of adults average 17–20cm, hatchlings 3–4 cm. Adult mass unknown. The shell is usually brown, sometimes with pale reticulations. The plastron is yellow (mostly) and black. The head is brown with fine yellow vermiculations, the body skin is yellow or yellow-brown.

Pelusios castanoides

HABITAT AND DISTRIBUTION: In still lakes, and swamps, along rivers; tends to be in backwaters and creeks, at altitudes between sea level and around 1,000m. In East Africa found along the Kenyan and Tanzanian coastal plain, extending inland to the Tsavo/Galana junction, Kilosa and the Kilombero Valley, also occurs in Lake Malawi/Lake Nyasa, on Pemba and Zanzibar. Elsewhere, south along the eastern seaboard to South Africa. Conservation Status: Lower Risk/Least Concern, it is widespread and hard to catch, so under no threat.

NATURAL HISTORY: Active by day in weedy water bodies, such as small lakes, backwaters and swamps; when these dry up it buries itself and aestivates. Such areas are prone to

Pelusios castanoides (Yellow-bellied Hinged Terrapin) Left: Bill Branch, Mozambique. Right: Dennis Rodder, Pemba.

bush fires, so the shells of adults often show fire damage. Clutches of up to 25 eggs, 3×2cm, recorded in Malawi in September. Eats insects, frogs, water snails and floating vegetation, including Nile Cabbage (*Pistia stratiotes*).

CONGO HINGED TERRAPIN *Pelusios chapini*

IDENTIFICATION: A fairly large hinged terrapin with a brown shell, with a hinged plastron and a blunt head with two little tentacles (barbels) under the chin. The beak is strongly bicuspid. The shell is smooth and domed (keeled in juveniles). The limbs are broad and webbed, with sharp black claws. The males have longer thicker tails than the females. Shells of adults up to 38cm, average unknown, hatchlings 3cm. Adult mass unknown. The shell carapace is usually grey, blackish-brown to brown or greenish-brown above, the plastron is black, edged or blotched with yellow.

Pelusios chapini

HABITAT AND DISTRIBUTION: Shallow, slow-flowing rivers and creeks, swamps, shallow pans and isolated temporary pools. Within East Africa known only from Lake Albert, the Semliki Forest Reserve and the environs, at 620m altitude, elsewhere west and north in the rivers and lakes of northern DR Congo.

NATURAL HISTORY: Diurnal, may bask on shorelines and in shallow water. Known to move out of rivers into flooded savanna during the rainy season. Congo specimens were observed resting in water vegetation and debris with the head and shell partially out of the water. Shy and nervous. Lays hard-shelled eggs in a pit dug by the female, a clutch of 7 eggs, roughly 3.5cm diameter were laid in a hole 60cm from the water's edge; a hatchling measuring 27×34mm was recorded in Gabon in February. Omnivorous, it eats plant material and fruit, fish and aquatic invertebrates, even carrion. Some museum specimens were caught in nets, fish traps or shallow pools.

Pelusios chapini (Congo Hinged Terrapin) Left: Laurent Chirio, Gabon. Right: Laurent Chirio Gabon.

FOREST HINGED TERRAPIN *Pelusios gabonensis*

IDENTIFICATION: A fairly large hinged terrapin with a broad, flat blunt head with two little tentacles (barbels) under the chin. The upper jaw has two toothed cusps. The shell is flat and oval, with a prominent median keel; this may disappear in big adults. The limbs are broad and webbed, with sharp claws. The males have longer thicker tails than the females. Shells of large adults up to 33cm long, 18cm wide and 10cm deep, but average length 18–25cm, hatchling size around 4cm. Shell carapace colour very variable, juveniles may be brown or orange-brown, adults tan, brown, yellow-brown, black or grey, often with a distinct black vertebral stripe which broadens to a dark V at the front; this fades in big adults. The shell margin is often speckled with black, and in large adults may be totally black. The plastron is black and tan, the plate margins outlined with dull yellow. Juveniles often have rufous or yellow temples, in adults the head is yellowish or warm brown above, there is often a broad black Y-shape between the eyes, which extends backwards centrally onto the neck. The limbs are black in juveniles, tan in adults.

Pelusios gabonensis

HABITAT AND DISTRIBUTION: Rivers, streams and swamps in forest. Adults said to prefer larger rivers, juveniles quiet waters. Widely distributed in the Guinean-Congolian forest blocks, in East Africa known only from the streams of the Bwamba Forest in north-west Uganda at around 600m altitude, south of Lake Albert; this forest represents the eastern extension of the Ituri Forest in the DR Congo. Elsewhere, west to Guinea and Angola.

NATURAL HISTORY: Virtually nothing known. Presumably similar to other hinged terrapins, i.e. diurnal, basks in exposed sites, lays hard-shelled eggs in a pit dug by the female near the water's edge. A clutch of 12 eggs, 25×53mm recorded. Diet aquatic insects, worms, snails and fish, it is attracted to fish traps baited with fish, will also eat water plants and fallen fruit and nuts. Eaten in West Africa by some forest people.

Pelusios gabonensis Forest Hinged Terrapin Left: Konrad Mebert, Easten Congo, juvenile. Right: Harald Hinkel, DR Congo, Adult and juvenile.

DWARF HINGED TERRAPIN *Pelusios nanus*

IDENTIFICATION: The smallest hinged terrapin, just reaching our area. The shell is elongate and oval and finely rounded, without a central keel, or with a very weak one. The plastron is large with a long broad anterior lobe; the hinge is weakly developed. The posterior lobe is broad, with a V-shaped anal notch, and an obvious constriction at the abdominal-femoral seam. The limbs are broad, powerful and webbed, with sharp claws, but there are no large transverse scales on the anterior surface of the front legs. The males have longer thicker tails than the females. Shells of large adults up to 12cm long, average size 8–10cm, hatchling size unknown. The shell is beautiful red-brown or olive brown, the bridge is black, mottled with yellow. The plastron is orange, pink or brown, with dark streaks and sometimes yellow on the seams. The head is grey or tan above, with dark vermiculations, the neck dark grey-brown, the chin, neck, shoulder and limb skins vivid yellow; the two chin barbels yellow and obvious.

Pelusios nanus

HABITAT AND DISTRIBUTION: Inhabits water bodies in moist savanna. In our area known only from Tatanda, southwest Tanzania, at altitude 1,800m. Elsewhere, west across northern Zambia and the Congo to central Angola.

Pelusios nanus Dwarf Hinged Terrapin. Top: Stephen Spawls, Northwest Zambia. Bottom: Frank Willems, Northern Zambia.

Family Pelomedusidae: African side-necked terrapins

NATURAL HISTORY: Lives in small ponds, swamps and marshes; in northern Zambia found in flooded grassland. Habits presumably similar to other hinged terrapins, i.e. diurnal, basks in exposed sites, lays hard-shelled eggs in a pit dug by the female near the water's edge, carnivorous. A Zambian female laid 5 eggs, roughly 3×15cm. Presumably aestivate if their ponds dry up in the dry season.

ZAMBIAN HINGED TERRAPIN   *Pelusios rhodesianus*

IDENTIFICATION: A small hinged terrapin with a dark shell, with a hinged plastron, a small head, two cusps on the beak and two little tentacles (barbels) under the chin. The shell is hard, fairly deep and rounded. The limbs are broad and webbed, with sharp claws. The males have longer thicker tails than the females. Shells of large adults up to 25cm long, average 15–20cm, hatchling size 3–4cm. Adult mass up to 1kg. The shell is usually black or dark grey, sometimes the carapace scutes outlined in white. The plastron is black, with pale seams, sometimes with cloudy yellow blotches near the centre, very occasionally uniform yellow. The head is brown or black above, but distinctly cream or yellow on the sides, chin and throat.

Pelusios rhodesianus

Pelusios rhodesianus Zambian Hinged Terrapin Below: Andre Zoersel, northern Zambia. Bottom left: Andre Zoersel plastron colours. Bottom right: Andre Zoersel head close-up.

HABITAT AND DISTRIBUTION: Widespread in Rwanda and Burundi, found in lagoons and weedy estuaries bordering parts of Lakes Victoria, Tanganyika and Malawi (Nyasa), probably also in Lake Rukwa; at altitudes between 400 and 1,500m. A sight record (not shown on map) from north of Kasese, Uganda. Elsewhere, west to northern Zaire, south to Angola and Mozambique. Conservation Status: Lower Risk/Least Concern, it is widespread and hard to catch, so under no threat.

NATURAL HISTORY: Little known. Presumably similar to other hinged terrapins, i.e. diurnal, basks in exposed sites. In Zimbabwe it frequents weed-choked dams and rivers. The female lays 11–14 hard-shelled eggs, 2×3.5cm, in a pit dug by the water's edge. In Zambia a number of females (and one male) were found walking around in a shallow valley during September during the first storm of the season, presumably intending to lay eggs. In southern Africa hatchlings were found in December and January. Known to eat insects, frogs and fish.

SERRATED HINGED TERRAPIN *Pelusios sinuatus*

IDENTIFICATION: A huge dark hinged terrapin with a hinged plastron. The head is very broad, the snout elongated, with a weakly bicuspid beak and two little tentacles (barbels) under the chin. The shell is hard, domed and fairly deep, juvenile plastron distinctly keeled with tent-like peaks; this may vanish in adults or persist as a few raised bumps. The rear marginal scales are distinctly serrated, thus the spiky rear end is an excellent field character, even in large adults. Unusually for a *Pelusios*, the plastron concavity is almost absent. The limbs are broad and webbed, with sharp claws. The males have longer thicker tails than the females. Shells of large adults up to 55cm long in upland Kenya, smaller elsewhere, most adults average 20–30cm, hatchling size 4–5cm or

Pelusios sinuatus

slightly larger. Adult mass in excess of 20kg. In juveniles the shell is grey or greenish, each scale with a lighter centre and fine radial markings, but this fades in adults to uniform dark grey or black (hence the common Kenyan name 'black water turtle'), sometimes with pale edging to the scales. In juveniles the top of the head and the limbs are grey or greeny-grey with fine black speckling, and the sides of the head and chin are light, fading in adults to uniform grey or black. The plastron is yellow or orange, with a black border.

HABITAT AND DISTRIBUTION: Rivers, lakes and waterholes. The most common hinged terrapin in Kenya and Tanzania, probably in most major water bodies in the eastern side of East Africa, but records lacking. Known from the Daua River, Lake Turkana and the Galoss waterhole east of it, the northern Uaso Nyiro, the Athi/Galana/Tsavo river, upstream to Nairobi National Park at 1,700m, the Tana River, lakes around Malindi and Watamu, the rivers around Dar Es Salaam, the Rufiji, Ruaha and Rovuma River systems, and Lakes Tanganyika, Rukwa and Malawi (Nyasa). No records for Uganda. Elsewhere, north to Somalia, south to South Africa.

NATURAL HISTORY: Diurnal, swimming in big rivers and lakes, often seen basking on shorelines, logs, exposed rocks and mudflats; big adults usually visible at the hippo pools in Nairobi National Park, with head and legs fully extended to catch the warmth. Juveniles are

Pelusios sinuatus Serrated Hinged Terrapin. Top: Stephen Spawls Nairobi National Park. Left: Stephen Spawls, juvenile plastron colours, Watamu. Right: Stephen Spawls juvenile showing keel.

sometimes active at night in shallow water, thus avoiding predation by fish eagles. Breeding details poorly known; clutches of 7–30 eggs recorded, a captive female laid a clutch of 27 eggs, roughly 3.5cm diameter, they hatched in two months. Near Athi River, hatchlings 5–6cm long were found heading towards the river 300m away; in Garissa in July, 8cm juveniles were active in an ox-bow lake. These hinged terrapins are largely carnivorous, eating mostly molluscs and snails, but also fish, amphibians and insects; they will scavenge on carrion, and when doing so are often themselves attacked by crocodiles. Occasionally known to take floating fruit. If handled, they will bite and scratch, and eject a foul-smelling fluid from glands near the rear limbs. The nests are sometimes excavated by Nile Monitors and mongooses.

PAN HINGED TERRAPIN/BLACK-BELLED HINGED TERRAPIN *Pelusios subniger*

IDENTIFICATION: A small brown terrapin with a hinged plastron, a big head, a single cusp on the beak, with two little tentacles (barbels) under the chin. The shell is domed and oval. The limbs are broad and webbed, with sharp claws. The tail is relatively short, although males have longer thicker tails than the females. Shells of large adults up to 20cm long, average 14–18 cm, hatchling size 3–4 cm. Adult mass unknown. The shell is usually warm brown to grey-brown, the bridge (between carapace and plastron) is yellow with black edging, giving an impression of vertical light and dark bars. The plastron is yellow, the scales edged in black. The head is brown or blue-grey, sometimes with lighter patches or dark speckling (not vermiculations), chin and throat lighter, cream or dirty white, the limbs and other skin is dark grey to black.

HABITAT AND DISTRIBUTION: Waterholes, flooded pans (hence the common name), swamps, lakes, rivers and streams. In East African known only from Tanzania and Burundi; at altitudes between 400 and 1,300m. In Tanzania known from waterholes in Tarangire National Park and around Dodoma, also Kilombero, Usangu Plains, and the area between Katavi National Park, Tatanda and west of Lake Rukwa; probably in the lagoons bordering Lake Tanganyika but not in the lake itself, which is occupied by large Serrated Hinged Terrapins. Known from the Malagarasi River tributaries near Kiharo, on the Tanzanian/Burundi border. Elsewhere, south to South Africa, west to Congo. Conservation Status: IUCN Lower Risk/Least Concern; it is widespread, even in arid habitats, and hard to catch, so under no threat.

Pelusios subniger

NATURAL HISTORY: Not well known. Presumably similar to other hinged terrapins, i.e. basks in exposed sites during the day, but there are reports of nocturnal activity. It lays hard-shelled eggs in a pit dug by the female near the water's edge, clutches of 8–12 eggs,

Pelusios subniger Pan Hinged Terrapin: captive

roughly 2×3.5cm, recorded; eggs incubated in captivity at 30°C hatched in 58 days. Females reach sexual maturity at 17cm length. Said to be omnivorous, taking vegetable matter, crabs, worms, insects, frogs and tadpoles. Known to wander across country to new water bodies in the rainy season in southern Africa, and will aestivate, burying itself in mud to wait for the next rains if its waterhole dries up.

WILLIAMS' HINGED TERRAPIN   *Pelusios williamsi*

IDENTIFICATION: A fairly large dark hinged terrapin with a hinged plastron, a broad head and a slightly projecting snout, two cusps on the upper jaw, with two little tentacles (barbels) under the chin. The shell is oval, with a low blunt keel (may be quite prominent in juveniles). The limbs are broad and webbed, with sharp claws. The males have longer thicker tails than the females. Shells of large adults up to 25cm long, average 15–20 cm, hatchling size unknown but probably 3–4 cm. Adult mass unknown. The head and neck are finely marked with black on yellow reticulations. The carapace is black or dark brown, the plastron is black with a yellow rim and midseam, or uniform yellow, with or without darker speckling, the limb sockets yellow. Three subspecies are known, distinguished by their plastron colour.

Pelusios williamsi

HABITAT AND DISTRIBUTION: Lakes, rivers and swamps. Western Kenya and Uganda, at altitudes between 400 and 1,300m, records include the Lake Victoria basin and the rivers draining into it on the Kenya side (Yala, Nyando and Sondu), the Victoria Nile, Lakes Kyoga, Salisbury, Albert, Edward and George and the Semliki River. An isolated record from Seronera, in the Serengeti, indicates it may occur west of there in the rivers flowing

Pelusios williamsi William's Hinged Terrapin Main: Entebbe. Inset: Andre Zoersel head close-up Entebbe.

into Lake Victoria. A single Rwanda record, from Akagera National Park. A near-endemic of East Africa, outside our area only known from the DR Congo shores of Lakes Albert and Edward.

NATURAL HISTORY: Poorly known. Presumably similar to other hinged terrapins, i.e. diurnal, basks in exposed sites, lays hard-shelled eggs in a pit dug by the female near the water's edge. Captive female laid clutches of 7–18 eggs, roughly 2×4cm, these hatched in 60–80 days; hatchlings weighed 8–11g. Diet insects and amphibians, possibly fish, but few details recorded. The Akagera specimens were all found in temporary pools on hilltops; it is not known how they survive the dry season.

ORDER SQUAMATA
SCALED REPTILES

Lygodactylus williamsi
Turquoise Dwarf Gecko
Michele Menegon

ORDER SQUAMATA

The order Squamata, or squamate or scaled reptiles, contains three sub-orders: lizards (sub-order Sauria); snakes (sub-order Serpentes); and the worm lizards (sub-order Amphisbaenia). When we originally published this book in 2002 about 2,920 species of snake, 4,450 species of lizard and 150 worm lizard species were known world-wide. Since then, field and taxonomic research have increased those numbers and there are now over 3,570 species of snake, 6,154 species of lizard and 188 species of worm lizard; squamate biodiversity has increased by nearly 32%.

The squamate reptiles are distinctive animals, they have scales, diapsid skulls, and their upper jaw shows kinesis, i.e. it can move relative to the cranium. The males have a pair of copulatory organs (the hemipenes), and the females lay eggs. Lizards and snakes occur world-wide, except in the Antarctic, and there are virtually none in latitudes above the Arctic Circle; their greatest diversity is in the tropics.

SUBORDER SAURIA: LIZARDS

Lizards are the most diverse, abundant and visible group of reptiles; no one visits East Africa without seeing a few lizards. The world's largest lizard is the Komodo Dragon *Varanus komodoensis*, a monitor lizard that can weigh over 150kg and grows to over 3m in length. East Africa's largest lizard, the Nile Monitor, *Varanus niloticus*, belongs to the same genus and can reach 2.5m, and possibly more. The smallest lizards in East Africa are the dwarf geckoes of the genus *Lygodactylus*, some of which are only 6–7cm long as adults. At present around 230 species of lizard are recorded in East Africa; there are almost certainly many more undiscovered species especially in the dry east and north, and in the forests and woodland of the Eastern Arc Mountains. Lizards in East Africa are found from sea level to montane moorland at 3,500m; in general they are more abundant at low altitude, especially in open country where there are rocks and opportunities to bask. The most numerous families in East Africa are the chameleons (60 species), geckoes (59 species) and skinks (53 species). No East African lizards are venomous, despite legends to the contrary, although most will bite in defence and big species can scratch and lash their tails. All lizards have scales and most have four limbs; there are a few with reduced limbs, some have no legs. Most lizards have external ears, and most have eyelids and can close their eyes. Many can shed their tails in defence (chameleons and monitor lizards cannot) and the regenerated tail may have a different shape to the original. Lizards usually lay eggs; a few species retain the eggs internally until they are about to hatch. Unlike snakes, some lizards are social animals, living in structured colonies, observing hierarchies, defending territory and interacting with other members of the colony. Lizards largely eat invertebrates, although a few take larger animals and some eat plant food. Their many enemies include snakes, small mammalian carnivores and predatory birds.

LIZARD IDENTIFICATION

When we originally published this book there were twenty families of lizard worldwide, eight in East Africa. Taxonomic work, based on DNA and other biological molecules, which is ongoing and often controversial, indicates that at present there are six infraorders of lizards and 37 families. Ten families, as presently understood, occur in East Africa. In our area, lizards can usually be identified to their family quite quickly. The geckoes are small, mostly nocturnal, with a soft granular skin, often with scattered enlarged dorsal scales, called tubercles. The nocturnal geckoes have curious vertical slit-like pupils; no other East African lizard has vertical pupils. The dwarf geckoes, active by day, are small, with round pupils and enlarged toe tips. The agamas are diurnal, relatively large lizards, with scaly, spiky bodies, big rounded heads, and thin necks; the head has spiky scales. They often perch in prominent places, and the males of several species have brightly-coloured heads. Chameleons are slow-moving, with rotating eyes set in little turrets, with grasping feet. Most are green, live in trees or bushes and have long prehensile tails; the pygmy chameleons are brown with stumpy tails. The monitor lizards are huge, with forked tongues, muscular limbs and tails that are flattened vertically. The skinks all have shiny bodies and fairly long tails, most have little limbs and there are a few limbless species. Many are striped. Any lizard found buried underground or pushing through leaf litter will be a skink. The lacertids or 'typical' lizards have rough scales; they usually have long tails, and most of the dry country species are striped or barred. Many are very quick moving. A small, very fast-moving ground lizard in dry country will be a lacertid. The plated lizards are relatively large; two species are striped, one is big and brown, all have a curious and distinctive skin fold along their sides. This family also contains a plated snake lizard, with a very long tail and no front limbs. The family of girdled lizards and their relatives contains three genera. A single species of flat lizard (*Platysaurus*) is found only in southern Tanzania. Two species of slim, limbless, striped brown grass lizard or snake lizard (*Chamaesaura*) occur in highland areas. The girdled lizards are brown, rough-scaled lizards with spiny heads and tails.

KEY TO THE EAST AFRICAN LIZARD FAMILIES

This key assigns all East African lizards to family level. Higher taxonomic orders (Infraorder and Superfamily) are listed in the relevant introductions, not being of significance here.

1a	Has both a vertical pupil and large eyes that can close with eyelids... Eublepharidae, eublepharid geckoes [p. 69]		4a	Head much wider than neck, dorsal scales keeled, overlapping and often spiky, tongue short and broad......Agamidae, agamas [p. 220]
1b	Does not have that combination of characters......2		4b	Head same width or slightly wider than neck, dorsal scales smooth or granular, not overlapping, never spiky, tongue long and slender, tail cannot be shed...... 5
2a	Top of head covered with granular or small, irregular scales 3			
2b	Top of head covered with large, symmetrical shields 6			
3a	Eyelids present, eye closes...... 4		5a	Digits fused into opposed bundles for grasping, tongue very long and tele-
3b	No eyelids, eye can't close.....9			

scopic, eye very tiny, in revolving turrets, tail cylindrical, no adult larger than 70cm ......Chamaeleonidae, chameleons [p. 242]

5b Digits separate, in a single plane, tongue long and forked, eye normal, not in revolving turret, tail laterally flattened, adults always larger than 70cm......Varanidae, monitor lizards [p. 308]

6a Shiny, the dorsal scales usually highly polished, overlapping, smooth or with keels, ventral scales similar but lacking keels and slightly larger, femoral pores absent, eyelids may be present or absent......Scincidae, skinks [p. 127]

6b Not shiny, dorsal scales usually matt, not overlapping, with a strong median keel or small and granular, ventral scales rectangular, larger than dorsals, femoral pores or a row of differentiated scales on posterior face of thigh, eyelids present...... 7

7a A lateral granular fold absent, limbs well-developed ......Lacertidae, lacertid or typical lizards [p. 179]

7b A lateral granular fold present and/or limbs vestigial...... 8

8a Tongue long, scales on the tail rectangular, not keeled or spiny......Gerrhosauridae, plated lizards and seps [p. 213]

8b Tongue short, scales on the tail keeled and/or spiny......Cordylidae, girdled lizards and relatives [p. 205]

9a Unregenerated tail of males with a curious serrated crest, pupil round, digits thin with claws, toe tip not expanded......Sphaerodactylidae, clawed geckoes [p. 71]

9b Does not have the above combination of four characters......Gekkonidae, geckoes [p. 73]

Holodactylus africanus
Somali-Maasai Clawed Gecko
Claudia Hemp, Captive

SUB-ORDER SAURIA: LIZARDS
FAMILY EUBLEPHARIDAE
EUBLEPHARID GECKOES

Eublepharid geckoes are small, slow-moving, nocturnal terrestrial lizards. Unlike most other geckoes, they have claws rather than specialised toe pads. Most live in arid regions. Many are attractively marked with spots and bands and are popular in the pet trade. There are 36 species in the family, in six genera; they are curiously widespread, with species in the south-western United States, Central America, East and West Africa, the Middle East and extreme South-East Asia. Four species are African, and one of these is found in Kenya and Tanzania.

NORTH-EAST AFRICAN CLAWED GROUND GECKOES *HOLODACTYLUS*

A genus of small, slow-moving nocturnal geckoes with eyelids that can close and thin digits with claws, not expanded toe tips. Two species are known, both from arid country in north-east Africa.

SOMALI-MAASAI CLAWED GECKO *Holodactylus africanus*

IDENTIFICATION: Unmistakeable; a stocky, ground-dwelling gecko with a rounded head, distinctive yellow moveable eyelids, a short stumpy tail and clawed feet. The body is covered with small granular scales, giving it a velvety appearance. Maximum size about 10.5 cm, average 6–8 cm. The back is marked with distinct dark, brown or purplish crescent-shaped crossbars on a lighter background, some individuals have a pale vertebral stripe, which may be indistinct; adult markings usually much duller than juveniles. The ventral surface is uniformly light white or yellow.

Holodactylus africanus

HABITAT AND DISTRIBUTION: Probably occurs almost throughout the dry savanna and semi-desert of northern and eastern Kenya and north-east Tanzania, largely in low country, at altitudes between 200m and 1000m, but records are lacking. A population occurs from Elangata Wuas, near Kajiado in Kenya, south into northern Tanzania, extending west to Olduvai Gorge, east to Mkomazi (where it was first recorded in East Africa, in 1971). There is a cluster of records around the south-western corner of Lake Turkana and the lower Kerio River, and also between the base of the Ndoto Mountains, across to Laisamis, south to the Samburu National Reserve and east to Kula Mawe. Isolated records include Dodoma, Kakuma in north-west Kenya, Mandera and Malka Murri; elsewhere up into the Horn of Africa, to Djibouti.

NATURAL HISTORY: A terrestrial, slow-moving nocturnal gecko, active at night in sandy areas, during the day hides in burrows. The tail stores fat and varies greatly in size; it has been suggested that in the wet season these geckoes gorge themselves on termites and aestivate during the dry season, living on the stored fat. Captive specimens lay 1–2 hard-shelled eggs. Diet invertebrates. It is greatly feared in Somali areas, since it is erroneously believed to be venomous and the bite causes your skin to drop off.

Holodactylus africanus Somali-Maasai Clawed Gecko. Stephen Spawls, Mkomazi Game Reserve.

Pristurus crucifer
Cross-marked Sand Gecko
Tomas Mazuch, Ethiopia.

SUB-ORDER SAURIA: LIZARDS
FAMILY SPHAERODACTYLIDAE
CLAWED GECKOES

A family of over 210 species, in 12 genera, of diverse, relatively small clawed geckoes, with a near world-wide distribution in tropical and temperature regions; absent only from Australasia and the southern half of Africa. In our area represented by a single species, *Pristurus crucifer*.

SAND GECKOES *PRISTURUS*

A genus of small, active, diurnal geckos with curiously rounded heads (they bear some resemblance to small agamas) and large eyes with round pupils. The body is cylindrical, the tail very long, the unregenerated tail (especially of the males) often has a curious serrated crest, the feet are clawed. They are tolerant of arid conditions and live in semi-desert and near desert. Over 25 species are known, mostly in north-east Africa, Socotra and the Middle East, one species just reaches extreme north-east Kenya.

CROSS-MARKED SAND GECKO/CROSS-MARKED SEMAPHORE GECKO *Pristurus crucifer*

IDENTIFICATION: A little grey or brownish gecko, with a round head, eye quite large, pupil circular. Body cylindrical, the legs are slim, the feet are clawed, not broad or flattened as in most geckos, the tail is long, about 60% of total length; the tail of the male is compressed laterally with a low, slightly serrated crest, the tail of the female has no crest and is scarcely compressed. Dorsal scales are homogenous and overlapping. Maximum size about 10cm, average 6–9cm, hatchling size not known. Colour grey, grey-brown or brown above, some have black spots on the neck. There is a pale vertebral stripe, with short, narrow tooth-shaped crossbars and a dark dorsolateral stripe punctuated by white or cream spots; the flanks may be speckled red. The stripe continues on to the tail, punctuated by pale blotches. Whitish below, chin yellow mottled with grey.

Pristurus crucifer

HABITAT AND DISTRIBUTION: In Kenya, known only from the semi-desert around Mandera, on Kenya's north-east border, at 230m altitude, might occur further south but not known from Wajir or El Wak. Elsewhere, to Somalia, eastern Ethiopia, Eritrea and the Arabian Peninsula.

NATURAL HISTORY: Diurnal and terrestrial. It usually lives in a small hole, dug in sand near the base of a bush. Curls the tail up into the air. Hunts from ambush, dashing out to snatch passing insects. In Mandera it was observed to be mostly active in the early morning and mid to late afternoon. It lays eggs, clutch details unknown. Diet: fond of beetles.

Pristurus crucifer Cross-marked Sand Gecko, Tomas Mazuch, Somalia.

Stenodactylus sthenodactylus
Elegant Gecko
Robert Sindaco, Lake Turkana

SUB-ORDER **SAURIA: LIZARDS**
FAMILY **GEKKONIDAE**
GECKOES

Geckoes are small, plump, mostly nocturnal lizards, often grey, pinkish or brown, with large heads. Their eyes are large, as they hunt by sight in near darkness. Their eyelids are fused and immoveable. Their eyes are covered by a transparent spectacle which they lick clean with their tongues; a number have vertical pupils that contract to a curious shape (usually three small holes) in bright light. They have a soft, granular skin, often with scattered, enlarged scales known as tubercles. Their tails are easily shed; in some species just a single touch will cause the tail to drop off; it then twitches for some time. The regenerated tail may be quite a different shape and colour from the original, particularly if it is storing a lot of fat. Male geckoes often have femoral and precloacal pores, the secretions from which may have a territorial or reproductive function. Unlike other lizards, some geckoes can communicate by sound, a useful skill for a nocturnal animal. Nocturnal East African geckoes mostly just squeak, a sound likely to be heard when the animal is seized, but some Asian geckoes have a strident grunt, some southern African species have a repertoire of clicks and barks, and the chirp of the Tropical House Gecko *Hemidactylus mabouia* is often heard in lowland homes in East Africa. Over 1,000 gecko species are known world-wide, mostly in tropical and warmer temperate regions; 63 are so far recorded from East Africa. The distribution of nocturnal forms is strongly limited by altitude; because animals that rely on external heat cannot survive where it is cold at night.

Geckoes are superb climbers. The toes are usually dilated at the tips into pads, called scansors, on the lower surface of which is a mass of hairlike, often branched structures (setae). There is often a claw on the toe, which may be retractable. When the gecko is climbing, a combination of the setae tangling with minute irregularities, van der Waals forces (tiny electrical forces) and a sharp claw allow the gecko to ascend walls (and glass) and hang upside down. Geckoes often enter buildings and hunt on the walls and ceilings, catching the insects that are attracted to lamps. Most eat arthropods exclusively, but a few are omnivorous; the day geckoes (*Phelsuma*) will eat pollen and flowers, and the dwarf geckoes (*Lygodactylus*) will take nectar and flowers. Some nocturnal geckoes can change colour fairly rapidly, between light and dark. Most geckoes lay two eggs, which are soft when laid; they stick to the surface (often a rock crevice or tree cavity) and harden, and are quite resistant to desiccation. Some species are communal egg-layers, and agglomerations of many eggs may be found in suitable spots. Several clutches may be laid in a season, and the females have curious neck pouches called endolymphatic sacs, where calcium for the egg shells is stored.

Due to their habit of entering buildings, rock and timber piles, etc., they are often accidentally translocated; some species have spread far around the tropical world in this fashion. In East Africa there are some curious superstitions about geckoes, for example they are thought to cause leprosy (a legend stemming from the way their skin sloughs in pale patches). And yet, geckoes are beneficial animals, eating many mosquitoes and other troublesome insects, they should be seen as the householder's friend.

KEY TO THE EAST AFRICAN GECKO GENERA AND SOME SPECIES

1a	Eye pupil vertical...... 2	6b	Rostral not bordering nostril, precloacal pores absent......*Chondrodactylus turneri*, Turner's Thick-toed Gecko [p. 117]
1b	Eye pupil round 8		
2a	Digits thin, not expanded at tips...... *Stenodactylus sthenodactylus*, Elegant Gecko [p. 75]	7a	Claws retractable, back covered with smooth, non-overlapping scales, all of similar size...... *Homopholis fasciata*, Banded Velvet Gecko [p. 121]
2b	Digits expanded at tips...... 3		
3a	Digits clawless or with minute claws...... 4	7b	Claws not retractable, back usually with some keeled scales, scales usually of varying sizes...... *Hemidactylus*, half-toed geckoes [p. 83]
3b	Digits with claws...... 7		
4a	Snout quite pointed, only on Pemba Island...... *Ebenavia inunguis*, Madagascan Clawless Gecko [p. 75]	8a	Has five obvious thin toes, and toes not expanded at tips...... *Cnemaspis*, forest geckoes [p. 77]
4b	Snout fairly rounded, not on Pemba Island...... 5	8b	Has four obvious large toes and one tiny toe, large toes dilated at tips and digital pads obvious...... 9
5a	Confined to forests of the Eastern Arc mountains of Tanzania, tail tip with a gripping (scansorial) pad ...... *Urocotyledon*, tail-pad geckoes [p. 124]	9a	Usually some green on the body, claws absent or minute, coastal East Africa and islands only, adults up to 15cm...... *Phelsuma*, day geckoes [p. 122]
5b	On mainland Tanzania and extreme southern Kenya, not in forest, no scansorial pad on tail tip...... 6	9b	Usually grey or brown, a small claw present on a free distal digital joint, widespread, adults rarely larger than 10cm...... *Lygodactylus*, dwarf geckoes [p. 98]
6a	Rostral bordering nostril, precloacal pores present in males......*Elasmodactylus*, East African thick-toed geckoes [p. 119]		

MADAGASCAN LONG-NOSED GECKOES *EBENAVIA*

A genus of slim little clawless geckoes with long pointed snouts. Two species are known, both on Madagascar, one species has colonised various Indian Ocean islands, including the Comoros, Mauritius and also Pemba Island.

MADAGASCAN CLAWLESS GECKO *Ebenavia inunguis*

IDENTIFICATION: A small gecko with a pointed snout. Eye pupil vertical, claws absent from digits, but the toes and fingers end in a pair of rounded tips. Maximum size around 8cm, of which the tail is 50%. Brown or pinky brown above, a clear dark brown line, white or yellow edged above, from the snout through the eye extends onto the flanks; the tail has light and dark bands; light grey below.

HABITAT AND DISTRIBUTION: In East Africa, known only from the island of Pemba. Originally described from both the north and the south of the island, it was recently confirmed to be still present in north-western Pemba at least, living on the ground under palm fronds in a coconut and banana plantation. Widespread in eastern Madagascar, it presumably reached Pemba either on a trading ship or by rafting. Conservation Status: IUCN Least Concern, and widespread in Madagascar, but the status of the Pemba population not well understood.

Ebenavia inunguis

Ebenavia inunguis Madagascan Clawless Gecko, Frank Glaw, Pemba.

NATURAL HISTORY: Both terrestrial and arboreal, often on trees, 1–3m above ground level. Nocturnal but shows some diurnal activity. They lay 1–2 eggs, roughly 7mm diameter. Diet invertebrates.

CLAWED DESERT GECKOES *STENODACTYLUS*

A genus of 13 species of slow-moving, nocturnal geckoes (none larger than 15cm) that occur in the flat and sandy deserts and near-deserts of the Middle East, Turkey, northern and north-east Africa; a single species reaches northern

Kenya through South Sudan. They have large heads and prominent eyes, with strangely-shaped vertical pupils; the body scales are small and homogenous. The toes are finger-like, undilated.

ELEGANT GECKO *Stenodactylus sthenodactylus*

IDENTIFICATION: A small gecko with a big head and a huge eye, with a vertical pupil. Body slightly depressed, limbs slender, with five thin clawed toes. Tail thin, less than 50% total length and can be shed. On either side of the base of the tail is a curious row of 2–4 white, tooth-like tubercles. Maximum size about 10cm, average 7–9cm, hatchling size 5cm. Colour very variable, usually brown or grey, with dorsal cross bars and/or speckling. Kenyan examples are usually brown, pinky-brown or rufous, with light ocelli and dark speckling and a banded tail; immaculate white or cream below.

Stenodactylus sthenodactylus

HABITAT AND DISTRIBUTION: Lives on gravel plains and small dunes in desert and semi-desert, mostly at low altitude, but up to 1,000m altitude in parts of its range. In Kenya known only in the vicinity of Lake Turkana and eastwards to the base of Mt Marsabit, localities include Koobi Fora, Alia Bay, Balesa Kulal, the Omo Delta and Kakuma; elsewhere through South Sudan and Sudan to much of North Africa and parts of the Middle East.

NATURAL HISTORY: Terrestrial and nocturnal, emerges at dusk on warm nights and creeps around looking for food. In the deserts of North Africa it is most active between April and September, but habits virtually unknown in Kenya. Hides mostly in holes, but also under ground cover. Captive males show territorial behaviour. It has a click-type call and if seized produces a range of call of differing frequencies and duration, which may confuse predators. If threatened, it may stand up high, twist its body into a circle or run. Males approaching a possibly receptive female lift the tail and wave it horizontally. Two spherical eggs about 1cm diameter are laid. No breeding details known for Kenyan examples but in North Africa the eggs hatch 65–80 days after laying. Diet insects, especially small desert beetles.

Stenodactylus sthenodactylus Elegant Gecko, Eduardo Razzetti, Lake Turkana.

FOREST GECKOES *CNEMASPIS*

A genus of poorly-known, small slim forest geckoes, they have large eyes with round pupils. Many have a chain of light, sub-triangular or rhomboidal blotches or chevrons down the spine, and several species (especially the juveniles) have a yellow underside. They do not have expanded digits, their thin elongate toes have sharp claws; the rear toes are often kinked. They climb superbly, living largely on trees and rocks, although they may descend to the ground. There is debate as to whether they are nocturnal, diurnal or crepuscular, although the yellow lenses of one species, *Cnemaspis africana*, indicate it is diurnal. Most occur in forest and woodland. Over 120 species are known, mostly in Asia (and many of these have been described in the last 20 years). Only about 15 species occur in Africa, in the forests and woodlands of tropical Africa, six in East Africa, but this number is likely to increase as East Africa's hill forests become better known; the situation in the Eastern Arc Mountains and the Albertine Rift forests is unclear, and several unusual forms are present in the hilltop forests of central, eastern and northern Kenya.

KEY TO EAST AFRICAN MEMBERS OF THE GENUS *CNEMASPIS*

1a Enlarged dorsal tubercles in 2–6 rows.....2
1b Enlarged dorsal tubercles in 8–12 rows.....3
2a Tail with tubercular scales, usually four rows of dorsal tubercles.....*Cnemaspis quattuorseriata*, Four-lined Forest Gecko [p. 81]
2b Tail smooth to the level at which it breaks, usually two rows of dorsal tubercles (occasionally four or six).....*Cnemaspis dickersonae*, Dickerson's Forest Gecko [p. 79]
3a Tail with tubercles, only in Udzungwa Mountains.....*Cnemaspis uzungwae*, Udzungwa Forest Gecko [p. 82]
3b Tail smooth.....4
4a Subcaudals with a broad continuous median row, 9–12 precloacal pores.....*Cnemaspis africana*, Usambara Forest Gecko [p. 77]
4b Subcaudals with a discontinuous median row, precloacal pores not 9–12.....5
5a Precloacal pores 6–8. In western Kenya and Uganda.....*Cnemaspis elgonensis*, Elgon Forest Gecko [p. 80]
5b Precloacal pores 14, in Uluguru Mountains, Tanzania.....*Cnemaspis barbouri*, Uluguru Forest Gecko [p. 79]

USAMBARA FOREST GECKO *Cnemaspis africana*

IDENTIFICATION: A medium-sized forest gecko with an elongate snout and long digits. Back and flanks covered with small granules, among which are scattered keeled, conical tubercles of variable size forming 10–14 irregular longitudinal rows. Males have 8–14 precloacal pores. Maximum size about 10cm, average 7–8cm, hatchling 3.4cm total length. Colour: Male, above mottled olive green, brown and grey, usually has a broad pale vertebral stripe with lateral projections; specimens from Endau lack a vertebral stripe and have vivid red and green banded tails. Below, throat white; belly, base of tail, thighs, groin, and anterior aspects of tibia, yellow to pale yellow.

HABITAT AND DISTRIBUTION: Woodland and hill forest, from sea level to about 2,200m. An East African endemic, known from wooded hills in eastern and south-east Kenya and north-east Tanzania; localities includes Endau Mountain east of Kitui, the Taita and Shimba Hills in Kenya, Mt Meru, southern slopes of Mt Kilimanjaro, North Pare Mountains, Maji Kununua Hill, Ibaya Hill and South Pare Mountains, near Mkomazi Game Reserve, Usambara Mountains, Magrotto Hill and the Nguru Mountains. Conservation Status: Assessed by IUCN as Least Concern, but being forest dependant it might be under threat from forest clearance and fragmentation.

Cnemaspis africana

NATURAL HISTORY: A diurnal, arboreal and rock-dwelling species. Basks in the sheltered interiors of hollow, rotting trees and in rock crevices. Often found in leaf litter which collects at the base of buttress roots of forest trees. Two females were in rocky outcrops and others on trees; will also utilise road cuttings with earth banks, where it shelters in the cracks. Able to swim for at least several metres. Breeding details: eggs 9.5×7.5mm, found beneath a log. Diet: invertebrates

Cnemaspis africana Usambara Forest Gecko. Left top: Michele Menegon, Usambara Mountains. Right: Patrick Malonza, Endau Hill. Bottom left: Michele Menegon, Usambara Mountains.

ULUGURU FOREST GECKO Cnemaspis barbouri

IDENTIFICATION: A small forest gecko; body slightly depressed, enlarged dorsolateral tubercles in 6 or more rows. Toes thin. Male precloacal pores 14. Maximum size about 6cm. Colour above, speckled and blotched, various shades of green olive to grey, with irregular pale chevrons extending the length of the body and onto the tail, some specimens have a thin irregular vertebral line. Below, pale grey, with a distinct pattern of three dark grey longitudinal lines running from the throat to the pectoral area.

HABITAT AND DISTRIBUTION: Hill and montane forest and woodland, from 610–1,300m altitude. A Tanzanian endemic, occurs in Uluguru, East Usambara, and related coastal forests; known localities include Vituri, Uluguru Mts; Ruvu South forest reserve, Tongwe forest reserve, Muheza District; Amani Nature Reserve, Muheza District; Bamba Ridge forest reserve, Semdoe forest reserve.

Cnemaspis barbouri

Cnemaspis barbouri Uluguru Forest Gecko. Left: Michele Menegon Uluguru Mountains.

NATURAL HISTORY: Poorly known, but presumably similar to other forest geckoes, i.e. diurnal, possibly crepuscular, arboreal, living on tree trunks and among buttress roots, hiding under bark, in tree crevices, laying eggs, eating insects and other small arthropods. The specific name honours Thomas Barbour, who with Arthur Loveridge, published a major study of the herpetofauna of the Uluguru and Usambara Mountains in 1928.

DICKERSON'S FOREST GECKO Cnemaspis dickersonae

IDENTIFICATION: A relatively large slender forest gecko. The body is depressed, the tail is just over half the total length. The toes are long and slim. Males have 7–8, exceptionally 9, precloacal pores. Maximum size about 9.1cm; average size about 7cm. Hatchling size unknown. Colour very variable; in some specimens the back is olive-grey, olive-green or brown, a series of light grey chevrons or rhombuses run from the back of the head to the tail; highlighted by the surrounding darker brown or grey pigmentation of the back. Other individuals may be uniform light or dark brown, and some have a fine light vertebral stripe; individuals from the Mathews Range are rich brown and orangey-pink. Below, grey or yellow; throat with diffuse speckles, not a distinct pattern. The limbs are lightly banded with darker brown or grey. This species has enlarged

Cnemaspis dickersonae

Cnemaspis dickersonae Dickerson's Forest Gecko. Left: Stephen Spawls, Lolldaiga Hills. Right: Stephen Spawls Ngaya Forest

dorsolateral tubercles in four or fewer rows. The absence of tubercles on the tail is noticeable and distinguishes this species from the closely related and similar Four-lined Forest Gecko *Cnemaspis quattuorseriata*. Taxonomic Note: It is suggested that the differing colour morphs of this species found at Ol Ari Nyiro, in the Mathews Range and the woodland of the Laikipia area may represent undescribed species.

HABITAT AND DISTRIBUTION: Low altitude rain forest, medium altitude hill forest, high woodland and riverine and valley forest; from 400m altitude to over 2,000m. The distribution is curiously disjunct. Originally described from the Ituri Forest in eastern Democratic Republic of the Congo. Known East African localities include Toro in western Uganda, Mt Elgon and Mt Kadam (Debasien), the Mathews Range, the Nyambene Hills and mid-altitude forests around Mt Kenya, Ol Ari Nyiro Ranch in Kenya, Mount Hanang and Mto-wa-Mbu (near Lake Manyara) in Tanzania, said to occur in the Udzungwa Mountains, but details lacking, so not shown on map. Elsewhere, found in eastern DR Congo, south-western Ethiopia and the Imatong Mountains in South Sudan.

NATURAL HISTORY: Poorly known. It is diurnal and arboreal; specimens from central Kenyan forests wait in ambush under the edges of bark, in tree cracks, etc., dashing out for food. Specimens will move onto buildings, especially wooden ones. They lay eggs, but no other breeding details known. They eat invertebrates. Often parasitised by tiny red mites. The specific name honours Mary Cynthia Dickerson, an Associate Curator of Herpetology in the American Museum of Natural History in 1919.

ELGON FOREST GECKO *Cnemaspis elgonensis*

IDENTIFICATION: A medium-sized forest gecko from Uganda and western Kenyan forests. Snout rounded, eyes fairly large, pupil round. Body slightly depressed, tail about half the total length, limbs with clawed digits. Back and base of tail with 10–14 irregular rows of enlarged tubercles. Maximum size about 11cm, average 8–10cm; tail slightly over 50% of total length. Colour quite variable; usually greyish, with light and dark speckling, and a vertebral line of lighter sub-triangular blotches; there is a dark line through the eye. The tail is distinctively marked with fine light and dark longitudinal lines. Newly-hatched young are faintly yellowish from nape to vent, tail pink below. Half-grown young

Cnemaspis elegonensis

are bright mustard-yellow below from chin to vent and to base of tail; the remainder of tail is grey.

HABITAT AND DISTRIBUTION: Montane and evergreen forest at 1,200–2,200m. Known from a chain of forests and mountains on the Kenya-Uganda border (Kakamega, Kaimosi, forests around Mt Elgon, Mt Ḳadam/Debasien), Amaler River, Cherang'any Hills, also from the Rwenzori Range (but apparently not in between). Conservation Status: IUCN assessed as Vulnerable; found in a small number of forest blocks, in an area where forests are generally threatened for unsustainable use, although some of these are in national parks.

Cnemaspis elgonensis Elgon Forest Gecko. Bio-Ken archive, Mount Elgon.

NATURAL HISTORY: Poorly known. Diurnal and arboreal, found on forested mountain slopes on trees, also on the sides of buildings in formerly forested areas. May live in crevices of hollow trees, in which it basks. Lays two eggs, roughly 10–11mm diameter. Diet presumably invertebrates.

FOUR-LINED FOREST GECKO *Cnemaspis quattuorseriata*

IDENTIFICATION: A medium-sized forest gecko from the Albertine rift valley. The body is flattened, the tail is long, over half the total length. There are four rows of small, sparse but distinctly raised dorsal tubercles, which continue onto the tail. Irregular flat nail-like tubercles are scattered over the tail, which also has a swollen tubercle on either side of the base. Underside of tail with a median series of irregularly enlarged scales. Males with 6–8 precloacal pores. Maximum size around 10cm, average 7–9cm, a hatchling on Idjwi Island was 2.7cm total length. Coloration quite variable, may be greyish, rufous, brown or green, mottled and speckled, but there is a ring of dark radial streaks around the eye and a dark line through it. Usually there is a chain of irregular, often dark-edged light sub-rectangular blotches along the spine. The tail is irregularly barred light and dark, and these bars may become sharply-pointed chevrons as they approach the rear limbs. Underside pale and throat markings absent.

Cnemaspis quattuorseriata

HABITAT AND DISTRIBUTION: Medium to high altitude forest and woodland of the Albertine rift, at altitudes of 1,000–2,200m. Localities include Kibale Forest and Bwindi Impenetrable Forest in Uganda, western Rwanda including Nyungwe Forest, also known from Idjwi Island, from just west of Lake Kivu and in the Virunga National Park in DR Congo. Distribution elsewhere uncertain (it was confused for a long time with *Cnemaspis dickersonae*).

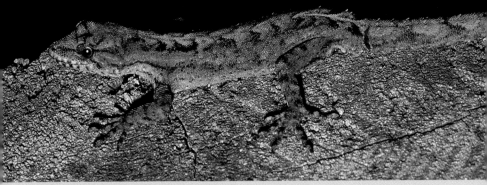

Cnemaspis quattuorseriata Four-lined Forest Gecko. Robert Drewes, Bwindi Impenetrable National Park.

NATURAL HISTORY: Virtually unknown, not helped by being confused with another species. Presumably similar to other forest geckoes, i.e. diurnal, arboreal, eats invertebrates. Twelve eggs, 0.7×0.6cm were found in bark crevices and among debris at foot of two large trees, on Idjwi Island. The origin of the specific name (*quattuor* = 4, *seriata* = series) refers to the usually four enlarged plates beneath the basal phalange of the median toe.

UDZUNGWA FOREST GECKO *Cnemaspis uzungwae*

IDENTIFICATION: A small forest gecko endemic to the Udzungwa mountain forests and some smaller low-altitude forests of eastern Tanzania. Unlike other species of *Cnemaspis*, it has a strongly tuberculated tail, and many flat button-like tubercles on the back. Maximum size presumably about 9cm, body length 4cm. Colouration, grey to brownish grey, uniform or speckled light and dark, usually with a series of dark brown-grey chevrons along the back; some individuals have a distinct light vertebral stripe. Below, a variable pattern of light grey to brown marks on a grey background. Similar species: Distinguished from *C. africana* and *C. barbouri* by the strongly tuberculated tail, the asymmetric subcaudals and two pairs of tubercles in the anal region.

Cnemaspis uzungwae

Cnemaspis uzungwae Udzungwa Forest Gecko. Michele Menegon, Udzungwa Mountains.

HABITAT AND DISTRIBUTION: Known only from the Udzungwa mountains, Tong'omba forest reserve and Kiwengoma Forest, south-eastern Tanzania; from 100 to over 2,000m altitude. Conservation Status: Forest-dependent in a country where forests are under threat, so vulnerable to habitat loss.

NATURAL HISTORY: Poorly known, presumably similar to other forest geckoes; i.e. arboreal and diurnal, lays two eggs, eats invertebrates. Commonly seen in the Udzungwa National Park at night, at heights of a metre or less, motionless on vegetation.

TROPICAL, HOUSE OR HALF-TOED GECKOES
HEMIDACTYLUS

A genus of over 130 species of nocturnal geckoes, widespread across the islands of the Pacific and Indian Oceans, in Central America, southern Europe, Asia and Africa; and introduced in a number of places. Some species are commensal with humanity, and may be seen near lamps at night. They have broad toes with retractile claws and prominent scansors; usually in pairs. Some species can change colour in a limited fashion. Many species can make sounds, clicks and grunts, and are territorial, living in structured colonies with a dominant male who keeps conspecific males away. Adaptable in semi-desert and dry savanna, the genus has radiated in north-east Africa; over 40 species are known from the area, 18 in East Africa. However, their taxonomy is not fully resolved and some species are very poorly known. There are almost certainly undiscovered and unrecognised species in East Africa. We have provided a key, but it must be used with caution; it may not work for females (which have no precloacal pores), and poorly known forms may show variation beyond the range given in the key. However, many species are readily identifiable by a mixture of general appearance, colour and range.

KEY TO THE EAST AFRICAN GECKOES OF THE GENUS
HEMIDACTYLUS

1a Body covered with granules or overlapping scales, no enlarged tubercles.....2
1b Body covered with granules or scales, among which are enlarged, keeled or smooth tubercles.....7

2a Body covered with overlapping (imbricate), smooth scales of roughly equal size.....3
2b Body covered with overlapping (imbricate), strongly- or weakly-keeled scales of equal or unequal size.....4

3a Males with 4–8 precloacal pores, in northern Kenya..... *Hemidactylus isolepis*, Lanza's Gecko [p. 89].
3b Males with 10–18 precloacal pores, in coastal Kenya.....*Hemidactylus modestus*, Tana River Gecko [p. 92]

4a Overlapping scales of dorsal surface roughly equal in size, strongly or weakly keeled, scales on head and neck tiny and granular.....*Hemidactylus funaiolii*, Archer's Post Gecko [p. 88]
4b Overlapping body scales of unequal size, head and neck scales not very different to dorsal scales.....5

5a Males with 6–10 precloacal pores, no femoral pores.....*Hemidactylus tropidolepis*, Ogaden Gecko [p. 98]
5b Males with 10 or more precloacal-femoral pores in a continuous row.....6

6a Males with 10–16 precloacal pores in a continuous row, usually in dry savanna.....*Hemidactylus squamulatus*, Nyika Gecko [p. 96]
6b Males with 16–23 precloacal pores in a continuous row, usually in coastal habitat.....*Hemidactylus barbouri*, Barbour's Gecko [p. 85]

7a Distance from tip of snout to eye greater than distance from eye to ear opening, snout elongate.....8
7b Distance from tip of snout to eye shorter than distance from eye to ear opening, snout rounded.....10

8a Enlarged tubercles on back arranged in 8–12 (usually 10) longitudinal rows, males with 45–50 precloacal-femoral pores in a continuous row, size to 19cm.....*Hemidactylus platycephalus*, Tree Gecko [p. 93]
8b Enlarged tubercles on back arranged in 11–18 longitudinal rows, males with 22–40 precloacal-femoral pores in a continuous row, size to 15cm.....9

9a Enlarged tubercles on back arranged in 12–18 longitudinal rows, size to 15cm, no enlarged tubercles on upper

part of the legs.....*Hemidactylus mabouia*, Tropical House Gecko [p. 90]
9b Enlarged tubercles on back arranged in 11–14 longitudinal rows, size to 11cm, enlarged tubercles present on upper legs.....*Hemidactylus mrimaensis*, Kaya Gecko [p. 92]

10a Original tail strongly constricted at base.....11
10b Original tail not strongly constricted at base.....12

11a Enlarged body tubercles strongly keeled, arranged in 14–18 more or less regular longitudinal rows, males with 28–34 precloacal-femoral pores in a continuous row.....*Hemidactylus ruspolii*, Prince Ruspoli's Gecko [p. 96]
11b Enlarged body tubercles strongly keeled, arranged in 18–20 more or less regular longitudinal rows, males with 42 precloacal-femoral pores in a continuous row.....*Hemidactylus tanganicus*, Dutumi Gecko [p. 97]

12a Innermost digit, particularly on the front foot, very short, with a short flat claw.....*Hemidactylus frenatus*, Pacific Gecko [p. 88]
12b Innermost digit normal in length compared to other digits, claw angled upwards.....13

13a Digits webbed at the base, conspicuous lateral fold of skin between fore and hind limbs, fringe on sides of tail looks like fur.....*Hemidactylus richardsoni*, Richardson's Forest Gecko [p. 94]
13b Digits not webbed at base, no conspicuous lateral skin fold, tail fringes (if present) do not look like fur.....14

14a Males with 20–46 precloacal-femoral pores in a continuous row, enlarged tubercles of back strongly keeled, in 14–25 more or less regular longitudinal rows... *Hemidactylus angulatus*, East African House Gecko [p. 84]
14b Males with less than 14 precloacal pores, no femoral pores.....15

15a Enlarged dorsal tubercles weakly keeled.....*Hemidactylus robustus*, Somali Plain Gecko [p. 95]
15b Enlarged dorsal tubercles strongly keeled.....16

16a Juveniles and adults not strikingly banded, either faint dark bar through eye or no dark bar, males with 6–11 precloacal pores.....*Hemidactylus macropholis*, Boulenger's Gecko [p. 91]
16b Juveniles strikingly banded, adults banded or retaining traces of bands and a dark bar through eye, males with seven or eight precloacal pores.....17

17a Only three dark crossbars between snout and insertion of hindlimbs..... *Hemidactylus bavazzanoi*, Somali Banded Gecko [p. 87]
17b At least four dark crossbars, or traces of them, between snout and insertion of hindlimbs.....*Hemidactylus barbierii*, Barbieri's Turkana Gecko [p. 86]

EAST AFRICAN HOUSE GECKO *Hemidactylus angulatus*

IDENTIFICATION: A fairly large gecko, body depressed, quite stocky, with a short snout. The unregenerated tail is long and not constricted at the base; regenerated tails are flattened and constricted at the base and may be turnip-shaped. The body and tail are covered with prominent tubercular scales, in large adults these can be almost pyramidal; there are 14–25 rows of these enlarged scales on the dorsum; and there are often 6–8 rows of long tubercles on either side of the tail, near the base. The males have 20–46 precloacal pores. A big gecko, up to 15cm total length, although 10–12 cm on average; unregenerated tail about half the total length. A dark line passes through the eye. The ground colour may be brown, rufous, pink or grey, with three chains of dark marks along the back, which may coalesce to form saddles; the tail is barred, light and dark. Taxonomic Note: Formerly known in East Africa as Brook's Gecko *Hemidactylus brooki*; that name is now restricted to Asian specimens, East African members of the group are a separate species.

Hemidactylus angulatus

HABITAT AND DISTRIBUTION: Found in coastal thicket, moist and dry savanna, semi-desert and woodland, and also in towns. Widespread throughout Kenya save the north-east, widespread across northern and north-eastern Tanzania and eastern Rwanda; seemingly absent from the eastern Lake Victoria shore and southern and south-western Tanzania and a part of north-central Uganda. Occurs from sea level to about 2,300m altitude. Elsewhere, west to Senegal, north to the Sudan.

NATURAL HISTORY: Nocturnal, equally at home on the ground, on trees, rocks and buildings, may be seen on walls at night near lamps hunting insects, although it is not as bold as the Tropical House Gecko, *Hemidactylus mabouia*. In natural settings it shelters under ground cover, in abandoned termitaria, under bark and in cracks in trees and rocks. Lays two hard-shelled eggs. Eats invertebrates.

Hemidactylus angulatus East African House Gecko, Stephen Spawls. Left: Watamu, juvenile. Right: Nakuru.

BARBOUR'S GECKO *Hemidactylus barbouri*

IDENTIFICATION: A small grey or brown coastal gecko with a big eye and short snout. Body fairly slim, and the unregenerated tail is also long and cylindrical, about 50% of total length. The scales are snake-like, being small and overlapping, and mostly the same size (there are a few slightly enlarged ones); there are no enlarged dorsal tubercles. The scales beneath the tail are enlarged. Males have 16–23 preano-femoral pores. Maximum size about 9cm, average 6–8cm. Colour quite variable; usually brown or grey, some specimens with light yellow speckling, sometimes with dark blotches or bars that may coalesce to form a series of dark crossbars; in some individuals dark lines through the eyes meet on the back of the head in a dark crescent. Lips paler, underside white.

Hemidactylus barbouri

HABITAT AND DISTRIBUTION: Coastal woodland, thicket, savanna and peri-urban areas, from sea level to about 400m altitude, replaced inland by the Nyika Gecko *Hemidactylus squamulatus*, of which it was formerly considered a subspecies. Occurs on the coastal strip, from the Tana Delta and Witu south to Bagamoyo, but there are no records from more than about 30km inland.

Hemidactylus barbouri Barbour's Gecko, Stephen Spawls, Shimba Hills.

NATURAL HISTORY: Nocturnal and terrestrial, sheltering by day under ground cover, under rocks, in coral rag, in logs or vegetable debris; often in piles of makuti (coconut palm) thatch. Emerges after dark and forages on the ground. Not uncommon in the Shimba Hills National Park, also tolerant of humanity and present in urban areas, provided there is some natural vegetation. Lays two hard-shelled eggs. Eats invertebrates.

BARBIERI'S TURKANA GECKO *Hemidactylus barbierii*

IDENTIFICATION: A small terrestrial stocky gecko with a flattened head and a large eye, from the Lake Turkana area. The unregenerated tail is long and thin. Back covered by large, trihedral, strongly-keeled tubercles intermixed with a few small, irregular shaped granules, forming 14 quite regular transverse rows. The male has eight precloacal pores, females have none. Both known adults had missing tails but had bodies of 4 and 5cm length, so maximum length probably around 10cm. Adult coloration pinky-brown, both known adults had four darker pink crossbars on the neck and body, these edged by a scattering of dark brown scales. A dark line through the eye extended around the back of the head in one specimen. White below. In the only known juvenile, the four crossbars are vivid yellow and dark brown, top of head orange, the tail with eight narrow chocolate brown bands on a bright yellow background.

Hemidactylus barbierii

Hemidactylus barbierii Barbieri's Turkana Gecko Left: Bill Branch, Koobi Fora. Right: Eduardo Razzetti, Lake Turkana.

HABITAT AND DISTRIBUTION: Known only from three specimens, from just north of Allia Bay and near Koobi Fora, on the east shore of Lake Turkana around 400m altitude; the adults were active at night, one in a dry river bed running through semi-desert, the other on open stony ground, the juvenile was under a rock.

NATURAL HISTORY: First collected in 2005 and described in 2007, the biology of this species is virtually unknown, but presumably similar to other *Hemidactylus* of the dry country, i.e. nocturnal and terrestrial, shelters under ground cover or in holes, lays eggs and eats invertebrates.

SOMALI BANDED GECKO   *Hemidactylus bavazzanoi*

IDENTIFICATION: A small, attractive terrestrial stocky gecko from north-east Kenya. Both known adults have regenerated tails, slightly shorter than body length and covered with dorsal keeled tubercles. Back covered by smooth scales, of unequal size, among which are large strongly-keeled tubercles in 16–18 rows. The males have seven precloacal pores. Both specimens had a body length of 4cm, and total length around 7cm. Colour orange or pinkish brown, top of head russet, with a dark brown crescent-shaped neck bar that extends round through the eyes to the snout, two dark brown cross bars on the body and several on the tail, underside pinky-white.

Hemidactylus bavazzanoi

HABITAT AND DISTRIBUTION: Known only from two specimens; one from just east of Lugh in south-west Somalia, east of the Webi Shebelle River, the other from the banks of the Daua River, Mandera, on Kenya's north-east border; at altitudes between 200 and 230m. These two localities are about 70km apart, in the Somali-Maasai ecosystem, but both specimens are from river valleys, so might be associated with riverine gallery forest.

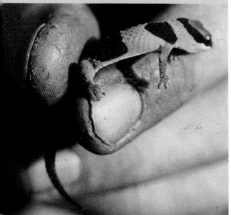

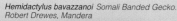

Hemidactylus bavazzanoi Somali Banded Gecko. Robert Drewes, Mandera

Conservation Status: Known from only two specimens so very hard to draw conclusions; its broad habitat is semi-desert, unlikely to be developed, but riverine valleys are potential sites for large-scale agriculture.

NATURAL HISTORY: Known from two specimens alone, collected in 1971 and 1977. One specimen was under a palm log in riverine forest at Mandera in the morning, the Somali specimen was collected at night; it dropped from a low bush and tried to crawl away slowly. Biology unknown but presumably similar to other *Hemidactylus* of the dry country, i.e. nocturnal and terrestrial, shelters under ground cover or in holes, lays eggs and eats invertebrates.

PACIFIC GECKO Hemidactylus frenatus

IDENTIFICATION: A fairly large gecko, body depressed, quite stocky. The tail is fairly long and not constricted at the base. The body and tail are covered with small granular scales, among which are a few larger, flat tubercles; there are 2–8 rows of these enlarged scales on the dorsum (although they are not obvious in most specimens, the dorsal scalation looks homogenous). The innermost digit (particularly on the front foot) is very short, and the claw lies flat. The males have 24–36 preano-femoral pores. Maximum length about 12.5cm, average 8–11cm, the tail is as long as the body. Elsewhere, may be grey or brown, with two diffuse darker dorsolateral stripes and a dark head stripe, but East African specimens are usually uniform pinky white; juveniles have a few fine yellow dorsal spots, that may coalesce to form narrow bars on the tail.

Hemidactylus frenatus

HABITAT AND DISTRIBUTION: In East Africa, known only from Lamu and coastal Somalia, at sea level or just above. It was probably transported there in cargoes, elsewhere widespread across the islands and shores of the Indo-Pacific Ocean; where it lives in coastal vegetation and in towns. Conservation Status: Not assessed by IUCN and widespread over a huge range outside our area, but its presence on the East African coast should be monitored; it has been shown in Australia to be an aggressive coloniser of urban habitats, driving other geckoes away.

Hemidactylus frenatus Pacific Gecko. Stephen Spawls, Djibouti.

NATURAL HISTORY: Nocturnal, terrestrial and arboreal, the Lamu specimens were found in a coconut plantation in the 1940s and it is not known if it still occurs there. Elsewhere it is often commensal with humans, and may be seen on walls at night near lamps hunting insects. It is a territorial species; a dominant male establishes a territory and drives away smaller males. Has two distinctive calls, a series of chirps and a 'chuck, chuck, chuck' call. Lays two hard-shelled eggs. Eats invertebrates.

ARCHER'S POST GECKO Hemidactylus funaiolii

IDENTIFICATION: A small terrestrial stocky gecko from northern Kenya. Head broad, body fairly stout. Both regenerated and undamaged tails taper smoothly, although regenerated tails are shorter. The scalation is unusual, with scattered smooth conical granular scales on the neck, becoming overlapping and strongly keeled scales on the body. The males have six precloacal pores. Maximum length about 7.5cm, tail slightly over half the total length. Colour brownish, pinkish on head, the dorsum marked by yellow spots, and brown streaks that may coalesce to form vague bars; there is a brown streak from the nostril to the eye and on the back of the head; underside uniform white.

HABITAT AND DISTRIBUTION: Semi-desert and dry savanna; an endemic of the Horn of Africa. In East Africa, known only from a handful of specimens from three widely dispersed localities in northern Kenya; several from Archer's Post, at the eastern entrance to Samburu National Reserve, it is also recorded from Lerata, just north of Samburu, at Laisamis and the Hurri Hills, and from Baidoa in southern Somalia; at altitudes between 500–1,000m.

Hemidactylus funaiolii

Hemidactylus funaiolii Archer's Post Gecko. Quentin Luke, Archer's Post

NATURAL HISTORY: Poorly known. The Laisamis and one Archer's Post specimens were under rocks. Biology presumably similar to other *Hemidactylus* of the dry country, i.e. nocturnal and terrestrial, shelters under ground cover or in holes, lays eggs and eats invertebrates.

UNIFORM-SCALED GECKO/LANZA'S GECKO
Hemidactylus isolepis

IDENTIFICATION: A small slim terrestrial gecko from northern Kenya. Head broad, eyes prominent. Tail slightly over 50% of total length and tapering smoothly (although regenerated tails may become swollen). The scalation is unusual; all the dorsal scales are roughly the same size, smooth and overlapping. The males have 4–8 precloacal pores. Maximum length about 7.5cm. Colour pink or pinkish brown, with a distinct dark line through the eye, some individuals are boldly marked with dark speckles and streaks and pale spots, in others the markings are subdued. Underside uniform white. Taxonomic Note: It has been suggested that true *Hemidactylus isolepis* are confined to Somalia and Ethiopia, and Kenyan animals assigned to this taxon are probably another, as-yet-unnamed species.

Hemidactylus isolepis

HABITAT AND DISTRIBUTION: Semi-desert and dry savanna of northern Kenya, at low altitude, between 300–1,000m. Probably very widespread but present records sporadic: occurs widely around Lake Turkana area, also known from Samburu National Reserve and Archer's Post, Baringo, parts of the Dida-Galgalu Desert, and Mandera.

NATURAL HISTORY: Poorly known. A Baringo specimen was under a rock. Biology presumably similar to other *Hemidactylus* of the dry country, i.e. nocturnal and terrestrial, shelters under ground cover or in holes, lays eggs and eats invertebrates.

Hemidactylus isolepis Lanza's Gecko Left: Robert Sindaco, Lake Turkana. Right: Stephen Spawls, Lake Baringo.

TROPICAL HOUSE GECKO *Hemidactylus mabouia*

IDENTIFICATION: A fairly large gecko, common in many places, often on buildings, near lamps at night. The body is depressed, quite stocky, with a long snout. The tails (even unregenerated ones) are virtually never constricted at the base. The body and tail are covered with small but prominent, weakly-keeled tubercular scales, looking like little spots, in 12–18 rows, and on the sides of the tail these scales can be quite large and pointed. The males have 22–40 preano-femoral pores. A big gecko, up to 15cm total length (and females tend to be larger than males), 10–12 cm on average, the unregenerated tail is slightly over half the total length. These geckoes can change colour rapidly; those in woodland tend to be dark, those near lamps at night are often pale pink, sometimes grey or yellow. Where visible, the pattern consists of around five irregular darker crossbars or pairs of dark blotches; in some individuals there is a series of light subrectangular blotches along the spine. White below. Taxonomic Note: Several authors have suggested that the name *Hemidactylus mabouia* is inappropriate for the African animals to which this name is applied, and they should perhaps be called *Hemidactylus mercatorius*, but for the time being we retain *Hemidactylus mabouia* for stability, until the situation is fully clarified.

Hemidactylus mabouia

HABITAT AND DISTRIBUTION: One of East Africa's most widespread and easily seen lizards, due to its habit of waiting in ambush near lamps. Found in coastal thicket, moist and dry savanna, semi-desert and woodland, and also in towns. Widespread throughout Kenya (records sporadic in the dry north), most of northern and eastern Tanzania, and most of Uganda, eastern Rwanda and Burundi. Occurs from sea level to about 2,500m altitude. Elsewhere: widespread all over sub-Saharan Africa, although records sporadic in the west; on islands in the Indo-Pacific; and by natural or human agency (it is suggested it reached the New World on slave ships from West Africa) it has dispersed along the east coast of South America, and through the Caribbean into Florida.

NATURAL HISTORY: Nocturnal, equally at home on trees, rocks or in caves; often in buildings where it may be seen on walls at night near lamps hunting insects, and will feed by day in darkened places if insects are abundant. Rarely on the ground. In natural settings it shelters under bark and in cracks in trees and rocks, hides in any crevice in human habitation, often behind pictures, for example. It lays two hard-shelled eggs, roughly 10mm in diameter, in any suitable refuge; large communal clutches in suitable refuges have been found. South American females laid clutches every 16 days, and the eggs hatched in around two months. Courting males produce a series of multiple chirps. It is a territorial species; a dominant male establishes a territory and drives away smaller males. Eats a wide range of invertebrates, and eaten itself by many small carnivores, including spotted bush snakes and boomslangs.

Hemidactylus mabouia Tropical House Gecko. Left: Daniel Hollands, Murchison Falls. Right: Stephen Spawls, Bagamoyo

BOULENGER'S GECKO *Hemidactylus macropholis*

IDENTIFICATION: A fairly large gecko, quite stocky. The tail is fairly long and not constricted at the base. The body and tail are covered with 12–16 rows of large, strongly-keeled tubercles, giving the animal a striped appearance; the scales between the rows are small and granular. The males have 6–11 precloacal pores. Maximum length about 14cm, average 9–12cm, the tail is usually slightly shorter than the body. Pink, pinky-brown or rufous in colour, some specimens have around five diffuse darker crossbars on the back. The tail is banded pink and brown and the dorsum is covered with small rufous or brown spots. Underside white.

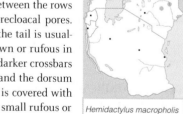

Hemidactylus macropholis

Hemidactylus macropholis Boulenger's Gecko, Stephen Spawls, Somalia.

HABITAT AND DISTRIBUTION: Dry savanna and semi-desert of eastern and northern Kenya, with a single dubious record from north of Arusha, northern Tanzania. Kenyan localities include Ngomeni, near Kitui, Sankuri on the Tana River, Laisamis, Wajir, Wajir Bor, Buna, eastern Lake Turkana, Rhamu and Mandera; thence north to eastern and north-east Ethiopia and Somalia; the Kenya records mostly at altitudes below 1,000m.

NATURAL HISTORY: Nocturnal and terrestrial; Kenyan specimens have been found under rocks or other ground cover in rocky areas, in termitaria and in rodent burrows. Like other members of the genus it lays two hard-shelled eggs and eats invertebrates.

TANA RIVER GECKO Hemidactylus modestus

IDENTIFICATION: A small terrestrial stocky gecko. Head broad, body fairly stout. Tail short and fairly thick. Unusually, the body is covered by small, overlapping scales of roughly equal size; there are no granular scales and no enlarged tubercles. The males have 10–18 precloacal pores. Maximum length about 8cm, tail about 40% of total length. The type was described as uniform brown above, whiter below, but that was probably a specimen in alcohol; a handful of more recent specimens are light brown with darker longitudinal markings forming irregular lines down either side of the back, with additional elongate dark markings intruding from these lines into the middle of the back. These animals have a dark bar through each eye, that coalesce on the back of the neck to form a crescent; in some specimens additional incomplete crescents occur behind the eyes.

Hemidactylus modestus

HABITAT AND DISTRIBUTION: Coastal forest and thicket. Originally described from Ngatana in the Tana delta, subsequent specimens were collected in the Arabuko-Sokoke Forest, at Kilifi and at Ukunda, on Kenya's south coast; at sea level or just above. Conservation Status: Assessed as Data Deficient by the IUCN.

NATURAL HISTORY: Essentially unknown. Ngatana specimens were found under piles of rubbish and rotting vegetation under mango trees. Biology presumably similar to other *Hemidactylus*, i.e. nocturnal, lays eggs and eats invertebrates.

KAYA GECKO Hemidactylus mrimaensis

IDENTIFICATION: A recently-described arboreal, spiky-looking gecko from the sacred forests of southern coastal Kenya. Small and fairly slim, with a long head, the tail (even regenerated ones) tapering smoothly, and slightly over 50% of total length. Dorsal scalation distinctive, consisting of 11–14 rows of enlarged, posteriorly elongate and highly visible tubercles interspersed by small granular scales; tubercular scales on the tail clearly enlarged and pointed, some resemble small curved thorns. Males have 32–34 precloacal pores. Maximum length about 11cm, average 8–10cm. Dorsal colour shades of grey or brown, some almost uniform, others with around five diffuse darker crossbars; body with light and dark blotches, tail and limbs with vague narrow dark barring. Pale below, in some specimens the underside of the tail is orange. Resembles the Tropical

Hemidactylus mrimaensis

House Gecko *Hemidactylus mabouia*, but it has enlarged tubercles on slim upper legs; *H. mabouia* has stouter legs without tubercles.

HABITAT AND DISTRIBUTION: A gecko of coastal forest; known from three kayas, sacred forests of southern coastal Kenya; Kaya Mrima, just northwest of the Shimoni peninsula, Kaya Kinondo, north-east of Gazi and Kaya Jibana, east of Kaloleni, at altitudes between sea level and 350m; also recorded from the southern Shimba Hills. Known also from the forest around Witu, north of the Tana Delta, so probably occurs the length of the coast in suitable undisturbed forest patches. Not found yet in disturbed forests or near habitation. Conservation Status: Not assessed by IUCN; the sacred coastal kayas where it occurs have traditional protection but at the same time coastal land is under threat for development for tourism and agriculture.

Hemidactylus mrimaensis Kaya Gecko, Patrick Malonza, Kaya Jibana.

NATURAL HISTORY: Active by day, and possibly by night; specimens were observed on small tree stems and stumps, a short distance above the ground. At night they have been found on banana plants and under fallen logs. They are described as shy, highly active and difficult to catch. Present in much smaller numbers than the sympatric *Hemidactylus mabouia*. Presumably similar to other *Hemidactylus* geckoes in laying two eggs and eating invertebrates.

TREE GECKO *Hemidactylus platycephalus*

IDENTIFICATION: A large, flat, arboreal gecko, similar in appearance to the Tropical House Gecko *Hemidactylus mabouia* (although genetically it has been shown to be quite different, Kenyan specimens are in a different clade). The body is depressed, with a long snout. The tail (even if unregenerated) is virtually never constricted at the base. Between the small granular scales on the back are a series of small but prominent, weakly-keeled tubercular scales, looking like little buttons, in 8–12 rows, and on the sides of the tail these scales are pointed and obvious. The males have 45–57 precloacal pores. A big gecko, up to 19cm total length, 12–17cm on average, the unregenerated tail is slightly over half the total length. Usually some shade of grey or brown in colour, with 4–5 diffuse darker crossbars; in Kenyan specimens these crossbars often resemble the lower half of a six-pointed star, with the lowest point towards the tail. White below.

Hemidactylus platycephalus

HABITAT AND DISTRIBUTION: In forest, woodland, moist and dry savanna and semi-desert, from sea level to around 1,500m altitude. Widespread in south-eastern Tanzania, including the islands, and eastern Kenya, through Tsavo and Ukambani to the northern bend of the Tana River and the northern Uaso Nyiro River at Samburu National Reserve. Sporadic records from northern Kenya include Loyengalani, Kalacha, South Horr, Koobi Fora, Laisamis, Wajir, Moyale, Rhamu and Mandera, also recorded from Tarangire and Lake Manyara in Tanzania. Elsewhere, south to Zimbabwe, and on the Comoros.

NATURAL HISTORY: Nocturnal. Often on large trees, also on rocks; will sometimes enter buildings, although on Kenya's north coast where the species overlaps with *Hemidactylus mabouia*, this species is usually found on large trees, and *H. mabouia* on buildings. Rarely descends. In natural settings it shelters under bark and in cracks in trees and rocks; in buildings hides in any suitable crevice. May congregate in places, a group of eleven were observed high on the trunk of a Mvumo Palm in the Shimba Hills. It lays two hard-shelled eggs, roughly 10mm in diameter, in any suitable refuge. Eats a wide range of invertebrates.

Hemidactylus platycephalus Tree Gecko, Stephen Spawls, Shimba Hills.

RICHARDSON'S FOREST GECKO *Hemidactylus richardsoni*

IDENTIFICATION: A large grey robust forest gecko from western Uganda. Quite stocky in appearance, but unmistakeable, with velvety skin and a conspicuous lateral skin fold between the front and rear legs, digits webbed at the base and a fringe of long tubercles on the side of the tail. Tail roughly 50–55% of total length. Scales mostly small and granular, among which a few scattered smooth tubercles largely form one or two rows on each flank; the tail is flattened and has several rows of long tubercles, those at the outer edge on each side are so long they form a lateral fringe. There are 40–48 precloacal-femoral pores in a continuous row. Maximum size about 16cm, average 10–14cm. Colour grey or grey-brown overall, with a broad dark line passing through the eye and along the flanks. In some specimens there is a series of four or five broad darker dorsal crossbars on a lighter background, in others the back is simply mottled dark and light grey. The underside is deep yellow.

Hemidactylus richardsoni

HABITAT AND DISTRIBUTION: A forest-dwelling species of central Africa, in East African known only from Ntandi, at 700m altitude in the Bwamba Forest of western Uganda, elsewhere extends westwards to Cameroon and Nigeria.

NATURAL HISTORY: Nocturnal. Lives on big trees in forest, where it is said to live in the canopy; one Ugandan specimen was on a dead tree. In parts of its range it is known to enter buildings, and Congo specimens were found on a brick wall and a tree stump. In natural settings it shelters under bark and in cracks in trees. It has strong claws and a powerful grip; said to be difficult to remove it from a branch without injuring the animal. Captive specimens have been observed leaping acrobatically from branch to branch. Two hard-shelled eggs were laid by a Nigerian specimen in May; start of the rainy season. Eats a wide range of invertebrates.

Hemidactylus richardsoni Richardson's Forest Gecko. Olivier Pauwels, Gabon.

SOMALI PLAIN GECKO *Hemidactylus robustus*

IDENTIFICATION: A slender terrestrial gecko, known only from two localities in Kenya. Body fairly slender, and slightly depressed, with a short snout and large and prominent eyes. Tail about the same length as body, tapers smoothly, not noticeably constricted at the base. There are 13–18 more or less regular rows of enlarged, feebly-keeled tubercles along the back. Males have 5–8 (usually 6) precloacal pores. Maximum size about 11cm, average 8–10cm. Quite variable in colour: pink, grey, yellowish or rufous brown or some intermediate shade. There is usually a dark line through the eyes, which tapers off on the neck; the back may have 4–5 diffuse, irregular darker cross bars, or vague longitudinal stripes, or even just some irregular darker mottling; tail with narrow diffuse but clear darker crossbars. White or pinkish below.

Hemidactylus robustus

HABITAT AND DISTRIBUTION: In Kenya, known only from the dry savanna and semi-desert of the low northeast; from two places only, Garissa and Mandera between 150–230m altitude. Elsewhere, to Ethiopia and Somalia, widespread around the Arabian Peninsula and parts of the Middle East.

NATURAL HISTORY: Nocturnal and terrestrial. In the Middle East it is often commensal with man and found in buildings. The Mandera specimens were found under rock and palm logs. Biology similar to other *Hemidactylus*, i.e. lays two eggs and eats invertebrates.

Hemidactylus robustus Somali Plain Gecko, Tomas Mazuch Garissa.

PRINCE RUSPOLI'S GECKO Hemidactylus ruspolii

IDENTIFICATION: A small chunky gecko from northern Kenya. Quite stout in appearance, unmistakeable, with a stocky body, a short snout and a tail with an obvious constriction at the base; the unregenerated tail is slightly less than 50% of total length. Between the small granular scales on the dorsal surface there are 14–18 rows of strongly-keeled, prominent and obvious tubercular scales, giving a distinctive spiky appearance to this gecko. The unregenerated tail is even more spiky, it looks like the tail of an Ankylosaurus, while regenerated tails are usually very swollen, fat and turnip-shaped, usually without tubercles. Maximum size about 10cm, average 6–8cm. Ground colour variable, may be pinkish, grey, brown or almost black. Usually there is a series of four or five dark sub-rectangular blotches along the spine, with two parallel rows of similar blotches on either side. Many of the dorsal tubercles are white or yellow, and in striking contrast to the ground colour, especially in specimens from north-east Kenya, which may be almost black. The underside is white. Some juveniles are almost uniformly black, with yellow or red tubercles, the tail appears black with fine yellow bands, and some adults retain this colour.

Hemidactylus ruspolii

Hemidactylus ruspolii Prince Ruspoli's Gecko. Eduardo Razzetti, Lake Turkana.

HABITAT AND DISTRIBUTION: An inhabitant of the dry savanna and semi-desert of north-central and north-east Kenya, at low altitudes, usually below 1,000m, although there is a record from the northern edge of Laikipia at 1,500m. Not known from west of Lake Turkana. Isolated records from Tsavo National Park and Ngomeni in Kitui District. Elsewhere into the Ogaden.

NATURAL HISTORY: Nocturnal and apparently largely terrestrial in Kenya, although some specimens have been found under tree bark. By day, shelters under rocks and in logs. Two hard-shelled eggs are laid. Eats a wide range of invertebrates.

NYIKA GECKO Hemidactylus squamulatus

IDENTIFICATION: A small terrestrial gecko; perhaps the terrestrial gecko most likely to be encountered in the low dry country of eastern Kenya and Tanzania. The body is slightly depressed, the head is box-like, with a short snout. The tail tapers smoothly, and is not constricted at the base. The back is covered by overlapping scales of unequal sizes, without enlarged tubercles, but the larger scales are usually keeled and form 10–16 longitudinal rows. The males have 10–20 precloacal pores. A little gecko, up to about 9cm total length, but 6–8 cm is more usual. The unregenerated tail is slightly over half the total length. Very variable in colour and appearance; there are probably at least two cryptic forms at present concealed within this species, as it is presently defined. Usually some shade of brown or grey; with a number of fine, vague darker crossbars, some specimens have dark patches behind the eyes, these may extend backwards and meet to form a crescent round the back of the head. Specimens from the west (Mara, Serengeti) are larger and vividly speckled brown and white. Hatchlings and juveniles may be brown or pink, finely speckled yellow or white.

HABITAT AND DISTRIBUTION: In semi-desert, dry and moist savanna and grassland, from just above sea level to over 2,200m, but most common below 1,400m. Widespread in north-east and north-central Tanzania, as far west as the Serengeti and the Mara, also in north-eastern and eastern Kenya (but not on the coast, where *Hemidactylus barbouri*, long regarded as a subspecies of this form but now as separate species, is found), as far north as Wajir and Laisamis, the only records from north-west Kenya so far are South Horr and Eliye Springs.

Hemidactylus squamulatus

NATURAL HISTORY: Nocturnal and terrestrial. Hunts on the surface at night; it sometimes emerges at dusk. By day shelters in holes, under ground cover such as rocks and logs. It lays two hard-shelled eggs, roughly 7mm in diameter, in any suitable refuge. Eats a wide range of invertebrates.

Hemidactylus squamulatus Nyika Gecko. Left: Stephen Spawls, Maasai Mara. Middle: Stephen Spawls, Ithumba Hill. Right: Michele Menegon, Tanzania.

DUTUMI GECKO *Hemidactylus tanganicus*

IDENTIFICATION: A large *Hemidactylus* gecko from north-eastern Tanzania. Poorly known and apparently never photographed alive. The body is stout, limbs relatively short, the tail is carrot-shaped and depressed, slightly over half the total length. Between the small granular scales on the back there are 18–20 longitudinal rows of enlarged keeled tubercles, and there are eight rows of similar tubercles on the tail. The males have 42 precloacal pores. Maximum length about 16cm. The type was described as buff with a pinkish tinge, with six dark blotches along the spine, and similar blotches on the flaks; uniformly white with a pinkish tinge below.

Hemidactylus tanganicus

HABITAT AND DISTRIBUTION: Known only from two localities, Dutumi and Mkomazi, in the savanna of north-east Tanzania. Dutumi is in Miombo woodland, and Mkomazi is in dry savanna; both at low altitude, 300m or less. Assessed by the IUCN as Data Deficient, probably because its status as a valid species is unclear.

NATURAL HISTORY: Essentially unknown. The Mkomazi specimens were on trees, one hiding under the bark, so presumably arboreal. Biology presumably similar to other *Hemidactylus*, i.e. nocturnal, lays eggs and eats invertebrates.

OGADEN GECKO *Hemidactylus tropidolepis*

IDENTIFICATION: A small terrestrial gecko. The body is slightly depressed, the head has a short snout. The tail tapers smoothly, and is not constricted at the base. The back is covered by overlapping scales of unequal sizes, all of which are keeled, the larger ones randomly scattered across the back. The males have 6–10 precloacal pores. A small gecko, up to about 7cm total length. The unregenerated tail is slightly over half the total length. Very variable in colour; some shade of brown or grey; some specimens have a number of fine, vague darker crossbars, or wavy lines on the back, others have extensive black mottling on the back, increasing towards the flanks, interspersed with white spots. Some specimens have a dark patch behind the eye; these may extend backwards to form a crescent round the back of the head.

Hemidactylus tropidolepis

HABITAT AND DISTRIBUTION: The wide range of habitats that this species is recorded from (river delta, dry savanna, semi-desert) and the curious variety of localities suggest that this species is not properly defined. Specimens are known from Olorgesaile, Voi, the Tana delta and Wajir, thence into Somalia and the northeast of the Ogaden, at a range of altitudes between sea level and 1,300m.

NATURAL HISTORY: Nocturnal and terrestrial. Hunts on the surface at night; may emerge at dusk. By day shelters in holes, under ground cover such as rocks and logs. It lays two hard-shelled eggs in any suitable refuge. Eats a wide range of invertebrates.

DWARF GECKOES *LYGODACTYLUS*

A genus of attractive, small, diurnal, largely African geckoes, though a few species occur on Madagascar and South America. Over 60 species are known, with 23 in our area, nine of which are endemic. Often highly visible, they are territorial, in colonies with a dominant male. Many are brown or grey but a number of species have brightly-coloured heads and necks; they have a limited ability to change colour. Mostly arboreal (some will use rocks), they will venture onto buildings, fences and ornamental trees; it has been shown that in savanna, trees felled by elephants improve available habitats for dwarf geckoes. They speciate rapidly, small climatic and vegetational changes isolate populations and their rapid rate of breeding mean that changes accumulate quickly to produce new forms. They are also prone to accidental translocation, as eggs or living animals, in loads of firewood and timber, produce and building stone. Consequently their taxonomy is not fully resolved. If you see one of these geckoes, photograph it and send the picture to the National Museum; preserved specimens are also needed.

KEY TO EAST AFRICAN MEMBERS OF THE GENUS *LYGODACTYLUS*

This key is based largely on males of species with reasonably clear-cut colour patterns and distributions. It excludes three problematic species, for reasons as follows: *Lygodactylus angolensis* (incomprehensible distribution patterns), *Lygodactylus broadleyi* (type description doesn't describe the colour), *Lygodactylus inexpectatus* (briefly described from a single female specimen, never rediscovered). However, we have described these three forms in the species accounts. We have also treated the form *Lygodactylus grzimeki* (regarded as a full species by some) as a subspecies of *Lygodactylus angularis*, thus *L. a. grzimeki*

1a Head some shade of yellow, cream or white, with dark markings. Head and/or neck in clear contrast to body colour.....2
1b Head not yellow, white or cream. Head colour similar to dorsal coloration.....9
2a Head vivid mustard yellow, often with diffuse dark markings.....*Lygodactylus picturatus*, Yellow-headed Dwarf Gecko [p. 112]
2b Head white, cream, pale or dull yellow.....3
3a Body with five long contrasting stripes that extend to the level of the rear limbs or beyond, 11 precloacal pores.....*Lygodactylus kimhowelli*, Kim Howell's Dwarf Gecko [p. 109]
3b Body lacking five clear-cut stripes that extend onto tail, less than 11 precloacal pores.....4
4a A broad single yellow stripe (bright or dull) extends down the centre of the back beyond the level of the rear limbs.....*Lygodactylus tsavoensis*, Tsavo Dwarf Gecko [p. 115]
4b No single broad yellow stripe down the centre of the back.....5
5a Yellow stripe in the centre of the head forms a Y-shape; in high savanna around Mt Kenya.....*Lygodactylus wojnowskii*, Mt Kenya Dwarf Gecko [p. 116]
5b Yellow stripe in the centre of head does not form a Y-shape.....6
6a One or more pale patches on top of head totally enclosed by black pigment.....*Lygodactylus mombasicus*, White-headed Dwarf Gecko [p. 111]
6b No pale patches totally enclosed by black pigment on top of head.....7
7a Head and neck very dull yellow, speckled with brown, low contrast between dorsal and head/neck colour; in savanna of western Kenya and extreme northern Tanzania.....*Lygodactylus manni*, Mann's Dwarf Gecko [p. 110]
7b Head and neck not dull yellow, fairly clear cut division between head/neck and body colour; in dry northern or eastern Kenya.....8
8a Three clear, sharp-edged yellow lines on head and neck, no light spot on anterior or chin, in East Africa only known from the northern Tana River.....*Lygodactylus scorteccii*, Scortecci's Dwarf Gecko [p. 113]
8b Head and neck yellow with black markings but no clear lines, usually a light spot on anterior chin, widespread in northern Kenya.....*Lygodactylus keniensis*, Kenya Dwarf Gecko [p. 108]
9a Adult males distinctly blue, green or turquoise.....*Lygodactylus williamsi*, Turquoise Dwarf Gecko [p. 115]
9b Adults not blue, green or turquoise.....10
10a Three fine brown or black parallel lines on chin, in extreme northern Kenya, joined dark spots above shoulder.....*Lygodactylus grandisoni*, Bunty's Dwarf Gecko [p. 104]
10b Does not have this combination of characters.....11
11a Light brown or grey-brown, with virtually no high contrast markings on dorsal surface or throat, only in northern Kenya...*Lygodactylus somalicus*, Somali Dwarf Gecko [p. 114]
11b Usually grey, sometimes brown, usually with some high contrast black markings, not in northern Kenya...12
12a Dorsal surface usually with conspicuous lateral or dorsolateral stripes.....13
12b Dorsal surface usually without conspicuous lateral stripes.....16
13a Eye stripe brown or light, rarely black.....14
13b Eye stripe dark brown or black.....15
14a Dorsolateral stripes with diffuse edging, tail rarely rufous.....*Lygodactylus capensis*, Cape Dwarf Gecko [p. 102]
14b Dorsolateral stripes with sharp edging, tail often rufous.....*Lygodactylus grotei*, Grote's Dwarf Gecko [p. 105]

- **15a** Maximum size about 9cm, unregenerated tail about 55% of total length.....*Lygodactylus gravis*, Usambara Dwarf Gecko [p. 104]
- **15b** Maximum size about 5cm, unregenerated tail roughly 50% of total length.....*Lygodactylus conradti*, Conradt's Dwarf Gecko [p. 103]

- **16a** Bold lateral discrete spots or blotches on flanks.....17
- **16b** No bold lateral spots on flanks.....18

- **17a** A large spot on the neck, sometimes spots on the flanks.....*Lygodactylus scheffleri scheffleri*, Scheffler's Dwarf Gecko [p. 113]
- **17b** Several fairly large, dark spots on the flanks, but neck spot absent or tiny.....*Lygodactylus laterimaculatus*, Side-spotted Dwarf Gecko [p. 109]

- **18a** Usually brown, grey or almost black, with three dark, often ocellate flank spots, known only from the western side of East Africa, in forest and woodland .....*Lygodactylus gutturalis*, Forest Dwarf Gecko [p. 106]
- **18b** Usually grey or brown, in southern or eastern East Africa, not in the west.....19

- **19a** In extreme southern Tanzania, usually brown, often with a yellow shoulder patch.....*Lygodactylus angularis*, Angulate Dwarf Gecko [p. 101]
- **19b** Endemic to eastern Tanzania.....20

- **20a** Usually grey with 5–7 precloacal pores, endemic to the Uluguru Mountains, Tanzania.....*Lygodactylus scheffleri uluguruensis*, Uluguru Dwarf Gecko [p. 113]
- **20b** Usually pale to dark brown to grey with seven precloacal pores, endemic to coastal forest of eastern Tanzania and Zanzibar.....*Lygodactylus howelli*, Copal Dwarf Gecko [p. 107]

ANGOLAN DWARF GECKO *Lygodactylus angolensis*

IDENTIFICATION: A small dwarf gecko, usually grey or brown; some Zambian specimens almost black. Tail about 50% of total length. Males have 7–10 precloacal pores. Maximum size about 6.5cm, average 4–5cm, hatchling size unknown, probably about 2.5cm. Light grey, or olive-brown, some specimens (not all) have a dark line through the eye. Dorsal surface marked with fine dark lines and a line of pale lateral spots, below that a lateral marking consists or alternate light and dark bars. Often the lateral spots are edged with black above and below, forming near-ocelli; this is a good field character. White below.

Lygodactylus angolensis

HABITAT AND DISTRIBUTION: Occurs in the savanna and woodland of south-eastern Africa, at altitudes between 1,000–1,800m. Records from East Africa are curiously disjunct; most are in low altitude savanna, and include the Kedong Valley in Kenya, eastern Serengeti, Monduli, Dodoma, Mahale Peninsula, Usangu Plains, the Selous Game Reserve, Tatanda and Nachingwea. Elsewhere, south to northern Zimbabwe, south-west to Angola.

Lygodactylus angolensis Angolan Dwarf Gecko. Left: Beate Roll, throat colours, female. Right: Beate Roll, captive.

NATURAL HISTORY: Diurnal and arboreal. Forages in low trees and bushes and on dead trees, but also in and around habitation. It is speculated that it does not live very long, less than two years. Two eggs, roughly 0.6cm diameter, are laid in a suitable refuge. Diet; ants and termites.

ANGULATE DWARF GECKO Lygodactylus angularis

IDENTIFICATION: A small dwarf gecko, usually brown. Tail about 50% of total length. Males have 5–8 precloacal pores. Maximum size about 9.5cm, average 6–8cm, hatchling size unknown. Some specimens uniform grey above, with a few fine dark wavy lines, others are mottled brown and grey, with broad pale dorsolateral stripes and a few black lateral spots. Males are pink below, with yellow throats, females yellow; on the chin two lines converge from the lips to the base of the throat.

HABITAT AND DISTRIBUTION: In low- to medium-altitude woodland and forest clearings of southern Tanzania, from the north end of Lake Malawi east across to the Kilombero Valley, west to the south end of Lake Tanganyika, thence into Malawi and northern Zambia. The subspecies *Lygodactylus angularis grzimeki* was described from Lake Manyara National Park. A couple of curious disjunct records from Mombasa, Kenya and Tongwe Forest Reserve near Tanga require investigation.

Lygodactylus angularis

NATURAL HISTORY: Diurnal and arboreal; living on tree trunks, occasionally ventures onto buildings, one was found on a telegraph pole. Two eggs, roughly 0.8cm diameter, are laid in a suitable refuge. Diet: small invertebrates, including caterpillars; one appeared to have eaten a gecko egg.

Lygodactylus angularis Angulate Dwarf Gecko. Left: Ignas Safari, Mbeya. Right: Colin Tilbury, northern Malawi.

BROADLEY'S DWARF GECKO Lygodactylus broadleyi

IDENTIFICATION: A small dwarf gecko, described in 1995. Tail long, about 60% of total length. Males have six precloacal pores. Dorsal scales are smaller than any known *Lygodactylus*, there are 212–235 scale rows between the rostral scale, on the nose, and the base of the tail; few other dwarf geckoes have more than 200. The type specimens were about 6cm total length. Colour undescribed; the type description says 'the colour is so variable that it cannot be used as a character', although the male had a red stripe under the tail and the underside is yellow, becoming orange towards the tail.

HABITAT AND DISTRIBUTION: A Tanzanian endemic, known from a widespread group of medium to low altitude forests in the east, including Kilulu Forest near the Kenyan border, Amani in the Usambaras, Kiwengoma Forest south of the Rufiji and Zaraninge Forest near Bagamoyo; from sea level to nearly 1,000m altitude.

Lygodactylus broadleyi

NATURAL HISTORY: Unknown, never knowingly observed alive, but presumably similar to other dwarf geckoes, i.e. diurnal and arboreal (the type specimen was on a tree), lays two small spherical eggs, eats arthropods.

CAPE DWARF GECKO *Lygodactylus capensis*

IDENTIFICATION: A small brown or grey-brown dwarf gecko. Tail about 50% of total length. Males have 4–8 precloacal pores. Maximum size about 8cm, average 5–7cm, hatchling size 2.5cm (the subspecies *L. c. pakenhami* has even smaller hatchlings, just over 2cm). Colour variable, usually shades of brown or grey-brown; in East African specimens there is usually a faint dark line from the snout through the eye that extends into a dorsolateral stripe. Cream or white below, without markings. Specimens from central Kenya and parts of Tanzania sometimes have a fine light mid-dorsal stripe. On Pemba, not all individuals of the subspecies *L. c. pakenhami* have the side stripe, in some the dark stripe is light beyond the eye, and the dark stripe is low on the flanks. Taxonomic Note: Some authorities believe that Kenyan and northern Tanzanian examples of this species may not be true *Lygodactylus capensis*, but a cryptic species that superficially resembles it.

Lygodactylus capensis

Lygodactylus capensis Cape Dwarf Gecko. Top: Stephen Spawls, Naivasha. Bottom: Frank Glaw, Pemba, mating pair.

HABITAT AND DISTRIBUTION: Coastal woodland and thicket, moist savanna at low and medium altitude, from sea level to around 2,000m. Widespread across much of northern and eastern Tanzania, and also along the Lake Tanganyika shore from Ujiji south to lake Rukwa, thence into Malawi. The subspecies *L. capensis pakenhami* occurs on Pemba, where it is widespread. In Kenya, sporadic records include Chemilil, Lake Naivasha, Athi River, the southwestern slopes of Mt Kenya and Tsavo West. Elsewhere, south to eastern South Africa, south-west to Angola.

NATURAL HISTORY: Diurnal and arboreal; living on tree trunks, branches, shrubs and bushes, will also live on fences and walls. South African specimens reached sexual maturity in eight months and lived from 15–18 months only. Gravid Tanzanian females were collected in September, in South Africa they appear to breed throughout the year. One to two eggs, roughly 0.6cm diameter, are laid in a suitable refuge; often under bark or in a deep tree crack. Communal nests with many eggs have been recorded. Diet: arthropods, fond of ants. If seized they are said to sham death, they do sometimes become immobile when held.

CONRADT'S DWARF GECKO *Lygodactylus conradti*

IDENTIFICATION: A very small dwarf gecko, the tail is roughly 50% of total length. Males have 5–7 precloacal pores. Maximum length about 5cm. Colour quite variable, may be olive, yellow or grey, a dark stripe through the eye. A broad dorsolateral line starts above the eye and runs the length of the body, joining with its fellow at the base of the tail, which is marked with small dark blotches or crossbars. The males are bright yellow underneath.

Lygodactylus conradti

HABITAT AND DISTRIBUTION: A Tanzanian endemic, known from the forests of eastern Tanzania (East Usambara Mountains and various coastal forests in their vicinity), at altitudes between 1,200–2000m. The identification of this species (type locality Derema, East Usambara Mountains) has been confused with that of the larger *Lygodactylus gravis* (also known from nearby Amani in the East Usambaras, and from Magamba in the West Usambaras). Because of this confusion, it is difficult to reliably describe the details relating to this species. Generally, the Usambara Mountain *Lygodactylus* of the *scheffleri*

Lygodactylus conradti Conradt's Dwarf Gecko. Stephen Spawls, Usambara Mountains.

species group, to which *L. conradti* belongs, all have six adhesive lamellae under the fourth digit, although some have seven.

NATURAL HISTORY: Poorly known but presumably similar to other members of the genus, i.e. diurnal and arboreal, lives in structured colonies, lays two eggs, eats arthropods.

BUNTY'S DWARF GECKO *Lygodactylus grandisonae*

IDENTIFICATION: A small dwarf gecko, described in 1962. Tail about 50% of total length, or slightly more. Males have 6–7 precloacal pores. The type specimens were about 5cm total length; the largest 5.4cm. The dorsum is grey-brown with fine rufous spots, a dark brown or black line runs from the snout through the eye; on either side of the forelimb it widens into three dark blotches and then breaks up. On the chin is a distinctive pattern of three fine, parallel dark lines.

HABITAT AND DISTRIBUTION: Described by Georges Pasteur from specimens collected by J.C. Battersby in 1951, in the dry savanna woodland at Malka Murri, on the Kenyan-Ethiopian border at 900m altitude. Assessed as Data Deficient by IUCN, but its habitat is protected by a National Park on the Kenyan side.

Lygodactylus grandisonae Bunty's Dwarf Gecko, Stephen Spawls, preserved, throat pattern.

Lygodactylus grandisonae

NATURAL HISTORY: Unknown but presumably similar to other dwarf geckoes, i.e. diurnal and arboreal, lays two small spherical eggs, eats arthropods. Not collected since its discovery and never photographed alive. The specific name honours Alice G.C. ('Bunty') Grandison, redoubtable former curator of herpetology at the Natural History Museum in London.

USAMBARA DWARF GECKO *Lygodactylus gravis*

IDENTIFICATION: A relatively large dwarf gecko. Tail about 55% of total length. Males have 7–9 precloacal pores. Maximum size about 9cm. Dorsal colour ranging from yellows to blackish grey; usually a dark line from the nostril, running through the eye to just above the opening of the ear; dorsum sometimes uniformly coloured, or with darker streaks and mottling; in most females a broad, dark-edged, light-brown or yellow line from just behind the eye to the base of the tail, and may merge with its partner from the other side of the animal; between the lines is a double row of vague dark chevrons. The males tend to be more uniform. The sides of the body are either uniformly coloured, or have a series of dark blotches. Below, lightly spotted or without spots.

Lygodactylus gravis

Lygodactylus gravis Usambara Dwarf Gecko. Left: Beate Roll, captive, female. Right: Beate Roll, male captive.

HABITAT AND DISTRIBUTION: A Tanzanian endemic, from the forests of the Usambara and South Pare Mountains, at altitudes below 2,200m; known localities include Magamba and Kwai, in the West Usambaras, Amani (East Usambaras) and Maji Kununua Mountain, Mkomazi National Park. Conservation Status: Assessed by IUCN as Vulnerable B1ab(iii).

NATURAL HISTORY: Diurnal and, unusually for a dwarf gecko, apparently terrestrial, found at the bases of wild banana plants and under logs and rocks. At Magamba, communal egg laying has been noted, with 102 eggs being found in forest soil. Feeds on insects.

GROTE'S DWARF GECKO *Lygodactylus grotei*

IDENTIFICATION: A small brown or grey-brown striped dwarf gecko. Tail about 55 % of total length. Males have 5–6 precloacal pores. Maximum size about 7.5cm, average 5–7cm, hatchling size 2.5cm. Colour variable, usually shades of brown or grey-brown; with two fairly sharply edged yellow dorsolateral stripes, and a vertebral stripe, this may be more light grey. The tail is often rufous. Cream or white below, without markings.

HABITAT AND DISTRIBUTION: Coastal woodland and thicket, moist savanna at low and medium altitude, from sea level to around 1,000m, in south-eastern Tanzania. Elsewhere, south to northern Mozambique.

Lygodactylus grotei

NATURAL HISTORY: Diurnal and arboreal; living on tree trunks, branches, shrubs and bushes, will also live on fences and walls. Pairs were observed mating in February in both Pemba and Morogoro. One to two eggs, roughly 0.6cm diameter, are laid in a suitable refuge; often under bark or in a deep tree crack. Diet: invertebrates, fond of ants.

Lygodactylus grotei Grote's Dwarf Gwecko. Left: Michele Menegon, Mozambique. Right: Beate Roll, captive.

CHEVRON-THROATED DWARF GECKO *Lygodactylus gutturalis*

IDENTIFICATION: A small dwarf gecko from the western forests of East Africa. Unregenerated tail about 50% of total length. Males have 6–9 precloacal pores. Maximum size is about 9cm. The dorsal colour is quite variable (and can change according to the animal's temperament); it may be predominantly black, grey or brown (Ugandan specimens quite light brown), with lighter blotching, and these blotches may form a diffuse dorsolateral band. There is usually a dark line through the eye, and above and on either side of the shoulder there are at least three dark blotches. Tail with diffuse lighter bars. Throat of male white with three dark chevrons, or a solid arrow-head or spot as a third basal mark; throat of female similar to almost immaculate. Subcaudals with a grey stippled median stripe. Ventrum vivid yellow or cream.

Lygodactylus gutturalis

HABITAT AND DISTRIBUTION: Forest, woodland and gallery forest of the west, between 700–2,000m altitude. Recently recorded in the Maasai Mara national reserve and Kakamega Forest in Kenya, sporadic records in Uganda from the forests of the east and the west (but not the centre) include the Sesse Islands, Wadelai, Mbale, Lake Albert, Katakwi, Bubula (Mt. Elgon), Greeki river, Mount Debasien, Budongo Forest, Bundibugyo. In Tanzania recorded from Seronera in the Serengeti, Rubondo Island, Ujiji, Burigi Game Reserve, south to the Ugalla River, widespread in Rwanda and Burundi, especially in the east. Elsewhere, north to Ethiopia and South Sudan, west to Senegal in the forest; it has the widest distribution of any *Lygodactylus* species.

Lygodactylus gutturalis Chevron-throated Dwarf Gecko, Left: Max Dehling, Rwanda. Right: Caspian Johnson, Ugalla.

NATURAL HISTORY: Diurnal and arboreal; living on trees, particularly the bases of large forest trees, although it will inhabit fences and buildings. In East Africa it appears to be active at all times during the day, although Nigerian specimens at lower altitude were most active between 5–7pm. In places where it is sympatric with other *Lygodactylus* species it inhabits larger trees and deeper shade, although Congo specimens were found in heaps of plant debris. One to two eggs, roughly 0.8cm diameter, are laid in a suitable refuge; often under bark or in a deep tree crack, gravid females found in western Tanzania in May, clutches found in December in Uganda. It probably breeds throughout the year. Diet: invertebrates, fond of ants.

COPAL DWARF GECKO/HOWELL'S DWARF GECKO
Lygodactylus howelli

IDENTIFICATION: A small dwarf gecko of eastern Tanzania. Unregenerated tail about 50% of total length or slightly longer. Males have seven precloacal pores. Maximum size is unknown. The body scales are apparently very small, even for a dwarf gecko. The dorsal colour is pale to dark brown or grey, with a dark bar between the eyes and a dark line from the nostril through the eye; there is a pair of light spots at the base of the tail. Some specimens are immaculate white below; others have many dots and streaks of dark brown, particularly males. Taxonomic Notes: This species was originally described from a sub-fossil specimen in copal resin, as *Lygodactylus viscatus*, then re-described as *Lygodactylus howelli*, from a modern Zanzibari specimen. Later, more live specimens were found on the Tanzanian coast. There is debate over its status, validity and nomenclatural precedence; at present the name *Lygodactylus howelli* is being used.

Lygodactylus howelli

HABITAT AND DISTRIBUTION: A coastal species, known from Jozani Forest on Zanzibar, from Mafia island, and from a handful of Tanzanian coastal forests including Kilulu, Zaraninge, Kiwengoma, Tong'omba and Namkutwa, all at low altitude. Conservation Status: Assessed by IUCN (as *Lygodactylus viscatus*) as Near Threatened.

NATURAL HISTORY: Diurnal and arboreal, lives on saplings and large trees in groundwater forests, in structured colonies. Apparently it is hard to detect, alert and agile, and if threatened and unable to ascend will jump to the ground and hide in leaf litter. Presumably lays two eggs, eats small arthropods.

DAR ES SALAAM DWARF GECKO *Lygodactylus inexpectatus*

IDENTIFICATION: A species described by Georges Pasteur in 1965, in 12 terse lines, from a single 8cm female specimen, and never located since. The type had the following characters: nine enlarged dimpled precloacal scales, two enlarged subcaudal scales per whorl of the tail, the head and neck colour did not contrast with the dorsal colouration; the chin had three lines converging at the front to form an arrowhead, and a series of dark scales under the tail. Taxonomic Note: It has been suggested that the type specimen is a slightly aberrant waif of some known species and that it reached the port on a ship.

Lygodactylus inexpectatus

DISTRIBUTION: Known only from the holotype, collected at Dar es Salaam, Tanzania. Conservation Status: Assessed by IUCN as Data Deficient.

NATURAL HISTORY: Unknown, not collected since its discovery and never observed or photographed alive.

KENYA DWARF GECKO *Lygodactylus keniensis*

IDENTIFICATION: A small dwarf gecko, the tail is roughly 50% or slightly less of total length. Males have 7–8 precloacal pores. Maximum length about 9cm, average 6–8cm, hatchlings 2.5cm. The head is cream or yellow, with distinctive black markings, sometimes just large spots, sometimes lines, usually there is a dark stripe through the eye. The head contrasts with the dorsal colour, which is usually some shade of grey, but may vary from very light to almost blue-grey, mottled to a greater or lesser extent with dark ocelli or blotches; the tail is sometimes uniform grey, or may have a series of dark chevrons. The underside is white, save the throat and chin, which have black reticulations and there is nearly always a distinctive white 'o'-shaped chin spot.

Lygodactylus keniensis

HABITAT AND DISTRIBUTION: Widespread in dry savanna and semi-desert of northern Kenya, from 200m altitude up to about 1,600m (and even higher altitudes in Ethiopia). The range spreads north, north-west and north-east from Lake Elmenteita to the northern border, although there are few records east of Lake Turkana. Extends into Uganda around Amudat and in Kidepo National Park; an isolated record from the Saka bend on the northern Tana River in Kenya. Elsewhere, to Ethiopia and southern Somalia. Conservation Status: Assessed by IUCN as Least Concern.

NATURAL HISTORY: Diurnal and arboreal, active in the morning and later afternoon at lower altitudes. It forages on all sizes of tree, but will also utilise buildings, walls and fences. A study of this species in the lower Kerio Valley found that it often shared its habitat with the Somali Dwarf Gecko *Lygodactylus somalicus*, but when both were on the same big tree, this species occupied the trunk, and was the only species on the larger trees. One tree had 46 individuals on it (15 females, 21 males, 10 juveniles), on smaller trees the ratio was one male to 2.1 females. Two eggs are laid in bark cracks and fissures; and in the Kerio Valley eggs from this species and the Somali Dwarf Gecko were found in the same crack. Diet: small invertebrates.

Lygodactylus keniensis Kenya Dwarf Gecko Left: Daniel Hollands, Lake Nakuru. Right: Daniel Hollands, throat markings, male.

KIM HOWELL'S DWARF GECKO/ZEBRA DWARF GECKO
Lygodactylus kimhowelli

IDENTIFICATION: A small dwarf gecko with five distinct dorsal stripes, from the coast. Unregenerated tail about 50% of total length. Males have 11 precloacal pores; no other East African species has this many. Maximum size is about 10cm. The head is striped cream or yellow against brown or black striping, and the head stripes extend the length of the body and onto the tail, the yellow becoming light grey and the brown or black becoming darker grey. Female colour is a subdued version of the male's. The chin is white, but with heavy black stippling and a light spot at the front; the ventral surface has a clear pale yellow central band that extends just beyond the cloaca. Juveniles have vivid orange-red ventrals, the colour extending to the rear limbs and tail base. Taxonomic Note: Molecular studies indicate that this species is very closely related to *Lygodactylus mombasicus*, the White-headed Dwarf Gecko, and could be considered just a striped morph of that species.

Lygodactylus kimhowelli

HABITAT AND DISTRIBUTION: An East African endemic, known from the coastal forests and associated agricultural plantations of the far northern Tanzanian coast, extending into Kenya, to the Shimba Hills; at altitudes between sea level and 350m.

Lygodacylus kimhowelli Kim Howell's Dwarf Gecko. Left: Lorenzo Vinciguerra, Tanga. Inset: Beate Roll, captive juvenile

NATURAL HISTORY: Diurnal and arboreal; living on trees. A popular species in the pet trade, breeders report they lay two eggs, roughly 0.8cm diameter, and those hatch in 2–3 months. Diet: invertebrates.

SIDE-SPOTTED DWARF GECKO *Lygodactylus laterimaculatus*

IDENTIFICATION: A small dwarf gecko, males have a distinctive row of black blotches on its flanks. Unregenerated tail about 55% of total length. Males have 6–7 precloacal pores. Maximum size is about 8cm, average 5–7cm. Dorsal colour grey, dark grey or grey-brown, the head the same background colour as the back, the back marked with light blotches that become crossbars on the tail; in dark individuals these blotches are pinkish. Males have 3–10 distinctive black blotches on the sides, contrasting strongly with the ground colour, and these may coalesce to form a partial stripe or a zigzag; females lack the spots, the flanks are lightly speckled. Pinkish-white below, with a few dark marks on the chin. Taxonomic Note: Some authorities regard this as a subspecies of Scheffler's Dwarf Gecko *Lygodactylus scheffleri*.

Lygodactylus laterimaculatus

Lygodactylus laterimaculatus Side-spotted Dwarf Gecko. Left: Stephen Spawls, Moshi, male. Right: Beate Roll, captive, female.

HABITAT AND DISTRIBUTION: An East African endemic, known from low- to medium-altitude savanna and wooded hills, from roughly 300–1,200m. Recorded from Voi and Buchuma in Tsavo, the lower slopes of the Taita Hills, westwards to the southern slopes of Mt Kilimanjaro. Conservation Status: Assessed by IUCN as Least Concern.

NATURAL HISTORY: Diurnal and arboreal; living on trees, observed to be common around Moshi town, living on small trees, fences and buildings; around Voi seen mostly on small trees, occasionally on big acacias. Presumably similar to other dwarf geckoes, laying two eggs and eating invertebrates.

MANN'S DWARF GECKO *Lygodactylus manni*

IDENTIFICATION: A medium-sized dwarf gecko. Unregenerated tail about 50% of total length. Males have six precloacal pores. Maximum size is about 8cm. Dorsal colour grey or grey-brown, the head dull yellow with brown markings; a series of diffuse darker ocellate markings form two dorsolateral stripes, the tail is grey. The dorsal colouration does not contrast strongly with the head markings although they are different colours. The throat has heavy black markings on a white background, the ventral surface has a clear pale yellow central band that extends just beyond the cloaca and the upper underside of the rear limbs is sepia.

Lygodactylus manni

HABITAT AND DISTRIBUTION: An East African endemic, known from medium altitude moist and dry savanna, between 1000–1,800m altitude. Found from Lake Baringo south and west onto the north shore of the Winam Gulf, and into northern Tanzania, the type specimen came from Saranda, northwest of Dodoma, also known from the east end of Lake Eyasi. Conservation Status: Assessed by IUCN as Least Concern.

Lygodactylus manni Mann's Dwarf Gecko. Stephen Spawls, Lake Baringo.

NATURAL HISTORY: Diurnal and arboreal; living on trees, will venture onto fences and buildings. Habits presumably similar to other dwarf geckoes, laying two eggs and eating invertebrates.

WHITE-HEADED DWARF GECKO Lygodactylus mombasicus

IDENTIFICATION: A dwarf gecko of coastal Kenya and north-eastern Tanzania. Unregenerated tail about 50% or slightly more of total length. Males have 6–10 precloacal pores. The head has a bold black pattern on a whitish, cream or pale yellow background. The dorsum is grey or pale grey. On the top of the head, between the eyes there is virtually always a black patch with two enclosed yellow or cream spots. Males usually have a uniform black throat, only rarely showing traces of chevrons; throat of female white or showing faint darker marking. The male underside is yellow around the cloaca. Hatchlings and juveniles are dark versions of the adult, looking predominantly brown but the head pattern is visible. Taxonomic Note: This gecko was frequently and erroneously called Lygodactylus picturatus in recent literature, but that name actually applies to the Yellow-headed Dwarf Gecko.

Lygodactylus mombasicus

HABITAT AND DISTRIBUTION: Coastal woodland and low-altitude savanna, from sea level to 500m altitude or maybe higher. Occurs northwards from the Gendagenda Forest on the northern Tanzanian coast, up the Kenya coast to Lamu. Extends up the Tana River as far as Meru National Park, across to Isiolo, and inland on the Galana River. Conservation Status: Assessed by IUCN as Least Concern.

Lygodactylus mombasicus White-headed Dwarf Gecko, Stephen Spawls. Top left: Throat colours, male. Bottom left: Watamu, dull male. Right: Watamu, displaying male.

NATURAL HISTORY: Diurnal and arboreal, active throughout the day. It forages on all sizes of tree, but will also utilise little ornamental shrubs, buildings, walls and fences. It thrives in suburbia, provided not too much insecticide is used; found in the grounds of virtually all Kenyan coastal plots, using the walls and the ornamental trees, each of which has a dominant male. They can become tame and confiding if not persecuted. A favourite prey of lizard-eating snakes like the Spotted Bush Snake Philothamnus semivariegatus, and the bark and sand snakes. Two eggs are laid in any suitable refuge. Diet: small invertebrates.

YELLOW-HEADED DWARF GECKO *Lygodactylus picturatus*

IDENTIFICATION: A strikingly-coloured dwarf gecko. Unregenerated tail about 50% or slightly more of total length. Males have 9–10 precloacal pores. Maximum size is about 9cm, average 6–8cm, hatchlings about 2.5cm. In males the head is vivid, mustard-yellow with black markings; the dorsal colour a strongly-contrasting grey or blue-grey, the back speckled to a greater or lesser extent with black markings, in some specimens the absence of speckling creates uniform grey dorsolateral stripes; specimens from inland Tanzania often have a distinctive white, black-ringed spot just above the insertion of the forelimb. Females are a subdued version of the male. The mature male's throat is black with yellow edging; the throat of the female is blotched and stippled black on a white background. The ventral surface has a clear pale yellow central band that extends just beyond the cloaca. Taxonomic Note: Until recently, this species was erroneously known as *Lygodactylus luteopicturatus*; recent work shows that this is almost certainly a junior synonym of *L. picturatus*. Two subspecies exist at present, *Lygodactylus picturatus picturatus* (mainland) and *L. p. zanzibaritis*, found on Zanzibar; this form has rather more sharply defined black head markings and ocelli on the flanks.

Lygodactylus picturatus

HABITAT AND DISTRIBUTION: Coastal forest, thicket, savanna and farmland, extending into urban areas, at altitudes from sea level to about 1,000m. Occurs from south-east Kenya and north-east Tanzania, east to Mombasa, south into Tanzania, to northern Mozambique. Inland on the Rufiji and Ruvuma Rivers; isolated records include the Tana Delta, around Lake Manyara and the western Serengeti. Conservation Status: Assessed by IUCN as Least Concern.

NATURAL HISTORY: Diurnal and arboreal; living on trees, an extremely agile species that makes vigorous use of the tail-tip scansors. If threatened, it dodges around the tree and may descent to the ground to hide. It will venture onto fences, walls and buildings and is tolerant of urbanisation; occurs in the heart of Dar es Salaam. In coastal East Africa this gecko may be active at night, feeding around electric lights. Males display to females by arching the head to display their black throats, and when two males approach, they arch their backs in a threat posture. Eggs are deposited in glued pairs under bark. Feeds not only on small arthropods but on plant juices and the honey of stingless bees.

Lygodactylus picturatus Yellow-headed Dwarf Gecko. Main image: Michele Menegon, Zanzibar subspecies. Inset: Stephen Spawls, adult male and juvenile, Dar es Salaam.

SCHEFFLER'S DWARF GECKO Lygodactylus scheffleri

IDENTIFICATION: A small chunky dwarf gecko, known from dry savanna. Unregenerated tail about 50% or slightly more of total length. Males have six precloacal pores. Maximum size is about 6cm. Dorsal colour olive to yellow-brown, the head and body the same colour. A diffuse narrow dark line runs through the eye. Males have a line of black or deep brown patches on the flanks, below a line of light diffuse dorsolateral patches; females lack these dark lateral spots. On the neck between the ear and the forelimb is a black or red-brown blotch; the tail is barred light and dark, these bars may be irregular or appear like coalescing yellow spots. The underside is white, sometimes the chin is flecked with brown. Taxonomic Note: Populations of the species Lygodactylus uluguruensis, sometimes regarded as a full species, are at present believed to be a subspecies of this form, Lygodactylus scheffleri uluguruensis. It has 5–7 precloacal pores and is grey, with dorsal darker grey blotches, each bordered anteriorly by a dark mark; the males do not have the dark lateral spots. The throat has dark median lines, flanked by converging lines.

Lygodactylus scheffleri

HABITAT AND DISTRIBUTION: An East African endemic, known from a handful of records in the dry savanna of the east; south-east Kenyan records include the Chyulu Hills, Kiboko, Kibwezi, Kitobo Forest, Voi and Maktau; it is probably widespread in eastern Ukambani. Tanzanian records include south-west of Mt Hanang, and between Arusha and Moshi; all lie between altitudes of 500–1,500m. The subspecies Lygodactylus scheffleri uluguruensis is confined to the forest and forest fringes of the Uluguru Mountains. Conservation Status: Assessed by IUCN as Data Deficient.

Lygodactylus scheffleri Scheffler's Dwarf Gecko. Lorenzo Vinciguerra, northern Tanzania.

NATURAL HISTORY: Diurnal and arboreal; living on trees, will venture onto fences and buildings. Habits presumably similar to other dwarf geckoes, laying two eggs and eating invertebrates.

SCORTECCI'S DWARF GECKO Lygodactylus scorteccii

IDENTIFICATION: A medium-sized dwarf gecko with a yellow head. Unregenerated tail about 50% of total length or slightly longer. Males have 6–8 precloacal pores. Maximum size is about 8cm. The head is bright yellow with brown and black striping, the stripes extend just past the shoulders and the yellow stripe then becomes light grey; the centre of the back is mottled grey-brown. Female colour is a subdued version of the male's. The underside is white; the male has black stippling on the throat.

Lygodactylus scortecci Scortecci's Dwarf Gecko. Tomas Mazuch, Garissa.

Lygodactylus scorteccii

HABITAT AND DISTRIBUTION: In Kenya, known only from riverine forest on the Tana River, between the northernmost bend and Garissa, at low altitude. Widespread in southern Somalia.

NATURAL HISTORY: Poorly known but presumably similar to other members of the genus, i.e. diurnal and arboreal, lives in structured colonies, lays two eggs, eats small arthropods.

SOMALI DWARF GECKO *Lygodactylus somalicus*

IDENTIFICATION: A small dwarf gecko, brown, found in northern Kenya. Unregenerated tail about 50% of total length or slightly longer. Males have 5–6 precloacal pores. Maximum size is about 6cm, males seem to be larger than females. The head and body are the same colour, brown or grey-brown. There is a fine diffuse darker stripe through the eye. In some individuals there is a suggestion of a broad grey mid-dorsal band, brown on the flanks, dorsal surface with irregular but large pale blotches. The tail is banded, brown with narrow light bands. Underside white, usually unmarked, sometimes some spotting on the chin and neck.

Lygodactylus somalicus

HABITAT AND DISTRIBUTION: Dry savanna and semi-desert of northern Kenya, at altitudes from 200 to around 600m Found from the lower Kerio River and southern Lake Turkana in a broad band across northern Kenya, not known south of Wajir. Elsewhere to south-east Ethiopia and south and central Somalia.

NATURAL HISTORY: Arboreal and diurnal, active in the morning and late afternoon in its hot habitat, on trees and bushes in the dry country. A study of this species in the lower Kerio Valley in Kenya found that it was sympatric with the Kenya Dwarf Gecko *Lygodactylus keniensis* and, being the smaller of the two, usually lived on the smaller shrubs; when on the same

Lygodactylus somalicus Somali Dwarf Gecko. Eduardo Razzetti, Lake Turkana.

tree as the Kenya Dwarf Gecko it was forced onto the higher, thinner branches. The males were highly territorial. However, dominance depended on individual size, not sex (in confrontations over food items); a female was seen to dominate a favourable position at a breach in a termite archway, she excluded the male within whose territory the breach occurred. They lay two eggs, often in the same bark cracks as eggs of the Kenya Dwarf Gecko. Diet: small invertebrates.

TSAVO DWARF GECKO   *Lygodactylus tsavoensis*

IDENTIFICATION: A medium-sized dwarf gecko with a distinctive yellow dorsal stripe. Unregenerated tail about 50% of total length or slightly longer. Males have 6–10 precloacal pores. Maximum size is about 9cm. The head is striped black and cream or dull yellow, the body grey or greyish-brown, the yellow central head stripe, bordered by a black or brown stripe, extends the length of the body to beyond the tail base. The flanks are speckled black, sometimes with ocellate markings. Ventrals surface white, with a central yellow band.

Lygodactylus tsavoensis

HABITAT AND DISTRIBUTION: A Kenyan endemic, recently described, it inhabits the dry savanna of eastern Kenya, at altitudes between 300–1,000m; from Hunter's Lodge and Makindu eastwards through Tsavo to as far as Buchuma on the eastern edge of Tsavo East.

NATURAL HISTORY: Diurnal and arboreal, lives on trees and stout bushes, in structured colonies, lays two eggs, eats small arthropods.

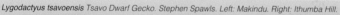
Lygodactyus tsavoensis Tsavo Dwarf Gecko. Stephen Spawls. Left: Makindu. Right: Ithumba Hill.

TURQUOISE DWARF GECKO/ELECTRIC BLUE DWARF GECKO   *Lygodactylus williamsi*

IDENTIFICATION: A spectacular dwarf gecko, and sadly much in demand in the pet trade. Unregenerated tail about 50% of total length. Males have 6–7 very small precloacal pores. Maximum size is about 9–10cm, average 5–8cm. The dorsal colour of the male may be deep blue, light blue, turquoise or green, usually with heavy black striping on the head;

the female may be greenish, bronze or brownish, both sexes may have black speckling on the flanks. The throat of the mature male is blue-black with lighter blue lines, and the body light blue, the female has a light blue throat with darker lines and is white below.

HABITAT AND DISTRIBUTION: A forest-dwelling Tanzanian endemic, currently only known from four forest patches; Kimboza Forest (the type locality), Ruvu Forest and two tiny outcrops just north of there, in Morogoro Region, Tanzania, in the foothills of the Uluguru mountains at altitudes between 170–480m. Conservation Status: Assessed by the IUCN as Critically Endangered B1ab(iii,v)+2ab(iii,v). It lives in two small areas and the population is rapidly decreasing due to collection for the pet trade (and to some extent deforestation); a recent study indicated that about 15% of the total population had been harvested for the pet trade. Specimens are collected in a destructive manner, with collectors simply chopping down the trees on which the animals are found and shaking them out of the leaves.

Lygodactylus williamsi

NATURAL HISTORY: Diurnal and arboreal; it appears to live only on the large leaves of the screw pines, distinctive monocotyledonous trees of the species *Pandanus rabaiensis*, which are found in about 20% of the forest. These geckoes live in structured colonies, usually with a single male and several females. Females appear to breed throughout the year, laying two eggs at a time, hatching times in captivity vary from 50–120 days.

Lygodactylus williamsi Turquoise Dwarf Gecko. Left: Michele Menegon, Kimboza Forest, male. Right: Michele Menegon, Kimboza Forest, female.

MT KENYA DWARF GECKO *Lygodactylus wojnowskii*

IDENTIFICATION: A medium-sized dwarf gecko of the foothills and central highlands round Mt Kenya. Unregenerated tail about 50% of total length or slightly longer. Males have eight precloacal pores. Maximum size is about 8cm. The head is striped black and cream or dull yellow, the body grey or greyish-brown, the head markings extend to just beyond the shoulders, where the sharp-edged black stripes become diffuse and gradually fade towards the tail The flanks are speckled black, sometimes with ocellate markings. Ventrals surface white or grey-white, with a central yellow band. Males and females have the same dorsal colouration, but the male has a black throat; the female has a black and white barred throat with a white spot at the front, like *Lygodactylus keniensis*.

HABITAT AND DISTRIBUTION: A Kenyan endemic, described in 2016. Known from forest fringes, woodland, savanna and suburbia, in central Kenya, at altitudes between 900–1,600m; originally described from around Chogoria town on the south-east slopes of Mt Kenya, specimens are also known from Igembe, in the Nyambene Hills, and there are sight records from Chiokarige on Tharaka, Sagana, Kindaruma and Masinga Dams in Machakos, near Machakos and also from Kenyatta University near Nairobi.

Lygodactylus wojnowskii

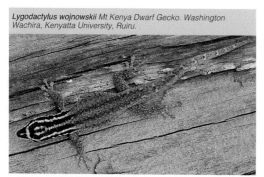

Lygodactylus wojnowskii Mt Kenya Dwarf Gecko. Washington Wachira, Kenyatta University, Ruiru.

NATURAL HISTORY: Diurnal and arboreal, lives on trees and bushes, observed on fences and fruits trees around Chogoria. Described as being highly agile, jumping from branch to branch. Presumably, like other dwarf geckoes, it lives in structured colonies, lays two eggs and eats small arthropods.

GIANT GECKOES *CHONDRODACTYLUS*

A genus with five species of large nocturnal geckoes. The single East African member of the genus was originally placed in the genus *Pachydactylus*, thick-toed geckoes, a diverse southern African genus; molecular analysis has rearranged the complex.

TURNER'S THICK-TOED GECKO *Chondrodactylus turneri*

IDENTIFICATION: A big, chunky gecko with a muscular head, rounded snout, the nostril points vertically upwards, a prominent eye with golden-brown iris and vertical pupil. The body is squat and powerful. Toe tips dilated. The tail is thick, slightly less than half total length. No precloacal pores. Body scales are of two types, little granular scales interspersed with rows of big-keeled spiky tubercular scales; when held the body feels distinctly rough due to these big scales. Maximum size about 18–20cm, average 12–15cm, hatchlings 6–7cm. Ground colour grey or brown, with 3–7 wavy, irregular black crossbars on the back and irregular black and white dorsal spots (often occupying a single coloured scale), often concentrated towards the shoulders. Usually two dark lateral stripes on the head, passing through the eye, and another two on the snout. Underside white.

HABITAT AND DISTRIBUTION: Moist and dry savanna, with big trees and rock outcrops, from sea level to 1,800m altitude. Distribution in East Africa patchy, probably due to under-collecting; it is a secretive species. Known from south-east Tanzania, from Saadani National Park west to Dodoma, south to the border, one record from north-west Rwanda, a cluster of records in the Kenya-Tanzania border country, from the escarpment north of Olorgesaille (maybe on the west of the Ngong Hills) south through Kajiado to west of Arusha, across beyond the Serengeti. A possible isolated sight record from Kora, in Kenya, so might be in the rocky hills of Ukambani. Elsewhere, south to Botswana and northern South Africa, south-west to Angola.

Chondrodactylus turneri

NATURAL HISTORY: Lives in rock cracks and under the bark of big trees. Will live on buildings, sheltering in roofs or under eaves and hunt insects around lamps. Nocturnal, but may ambush prey from rock cracks during the day. In suitable habitats it can be found by shining a light into deep cracks. When seized it makes a noisy squeak. It has strong jaw muscles, it will bite hard and hang on ferociously and may be reluctant to let go. Two eggs, about 1.5×2cm in diameter are laid in deep rock fissures or under bark, possibly in holes. Breeding season not known in East Africa, but eggs are laid between August and December in southern Africa, where the incubation time is 60–80 days. Its diet includes invertebrates, fond of termites, will take strong-bodied insects such as beetles.

Chondrodactylus turneri Turner's Thick-toed Gecko. Left: Stephen Spawls, Olorgesaille. Right: Michele Menegon, South Africa.

EAST AFRICAN THICK-TOED GECKOES *Elasmodactylus*

A genus with two species of large nocturnal terrestrial geckoes, both found in the south of East Africa, extending into south-central Africa. Like *Chondrodactylus* the genus emerged as a monophyletic clade following molecular analysis of the

Pachydactylus group. Both members of the genus are large, stocky geckoes, with broad dilated toe tips that are undivided; the body scales are small, granular and non-overlapping; scattered among them are large-keeled tubercles.

KEY TO EAST AFRICAN MEMBERS OF THE GENUS *ELASMODACTYLUS*

1a Rostral with a median cleft above, no swollen nasal ring, each caudal verticil with a transverse row of six enlarged dorsal tubercles......*Elasmodactylus tuberculosus*, Tuberculate Thick-toed Gecko [p. 119]

1b Rostral without a median cleft above, a greatly swollen nasal ring present, each caudal verticil with a pair of slightly enlarged scales......*Elasmodactylus tetensis*, Tete Thick-toed Gecko [p. 119]

TETE THICK-TOED GECKO *Elasmodactylus tetensis*

IDENTIFICATION: A big, strongly-built gecko with a large short head, rounded snout and prominent eye, with a vertical pupil. Body squat and muscular. Toes broadly dilated at the tips. Tail slightly less than half total length. Precloacal pores 8–14. The back is covered with large keeled tubercles, usually in eight longitudinal rows, interspersed with small granular scales. Maximum size is about 18cm, average 13–15cm, hatchling size unknown. Ground colour grey, with vague darker crossbars, the dorsal surface speckled with white tubercular scales. The tail has light and dark crossbars, the belly is white.

Elasmodactylus tetensis

HABITAT AND DISTRIBUTION: Savanna and woodland with rock outcrops at low altitude. In our area found in extreme south-east Tanzania, localities include Liwale, at 500m altitude, the Lumesule River and Nachingwea. Elsewhere, south through eastern Mozambique to the lower Zambezi valley. Conservation Status: Assessed by IUCN as Least Concern.

Elasmodactylus tetensis Tete Thick-toed Gecko. Bill Branch, Mozambique.

NATURAL HISTORY: Poorly known. Nocturnal, lives in rock cracks and hollow tree trunks, said to favour baobabs. Said to be gregarious, living in colonies, as do other big thick-toed geckoes. The skin is thin and tears easily. No breeding details known, presumably lays two eggs. Diet probably invertebrates.

TUBERCULATE THICK-TOED GECKO
Elasmodactylus tuberculosus

IDENTIFICATION: A fairly big, stocky gecko with a large head, rounded snout and big eye, with a vertical pupil. Body strong and muscular. Toes broadly dilated at the tips. Tail about half total length. Six to eight precloacal pores are present. The back is covered with large keeled tubercles, interspersed with small granular scales. Maximum size about 17cm, average 12–15cm, hatchling size 4–5cm. Adults various shades of brown, with irregular darker cross-bars, sometimes with a darker brown stripe from the nostril through the eye to the ear opening. Irregular dark bands on the tail. Juveniles have six pairs of dark blotches on either side of the spine, and tails with 11 dark crossbars. Uniform white or cream below.

HABITAT AND DISTRIBUTION: Open and wooded savanna, from sea level to 1,700m. Widely distributed across central Tanzania, west from Tanga across to Mwanza, south-west to Lake Rukwa and the shores of Lake Tanganyika. Elsewhere, in extreme northern Zambia and eastern DR Congo. Conservation Status: Assessed by IUCN as Least Concern.

Elasmodactylus tuberculosus

NATURAL HISTORY: Poorly known. Nocturnal, will live on trees and buildings, possibly on rocks. Known to descend to the ground and enter holes. Said to be very timid, emits a strident squeak if seized. Lays clutches of hard-shelled eggs throughout the wet season. Diet: beetles and bugs recorded, presumably any suitable arthropod taken. Feared in some parts of its range, people believe that if it emits a noise in a house then one of the human occupants will die.

Elasmodactylus tuberculosus Tuberculate Thick-toed Gecko. Michele Menegon, Tanzania.

VELVET GECKOES *HOMOPHOLIS*

A genus of medium to large arboreal and rock-dwelling geckoes, some are nocturnal, others diurnal. Usually shades of grey, brown or black. The eye pupil is vertical. They have soft, velvety skins, with small granular scales. The toes are dilated, with 8–12 V-shaped scansors and a retractile claw. The tail is thick and relatively short. Precloacal pores are present. Four species are known, three from southern Africa, one in our area.

BANDED VELVET GECKO *Homopholis fasciata*

IDENTIFICATION: A fairly large, stocky gecko, with a short head and rounded snout, large pale eyes with vertical pupils. The toes are broadly dilated, with big claws. The tail is stout, cylindrical and relatively short, it is easily shed. The scales are small, smooth, roughly homogeneous and granular, not overlapping. Males have two precloacal pores. Maximum size about 16cm, average 9–13cm, hatchling size unknown. Colour grey, beige, olive or purplish-brown. A broad dark band forms a crescent at the back of the head, there are 3–6 dark-edged, wavy brown crossbars on the back (in some cases, the light bands are very narrow), the tail is banded and blotched grey, brown and black. Lips and chin pale, often with fine brown reticulations, whitish below.

Homopholis fasciata

HABITAT AND DISTRIBUTION: Dry and moist savanna with big trees, from sea level to about 1,300m altitude. Sporadically recorded from central and northern Tanzania and east and northern Kenya; localities include Bukoba, Kakoma (south of Tabora), Dodoma and environs, between Lakes Manyara and Eyasi. Kenyan localities include Taita District and the southern sector of Tsavo West, Mombasa, Ngare Ndare, Kula Mawe, Laisamis, Lothagam and Lokitaung. Elsewhere, to Somalia and Ethiopia.

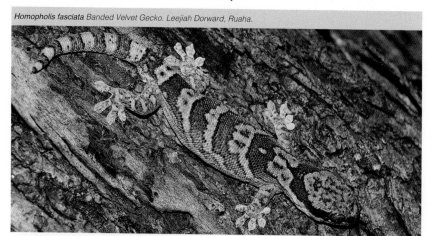

Homopholis fasciata Banded Velvet Gecko. Leejiah Dorward, Ruaha.

NATURAL HISTORY: Poorly known. Lives mostly on trees, fond of baobabs, big fig trees and acacias. Nocturnal on the Kenya coast, might show diurnal activity. Hides in holes and cracks in big old trees. Two eggs, roughly 1.5–2cm in diameter are laid; several clutches a year are laid. They have a powerful bite if handled. They eat insects and other arthropods.

DAY GECKOES *PHELSUMA*

A genus of attractive diurnal, arboreal geckoes with pointed snouts and big eyes with round pupils. Most species are vivid green, with a speckling of other bright colours; reds, blues and yellows. More than 50 species are known, largely on the islands of the western Indian Ocean; they have radiated spectacularly on Madagascar. Two species occur in East Africa and their ancestors probably reached our coast by rafting. They are popular pets in the west and hence have been over-exploited, all are on CITES Appendix II.

KEY TO EAST AFRICAN MEMBERS OF THE GENUS *PHELSUMA*

1a Subcaudals with median scale rows enlarged transversely, preano-femoral pores 32–38 in males, only on Pemba......*Phelsuma parkeri*, Pemba Day Gecko [p. 122]

1b Median subcaudals not enlarged, preano-femoral pores 22–29 in males. Range: coast of southern Kenya and Tanzania and Zanzibar......*Phelsuma dubia*, Dull Green Day Gecko [p. 123]

PEMBA DAY GECKO *Phelsuma parkeri*

IDENTIFICATION: A large green day-gecko, with a rounded snout, eye with a round pupil. Body slim. Toes broadly dilated at the tips. The back is covered with smooth granular scales. Males have 32–38 preano-femoral pores. Maximum size about 16.5cm, average 12–14cm, hatchling size 3cm. Adults on Pemba are vivid green, with fine slightly darker vermiculations, but the field impression is of a bright green gecko. The eye pupil is greenish-bronze. Below, whitish, immaculate, even on throat. Taxonomic Note: Originally regarded as a subspecies of the Madagascan species *Phelsuma abbotti*.

Phelsuma parkeri

HABITAT AND DISTRIBUTION: This species is endemic to Pemba Island, Tanzania, and recent work indicates it occurs almost throughout the island where there is suitable vegetation cover, including both farmland, peri-urban areas and natural woodland; although it is not found in the dry coral-rag regions. Conservation status: Assessed by IUCN as Least Concern, although on CITES Appendix II.

NATURAL HISTORY: Diurnal and arboreal, a recent survey found them living on coconut palms, bananas and Raphia palms, at heights of 1.5–1m; apparently most active during the afternoon. Although its closest relative in East Africa, the Dull Green Day Gecko *Phelsuma dubia*, has not yet been recorded on Pemba, given the frequency of trade by boat and the potential for adults to be transported in cargo such as coconuts, thatch and timber, it could be expected to occur there. It is equally likely that the Pemba Day Gecko occurs on Zanzibar but is overlooked, or will be accidentally introduced there. It would be of interest to see how the

Phelsuma parkeri Pemba Day Gecko. Dennis Rodder, Pemba.

two species interact. Diet: arthropods, pollen and nectar, will lick sweet fruit. Two eggs are laid, but the females are communal nesters, and often many eggs may be found in a suitable spot.

DULL GREEN DAY GECKO *Phelsuma dubia*

IDENTIFICATION: A large dull green diurnal gecko known from the coast and Zanzibar. Snout rounded, eye with a round pupil and orange iris. The toes are broadly dilated. Back covered with obtusely keeled granules; ventral scales smooth; males with 19–29 precloacal and femoral pores. Adults about 15cm. Hatchling 4–5cm. The dorsal surface is usually dull green, spotted and vermiculated with maroon (Kenya specimens) to dull orange (Tanzanian specimens), limbs speckled red-brown and green. They can change colour quite dramatically, the green fading to brown or red-brown. Below, whitish, immaculate, or a dusky upside-down U-shaped mark following contour of lower jaw to shoulder. Young minutely speckled with brown edged white spots and a red tail; the spots disappear when a length of about 5cm is attained.

Phelsuma dubia

Phelsuma dubia Pemba Dull Green Day Gecko. Michele Menegon, Dar es Salaam. Inset: Stephen Spawls, Nyali.

HABITAT AND DISTRIBUTION: The East African coastal plain, in farmland and suburbia, from sea level to 100m altitude. Has a patchy distribution; known localities include Kilifi, Nyali, Mombasa, Dar es Salaam, Bagamoyo, Singino, Zungomero and on Zanzibar. A couple of odd inland records (not shown on map) in Tsavo National Park in Kenya and Songea in Tanzania. Known mostly from coconut palms and other trees in coastal areas, at sea level or very low altitude, rarely found off the coastal plain. Described from

Madagascar, and also in northern Mozambique, on the Comoros and Mozambique Island, it was probably introduced from Madagascar in cargoes onto the East African coast, which would explain its sporadic distribution. Conservation Status: Assessed by IUCN as Least Concern.

NATURAL HISTORY: Diurnal, arboreal and relatively fast moving for a *Phelsuma*, in the Nyali area adults were on palm branches more than 2m above the ground and on the upper parts of buildings, not at ground level. Where coconut plantations are cleared to make way for housing, these geckoes and *Hemidactylus* geckos 'share' accommodation in buildings. Females taken in March contained two well developed eggs, in Madagascar after a single copulation females produce up to six clutches of two eggs, which took about 40–45 days to hatch. The eggs are sometimes laid in coconut palms 2m from ground, and may be deposited in the crowns of coconut trees. Diet: arthropods, including ants and beetles but also may take soft fruit; in Dar es Salaam they may dash from a hiding place to lap up sugary tea which has been spilt, much to the surprise of some householders.

TAIL-PAD GECKOES *UROCOTYLEDON*

A curious genus of gecko, with five species, two of which occur in Cameroon, two in Tanzania and one in the Seychelles; a distribution pattern shared by no other reptilian genus They have an unusual feature, adhesive papillae at the tip of the unregenerated tail, helping them to grip thin branches.

KEY TO EAST AFRICAN MEMBERS OF THE GENUS *UROCOTYLEDON*

1a	Four undivided lamellae under fourth toe, convex snout profile, dorsal pattern of whitish-grey diamonds or chevrons, in Udzungwa Mountains...... *Urocotyledon rasmusseni*, Udzungwa Tail-pad Gecko [p. 126]	1b	Eight to twelve undivided lamellae under fourth toe, concave snout profile, dorsal pattern of transverse diffuse pale bands, in Usambara or Uluguru Mountains ......*Urocotyledon wolterstorffi*, Uluguru Tail-pad Gecko [p.124]

ULUGURU TAIL-PAD GECKO *Urocotyledon wolterstorffi*

IDENTIFICATION: A forest gecko, known from the Eastern Arc Mountains. It has a wide, flat head, snout rounded and wide, with a prominent eye with a vertical pupil and a conspicuous dark orange iris. Body depressed, tail about 50% of total length, the tail tip has adhesive lamellae, like the dwarf geckoes. Granular scales on snout larger than those on the back of the head; the back is covered with small, subequal, smooth, granules or subimbricate scales, uniform or intermixed with large tubercles. Back, flanks, limbs and tail covered with small hexagonal granules. The ventral scales slightly larger than the dorsal scales. The toe tips are greatly expanded. Top of head mottled light grey/dark brown. A poorly-defined dark-brown crescent originates behind the eye and

Urocotyledon wolterstorffi

meets its fellow on the occiput. Dorsum dark brown with 4–6 indistinct lighter grey saddles. Dorsal section of the tail dark brown with a number of equally-spaced elongate dirty ivory patches. Fore and hind limbs dark with light blotching.

HABITAT AND DISTRIBUTION: Hill forest. This Tanzanian endemic is known from relatively few specimens collected in the Usambara, South Pare, Nguu, Nguru Uluguru and Udzungwa Mountains, at altitudes between 500–1,800m. Recent unpublished analysis suggest the presence of several cryptic species within the group. The type specimen from 'Tanga' could also have come from the nearby East Usambara Mountains. Some curious specimens from Arusha (not shown on map) might have been transported there in timber.

NATURAL HISTORY: Little known. Nocturnal and arboreal. Individuals were found at night climbing on bushes and small trees in the forest understory. One has been collected from the inside of a house in the Uluguru Mountains at Bunduki, another was taken on a sapling at night in forest near Amani, East Usambara Mountains. In the Kanga Mountains this species and the Usambara Forest gecko *Cnemaspis africana* were observed sharing an egg-laying site on dry sandy soil under a large overhanging rock. Two eggs of *Urocotyledon wolterstorffi* were found under wood debris. One egg was 1.2cm diameter, after 16 days in controlled condition a hatchling of about 40mm emerged. Feeds on insects and spiders.

Urocotyledon wolterstorffi Uluguru Tail-pad Gecko. Top: Michelel Menegon, South Pare Mountains. Below left: Michele Menegon, Mt Kanga. Below right: Udzungwa Mountains.

UDZUNGWA TAIL-PAD GECKO/JENS RASMUSSEN'S TAIL-PAD GECKO *Urocotyledon rasmusseni*

IDENTIFICATION: A small gecko with a short, slightly depressed head with a clearly convex snout profile. All digits bear slender, stylet-like claws and a single pair of large, leaf-like terminal scansors separated by a groove within which the claw is located. Known length is 42mm, it might get larger. The eyes are relatively large with a vertical pupil and crenulated margins, there are no enlarged superciliary scales. The scales on the snout and crown are small, granular and slightly heterogeneous. The tail of the only known specimen is partly regenerated, so the typical adhesive lamellae at the tip of the unregenerated tail haven't yet been observed. The original portion of the tail is slightly depressed in cross section, with scattered enlarged scales forming weakly denticulate margins, the dorsa scales of the tail are granular and arranged in regular rows. The ground colour is russet-brown with darker brown markings and a prominent mid-dorsal series of whitish-grey irregular diamonds or chevrons. Recent photographs of some individuals in the wild show a fairly constant colour pattern.

Urocotyledon rasmusseni

HABITAT AND DISTRIBUTION: Known only from the dry montane woodlands on the western slopes of the Udzungwa Mountains, within the Udzungwa Mountain National Park in Tanzania.

NATURAL HISTORY: Unknown. Individuals were observed on Commiphora trees about 3m above ground level It is assumed that this species is nocturnal and feeds on arthropods, as does *Urocotyledon wolterstorffi*. Like other members of the genus it is probably exclusively arboreal.

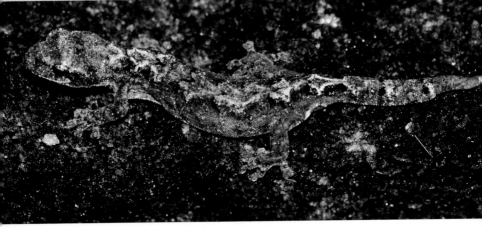

Urocotyledon rasmusseni Udzungwa Tail-pad Gecko. Michelel Menegon, Udzungwa Mountains.

Trachylepis dichroma
Bicoloured Skink
female, Stephen Spawls

SUB-ORDER SAURIA: LIZARDS
FAMILY SCINCIDAE
SKINKS

A family of shiny-bodied lizards, many with reduced or absent legs, varying in size from 5–60cm in East Africa. Their flat shiny scales are toughened by osteoderms, giving skinks a stout, fairly rigid but flexible coating which is resistant to wear, useful for burrowing or rock-dwelling animals. Most lose their tails easily. They inhabit a variety of habitats; on rocks and trees, on the ground or under it. Some are territorial and live in structured colonies. They occur throughout the tropical and part of the temperate world. Their taxonomy has changed a lot; when we originally wrote this book, there were about 80 genera and 700 species of skink worldwide, now there are over 150 genera and 1,618 species, an increase reflecting both improvements in molecular taxonomy and an increased field research effort. In 1998, 44 species in 14 genera were known from East Africa, now there are five subfamilies, with 51 species in 16 genera. Some 19 species are East African endemics.

KEY TO THE EAST AFRICAN GENERA AND SOME SPECIES OF SKINK

1a	Nostril pierced in or bordered by the rostral......2		and connected to its posterior border by a groove, limbs absent......7
1b	Nostril pierced between 2 or 3 nasals, well separated from the rostral......10	3a	Interparietal small and subtriangular, narrower than frontal, well-separated from the posterior supraoculars, limbs present......4
2a	Nostril pierced between the rostral and a small nasal, or between rostral, supranasal, postnasal and first labial, limbs present or absent......3	3b	Interparietal large, broader than frontal, in contact laterally with posterior supraoculars, limbs present or ab-
2b	Nostril pierced in a very large rostral		

sent......6

4a Limbs with four digits......*Sepsina tetradactyla*, Four-toed Fossorial Skink [p. 167]
4b Limbs with five digits......5
5a Ocelli on back, digits relatively long, in East Africa, only in northern Kenya...... *Chalcides bottegi*, Ocellated Skink [p. 166]
5b No ocellate markings on back, digits very short, in East Africa, only in the Usambara Mountains in eastern Tanzania......*Proscelotes eggeli*, Usambara Five-toed Fossorial Skink [p. 175]
6a Nostril pierced between the rostral, supranasal, nasal and first labial, pentadactyle limbs present......*Scelotes uluguruensis*, Uluguru Fossorial Skink [p. 176]
6b Nostril pierced between the rostral and first labial, limbless......*Melanoseps*, limbless skinks [p. 170]
7a Rostral bordered posteriorly by a single broad scale......8
7b Rostral bordered posteriorly by a pair of internasals......9
8a Third upper labial in contact with the orbit, a narrow yellow vertebral stripe, in south-western Tanzania.....*Typhlacontias kataviensis*, Katavi Blind Dart Skink [p. 168]
8b Upper labials separated from the orbit by subocular scales, vertebral stripe broad, in south-eastern Kenya and north-eastern Tanzania......*Acontias percivali*, Percival's Legless Skink [p. 129]
9a Eye exposed, nostril connected to the posterior border of rostral by a long straight sulcus......*Scolecoseps*, sand skinks [p. 177]
9b Eye covered by skin, nostril connected to the posterior border of rostral by a short curved sulcus......*Feylinia currori*, Western Forest Limbless Skink [p. 169]

10a Eyelids fused, immovable, the lower one with a large transparent disc that completely covers the eye......15
10b Eyelids moveable, the lower one with scaly or transparent disc......11
11a Lower eyelid with a large transparent disc, dorsal scales usually keeled, rarely smooth, limbs well-developed......*Trachylepis*, typical skinks [p. 131]
11b Lower eyelid scaly or with a small transparent disc, dorsal scales smooth, limbs short or vestigial......12
12a Prefrontals large and usually in contact, no frontonasal, limbs vestigial, two or three digits on forelimb and three on hindlimb......*Eumecia anchietae*, Western Serpentiform Skink [p. 130]
12b Prefrontals small and widely separated, a frontonasal present, limbs short but pentadactyle......13
13a Supranasals usually absent, widely separated if present......*Leptosiaphos*, five-toed skinks [p. 157]
13b Supranasals present and in broad contact......14
14a Relatively large, with a stout body and rounded snout, ear opening present and tympanum clearly visible, often vivid red pigment on flanks......*Lepidothyris hinkeli*, Hinkel's Red-flanked Skink [p. 155]
14b Relatively small, with a slim body and pointed snout, ear opening present but small, tympanum not visible, without vivid red pigment on flanks......*Mochlus*, writhing skinks [p. 149]
15a Interparietal fused with frontoparietals into a single shield, supraoculars four, in intertidal zone......*Cryptoblepharus africanus*, Coral Rag Skink [p. 163]
15b Interparietal distinct, frontoparietals paired or fused, supraoculars three, on land......*Panaspis*, snake-eyed skinks [p. 164]

SUBFAMILY **ACONTINAE**

A group of legless burrowing skinks, and the sister group to all other skinks; with around 25 species in two genera, all save one confined to the southern third of Africa.

LEGLESS SKINKS *ACONTIAS*

An African genus of 21 species of slim, smooth-bodied legless burrowing skinks with small recessed eyes. The scales are smooth and shiny. One species is found in eastern Kenya and Tanzania.

PERCIVAL'S LEGLESS SKINK *Acontias percivali*

IDENTIFICATION: A legless burrowing skink. The nose is rounded and distinctively light-coloured, the eye is small, dark and deeply recessed, with an elongate moveable lower eyelid. There are no external ears; careful examination will reveal a distinctive groove running back from the nostril to the rear edge of the rostral scale. Males tend to have slightly bigger heads than females. The body is cylindrical, the tail short and stubby, about 10–12% of total length. The scales are smooth and shiny, in 16–18 rows at mid body. Maximum size about 28cm, average 18–25cm, hatchling size unknown but probably 8–10cm. The adults have a purply-brown dorsal stripe, the flanks and underside are orange or pink; the distinctive rostral scale is pinkish or dull white. The type description describes the dorsum as 'glossy black', presumably in preservative.

Acontias percivali

HABITAT AND DISTRIBUTION: An East African endemic, it occurs in low dry savanna, from Ngulia and the Tsavo River in south-east Kenya, south through Voi and Kilibasi, thence into north-eastern Tanzania, also known from the vicinity of Dodoma, at altitudes between 400–1,200m. Conservation Status: Least Concern. Part of its range lies within two protected areas, the Tsavo ecosystem and Mkomazi National Park.

NATURAL HISTORY: A burrowing species, living below the surface, and can be found under deep-seated rocks, logs, etc. Activity patterns poorly known; sometimes on the surface at dusk, although an Ngulia specimen suddenly emerged from beneath the sand on a road after being run over at midday. Several specimens were found by following a plough at Voi. They squirm when handled but do not try to bite. Male combat has been observed in other species in the genus. Diet probably arthropods. Breeding details unknown but other members of the genus give live birth to between one and fourteen relatively large offspring. This skink is the northernmost representative of a southern African genus. Named for Arthur Blayney Percival, appointed Kenya's first 'Ranger for Game Preservation' in 1901.

Acontias percivali Percival's Legless Skink. Stephen Spawls. Left: Northern Tanzania. Right: Ngulia Lodge.

SUBFAMILY **MABUYINAE**

A widespread subfamily of over 190 species of skink, in more than 20 genera. Two genera occur in East Africa: one species of *Eumecia*, a virtually legless grassland animal, and 17 species of 'typical African skinks', genus *Trachylepis*.

SERPENTIFORM SKINKS *EUMECIA*

An African genus of elongate-bodied skinks with very reduced, bud-like limbs, which are not used in locomotion. They have smooth shiny scales. They are diurnal, inhabiting high grassland or woodland, where they slide through the grass like snakes. Little is known of their habits. Two species are known, one of which occurs in East Africa.

WESTERN SERPENTIFORM SKINK *Eumecia anchietae*

IDENTIFICATION: A long, almost legless striped skink. The snout is quite pointed, head short, the eye is fairly large, pupil round. The earhole is small and oval. The neck is slightly thinner than the head. The body is cylindrical, long and slender, with little bud-like limbs, with two or three minute digits on the front limbs and three on the hind limbs; the limbs are visible when the animal pauses. The tail is 55–65% of the total length. The dorsal scales are smooth. Maximum size about 55cm in East Africa (even larger specimens known from Angola), average 30–45cm, hatchling size 16–17cm in Kakamega specimens. Colour brown or olive above, with a broadish brown vertebral stripe, bordered by fine or broad lighter dorsolateral stripes; with fairly regular dark bars between the stripes; the flanks are also barred light and dark, although these may fade posteriorly. The belly is cream or grey-white.

Eumecia anchietae

HABITAT AND DISTRIBUTION: Forest clearings, high woodland, grassland, swamps and well-wooded savanna of western Kenya and western Tanzania, at altitudes between

Eumecia anchietae Western Serpentiform Skink. Left: Sjoered Henegouwen, Maasai Mara. Right: Frank Willems, northern Zambia.

about 1,400–2,500m. Three main populations exist; one in the high country of western Kenya from Mt Elgon south to Njoro, another from the Soit Ololo escarpment and the Maasai Mara into the northern Serengeti; an isolated record from Tatanda in south-west Tanzania; may well occur more widely in western Tanzania. Probably in suitable areas of Uganda, Rwanda and Burundi but no records.

NATURAL HISTORY: Not well known. A diurnal, terrestrial skink, it may be locally common in suitable habitat. It lives in grassland, in open country and glades, where it slides through vegetation. A short series from Kakamega were taken during the late morning and afternoon. These skinks have a curious mode of reproduction, almost mammalian; they give live birth to up to five neonates; before birth the developing embryos are initially fed by uterine secretions which are absorbed by the yolk sac and transferred to the foetus, in the final stages a placenta develops. A 52cm Kakamega female had near-full term embryos in September. Diet insects and other arthropods.

TYPICAL SKINKS *TRACHYLEPIS*

One of East Africa's most visible groups of lizards, *Trachylepis* are a genus of small to large diurnal skinks, living in a range of habitats. Originally in the genus *Mabuya*, a huge pan-tropical genus, typical skinks from the Afro-Malagasy region are now assigned to the genus *Trachylepis*. They have large eyes, a visible external earhole. They have no preanal or femoral pores. They have cylindrical, shiny bodies with mostly keeled, overlapping cycloid scales; the number of keels on the scales appears to increase with age in some species. The long tail is easily shed. Most are shades of grey or brown, with stripes, but some are vividly coloured and in some species the male and female (and in some species the juveniles) have different colour patterns. Most lay eggs, some give live birth, some do both. Over 80 species of *Trachylepis* are known, most in Africa, but others occur in Madagascar, on some western Indian Ocean islands, south-western Asia and an island off Brazil. Seventeen species are known from East Africa.

KEY TO THE EAST AFRICAN MEMBERS OF THE GENUS *TRACHYLEPIS*

1a	Scales on soles of the feet usually non-spinose, smooth or tubercular (except in some juvenile specimens of *Trachylepis quinquetaeniata*)......2	3a	Midbody scale rows 22–27, dorsal scales smooth or with 2–3 weak keels, unregenerated tail more than 2.3 times the total length......*Trachylepis megalura*, Grass Top Skink [p. 143]
1b	Scales on soles of the feet usually keeled and spinose.....13		
2a	Frontoparietals usually fused, subocular much narrowed inferiorly, sometimes not reaching lip......*Trachylepis bayoni (keniensis)* Bayon's Skink [p. 134]	3a	Midbody scale rows 27 or more, dorsal scales keeled, unregenerated tail less than 2.3 times the total length......4
		4a	Midbody scale rows 32–52, juveniles with blue tails 5
2b	Frontoparietals not fused, subocular not or scarcely narrowed inferiorly......3	4b	Midbody scale rows 27–36, juveniles without blue tails......6

5a Females and juveniles with five black-bordered blue or cream longitudinal stripes, midbody scale rows 32-42......*Trachylepis quinquetaeniata*, Five-lined Skink [p. 146]

5b Females and juveniles with three cream longitudinal stripes, midbody scale rows 38-52......*Trachylepis margaritifer*, Rainbow Skink [p. 142]

6a Dorsal scales of adults with 5-11 keels......7

6b Dorsals largely tricarinate......9

7a A broad pale, low lateral stripe from snout to groin, dorsal scales usually with five keels, Pemba Island......*Trachylepis albotaeniata*, Pemba Speckle-lipped Skink [p. 133]

7b No broad white lateral stripe, dorsal scales usually with 7-11 keels in adults, mainland East Africa and Zanzibar......8

8a Midbody scale rows 30-38, supraciliaries usually five, build robust, head length in adults more than 20% of snout-vent length, flanks usually darker than back, dorsum with dark and/or light flecks......*Trachylepis maculilabris*, Speckle-lipped Skink [p. 141]

8b Midbody scale rows 28-32, supraciliaries usually four, build slender, head length in adults less than 20% of snout-vent length, dorsum uniformly grey-brown or with a few scattered black flecks......*Trachylepis boulengeri*, Boulenger's Skink [p. 135]

9a Build moderate, body slim, unregenerated tail over 1.6 times snout-vent length, a broad dark lateral band from behind the eye to at least beyond the shoulder......10

9b Build robust, body thick-set, unregenerated tail less than 1.6 times snout-vent length, no broad dark lateral band......11

10a Ear opening small and horizontal, supranasal in broad contact, in savanna east of eastern Uganda......*Trachylepis planifrons*, Tree Skink [p. 145]

10b Ear opening moderately large and near-vertical, supranasals separated by the rostral, in our area in forested country of western Uganda only......*Trachylepis albilabris*, White-lipped Forest Skink [p. 133]

11a Midbody scale rows 30-38, paravertebral scales 52-57, flank scales with two keels......*Trachylepis dichroma*, Bi-coloured Skink [p. 138]

11b Midbody scale rows 30-34, paravertebral scales 40-55, flank scales with three keels......12

12a Paravertebral scale rows 46-55, flanks often orange or red, range north-west Uganda......*Trachylepis perroteti*, Orange-flanked Skink [p. 144]

12b Paravertebral scales 40-49, flanks not usually orange or red, range Kenya and Tanzania......*Trachylepis brevicollis*, Short-necked Skink [p. 137]

13a Lower border of subocular usually at least half the length of upper, usually a conspicuous white lateral longitudinal stripe......*Trachylepis varia*, Variable Skink [p. 148]

13b Lower border of subocular less than a third of upper, no conspicuous white lateral longitudinal stripe......14

14a Dorsal scales bicarinate or with a poorly defined median keel, lamellae beneath fourth finger 11-14......*Trachylepis brauni*, Ukinga Mountain Skink [p. 136]

14b Dorsal scales with at least three well-defined keels, lamellae beneath fourth finger 14-20......15

15 A double vertebral stripe, lamellae beneath fourth finger 14-15, in high altitude montane grassland......*Trachylepis irregularis*, Alpine Meadow Skink [p. 139]

15b No vertebral stripe, lamellae beneath fourth finger 15-20, not in high altitude montane grassland......16

16a Two broad light dorsolateral stripes with precise edges always present, dorsal surface immaculate or with fine pale speckling, midbody scale counts 33-42, but 34-39 in vicinity of Lake Victoria; widespread throughout East Africa......*Trachylepis striata*, Striped Skink [p. 147]

16b Light dorsolateral stripes usually absent, if present then narrow and without precise edges, dorsal surface heavily blotched black and gold/white, not speckled, midbody scale counts 38-41, only on Lolui Island, Lake Victoria......*Trachylepis loluiensis*, Lolui Island Skink [p. 140]

WHITE-LIPPED FOREST SKINK Trachylepis albilabris

IDENTIFICATION: A medium-sized skink with a long head. The snout is rounded. The body is cylindrical, the toes long and thin, the tail is surprisingly long and slender for a largely terrestrial skink, 65–70% of total length. Dorsal scales are keeled, with three keels; midbody scale count 27–32. Maximum size about 23cm, average 15–22cm, hatchling size 7–8cm. Quite variable in colour: the back is various shades of brown or rufous, the flanks may be dark brown (particularly in juveniles); adult females have a broad yellow stripe extending from the lips to the forelimbs, males have a yellow blotch in front of the forelimbs. The belly and throat are greenish yellow in females, in males the throat is white with black spots and the belly is deep yellow.

Trachylepis albilabris

HABITAT AND DISTRIBUTION: A forest species. In our area, only known from the Semliki National Park in western Uganda, elsewhere west to Guinea.

NATURAL HISTORY: Diurnal and terrestrial, living in leaf litter, in Ghana was often observed between the buttresses of large trees; adults were not seen in trees although a juvenile was found on a fern. They lay eggs, a juvenile was collected in May in Ghana. Eats a wide range of arthropods.

Trachylepis albilabris White-lipped Forest Skink. Bill Branch, Liberia.

PEMBA SPECKLE-LIPPED SKINK Trachylepis albotaeniata

IDENTIFICATION: A fairly large, long-tailed arboreal skink. The snout is pointed, the eye is large for a skink. The body is cylindrical, the toes long and thin, the tail is very long and slender, 60–70% of total length, and it is used as a climbing aid. Dorsal scales are keeled, with five keels; midbody scale count 28–30. Maximum size about 28cm, average 15–25cm, hatchling size unknown but probably similar to typical speckle-lipped skinks, 5–8cm. Reddish-bronze dorsally, although this may either fade to grey or darken towards the rear limbs; with a dark lateral stripe; the lips are cream or yellow, and this broadens down the body into a light stripe low on the flanks. The tail is brown or bronze. Breeding males have a vivid yellow underside, females light yellow or cream below. The ear opening is obvious, near vertical, the ear lobules and scales bordering the opening often butter yellow. Taxonomic Note: Regarded for a long time as a subspecies of the Speckle-lipped Skink *Trachylepis maculilabris*.

Trachylepis albotaeniata

HABITAT AND DISTRIBUTION: Only known from Pemba Island, where it lives at sea level or just above, in natural forest and woodland, in agricultural land, coconut plantations and in urban areas, provided there are trees or bushes. Conservation Status: Not assessed, but this is because IUCN still regard it as a subspecies of *T. maculilabris*.

NATURAL HISTORY: Diurnal and arboreal, active on trees and bushes. It climbs expertly, using its tail; it will climb along even very thin twigs and on leaves. Sleeps at night in a leaf cluster, cracks and holes in trees, under palm fronds etc. No breeding details known, but its relative, the Speckle-lipped Skink *Trachylepis maculilabris*, has clutches of 6–8 eggs. It probably breeds throughout the year. Eats a wide range of arthropods.

Trachylepis albotaenia Pemba Speckle-lipped Skink Frank Glaw, Pemba.

BAYON'S SKINK *Trachylepis bayoni*

IDENTIFICATION: A medium-sized brownish skink, heavily built, with a cylindrical body, a short head and a round snout, The eye is small, the pupil round. There are 2 or 4 large lobules over the ear opening. The tail is slightly more than half the total length. Body scales with three keels, in 34–36 rows at midbody. Maximum size about 17cm, average 12–16cm, hatchling size unknown but probably about 6–7cm. Ground colour shades of brown, a thin white or yellow flank stripe from the upper lip extends to the level of the hind limbs, there is a similar white dorsolateral stripe from the nape to the level of the hind limb; in some specimens these stripes are faded. Back often marked with fine irregular black longitudinal lines. The chin, throat and underside are white or cream. Taxonomic Note: East African specimens are the subspecies *Trachylepis bayoni keniensis*, the nominate subspecies was originally described from Angola.

Trachylepis bayoni

HABITAT AND DISTRIBUTION: Medium to high savanna, grassland and alpine moorland, from 1,400–3,200m altitude. This subspecies is an East African endemic, known from a number of isolated localities in high central Kenya (Mt Elgon, Cherang'any Hills, Lake Sirgoit, Sotik, Aberdares, Mt Kenya, southern Ngong Hills) and northern Tanzania (Shira Plateau on Mt Kilimanjaro). Other subspecies occur in DR Congo and Angola. Conservation Status: Data Deficient. Occurs in a very restricted habitat but the East African form lives largely within national parks, so has adequate protection.

Trachylepis bayonii Bayon's Skink. Stephen Spawls, Aberdares.

NATURAL HISTORY: Very poorly known. Terrestrial and diurnal, living in grassland where it shelters within grass clumps; has also been found under rocks and roadside debris. Specimens from very high altitude areas presumably get right into grass clumps at night to avoid freezing. Lays eggs, clutch details unknown. Eats insects and other arthropods. Judging by its paucity in museum collections, it doesn't seem to be common in any parts of its range, but might just be highly secretive.

BOULENGER'S SKINK   *Trachylepis boulengeri*

IDENTIFICATION: A long slender skink with a very long tail, a short head, a rounded snout and small eye, the pupil is round. The tail is about two-thirds of the total length. Body scales smooth and shiny, with multiple keels (3–11, usually 7–9, the number increases with age), in 28–32 rows at midbody. Maximum size about 30cm (of which the tail is 20cm), average 15–27cm, hatchling size unknown. Ground colour warm brown, pinkish or grey-brown above, sometimes with scattered black flecking, some specimens may have a faint, pale dorsolateral stripe, others have a black streak from the eye to the ear. The chin, lips and underside are yellow.

Trachylepis boulengeri

HABITAT AND DISTRIBUTION: Low altitude savanna and coastal woodland, from sea level to about 1,500m. Within East Africa, found in south-east Tanzania thence south and west to southern Malawi and central Mozambique; there is also a single record from the Manga forest reserve, near Pangani, north-east Tanzania. Conservation Status: Not assessed, but widespread and adapts well to agriculture, not under threat.

NATURAL HISTORY: Diurnal. Mostly arboreal in the northern part of its range, it climbs coconut palms, seems to be more terrestrial in the south, basking on horizontal logs and hunting in leaf litter or tall grass, known to sleep on reeds at night. Sometimes basks on thatched roofs and road verges. Lays eggs, clutch details unknown. Diet: insects and other arthropods.

Trachylepis boulengeri Boulenger's Skink. Bill Branch, Mozambique.

UKINGA MOUNTAIN SKINK *Trachylepis brauni*

IDENTIFICATION: A small robust skink with a cylindrical body, a short head and small eye, the pupil is round. The tail is long, about 60% of the total length. Dorsal scales usually have two keels, in 38 rows at midbody. Maximum size about 15cm, average 10–13cm, hatchling size unknown. Ground colour brown above, with a distinct pale vertebral and dorso-lateral strip and many small pale spots on the back and sides; these may coalesce to form broken crossbars. In some specimens the top of the head is green. Flanks brown with extensive white spotting. White below, sometimes with black speckling.

Trachylepis brauni

HABITAT AND DISTRIBUTION: High altitude savanna and grassland above 2,200m in the Ukinga mountains of southern Tanzania and the Nyika plateau in northern Malawi. Conservation Status: Not assessed. Very restricted range, protected in the Nyika National Park in Malawi, but not in Tanzania.

Trachylepis brauni Ukinga Mountain Skink. Left: Colin Tilbury, Kitulo Plateau, Tanzania. Right: Gary Brown, northern Malawi.

NATURAL HISTORY: Very poorly known. Diurnal. It lives in grassland, sheltering in holes (including rodent burrows) but is also active on rocks. No breeding details known, might give live birth (some high-altitude skinks do). Diet presumably insects and other arthropods.

SHORT-NECKED SKINK *Trachylepis brevicollis*

IDENTIFICATION: A large robust skink with a cylindrical body. Despite its common name, it doesn't have a particularly short neck, but its head is short, eyes large, the pupil is round. The tail is stout at the base and about half the total length. Dorsal scales usually have two keels; the flank scales three keels, scales in 30–34 rows at midbody. Maximum size about 32cm, average 18–26cm, hatchling size about 7–9cm. The colour is tremendously variable, hatchlings are black with bright yellow barring on the anterior flanks, this changes to clusters or crossbars of fine yellow dots as they grow, they then become brown or grey, dark brown with light crossbars or dark brown with light crossbars. Adult females are usually brown, with irregular dark specks or crossbars, light or heavy, sometimes with a poorly-defined light dorsolateral stripe, sometimes uniform brown, Adult males usually distinctively marked in longitudinal black and brown stripes, sometimes vivid white speckling on the front half of the body. Uniformly paler below.

Trachylepis brevicollis

HABITAT AND DISTRIBUTION: Low altitude moist and dry savanna, woodland, semi-desert and coastal thicket, from sea level to about 1,500m in East Africa (sometimes slightly higher, for example on Lukenya Hill). Occurs virtually throughout east and northern Kenya, not in the centre or west. Reaches eastern Uganda in the Amudat area; sporadically distributed across north-central Tanzania. Elsewhere, north and east to Sudan, Somalia, Ethiopia, Eritrea and the southern Arabian peninsula.

NATURAL HISTORY: Mostly terrestrial but will climb rocks, fallen trees etc. Diurnal. Fond of basking in prominent positions (often on path-side rocks at Mzima Springs), but quite wary, It will utilise any suitable habitat, in holes on flat plains, squirrel warrens, termite hills, under logs, rocks, among boulder clusters on hillsides, in rock cracks, tree holes and so on. Often in large groups (which might be colonies), a big rotting acacia trunk in riverside bush at Garissa housed a group of 27 individuals. A clutch of four embryos

Trachylepis brevicollis Short-necked Skink. Stephen Spawls. Left: Voi, male. Centre: Ethiopia, female. Right: Hatchling.

recorded, they give live birth in Ethiopia. Hatchlings recorded in April in eastern Kenya. Eats a wide range of insects and other arthropods, sometimes larger prey, an Athi River specimen had eaten a mouse.

BI-COLOURED SKINK/ZEBRA SKINK *Trachylepis dichroma*

IDENTIFICATION: A large robust skink with a thick-set cylindrical body. The head is short, eyes large and prominent, the pupil is round. The tail is stout at the base and about half the total length. Dorsal scales usually have two keels, the flank scales two keels (some have a weak third keel), scales in 30–38 rows at midbody. Maximum size about 38cm, average 20–34cm, hatchling size about 7–9cm. Colour variable. Some adult males are finely striped light and dark brown dorsally, with fine white spots on the flanks, others are uniform brownish-bronze. In some (possibly breeding) males the lower flanks and side of the tail are vivid orangey-red. The throat may be black, blue or pale (possibly also connected with breeding). The belly is cream shading outwards to yellow. Adult females are brown, the scales black-edged, the body and tail have narrow cross-bars consisting of black and white scales; juveniles resemble the females but the black crossbars are finer, with fewer white scales. In some hatchlings, the upper dorsal surface is grey, with extensive white spotting, the flanks are barred black, grey and white, other hatchlings are finely barred black on brown. Sometimes the crossbars are interrupted in the centre of the back, giving the impression of a paler vertebral stripe. The lips and nose may be bright yellow or orange. Taxonomic Notes: in 2005 this species was elevated from the synonymy of the highly variable *Trachylepis brevicollis*. Subsequent work indicates that *Trachylepis dichroma* is much more varied than originally narrowly defined; the two species may live together in the same termitarium, some specimens appear to be intermediate between the two species; whether it is a genuine species rather than a colour morph remains to be confirmed.

Trachylepis dichroma

HABITAT AND DISTRIBUTION: Due to earlier confusion with Short-necked Skink *Trachylepis brevicollis* records, the distribution of the Bi-coloured Skink is not yet clarified. It inhabits low to medium high altitude savanna, and semi-desert, at altitudes between 200–1,800m. Known from the Maasai Mara; and from Lobo and Klein's Camp area in northern Serengeti, low altitude savanna west of Mt Kilimanjaro, in an area between Nairobi National Park, the upper Athi River and Kajiado, the Baringo, Garissa and Isiolo areas and in an area of the north-east between Buna, Rhamu and Mandera in extreme north-east Kenya; thence into Somalia and south-east Ethiopia. The type specimens were believed to originate from Dodoma. Probably much more wide-

Trachylepis dichroma Bi-coloured Skink. Top: Stephen Spawls, Dodoma, male. Bottom: Lorenzo Vinciguerra, Mt Meru.

spread in northern and eastern Kenya but unidentified. Conservation Status: Assessed by IUCN as Least Concern.

NATURAL HISTORY: Similar to *Trachylepis brevicollis*, and often found in the same habitat. Mostly terrestrial but will climb rocks, fallen trees, etc. Diurnal. Fond of basking in prominent positions, but quite wary, It will utilise any suitable habitat, in holes on flat plains, squirrel warrens, rodent holes, termite hills, under logs, rocks, among boulder clusters on hillsides, in rock cracks, tree holes and so on. Often in large groups (which might be colonies). Breeding details poorly known but presumably gives live birth. Eats a wide range of insects and other arthropods, might take small vertebrates.

ALPINE MEADOW SKINK *Trachylepis irregularis*

IDENTIFICATION: A large heavily-built skink with a short, rounded head and a large dark eye. The tail is slightly less than half the total length. Dorsal scales have 2–5 keels, scales in 31–34 rows at midbody. Maximum size 22–24cm, average 15–18cm, hatchling size unknown. Head speckled tan and black, upper labials black-edged, the back is black or dark brown, with a fine yellow double-stripe along the spine and a fine yellow single dorsolateral stripe, with yellow speckling (that may form dotted lines) between them. Yellow speckling on the flanks shades to white or pink. The belly is pink, white or whitish-blue, sometimes with irregular black stripes, the soles of the feet and the anal plate are pink.

Trachylepis irregularis

Trachylepis irregularis Alpine-meadow Skink. Stephen Spawls, Aberdares.

HABITAT AND DISTRIBUTION: High altitude moorland at 3,000m and above. An East African endemic, known only from the montane grassland of Mt Kenya, the Aberdares, the Mau and Mt Elgon. The holotype was described from 'Soy, slope of Mt Elgon' which is confusing as Soy, at 1,950m, isn't on the slope of Mt Elgon, it is more likely that Soy was where the collector lived. Conservation Status: Assessed by IUCN as Near Threatened, lives in a very restricted habitat but virtually all of it is within national parks.

NATURAL HISTORY: Becomes active on sunny days, it basks between 9–10am and then hunts until 3–4pm, or until it clouds over, then shelters as its montane habitat freezes at night. In the Aberdares it is known to live under rocks, where the temperature falls below freezing at night, so presumably it can supercool. Secretive and uncommon even in prime habitat; national park rangers in the Aberdares were surprised to be shown a specimen. Breeding details poorly known, although a female contained 'well-developed ova with embryos'. Might give live birth at such high altitude. A juvenile was collected in February. Diet insects and other arthropods, one specimen contained beetles and large ants.

LOLUI ISLAND SKINK *Trachylepis loluiensis*

IDENTIFICATION: A large robust, gregarious skink with a short, rounded head. The limbs are short and the rear limbs stocky, with long claws. The tail varies in length between 45–60% of total length, and is often noticeably stout at the base. Dorsal scales usually have three keels, although a few have two, four or five. Scales in 38–41 rows at midbody. Maximum size about 23cm, average 15–20cm, hatchling size unknown. Mottled black and gold, with more black on the body and more gold on the tail. The head is brown with black mottling. The chin and throat are heavily streaked with black on grey-white, giving the impression of incomplete bars running from the lips down onto the throat. Underside grey-white, with light black speckling; the tail is almost uniform grey-white below. Some specimens have two fine, irregularly-edged gold dorsolateral stripes. From a distance it may appear striped.

Trachylepis loluiensis

HABITAT AND DISTRIBUTION: Uganda's sole endemic lizard, known only from rocky outcrops on Lolui Island, in the Ugandan waters of Lake Victoria, at 0°08'S, 33°41'E; altitude 1,162m or just above. Conservation Status: Vulnerable D2. Although present in huge numbers, this skink might be vulnerable. Its distribution is very restricted. In addition, this skink occupies a niche that in the areas surrounding Lake Victoria is dominated by agamas; if agamas were introduced to the island, they might out-compete this species. It needs monitoring.

NATURAL HISTORY: Diurnal and rupicolous, living on large granite outcrops, where it utilises vertical, steeply sloping and flat rock faces within the outcrops. Unusually, it lives in huge groups, some numbering over a thousand individuals. The proportion of males and females within these colonies is not known, and it is uncertain if the colonies are socially structured, or simply a random aggregation of individuals utilising a favourable spot. Juveniles are present within the colonies, and individuals moved freely from one group to another. The lack of minimal space between individuals might indicate that they are not territorial, although combat has been observed, one individual seizing another by the leg. At night, they retreat into cracks. No breeding details are known, presumably lays eggs. They feed on a wide variety of invertebrates, but a major diet item are lake flies of the genus *Chaoborus*. It is possible that the massive concentrations of these tiny flies are the reason that such huge colonies of these skinks can exist.

Trachylepis loluiensis Lolui Island Skink. Steve Russell, Lolui Island.

SPECKLE-LIPPED SKINK   *Trachylepis maculilabris*

IDENTIFICATION: A fairly large, long-tailed arboreal skink. The snout is pointed, the eye is fairly large with distinct yellow eyelids. The body is cylindrical, the toes long and thin, the tail is very long and slender, 60–70% of total length, and it is used as a climbing aid. Dorsal scales are keeled (3–9, usually 5–8 keels), arranged in 29–38 (usually 29–35) rows at midbody. Maximum size about 30cm, average 15–25cm, hatchlings 5–8cm. Colour very variable, changes with location; although most specimens have distinct white, black-speckled lips, hence the common name. Kenyan coastal specimens are usually uniform brown above (sometimes with black speckling), with a broad orange flank stripe, the tail is brown, black speckled, the side of the neck is speckled white,

Trachylepis maculilabris

underside yellow, chin white with fine black dots. Specimens from Uganda and around Lake Victoria are brown above, with white speckling on the neck and flanks (decreasing towards the tail) and a broad dull red or orange flank stripe (not always present); some have vivid orange flanks, chin and throat. Occasional individuals have dark vertical flank bars. In animals from the Albertine Rift, the orange flank stripe only extends halfway to the hind limbs.

Trachylepis maculilabris Speckled-lipped Skink. Top left: Ignas Safari, Bagamoyo. Bottom left: Stephen Spawls, Entebbe. Right: Michele Menegon, Rwanda.

HABITAT AND DISTRIBUTION: Forest clearings, woodland, coastal thicket, farmland, and gardens, often near natural water sources. From sea level to about 2,300m altitude, possibly higher. Occurs the length of the East African coast, in the Taita Hills and Kitobo Forest in eastern Kenya, in Tanzania inland to the Usambaras and up the Rufiji River. A population also occurs along virtually the length of the western rift valley, from northern Lake Malawi (Lake Nyasa) up along Lake Tanganyika to Rwanda and Burundi, up through parts of west and north-west Uganda, also found all around Lake Victoria and on its islands, and the western Kenya/south-east Uganda border country. Elsewhere, west to Senegal, south to Angola and central Mozambique.

NATURAL HISTORY: Diurnal and arboreal, active on trees and bushes. It climbs expertly, using its tail, it will climb along even very thin twigs and on leaves. It will also utilise buildings, walls, thatched roofs and rock outcrops, at the coast it is often on coconut palms, around Lake Victoria often in waterside bushes. In dense forest it is usually found around clearings. Sleeps at night in a leaf cluster, cracks and holes in trees, under palm fronds, etc. Clutches of 6–8 eggs, roughly 1×1.5cm recorded, and in central Africa it appears to have no definite breeding season, 5 or 6 clutches laid per year; Tanzanian female gravid in May and October. Eats a wide range of insects, other arthropods and snails, observed to eat a juvenile Tropical House Gecko *Hemidactylus mabouia*.

RAINBOW SKINK *Trachylepis margaritifer*

IDENTIFICATION: A large, conspicuous rock-dwelling skink with a short head, adult males are very differently coloured to female and juveniles. The body is cylindrical, the tail is fairly long, 60% of total length. Dorsal scales have three strong keels, scales in 38–52 rows at midbody. Maximum size about 30cm, average 22–28cm, hatchlings 7–8cm. Females and juveniles have three conspicuous cream, ivory or yellow stripes (may be reddish near the head) on a black or brown background, the tail is blue with black stripes anteriorly, it may fade in a big adults to dull olive-brown. Adult males are brown or bronze above, speckled white, with three big black blotches on each side of the neck, sometimes red on the lower neck, limbs greenish with black vermiculations, tail bronze or turquoise, chin and throat lilac with white speckling, underside pearly white. Taxonomic Note: Originally regarded as a subspecies of the Five-lined Skink *Trachylepis quinquetaeniata*, but they appear to be sympatric (which subspecies cannot be) at a couple of localities in south-eastern Kenya, Ngulia Hill in Tsavo West and Kiambere, south-east of Mt Kenya near the Tana River. The females and juveniles can be distinguished by the number of stripes (three in the rainbow skink, five in the Five-lined Skink), males by the presence (rainbow) or absence (five-lined) of spots on the back.

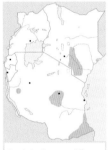

Trachylepis margaritifer

HABITAT AND DISTRIBUTION: Rocky hills and rock outcrops, particularly igneous and metamorphic rocks such as granite, schist and gneiss, in low- to medium-altitude savanna, from sea level to about 1,600m. In Kenya, occurs from Kiambere Hill, south-east of Embu,

Trachylepis margaritifer Rainbow Skink. Left: Stephen Spawls, Voi, male. Right: Andre Zoersel, northern Zambia, female.

south through Machakos and Sultan Hamud to Tsavo West and in the low country around the Taita Hills. Separate populations also occur in Tanzania, at Ugalla, around Dodoma, and also in south-eastern Tanzania, from Kilwa south to the Ruvuma River, thence south to south-east South Africa. Might be more widespread; this species was split from *Trachylepis quinquetaeniata*, but the true situation may be complex.

NATURAL HISTORY: A diurnal, rock-dwelling skink. Highly visible in suitable habitat, males bask in a prominent position and display their beautiful colours. They are territorial, big males dominate an area and do not tolerate other adult males. Males approach a possibly receptive female in a circuitous fashion and bob their heads and necks up and down in a curious horizontal movement. They hide and sleep in rock fissures, exfoliation cracks and similar places. They will live on buildings if they are close to or on rock, for example at Voi and Ngulia safari lodges in Tsavo. In such places there may be large numbers of individuals, due to the variety of artificial habitats. They lay 3–10 eggs, roughly 1.5×2cm. No further details known in East Africa but incubation times in South Africa about two months. They eat a wide range of insects and other arthropods, known also to attack and eat smaller lizards; the blue tail of a juvenile was found in the stomach of an adult. Eaten themselves by birds of prey and snakes, in Tsavo the Speckled Sand Snake *Psammophis punctulatus* is often encountered where these lizards are found. There are various theories concerning the bright blue tail; one is that it deflects predator attacks to the expendable tail, another that it indicates the lizard tastes horrible, another that it indicates a juvenile or female and thus not a rival to the dominant male.

GRASS TOP SKINK/LONG-TAILED SKINK *Trachylepis megalura*

IDENTIFICATION: A fairly large, striped, slim skink with little limbs and a very long tail. The snout is rounded, the eye is small, the legs are thin and spindly. The tail is very long and slender, 60–75% of total length and it is used as a climbing aid. Dorsal scales have three weak keels; they may be smooth in large adults, in 24–27 rows at midbody. Maximum size about 25cm (7cm body, 18cm tail), average 15–22cm, hatchlings 5–7 cm. Colour very variable, usually various shades of brown, golden or rufous; a prominent white flank stripe runs from the upper lip along the entire body and may extend onto the tail. Beneath this is a narrow black stripe, above it usually a broad black stripe. The back may be uniform brown, or have any number of fine or broad longitudinal black or white stripes; lowland specimens seem to have very pale stripes. Some females have broad red dorsal or dorsolateral stripes. Below white or yellow, sometimes up to nine fine black ventral stripes.

Trachylepis megalura

HABITAT AND DISTRIBUTION: Grassland and open grassy savanna, mostly above 1,500m altitude, but found from sea level to over 3,000m. Sporadic records from Tanzania, including the Matema massif, Udzungwa mountains, Usambara, Dar es Salaam, Bagamoyo, Longido and the crater highlands. Widely distributed in high western Rwanda, Burundi, and the Kenya highlands, also in the Chyulu and Taita Hills. A few records of rather unusual specimens from the Kenyan coast. The only Ugandan records are from Pelabek in the north and Bwindi in the south-west, but probably more widespread. Elsewhere the montane region of DR Congo, north to Ethiopia, south to central Mozambique.

NATURAL HISTORY: Diurnal, it lives in grassland and ascends up into clumps; using its long tail it slides swiftly through thick grass. In high-altitude areas it basks in the morning up in clumps before hunting It is shy and secretive, sliding rapidly away or hiding if approached. At night it hides under rocks, in grass clumps or in holes. It gives live birth, up to 15 young recorded in an Athi River specimen. Eats insects, especially grasshoppers, also fond of spiders; a series from southern Tanzania had eaten nothing but spiders. Its sliding movements mean it is liable to be mistaken for a snake if not seen clearly.

Trachylepis megalura Long-tailed Skink. Left: Stephen Spawls, Gravid female, Ethiopia. Top right: Florian Finke, Diani Beach. Bottom right: Stephen Spawls, Gilgil.

ORANGE-FLANKED SKINK *Trachylepis perroteti*

IDENTIFICATION: A big robust skink with a cylindrical body, a short head and small eye, the pupil is round. The tail is fairly long, about 55% of the total length. Dorsal scales usually have three keels, in 32–34 rows at midbody. Maximum size about 40cm, average 25–35 cm, hatchling size unknown. Ground colour dull green or bronze above, sometimes with fine longitudinal lines of brown dashes. Flanks dull orange (becoming vivid orange in males during the breeding season) or brown, shading down to orange or rufous-brown with extensive white spotting. Immaculate white or ivory below. Juveniles are less vividly coloured.

Trachylepis perroteti

HABITAT AND DISTRIBUTION: A species of the Sudan and Guinea Savanna of West and Central Africa, recently recorded in our area at Murchison Falls National Park, probably widespread in north-west Uganda as it occurs in Garamba National Park, Congo.

NATURAL HISTORY: Diurnal and terrestrial (although they will climb fallen logs and on termitaria), they live in holes, excavated under a bush, and mostly hunt by ambush, watching from cover until suitable prey goes by. Lays eggs but no further breeding details known. Diet: insects and other arthropods; Garamba specimens had eaten mostly grasshoppers, beetles and termites.

Trachylepis perroteti Orange-flanked Skink. Stephen Spawls, Ghana.

TREE SKINK   *Trachylepis planifrons*

IDENTIFICATION: A large, long-tailed arboreal skink. The eye is medium sized, body slightly depressed. The tail is very long and thin, two-thirds of total length. Dorsal scales mostly have three (sometimes five) weak keels, scales in 25–32 rows at midbody. Maximum size about 35cm, average 22–30cm, hatchlings 6–7cm. The back is grey or brown, specimens from coastal, northern and eastern Kenya have a distinct bronze, white or cream dorsolateral stripe and a black flank stripe extending from the snout through the eye to the hind limb. In southern Kenyan and northern Tanzanian specimens the dorsolateral stripe is poorly defined and the flank stripe fades to grey posteriorly. There are usually several rows of indistinct black speckles down the centre of the back, outside of these there are often one or two rows of white dots. The belly is grey or white, often with dark grey speckling, the tail is usually grey. Liable to be confused with the Striped Skink *Trachylepis striata*, but it has a much longer tail.

Trachylepis planifrons

HABITAT AND DISTRIBUTION: Coastal thicket, moist and dry savanna and semi-desert, provided there are reasonable-sized bushes present. Found from sea level to about 1,500m altitude. A single Ugandan record, Moroto, widely distributed in north and eastern Kenya and the Kenyan coast, and in a narrow band of the dry central Masai steppe in Tanzania, south to the Ruaha River; isolated records from Saadani National Park, Kilwa, Liwale, also known from the south end of Lake Tanganyika and Katavi National Park; it is probably more widespread in remote areas of Tanzania. Elsewhere, to Somalia, eastern Ethiopia, south-west to Malawi northern Zambia and south-eastern DR Congo.

NATURAL HISTORY: A diurnal, arboreal skink, living in both trees and bushes, on the trunks and even on quite thin branches. It climbs expertly, aided by its long tail. In dry areas it basks in the early morning sun, then hunts, retiring into shade when the day becomes hot. It is quick moving, when approached it sidles around the trunk, keeping the

Trachylepis planifrons Tree Skink. Stephen Spawls. Left: Samburu National Reserve. Right: Watamu.

wood between itself and danger, if pursued along a branch it will jump if threatened. It will descend to move to another tree, and sometimes hunts on the ground. Hides in tree holes and cracks, will sleep out on the ends of branches, but will also live in burrows or cracks in rock, on the Kenyan coast it sometimes lives in the coral rag. Lays eggs, but no breeding details known. Diet insects and other arthropods.

FIVE-LINED SKINK *Trachylepis quinquetaeniata*

IDENTIFICATION: A large, conspicuous robust rock-dwelling skink with a short head, adult males are very differently coloured to female and juveniles. The body is cylindrical, the tail is fairly long, about 55–60% of total length. Dorsal scales have three keels, scales in 32–42 rows at midbody, usually 36–40. Maximum size about 25cm, average 18–22 cm, hatchlings 6–7cm. Juveniles are black, females are brown, with five yellow or cream stripes on the body, and a blue tail, the head may be reddish or bronze. Adult males are various shades of brown or black, with white speckling, they have two or three big black blotches, separated by white dots, on each side of the neck. The lateral stripe persists anteriorly, where it may be yellow on the neck and blue or blue-white on the chin. Breeding males develop a black-speckled chin.

Trachylepis quinquetaeniata

HABITAT AND DISTRIBUTION: Usually on rocky hills, rock outcrops or lava fields, in low- to medium-altitude savanna, from 200m to about 1,600m, elsewhere to sea level. In East Africa occurs from Ngulia in Tsavo west and north through Ukambani and Tharaka, around the north side of Mt Kenya, across to the western border, and into parts of eastern and north-eastern Uganda and northwest Kenya. It is also on Central Island, Lake Turkana,

Trachylepis quinquetaeniata Five-lined Skink. Left: Bio-Ken archive, Tsavo West. Right: Steve Russell, Kidepo National Park

and along the Ethiopian border. A population occurs in north-western Uganda (from Murchison Falls north to Nimule, Arua and Moyo area), recorded from the north-east Victoria lakeshore and Lolui Island, and a single, anomalous Tanzanian record from Kwa Mtoro, north of Dodoma. Elsewhere, north to Egypt, west to Senegal.

NATURAL HISTORY: A diurnal skink, mostly rock-dwelling in Kenya but will move onto houses, huts and other man-made objects like bridges. In non-rocky areas may live on tree trunks, hiding under bark, sometimes found in holes in the ground; in the Dida-Galgalu desert in northern Kenya they live among lava boulders and tolerate diurnal temperatures of 42–44°C. Often visible in suitable habitat, males bask in a prominent position. They are territorial; big males dominate a space and do not tolerate other adult males. They are quite wary, if approached they move away; in Egypt they have been seen to jump off boulders into water and swim to escape. They lay 3–10 eggs, roughly 1.7×1cm, in a suitable spot such as in leaf litter (especially leaf drifts in rock cracks), under logs, deep in rock fissures, etc. Communal egg clutches of up to 60 eggs have been found. In west Africa a clutch of seven eggs hatched after 43 days incubation. They eat a wide range of insects and other arthropods; a Congo specimen had seeds, small fruit and an earthworm in its stomach, a Tanzanian animal had eaten the tail of a juvenile conspecific.

STRIPED SKINK *Trachylepis striata*

IDENTIFICATION: A medium-sized, robust, striped skink, very common in suitable habitats, often seen in towns. The head and body are slightly depressed. The tail is just over half the total length. Dorsal scales have 3–7 keels, scales arranged in 30–42 rows at midbody. Maximum size about 25cm, average 18–22cm, hatchlings 6–7cm. Ground colour shades of brown, olive, rufous or dull green, with two cream, yellow or light greenish dorsolateral stripes; these fade posteriorly in some animals. The top of the head and snout are reddish, the flanks usually speckled white or yellow, underside cream or white with some grey or black speckling. There may be white or black dorsal speckles. In dry northern Kenya some specimens are bronze-coloured. Similar in appearance to the Tree Skink *Trachylepis planifrons* but has a much shorter tail.

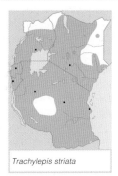

Trachylepis striata

HABITAT AND DISTRIBUTION: Forest clearings, coastal thicket, moist and dry savanna, semi-desert and urban areas. Found from sea level to about 2,300m altitude, or even higher. Probably throughout East Africa below 2,300m, but records sporadic in north-eastern and eastern Kenya and records lacking from northern Uganda and parts of south-central Tanzania.

NATURAL HISTORY: Diurnal and essentially arboreal, but it tolerates urbanisation very well, living on walls and fences and in gardens and plantations, often seen in East African towns. Takes cover in any suitable refuge. Quickly becomes tame and confiding, in urban gardens it rapidly learns to tolerate people. Specimens colonised a new house in rural Gilgil within three months of completion, moving in from the surrounding bush where they were uncommon. In suitable habitat it may occur in large numbers. Gives live birth; clutches of up to nine recorded. In southern Tanzania (Lake Rukwa) most young were born

Trachylepis striata Striped Skink. Left: Stephen Spawls, Kapkuro, Mt Elgon. Right: Michele Menegon, Selous Game Reserve, patternless.

between May and September. Diet: insects and other arthropods, particularly beetles, but seen to eat a gecko; also recorded taking vegetable matter, fruit and carrion. Their main enemies are birds of prey and snakes. Several dicephalic (two-headed) examples of this lizard have been recorded.

VARIABLE SKINK *Trachylepis varia*

IDENTIFICATION: A small to medium-sized striped skink, variable in colour, hence the name. The snout is quite pointed. The body is slightly depressed. The tail is long, two-thirds of total length, and thin. Dorsal scales have three keels and are arranged in 27–36 rows at midbody. Maximum size about 18cm (6cm body, 12cm tail), possibly slightly larger, average 10–15cm, hatchlings 5–6cm. Colour variable, usually some shade of brown, bronze or grey. Nearly all specimens have a white flank stripe, extending from the upper lip to the hind limbs, sometimes onto the tail; this stripe is a good field character. The back is brown, uniform or speckled to a greater or lesser extent with black, or longitudinally striped (especially specimens from montane areas, such as Usambara, Kilimanjaro, Aberdares) with up to nine fine stripes of black, red or yellow, there is often a dark or black, heavily speckled flank stripe. Animals from northern Serengeti and the Mara are a beautiful bronzy colour, with heavy black speckling on the back and tail, Mara juveniles have an orange head and dark vertical cross bars, interrupted by the dorsal and dorso-lateral stripe. The underside is usually white or cream, yellow or orange in high-altitude animals.

Trachylepis varia

HABITAT AND DISTRIBUTION: Coastal thicket, woodland, moist and dry savanna and high-altitude grassland, from sea level to 3,600m altitude. Not common in semi-desert. Occurs throughout south-east and highland Kenya, including the montane moorlands, goes north through the highlands around Maralal to the Ndoto Mountains and the southern end of Lake Turkana. Few records from the low altitude arid country of north and eastern Kenya save Wajir Bor. Occurs through Tanzania, Rwanda and Burundi, although records lacking from parts of western Tanzania. Curiously, there are virtually no Ugandan records, apart from around Mt Elgon and Nimule, it should be widespread there. Elsewhere, south to South Africa, west to the eastern DR Congo, north-east to Somalia and Ethiopia.

NATURAL HISTORY: Diurnal and terrestrial, sometimes climbs. Active on the ground, often in rocky areas, fond of broken country around the base of small rocky hills, ridges, etc. Hides in holes, under rocks, and under tree bark. Quite tolerant of urbanisation, in most Nairobi suburbs. Gives live birth in East Africa, with 3–10 babies recorded. Hatchlings collected in December, March and April in Kenya, in July in southern Tanzania, probably breeds throughout the year in moist areas. Diet: insects and other arthropods.

Trachylepis varia Variable Skink. Left: Stephen Spawls, Nairobi. Right: Stephen Spawls, Mt Elgon.

SUBFAMILY LYGOSOMINAE

A subfamily of just over 50 species of skink, in five genera. Two genera occur in East Africa; one species of *Lepidothyris*, an unusual large skink, and seven species of 'writhing skinks', genus *Mochlus*. Writhing skinks have a chequered taxonomic history; various species having been placed in the genera *Riopa*, *Mochlus* and *Lygosoma*; all East African forms are now in *Mochlus*, the name *Lygosoma* is reserved for Asian species.

WRITHING SKINKS *MOCHLUS*

A group of African burrowing skinks with small legs. As the name implies, when seized or handled they writhe and struggle fiercely, often jabbing the sharp snout into the restraining hand. They have scaly, mobile lower eyelids and small to medium sized eyes, most have sharp pointed snouts. Their smooth, shiny and elongated bodies are adapted for pushing through sand, soft soil or leaf litter, or for dwelling in holes. Some species lay soft-shelled eggs, others give live birth. They feed on small insects and their larvae. Thirteen species are known, seven occur in East Africa, three are endemic. They occupy a range of habitats, from semi-desert to woodland. There is debate about the supposed differences between the two most common forms; see the taxonomic note for Peters' Writhing Skink *Mochlus afer*.

KEY TO EAST AFRICAN MEMBERS OF THE GENUS *MOCHLUS*

1a	Unregenerated tail more than 55% of total length......2		snout slightly depressed......*Mochlus tanae*, Tana River Writhing Skink [p. 155]
1b	Unregenerated tail less than 55% of total length......3	5a	Prefrontals fused with frontonasal, midbody scale rows 22–24......*Mochlus pembanum*, Pemba Island Writhing Skink [p. 153]
2a	Adult up to 36cm total length, juveniles with a conspicuous vertebral stripe ......*Mochlus mabuiiforme*, Mabuya-like Writhing Skink [p. 152]	5b	Prefrontals distinct, midbody scale rows 24–30......6
2b	Adult up to 15cm total length, juveniles without a conspicuous vertebral stripe...... *Mochlus somalicum*, Somali Writhing Skink [p. 154]	6a	Adults 80–137mm excluding tail, back usually speckled black and white...... *Mochlus afer*, Peters' Writhing Skink [p. 150]
3a	Only on Mafia Island......*Mochlus mafianum*, Mafia Writhing Skink [p. 153]	6b	Adults usually 60–80mm excluding tail, back usually uniform brown or each scale with a dark spot at the base...... *Mochlus sundevalli*, Sundevall's Writhing Skink [p. 151]
3b	Not confined to Mafia Island......4		
4a	Lower eyelid scaly, snout strongly depressed, wedge-shaped......5		
4b	Lower eyelid with a transparent brille,		

PETERS' WRITHING SKINK *Mochlus afer*

IDENTIFICATION: A large, shiny-bodied, small-limbed writhing skink with a short snout and small eyes. The ear openings are tiny and deeply sunk. The body is stout and cylindrical, the tail is thick, about half the total length. The scales are smooth, in 25–29 rows at midbody. Largest male 23.0cm, largest female 22.0cm, average size 15–20cm, hatchling size unknown. Colour very variable. The back is usually pale to dark brown, or grey, with usually regular dark brown and white speckles. Animals which are about to slough may have uniform brown bodies, but light and dark speckling is usually visible on the tail. The ventrum may be spotted brown (especially in highland specimens) or immaculate; in females it is creamy white or blue-white, in mature males the ventrum and the flanks are bright yellow. Juveniles often have a bright red tail, this fades with maturity. Taxonomic Note: For a long time, this species and the following one, Sundevall's Writhing Skink *Mochlus sundevalli*, were regarded as conspecific. They were recently separated, but are usually identified by size (Peters' is larger, see the key) and colour; Peters' is usually heavily speckled, Sundevall's usually uniform. But both species vary a lot in colour, and a key based on size runs the risk of separating small and large individuals of the same species into two. Molecular taxonomy also appears to indicate these two taxa are actually the same.

Mochlus afer

HABITAT AND DISTRIBUTION: Occupies a wide range of habitat, including coastal savanna and woodland, dry and moist savanna, semi-desert, and medium to high altitude woodland. Ranges from sea level to 2,300m altitude. Widespread in East Africa, although unrecorded in Rwanda and Burundi; there are few records from the western side of Tanzania; and very sporadic in northern Kenya and most of Uganda save in the south-east

Family Scincidae: Skinks   Subfamily Lygosominae   151

Mochlus afer Peters' Writhing Skink. Left and Centre Stephen Spawls. Left: Nairobi National Park, male and female. Centre: Bagamoyo. Right: Eduardo Razzetti, Lake Turkana.

but this is probably due to undercollecting. Known from Lolui Island. Elsewhere, north to Ethiopia, west to eastern DR Congo, south to Mozambique.

NATURAL HISTORY: Burrowing, usually to be found below rocks, under logs, in soft soil and sand; often taken in pitfall traps. It may be crepuscular, sometimes seen on the surface at dusk; often mistaken for a snake as it moves across the ground. Males pursue the females in the mating season, copulation observed in April in southern Tanzania. They lay 4–7 eggs, roughly 1×1.5cm. Diet: Insects and other invertebrates.

SUNDEVALL'S WRITHING SKINK   *Mochlus sundevalli*

IDENTIFICATION: A medium to large, shiny-bodied, small-limbed writhing skink with a short, flat, wedge-shaped snout and small eyes, the lower eyelid is scaly. The ear openings are tiny and deeply sunk. The body is stout and cylindrical, the tail is thick, about half the total length. The scales are smooth, in 24–30 rows at midbody. Maximum size about 18cm, average size 12–16cm, hatchling size unknown. Colour very variable; usually brown to grey, uniform or with a dark spot at the base of each scale; the overall appearance of this colour form is of a finely speckled animal. Belly cream or white, sometimes speckled.

Mochlus sundevalli

HABITAT AND DISTRIBUTION: Occupies a wide range of habitat, including coastal savanna and woodland, dry and moist savanna, semi-desert, and medium to high altitude woodland. From sea level to 2,000m altitude. Widespread throughout Kenya and Tanzania. Enters Uganda in the Moroto area, also known from the south end of Lake Albert, but probably more widespread. Elsewhere, north to Somalia, south-west to Angola, south to northern South Africa. Conservation Status: Least Concern.

NATURAL HISTORY: Burrowing, usually to be found below rocks, under logs, in soft soil and sand. Like Peters' Writhing Skink *Mochlus afer*, it appears to be crepuscular, sometimes seen on the surface at dusk. It lays 2–7 eggs, roughly 1×1.5cm. Diet: insects, insect larvae and other invertebrates. It is vulnerable to flooding, especially in flat areas. During the El Niño floods of December 1997, a group of ground hornbills in Tsavo East were observed in the early morning picking these skinks off from the verges of the Aruba road where they had been forced to the surface by heavy rain, over 20 were eaten in ten minutes.

Mochlus sundevalli Sundevall's Writhing Skink. Left: Robert Sindaco, Lake Turkana. Right: Ignas Safari, Dodoma.

MABUYA-LIKE WRITHING SKINK *Mochlus mabuiiforme*

IDENTIFICATION: A very large writhing skink from the north Kenya coast. The snout is rounded, the ear-opening large with two small rounded lobules at the front. Limbs well developed with five digits. Scales smooth, in 26–30 rows at midbody. Males may reach a maximum size of 36.3cm, with the tail about 60% of the total length, average size unknown. Adults are faintly striped or uniformly plumbeous-grey; each individual scale is lighter at its base, the light area increasing in size towards the tail so that the scales on the tail are light centred with dark edges. The lips are white with brown barring. Ventrum uniformly white. The juveniles are more vivid; black crown of head mottled with pale brown, occipital scale conspicuously white with dark centre; a vertebral line of pale brown, one scale in width, begins behind the occipital scale and continues to the base of the tail where it disappears. The vertebral line is flanked on either side by a dorsolateral stripe of the same colour but two scales in width and commencing at the last supraocular. The limbs are uniformly black in some specimens; tip of tail transparent red, each scale edged with brown. Below, white, the internal organs visible through the scales; tail clear coral pink.

Mochlus mabuiiforme

Mochlus mabuiiforme Mabuya-like Writhing Skink. Stephen Spawls, Preserved specimen.

HABITAT AND DISTRIBUTION: In Kenya, known only from drier habitats in the Tana River delta at sea level or just above Elsewhere, known from southern Somalia.

NATURAL HISTORY: Little known. Diurnal and surface dwelling; seen in the shade of thornbush during the day. Presumably lays eggs, eats insects and other arthropods. The specific name refers to the fact that the author felt that this form resembles the typical skinks of the genus *Mabuya*.

MAFIA WRITHING SKINK Mochlus mafianum

IDENTIFICATION: A small to moderate-sized writhing skink, described in 1994 in Tanzania. Head moderate size, with a rather flat dorsal profile; rostral projecting onto snout along a broad, slightly curved, posterior edge. The body is cylindrical, tail subcircular in cross section, slightly more than half the total length. Limbs short with five toes. Dorsal scales tricarinate (three-keeled) in middle of back, smooth on nape, flanks and on base of tail; midbody scale rows 26–28. Total length about 12.3cm. Grey-brown above, paler on the flanks, uniform in subadults, but irregularly streaked with dark brown in adults; white below. A large dark median streak may be present on each dorsal scale, and flanks may have numerous dark and light streaks.

Mochlus mafianum

HABITAT AND DISTRIBUTION: Coastal woodland and savanna. Endemic to Tanzania on Mafia and Kisiju Islands, Coast Region. Conservation Status: Least Concern, but its precise microhabitat requirements are not known, so might be under threat from habitat destruction if it requires forest or closed vegetation cover; its limited range may also make it vulnerable.

NATURAL HISTORY: Unknown; the specimens were collected in a pitfall trap. Presumably similar to other writhing skinks; i.e. diurnal, burrowing, lays eggs, eats invertebrates.

PEMBA ISLAND WRITHING SKINK Mochlus pembanum

IDENTIFICATION: A small writhing skink. The head is of moderate length, the snout is depressed and wedge-shaped. The eye is small, with a scaly lower eyelid. The body is cylindrical, tail subcircular in cross section, slightly more than half the total length. Limbs short with five toes. Midbody scale rows 22–24. Total length about 12cm. Adults are iridescent bronzy or brown in colour, uniform or with linear black speckling, often with a vague dark lateral bar, this is speckled white, the speckling may be more pronounced on the tail; juveniles are more yellowish.

HABITAT AND DISTRIBUTION: Coastal and island woodland, at sea level. An East African endemic, known from Pemba Island in Tanzania and several localities on the Kenyan coast, including Ukunda, Likoni, Takaungu, Kilifi and the Arabuko-Sokoke Forest. Conservation Status: IUCN Least Concern. The exact requirements for this species are not known. It may be under threat from habitat destruction if it requires forest or closed vegetation cover; its limited range may also

Mochlus pembanum

make it vulnerable.

NATURAL HISTORY: Unknown. Presumably similar to other writhing skinks, i.e. diurnal, burrowing in sandy soil,, lays eggs, eats insects, insect larvae and other arthropods. The common and specific name refer to the type locality, Pemba island, coastal Tanzania.

Mochlus pembanum Pemba Island Writhing Skink. Frank Glaw, Pemba.

SOMALI WRITHING SKINK *Mochlus somalicum*

IDENTIFICATION: A small writhing skink. The head is of moderate length, the snout is depressed and wedge-shaped. The eye is small, with a scaly lower eyelid. The body is cylindrical, tail subcircular in cross section, about 60% of the total length. Limbs short with five toes. Midbody scale rows 26–28, the dorsal scales are iridescent. Total length about 15cm. Dorsal colour usually shades of brown (golden-brown, grey-brown), with a series of fine longitudinal black stripes. The head is flecked with black, the lips barred black and white. There is a broad dark lateral stripe. The tail is usually bright orange-red, pink or pinkish-brown. Below, cream or white.

Mochlus somalicum

HABITAT AND DISTRIBUTION: Dry savanna at low altitude. From Mkomazi in north-eastern Tanzania it spreads north through the eastern side of the Tsavo ecosystem; also on the coast, with an isolated record in Meru National Park. Elsewhere, in northern and central Somalia.

NATURAL HISTORY: Poorly known. The Meru National Park specimen was in a pitfall

Mochlus somalicum Somali Writhing Skink. Stephen Spawls. Left: Watamu. Right: Mzima Springs.

trap, Mzima Springs specimen was under a rock in dense acacia woodland, the Sala animal under a rock on a dry, barren plain. Behaviour presumably similar to other writhing skinks, i.e. diurnal, burrowing in sandy soil, lays eggs, eats insects, insect larvae and other arthropods.

TANA RIVER WRITHING SKINK *Mochlus tanae*

IDENTIFICATION: A long, very slender writhing skink. The head is of moderate length, the snout is slightly depressed. The eye is small, the lower eyelid has a transparent brille. The body is cylindrical, tail subcircular in cross section, 40–50% of the total length. The legs are small and very short, with five toes. Midbody scale rows 22–24. Total length about 15cm. Uniformly brown or plumbeous grey above, the upper labials flecked with lighter brown, similar light flecks on the side of the neck. Throat and lower labials white, heavily spotted with dark brown.

Mochlus tanae

HABITAT AND DISTRIBUTION: Coastal and dry savanna, at low altitude. Sporadic records from Witu and the Tana Delta, the Arabuko-Sokoke Forest and the Taita Hills in Kenya, and Ndungu at the base of the South Pare Mountains in Tanzania. Elsewhere, in river valleys in southern Somalia. Conservation Status: not assessed. Probably more widespread than it appears, suitable habitat is extensive and thus it is not under threat.

NATURAL HISTORY: Unknown. Presumably similar to other writhing skinks, i.e. diurnal, burrowing in sandy soil; its slim body indicates it probably favours soft silty sand. Probably lays eggs, eats insects, insect larvae and other arthropods.

RED-FLANKED SKINKS/FIRE SKINKS *LEPIDOTHYRIS*

A genus of three unusual, large central African skinks. Originally all three were regarded as a single variable species, known as the Fire or Red-flanked Skink, *Lygosoma fernandi*. A single species occurs in East Africa.

HINKEL'S RED-FLANKED SKINK *Lepidothyris hinkeli*

IDENTIFICATION: A spectacular, black and red forest skink. It has a short deep head, with a rounded snout. The eye is relatively large and obvious, the pupil round. The ear opening is large and oval. The body is stout and sub-cylindrical, the four limbs are short and strong, with five thin toes. The tail is stout, tapering smoothly, slightly over half the total length. The scales are smooth, in 31–34 rows at midbody. Maximum size about 34cm, average 22–30cm, hatchling size unknown but specimens from other members of the group were 6–8cm. Females are generally smaller than males. Vividly coloured, the lips are red, the back may be rufous, brown or black above, marked by irregular and often broken black, white speckled bars that extend down the vivid red or orange flanks. The tail is black, or barred brown and black, in some juveniles the tail is barred with vivid light blue. Females are usually less brightly coloured than males. The underside is white, unmarked. Two subspecies are known, *L. hinkeli hinkeli* and *L. h. joei*, East African specimens belong to the former.

HABITAT AND DISTRIBUTION: Woodland, forest and grassland at the forests' edge, at altitudes between 600–2,100m in East Africa. In Kenya, known only from the Kakamega Forest area, in Uganda known from Kampala and Entebbe and a series of localities along the Albertine Rift including Kibale National Park and the Bwindi Impenetrable Forest, Rwandan localities include Nyungwe Forest and east of Lake Kivu. Elsewhere, west to Mulungu, in eastern Congo, west of there the subspecies *L. h. joei* extends to Angola and northern DR Congo.

Lepidothyris hinkeli

NATURAL HISTORY: Secretive and rarely seen. Congo specimens have been described as active at night, but recent observations indicate that it is diurnal and terrestrial, and basks in the sunlight; a Nigerian study of a closely-related species found that almost all activity took place in the late afternoon, but some specimens were active in the early morning. Shelters in holes, rock crevices, termite hills and under ground cover, one Kakamega specimen was found lying motionless in short grass in mid-afternoon, another was seen on a forest path, one was caught in a pitfall trap. Older data indicate it digs extensive burrows, on one occasion a true pair were found in a burrow. Captive specimens show territorial behaviour, with males aggressive towards other mature males. Captive specimens lay clutches of up to 12 eggs. Diet: invertebrates. There are several curious legends attached to the various skinks in this genus; it is believed that they are highly venomous, it is bad luck to see one, and also that they can produce a fearsome shriek audible over several kilometres.

Lepidothyris hinkeli Hinkel's Red-flanked Skink. Left: Fabio Pupin, Nyungwe National Park, Rwanda. Right: Harald Hinkel, Rwanda, juvenile.

SUBFAMILY EUGONGYLINAE

A subfamily of about 430 species of skinks, in 45 genera, most occur in Australasia and the islands of the Pacific Ocean, but some ten species, in three genera, occur in East Africa. Some are large, but most of the East African forms are small, nondescript burrowers.

LEAF LITTER SKINKS *LEPTOSIAPHOS*

A genus of 18 species of small (usually less than 15cm total length), slim, smooth-bodied burrowing skinks, with very long and yet stout tails; a good field character. Most occur in tropical central Africa, largely in forest. The scales are smooth and shiny. Seven species are found in East Africa, all but two confined to the montane forest of western East Africa. In the past, they have been variously placed in the genus *Lygosoma*, *Panaspis* and *Siaphos*.

KEY TO EAST AFRICAN MEMBERS OF THE GENUS *LEPTOSIAPHOS*

1a Foot with five toes, lower eyelid with a central window or scaly......2
1b Foot with three or four toes, lower eyelid scaly......5
2a Frontoparietals fused with the interparietal, paravertebral scale rows transversely enlarged, less than 50 between nuchals and base of the tail, more than ten lamellae beneath fourth finger, more than 15 beneath fourth toe, endemic to the Udzungwa Mountains...... *Leptosiaphos rhomboidalis*, Udzungwa Five-toed Skink [p. 162]
2b Frontoparietals not fused with the interparietal, paravertebral scale rows not transversely enlarged, more than 50 between nuchals and base of the tail, 7–9 lamellae beneath fourth finger, 9–15 beneath fourth toe, Uluguru Mountains northwards......3
3a Lower eyelid with a central window, 10–12 lamella beneath fourth finger, 13–15 beneath fourth toe......4
3b Lower eyelid scaly, 7–8 lamellae beneath fourth finger, 9–11 beneath fourth toe...... *Leptosiaphos graueri*, Rwanda Five-toed Skink [p. 159]
4a Five digits on forefoot...... *Leptosiaphos kilimensis*, Kilimanjaro Five-toed Skink [p. 160]
4b Four digits on forefoot...... *Leptosiaphos aloysiisibaudiae*, Uganda Five-toed Skink [p. 158]
5a All limbs with three digits (tridactyle)...... *Leptosiaphos blochmanni*, Kivu Three-toed Skink [p. 158]
5b All limbs with four digits (tetradactyle)......6
6a Foot tetradactyle due to loss of first toe (hallux)...... *Leptosiaphos meleagris*, Rwenzori Four-toed Skink [p. 161]
6b Foot tetradactyle due to loss of fifth toe...... *Leptosiaphos hackarsi*, Virunga Four-toed Skink [p. 160]

UGANDA FIVE-TOED SKINK Leptosiaphos aloysiisabaudiae

IDENTIFICATION: A small, small-limbed skink with a short, rounded snout and medium-sized eyes, the lower eyelid has a large window. The ear openings are tiny and the tympanum is deeply sunk. The body is slim and cylindrical, the tail is thick and long, about 60% of the total length. The scales are smooth, in 22–24 rows at midbody. Limbs small and slender, four digits on forelimb, five on hindlimb, lamellae beneath fourth finger 10–12, beneath fourth toe 13–15. Average size about 14cm, hatchling size unknown. Brown, becoming reddish posteriorly, with darker median blotches on most scales, darker laterally, labials with light and dark vertical barring, chin and throat white, becoming yellow or reddish posteriorly. Similar species: Within its limited range, only likely to be confused with other small five-toed skinks; see our key.

Leptosiaphos aloysiisabaudiae

HABITAT AND DISTRIBUTION: In Uganda, recorded from Ajai and Katonga game reserves (the former specimen in riverine woodland, the latter in riverine swamp), Toro and Bukalasa and Mityana, in high savanna near Kampala; at altitudes between 800–1,400m altitude. Presumably an inhabitant of moist savanna or recently deforested areas. Elsewhere, recorded from the Imatong Mountains in South Sudan, Albert National Park in DR Congo, Cameroon and Nigeria. Conservation Status: IUCN Least Concern.

NATURAL HISTORY: Virtually nothing known, but presumably burrowing and diurnal, living beneath the surface of the soil or in leaf litter, lays eggs, eats small insects, insect larvae and other small arthropods that it finds beneath the leaf litter.

KIVU THREE-TOED SKINK Leptosiaphos blochmanni

IDENTIFICATION: A small, small-limbed skink with a short, rounded snout and medium-sized eyes, the lower eyelid is scaly. The ear openings are tiny. The body is slim and cylindrical, the tail is slender and long, about two-thirds of the total length. The scales are smooth, in 20–24 rows at midbody. Limbs very small and slender, tridactyle, lamellae beneath third finger 8–9, beneath third toe 9–11. Maximum length about 14cm, hatchling length 4.4mm. Warm or golden brown, the intense fine black speckling on the back may form narrow crossbars. A broad dark, yellow-edged stripe runs from the snout through the eye to just behind the shoulder. The tail is marked with fine dark longitudinal stripes. Ventral surface yellow to cream, but males may have the throat heavily spotted with black or uniform black, the rest of the ventrum may be white or each scale may have a black centre.

Leptosiaphos blochmanni

HABITAT AND DISTRIBUTION: Within our area, known only from the Nyungwe Forest in south-west Rwanda, at around 2,000m altitude, probably in northern Burundi, elsewhere known from the high country along the eastern border of DR Congo. Conservation Status: Data Deficient.

NATURAL HISTORY: Virtually nothing known, but presumably burrowing and diurnal, living beneath the surface of the soil or in leaf litter, lays eggs, eats small insects, insect larvae and other small arthropods that it finds beneath the leaf litter.

Leptosiaphos blochmanni Kivu Three-toed Skink. Harald Hinkel, Rwanda.

RWANDA FIVE-TOED/FOUR-TOED SKINK *Leptosiaphos graueri*

IDENTIFICATION: A small, small-limbed skink with a short, rounded snout and medium-sized eyes, the lower eyelid is scaly. The ear openings are tiny. The body is slim and cylindrical, the tail is thick and long, about 65% of the total length. The scales are smooth, in 22–24 rows at midbody. Limbs very small and usually pentadactyle, but forelimb sometimes with only four digits, lamellae beneath fourth finger 7–8, beneath fourth toe 9–11. Maximum length 20cm, average length 16cm, hatchling length 5cm. Coppery or metallic yellow-brown to grey, the side of the head and the flanks are grey-white, heavily speckled and blotched with black, this speckling may form fine dark bars on the mid-body section. Ventral surface grey-white, the scales often black-centred, becoming salmon-pink in the preanal region and on the hind limbs and tail.

Leptosiaphos graueri

Leptosiaphos graueri Rwanda Five-toed Skink. Left: Max Dehling, Rwanda. Right: Paul Freed, Rwanda.

HABITAT AND DISTRIBUTION: Known from a handful of localities along the Albertine Rift, Uganda records include the Bihunga escarpment in the Rwenzoris, Bwindi Impenetrable National Park (where sympatric with *Leptosiaphos hackarsi*) and Kabale, Sabinyo Volcano and Nyungwe Forest in Rwanda, and the Virunga National Park in DR Congo, at medium to high altitude, between 1,200–2,500m. A curious early record from Entebbe. Elsewhere on the eastern escarpment in DR Congo.

NATURAL HISTORY: Diurnal and terrestrial, burrowing into soil and leaf litter. Specimens were found among the roots of ferns growing between the buttress roots of trees in wet forest; in Rwanda specimens were found on moss-covered rock ledges, on a hummock in a swamp and crossing paths. Two eggs (11×6mm) are laid, diet small insects, insect larvae and other small arthropods that it finds in the leaf litter.

VIRUNGA FOUR-TOED SKINK Leptosiaphos hackarsi

IDENTIFICATION: A small, small-limbed skink with a short, rounded snout and medium-sized eyes, the lower eyelid is scaly. The ear openings are tiny. The body is slim and cylindrical, the tail is slender and long, about 60% of the total length. The scales are smooth, in 22–24 rows at midbody. The limbs are very small and slender, with four digits, the hind limb tetradactyle due to loss of the fifth toe. Lamellae beneath fourth finger 7–8, beneath fourth toe 9–12. Maximum length about 14cm, average length 13cm, hatchling size unknown. Adult male with head mottled light and dark brown, anterior three quarters black with scattered yellow spots, posterior dorsum and flank orange brown, chin and throat black or yellow blotched with black, rest of ventrum yellow, becoming orange posteriorly.

Leptosiaphos hackarsi

HABITAT AND DISTRIBUTION: Appears to be an inhabitant of high moist savanna or recently deforested areas along the Albertine Rift, recorded from Bwindi Impenetrable National Park in Uganda (where sympatric with *Leptosiaphos graueri*), Mt Karissimbi in northern Rwanda and the Virunga National Park in the eastern DR Congo, at altitudes over 1,500m.

NATURAL HISTORY: Virtually nothing known, presumably as for other members of the genus.

Leptosiaphos hackarsi Virung a Four-toed Skink. Michele Menegon, Rwanda.

KILIMANJARO FIVE-TOED SKINK Leptosiaphos kilimensis

IDENTIFICATION: The only leaf litter skink from Kenya and much of Tanzania; a small, slim, tiny-limbed skink with a very short, deep head, rounded snout and fairly large eyes, the iris is golden brown, the lower eyelid is scaly with a small window. The ear openings are small. The tail is thick and long, but often regenerated, usually 50–70% of the total length, tapering gently to a blunt tip. The scales are smooth, in 22–26 rows at midbody. Maximum size about 18cm, average size 10–15cm, hatchling size 56–58mm. Iridescent golden brown or bronze above, with fine black flecks that form broken longitudinal lines. There is an irregular dark line through the eye, and the lips are clearly barred black and white, the head scales finely mottled with black. The anterior flanks are pink or pinky-orange, sometimes finely speckled black and white. The tail is darker, grey or blue-grey, the underside uniform pinky-white.

Leptosiaphos kilimensis

HABITAT AND DISTRIBUTION: Widespread in rain forest, moist savanna, high grassland and woodland, from near sea level up to about 2,000m. Known from eastern Tanzania (Usambara, Nguu, Nguru, Ukaguru, Uluguru and Udzungwa Mountains), all along the high country of the north-east Tanzanian border and also in the Taita Hills in Kenya, widely distributed in the Kenya highlands from Nairobi (where still common in wooded suburbs such as Parklands, the Hill, Spring Valley) north to Ngaia Forest and the Nyambene Hills round the eastern flank of Mt Kenya, (in this area it does not occur below 1,400m), Subukia, and the wooded areas of Laikipia, across to Ol Ari Nyiro Ranch, north-west of Rumuruti. Ugandan records sparse; they include Pelabek, Kibale and the Bwindi Impenetrable National Park. Elsewhere, south-west to Angola.

NATURAL HISTORY: Diurnal, sometimes found basking in the forest in patches of sunlight, but often in leaf litter between buttress root of forest trees or under logs. It might also be crepuscular; in Nairobi often on the surface at dusk and not long after dawn. Lays a clutch of 2–4 eggs, measuring about 1.4×0.8cm, many clutches may be laid together beneath a suitable log or in leaf litter between buttress roots. Eats small insects, insect larvae and other small arthropods that it finds beneath the leaf litter.

Leptosiaphos kilimensis Kilimanjaro Five-toed Skink. Michele Menegon, Nguu North Forest Reserve.

RWENZORI FOUR-TOED SKINK *Leptosiaphos meleagris*

IDENTIFICATION: A medium sized skink, with small limbs and a very short head with a rounded snout, the lower eyelid is scaly. Scales smooth, in 22–24 rows at midbody. The largest known male was 19.6cm; the largest female 19.3cm, hatchling 5.3cm. Males and females have different colours. The dorsal colour in males is dark brown or black, with each dorsal scale lightly speckled; females are iridescent reddish-brown anteriorly, shading gradually to grey towards the tail which is plumbeous grey. A dark stripe from the nostril through the eye continues to just past the front legs; this may be diffusely edged with yellow on its upper surface. The back is speckled with black behind the front limbs. The male throat is mottled black and bluish white; chest, belly, hind limbs and base of tail salmon red; rest of tail bluish white, heavily mottled with black. The female throat is yellowish white, the belly pink, the tail under-surface bluish-white. Juvenile males resemble

Leptosiaphos meleagris

Leptosiaphos meleagris Rwenzori Four-toed Skink. Eli Greenbaum, eastern Congo.

adult females, yellowish white or uniform pink from chin to base of tail, this colour turns to salmon red as they reach maturity.

HABITAT AND DISTRIBUTION: A high-altitude species, in East Africa known only from the Mubuku Valley, 2,100m, Rwenzori Mountains, Uganda; elsewhere known only from the Bugongo Ridge at 2,700m and Kalonge, in the Virunga National Park and 'Ituri' in DR Congo. Conservation Status: Vulnerable B1ab(iii). Apparently restricted to montane habitats and known from only a single area in East Africa. Little information appears to be available as to its habitat requirements.

NATURAL HISTORY: Poorly known. Burrows in leaf litter and soft soil. Several specimens were found among the roots of ferns growing at the bases of buttress roots in wet forest. Presumably diurnal. Females lay two eggs, 12×8.5mm, sometimes several females lay in the same spot, up to 22 eggs recorded in a small area. Diet: centipede, spiders, mites, isopods and insects.

UDZUNGWA FIVE-TOED SKINK *Leptosiaphos rhomboidalis*

IDENTIFICATION: A small, small-limbed skink, with the frontoparietals fused with the interparietal. It has a short, rounded snout and medium-sized eyes, the lower eyelid has a horizontally oval window. The ear openings are tiny, with tympanum deeply sunk. The body is moderate and cylindrical, the tail is thick and long (regenerated in the type). The scales are smooth, in 23 rows at midbody, the paravertebral rows transversely enlarged and only 46 between nuchal and base of tail. Back brown with scattered dark flecks, a thin cream dorsolateral stripe from snout to base of tail, becoming faint posteriorly, a dark brown lateral band from snout to base of tail, a thin light stripe from the posterior supralabial through ear to forelimb, black spots along the lower labials extending towards the forelimb, belly white.

Leptosiaphos rhomboidalis

HABITAT AND DISTRIBUTION: A Tanzanian endemic, known only from a single specimen caught in a pitfall trap in Mwanihana forest reserve in the Udzungwa Mountains, south-east Tanzania. Presumably a forest endemic. Conservation Status: Data Deficient.

NATURAL HISTORY: Nothing known, but presumably similar to other members of the genus.

COASTAL SKINKS (SHINING SKINKS)
CRYPTOBLEPHARUS

A genus of small, egg-laying skinks with well-developed limbs and immoveable lower eyelids. They are largely terrestrial, some are arboreal and some climb rocks. Many species occur around coastlines. Originally only a single species was recognised, with many races/subspecies, but now the Australasian and Oceanic Island forms have been elevated into dozens of species. One species occurs along the Eastern African coast.

CORAL RAG SKINK *Cryptoblepharus africanus*

IDENTIFICATION: A small skink with well-developed limbs, five long clawed toes. Head distinct. Immoveable eyelids, each with a transparent spectacle. Scales smooth, close fitting, in 26–29 midbody scale rows. The tail is cylindrical, tapering to a fine point. Maximum size about 15cm, half of which is tail. The dorsum is blackish-bronze, with two speckled gold dorsolateral stripes; paler below.

Cryptoblepharus africanus

HABITAT AND DISTRIBUTION: On coral rag (old coral formations, on and above the waterline) all along the East African coast and the islands. Does not move away from the immediate vicinity of the intertidal zone. Elsewhere, from Somalia to Southern Africa.

NATURAL HISTORY: Diurnal. Quick-moving and active, forages on the intertidal zone of coastal coral formations, where it may be seen darting up and down, following wave movement, catching the insects and small crustaceans on which it feeds. It may also take small fish. It swims well, and if threatened and unable to reach a hole will jump from the coral and swim to safety. It is salt-tolerant, and the ancestral species dispersed across the

Cryptobleparis africanus Coral-rag Skink. Left: Daniel Hollands, Watamu. Right: Frank Glaw, Pemba .

Pacific by rafting. One or two eggs are laid in a crevice above the high water mark in the coral rag. Often occurs in large densities, even on quite small offshore coral islets; there were several hundred on a small (15m diameter) isolated coral outcrop in the bay at Watamu. May come ashore a short distance to scavenge; at the Slipway in Dar es Salaam they have colonised the walls.

SNAKE-EYED SKINKS *PANASPIS*

An African genus of small, slim, cylindrical-bodied egg-laying skinks with well-developed five-toed limbs; some have immoveable lower eyelids, in some the eyelids move. About 15 species are known, mostly in the great forests, but two savanna species occur in East Africa.

KEY TO EAST AFRICAN MEMBERS OF THE GENUS *PANASPIS*

1a Unregenerated tail two to three times the snout-vent length, midbody scale rows 20–22......*Panaspis megalurus*, Blue-tailed Snake-eyed Skink [p. 164]

1b Unregenerated tail one to two times snout-vent length, midbody scale rows 24–28......*Panaspis wahlbergi*, Wahlberg's Snake-eyed Skink [p. 165]

BLUE-TAILED SNAKE-EYED SKINK *Panaspis megalurus*

IDENTIFICATION: A small slim skink with a very long tail. Scales smooth, in 20–22 midbody scale rows. The tail is cylindrical, tapering to a fine point, slightly over half the total length. The tail is very long, 75% of total length, although regenerated tails are shorter. Maximum size about 17cm, average and hatchlings size unknown. Iridescent brown or olive above, with a pale, black-edged dorsolateral stripe from the neck to beyond the hindlimbs. A line of white blotches extends from behind the eye along the flanks to just behind the forelimbs. The upper flanks are grey to orange, becoming paler below, the underside of the limbs is orange pink, the belly white. The type specimen, a female, had a blue tail, a recently-reported male had an orange tail.

Panaspis megalurus

Panaspis megalurus Blue-tailed Snake-eyed Skink. Stephen Spawls, Maasai Mara.

HABITAT AND DISTRIBUTION: Known from the mid-altitude central plains of Tanzania, north and north-west of Dodoma, the Usangu Plains, at just over 1,000 m altitude and the Maasai Mara in Kenya. A specimen possibly from 'Arusha' (not shown on map) might be of this species.

NATURAL HISTORY: Unknown, but presumably similar to other small terrestrial skinks. The very long tail suggests that it might live in long grass and use its tail to assist with rapid movement, as do other long-tailed lizards. The Mara specimen was under a rock in grassland.

WAHLBERG'S SNAKE-EYED SKINK *Panaspis wahlbergi*

IDENTIFICATION: A small skink with well-developed limbs, five clawed toes. Head distinct. Immoveable eyelids, each with a transparent spectacle. Scales smooth, close fitting, in 22–28 midbody scale rows, the scales on the soles are rounded. The tail is cylindrical, tapering to a fine point, slightly over half the total length. Maximum size about 14cm, average 8–12cm, hatchlings about 3cm. The dorsum is grey, brown or bronze, often with up to six fine dark lines. There is usually a distinct dark side-stripe, edged above by a fine white stripe. The belly is white or blue-white, but may be pinkish-orange in breeding males. In Ethiopian males the tail is sometimes coral-pink. Taxonomic Note: used to be *Afroablepharus wahlbergi*, but recent taxonomic work shows that *Afroablepharus* and *Panaspis* are monophyletic; *Panaspis* has priority. In future, the East African animals are probably going to be named as a new species.

Panaspis wahlbergi

HABITAT AND DISTRIBUTION: Coastal bush, moist and dry savanna and high grassland, from sea level to about 2,200m. Widespread in eastern and parts of western Tanzania and south-eastern and central Kenya, sporadic records from other areas including South Horr, Marsabit and around Mandera in north-eastern Kenya; it is probably much more widespread but overlooked. Elsewhere, south to eastern South Africa.

NATURAL HISTORY: Diurnal. Quick-moving and active, forages on the ground, living in grass tufts, leaf litter and in recesses in broken ground. In South Africa, males live only 10–12 months, females a few months longer, but lifestyle unstudied in East Africa. They reach sexual maturity in 8–9 months. A clutch of 2–6 eggs, roughly 4×7mm, is laid. Diet: termites and other small insects.

Panaspis wahlbergii Wahlberg's Snake-eyed Skink. Stephen Spawls. Left: Nairobi. Right: Ethiopia.

SUBFAMILY **SCINCINAE**
'TYPICAL' SKINKS

A subfamily of some 280 species of skink, in over 30 genera, widespread in the Americas, Africa, Madagascar and Asia; some 14 species, including seven species of *Melanoseps*, in eight genera, occur in East Africa, the African forms are all terrestrial and burrowing.

BRONZE SKINKS *CHALCIDES*

An extremely variable group of shiny, elongate skinks with cylindrical bodies, and smooth scales, they are terrestrial and most are burrowers, living in holes or pushing through soft soil and sand. The snout is conical, not markedly wedge-shaped; the eyelids are functional, the lower with an undivided transparent disc; the limbs are small but functional and the nostril opening is located within the rostral scale, bordered behind by a small nasal. A Palaearctic genus, *Chalcides* has radiated widely in the northern half of Africa, southern Europe and the Middle East, a single species just reaches northern Kenya.

OCELLATED SKINK *Chalcides bottegi*

IDENTIFICATION: A stocky skink of northern Kenya. The head is broad, with a slight neck and a pointed nose; the body stout and cylindrical. The tail is thick and tapers smoothly, it is roughly 50% of total length. The scales are smooth and the dorsal scales are broader than they are deep, in 22 (very occasionally 24) rows at midbody. Maximum size about 27cm, average 18–25cm, hatchling size unknown. The colour pattern is distinctive: the top of the head is brown; the lips barred black and white. A broad, sharply defined darker brown, mid-dorsal band, about 3.5–4 scales wide, extends from the back of the head to the base of the tail; it is bordered on either side by a light tan stripe and below this a fine darker stripe. Arranged across these dorsal bands are more or less regular transverse rows of highly contrasting dark ocellate spots, with a white centre and dark edges. Underside uniform whitish to light tan in colour.

Chalcides bottegi

HABITAT AND DISTRIBUTION: Dry savanna and semi-desert of north-west Kenya in the vicinity of Lake Turkana; known only from a handful of localities, including South Horr, Lodwar, Kakuma, Lokitaung and Loarengak in northern Turkana District, Kenya, at low altitudes mostly below 1,100m, recorded also from Lokomarinyang, in South Sudan. It is apparently absent from north-east Kenya, although a similar species, *Chalcides ocellatus*, is known from Lugh in Somalia, not far from Mandera.

NATURAL HISTORY: All East African specimens were taken under logs, rocks and other debris in sandy areas during the daylight hours; their habits remain unknown as they have never been observed actively foraging, either at night or in the daytime. When uncovered, the Ocellated Skink thrashes about in a serpent-like manner in attempting to escape; when held, a specimen will push its snout against its captor's hand as if attempting to burrow. These skinks give live birth, details unknown but similar species may have between 2–20 offspring. Captive specimens appear to be territorial, with males fighting in the breeding season.

Chalcides bottegi Ocellated Skink. Eduardo Razzetti, Lake Turkana.

SAVANNA BURROWING SKINKS *SEPSINA*

An African genus of small burrowing skinks with reduced limbs, with 3 or 4 toes on each limb. Little is known of their lifestyle. There are five species, all occurring in the southern half of Africa, with a distribution centred on Angola, one species reaches Tanzania.

FOUR-TOED FOSSORIAL SKINK *Sepsina tetradactyla*

IDENTIFICATION: A small skink with a blunt head, a rounded snout and lidded eyes. The body is cylindrical, scales smooth and shiny, midbody scale rows 24, supraciliaries 4–5. The feet are tiny, with four fingers and five toes. The tail is elongate, fat and easily shed. Total length up to 15cm, of which the tail is usually less than 40%. Body grey, brown, bronze or pinkish above, if not pink then with suffusions of pink around the limb insertions; pink below. The two subspecies can be distinguished by tail colour; the eastern form has a blue-grey spotted tail, the tail of the western form is bronze.

Sepsina tetradactyla

HABITAT AND DISTRIBUTION: Moist savanna at low altitude. The eastern subspecies, Eastern Four-toed Fossorial Skink *S. tetradactyla tetradactyla*, is quite widespread in south-eastern Tanzania, known localities include Kiwengoma forest reserve and the Rondo and Litipo forest reserves, at altitudes below 1,000m, elsewhere south to Malawi and (presumably) northern Mozambique. The western subspecies, the Western Four-toed Fossorial Skink *S. tetradactyla hemptinnei*, is known from Tabora area and Ujiji in the west, between 700–900m altitude, thence west to the south-eastern DR Congo. Conservation Status: probably not threatened by human activities.

NATURAL HISTORY: Almost unknown. Burrowing in leaf litter and loose soil, probably diurnal. One specimen was found in the dry, somewhat sandy soil underneath roots of an enormous tree stump in gallery forest along a river, others were associated with soil around termitaria. One specimen contained four eggs, but it isn't known if they lay the eggs or the eggs develop within the mother. Diet: worker termites, other insects.

Sepsina tetradactyla Four-toed Fossorial Skink. Left: Michele Menegon, southern Tanzania. Right: Michele Menegon, southern Tanzania.

WESTERN BURROWING SKINKS *TYPHLACONTIAS*

A small genus of burrowing skinks, all but one totally legless. Most are striped and live in sand. They have a huge rostral scale, with a groove leading back from the nostril to the rear edge of the scale. Seven species occur, in southern Africa, a new species, the first recorded from East Africa, was described in 2006 from south-western Tanzania.

KATAVI BLIND DART SKINK *Typhlacontias kataviensis*

IDENTIFICATION: A small, striped legless, elongated burrowing skink with a pointed snout and a maximum recorded length not exceeding 12cm. Body cylindrical, the tail is fairly short, about one third of the total length. The eyes lack eyelids and there is no trace of external ear. There is a fine orange, black-edged vertebral stripe, on either side of this a broad yellow-brown stripe, the flanks have fine light and dark stripes. The vertebral stripe commences as a blackish blotch on the head. The tail tip is marbled grey and white. The colour pattern is very similar to that of the Iranian Fossorial Skink *Ophiomorus brevipes*. The underside is white, darker at the centre.

Typhlacontias kataviensis Katavi Blind Dart Skink. Toby Gardiner, Katavi National Park.

Typhlacontias kataviensis

HABITAT AND DISTRIBUTION: Known only from sand ridges bordering the flood plains in Katavi National Park, in the savanna of western Tanzania, at 850m altitude. Conservation Status: Endangered B2ab(iii), as it only exists (as far as we know) in a very small area.

NATURAL HISTORY: Largely unknown. The type specimens were collected in pitfall traps and by digging, so presumably a burrowing species. Other members of the genus have been found under ground cover during the day, or foraging on the surface at dusk or in darkness. The southern African species give live birth. Diet: presumably small arthropods including termites and ants.

FOREST LIMBLESS SKINKS Feylinia

An African genus of small, cylindrical-bodied, blunt-tailed skinks with pointed snouts. Their eyes are primitive, concealed beneath the head skin. They have no legs. They are usually brown or grey. Their scales are smooth and shiny. They inhabit the forest, woodland and moist savanna of central Africa, where they slide through the undergrowth, hunting by day. Their habits are poorly known. Six species are known, one occurs on Príncipe, an island off the central African coast, the others occur in central Africa, one reaches Uganda, western Tanzania and Kenya.

WESTERN FOREST LIMBLESS SKINK Feylinia currori

IDENTIFICATION: A small skink with a cylindrical body, no limbs, external eyes or ears. The head is very short, snout pointed, the tail varies from one-third to one-quarter of the total length, it is broad and does not taper very much, coming to a rounded end. The scales are smooth, in 20–28 rows at midbody. Maximum size about 36cm, average 28–33cm, hatchling size unknown. Dark bluish-grey, appearing black from a distance, the scales have a pale edging. Underside the same hue or slightly lighter than the back, the snout is light grey. Liable to be confused with a blind snake but the combination of absence of limbs, absence of eyes and pointed snout should identify it.

Feylinia currori

HABITAT AND DISTRIBUTION: Woodland and high savanna at altitudes between 700–1,500m, elsewhere in forest, to sea level. Only a few East African records: four from around Lake Victoria; these are Geita, south of Smith Sound and Bukoba, both in Tanzania; Entebbe and the Sesse Islands, in Uganda, also recorded along the Albertine Rift between Lake George and Albert; known localities include Kasese, Queen Elizabeth National Park and Toro game reserve. In Kenya, recently recorded from Kakamega Forest. Elsewhere, west to Cameroon (possibly further), south-west to Angola.

Feylinia currori Western Forest Limbless Skink. Kate Jackson, eastern Congo.

NATURAL HISTORY: Diurnal and terrestrial. Hunts on the ground, sliding rapidly through vegetation; the Kakamega specimen was found on a path in the afternoon. Takes shelter under logs, in leaf litter and other ground cover. Although unable to see, they appear to have good hearing, and will freeze if they hear or feel footsteps Presumably lays eggs but no details known; other members of the genus found with eggs in oviducts. Diet mostly termites, it is speculated that their hearing is so good that they can detect termites by the noise these insects make. Also known to take centipedes. If held they may sham death, or double up and jerk either half of the body when touched. In central Africa there are some curious legends about this skink. One is that they have two heads and if picked up can escape and enter the human body unseen, leaving during the night and causing the victim's death. Another is that if discovered under cover or in a hole, they must not be disturbed, and thus the finder ensures that the skink will never visit them and harm them!

LIMBLESS SKINKS *MELANOSEPS*

An African genus of small, shiny-bodied, burrowing skinks. They have no legs; several species were originally classified in the genus *Herpetosaura*, literally 'creeping lizards'. They are small, usually between 10–20cm. They live mostly in the leaf litter of evergreen forests and they are thought to originate from an ancestor occurring in the montane and lowland forests of central Africa. Little is known of their biology and lifestyle. Recent observation in captivity has confirmed that *M. emmrichi* has an ovoviviparous reproductive mode, as previously suspected for other species in the genus by several authors. They give birth to two to six fully-formed juveniles of relative large size, compared to the adults, suggesting a remarkably high energy investment by females. They are usually found under logs, deep in the leaf litter or by digging in the soft soil of the forest floor. Eight species are known, seven of which occur in Tanzania, five of these are Tanzanian endemics. The generic term *Melanoseps* translates as 'black; rotting'; the seps was a small mythical lizard whose bite caused putrefaction, the black refers to the colour of the original species.

KEY TO THE EAST AFRICAN MEMBERS OF THE GENUS *MELANOSEPS*

1a A small nasal separates the nostril from the first labial. Midbody scale rows 24-28 ......*Melanoseps uzungwensis*, Udzungwa Mountain Limbless Skink [p. 174]

1b Nostril in contact with first labial, midbody scale rows 18-26......2

2a Snout-vent length more than 142mm, midbody scale rows 21-26......3

2b Snout-vent length less than 142mm, midbody scale rows 18-22......4

3a Midbody scale rows 24-26, subcaudals 36-45, Uluguru Mountains. only...... *Melanoseps emmrichi*, Uluguru Limbless Skink [p. 171]

3b Midbody scale rows 22-24, subcaudals 43-61, Southern Highlands of Tanzania, Mulanje Mountain in Malawi and mountains in Mozambique......*Melanoseps ater*, Black Limbless Skink [p. 171]

4a Scales between chin and vent 139-150; adults exceed 95mm in snout-vent length, widespread......*Melanoseps loveridgei*, Loveridge's Limbless Skink [p. 173]

4b Scales between chin and vent 113-141; adults less than 95mm in snout-vent length......5

5a Scales between chin and vent 125-141;

5b	range north-eastern Tanzania......*Melanoseps pygmaeus*, Pigmy Limbless Skink [p. 173] Scales between chin and vent 120. Tail length more than half the snout-vent length......*Melanoseps longicauda*,	5c	Long-tailed Limbless Skink [p. 172] Scales between chin and vent 113–128. Tail length less than half the snout-vent length......*Melanoseps rondoensis*, Rondo Limbless Skink [p. 174]

BLACK LIMBLESS SKINK *Melanoseps ater*

IDENTIFICATION: A relatively large legless skink, midbody scale rows 22–28, maximum length about 21cm (possibly longer). Colour of the dorsum variable, but usually dark grey, sometimes completely black or brown. Individuals with bright red bellies are known from Mt Rungwe, although usually chin to vent is whitish, yellow or pink, sometimes with blackish brown lines.

Melanoseps ater

HABITAT AND DISTRIBUTION: Two subspecies have been described (some doubt their validity), with distributions as follows: *Melanoseps ater ater* from Mt Rungwe, southern Tanzania, midbody scale rows usually less than 22, ventrum usually black or heavily streaked black; and *Melanoseps ater matengoensis* the Matengo Limbless Skink, a doubtfully distinct form from the Matengo Highlands, Tanzania, with midbody scale rows usually 24, ventrum predominantly white. Elsewhere, the nominate subspecies is also known from the Shire Highlands, Malawi. Individuals tentatively assigned to this species were recently found on Mt Mabu and Mt Namuli, in Mozambique. Conservation Status: As far as is known this species is not threatened by human activities.

NATURAL HISTORY: Unknown. Presumably diurnal, living in leaf litter, probably live bearing, eats invertebrates.

Melanoseps ater Black Limbless Skink. Left: Michele Menegon, Mount Rungwe. Right: Michele Menegon, Mt Rungwe.

ULUGURU LIMBLESS SKINK *Melanoseps emmrichi*

IDENTIFICATION: A recently-described, large limbless skink closely related to *Melanoseps ater* with a blunt tail tip. The type specimen has 24 scale rows at midbody, 146–154 between mental and vent and 36–45 subcaudals. The largest known individual is 20.3cm of which 3.5cm of tail. Uniform black or grey, iridescent, above and below, the mental and two scales posterior to it can be white.

HABITAT AND DISTRIBUTION: Known only from the montane evergreen forest between 1,500–1,650m in the northern Uluguru Mountains, possibly more widespread across the entire Uluguru range. Conservation Status: As far as is known this species is not threatened by human activities.

Melanoseps emmrichi

Melanoseps emmrichi Uluguru Limbless Skink. Michele Menegon, Uluguru Mountains.

NATURAL HISTORY: Poorly known. Burrowing and active in soft soil and leaf litter. This species is most active during daylight, but has also shown night activity. They give live birth. Diet probably invertebrates. It occurs in sympatry with *Melanoseps loveridgei*.

LONG-TAILED LIMBLESS SKINK *Melanoseps longicauda*

IDENTIFICATION: A small, problematic limbless skink. It was originally described in 1900 by Gustav Tornier from two Tanzanian specimens. One specimen, from Korogwe, on the Pangani river, is now assigned to a different species, *Melanoseps pygmaeus*. The other type, from 'Maasailand', is lost and, according to Donald Broadley, Tornier's type description, in proportions, scale counts and especially colour pattern more closely matches the skink genus *Typhlacontias*, than *Melanoseps*. The two original types were described by Tornier as 'similar in colouration to the European limbless lizard *Anguis fragilis*'. According to Tornier, the ventral and dorsal scales are black with white borders, the back becomes bright olive along the dorsal midline, with two longitudinal rows of black dots. The tail is longer than body. Largest specimen 11.2cm, 18–19 midbody scale rows, 118–125 scales between mental and vent. The lack of a type specimen and the results of recent investigations pose some doubts on its actual taxonomic position. It is sometimes treated as a subspecies of *M. ater* or, as suggested, may actually be a *Typhlacontias*. Its true identity awaits investigation.

Melanoseps longicauda

HABITAT AND DISTRIBUTION: Endemic to Tanzania, from 'Maasailand', this presumably in north-eastern Tanzania. Conservation Status: No detailed information is available as to the habitat requirements of this species, it is probably overlooked; its habitat, dry woodland, is unlikely to be greatly altered except by extensive and repeated burning and/or overgrazing.

NATURAL HISTORY: Virtually nothing known. Presumably lives in soft soil, diurnal, probably live bearing, eats invertebrates. The specific name refers to the relatively long tail (long = long, cauda = tail) of this species.

LOVERIDGE'S LIMBLESS SKINK Melanoseps loveridgei

IDENTIFICATION: A large limbless skink. The type specimen has 18 scale rows at midbody, 135–150 between mental and vent. Largest individual is 13.6cm of which 1.7cm are of a regenerated tail. Gravid female measures 16cm, of which 4.1cm are tail. Grey above with light speckling. Below, lighter. Specimens from south-western Tanzania may be all black.

HABITAT AND DISTRIBUTION: Moist savanna at low altitude, submontane and montane forest, also in rural areas at relatively high altitude. Known from south-western and southern Tanzania, south from the Uluguru Mountains, localities include Tatanda; Vituri (Ulugurus), Tunduru, Songea, Liwale, Mpwapwa, Mkata and Lindi. Kenyan records tentatively assigned to this species include Kitobo Forest and Mt Kasigau. Elsewhere, found in north-eastern Zambia and possibly northern Mozambique. Conservation Status: A species with a relatively wide distribution and which is unlikely to be threatened.

Melanoseps loveridgei

NATURAL HISTORY: Poorly known. Presumably burrowing, diurnal, probably live bearing, diet invertebrates. This species and the Udzungwa Mountain Limbless Skink *Melanoseps uzungwensis* are sympatric in the Uzungwa Scarp Nature Reserve and at least parapatric at Lugoda Tea Estate, Mufindi. The specific name honours Arthur Loveridge, eminent herpetologist.

Melanoseps loveridgei Loveridge's Limbless Skink. Michele Menegon, southern Tanzania.

PIGMY LIMBLESS SKINK Melanoseps pygmaeus

IDENTIFICATION: A recently-described, dwarfed limbless skink with 18–20 scale rows at midbody like *Melanoseps rondoensis*. It is distinguished from the latter by its higher number of ventral scales (125–141) and subcaudals (44). Known largest individual is less of 112cm, of which 2.5cm is tail. Uniform dark brown above and below, with a white patch on the chin, looks speckled from above.

Melanoseps pygmaeus Pygmy Limbless Skink. Victor Wasonga, Shimba Hills.

Melanoseps pygmaeus

HABITAT AND DISTRIBUTION: Known from coastal and lowland rainforest up to 580m altitude in Tanga region and the vicinity of the Usambara Mountains in Tanzania, also recorded from the Shimba Hills in Kenya. Individuals were collected in pitfall traps or by digging in the leaf litter.

NATURAL HISTORY: Unknown. Presumably burrowing and active in soft soil and leaf litter, diurnal. The holotype contained two eggs, probably live bearing; diet probably invertebrates.

RONDO LIMBLESS SKINK *Melanoseps rondoensis*

IDENTIFICATION: A small limbless skink from southern Tanzania. Midbody scale rows 18–20; there are 114–118 scale rows between mental and vent. Size small, maximum length from snout to vent 93mm. The tail is 17–27% of total length. Colour reddish-brown dorsally, lighter below with rows of dots longitudinally.

HABITAT AND DISTRIBUTION: Moist savanna and woodland at low altitude. Endemic to Tanzania, found in the Rondo Plateau area and Nchingidi, Lindi and Newala. Conservation Status: If restricted to the Rondo Plateau, then this species is potentially vulnerable to forest destruction and habitat alteration and fragmentation.

Melanoseps rondoensis

NATURAL HISTORY: Poorly known. Presumably diurnal, burrows in soft soil or leaf litter, lays eggs, eats invertebrates. A female had developing ova in June, probably live bearing. The specific name refers to the Rondo Plateau, Lindi Region and District, south-eastern Tanzania, the type locality.

UDZUNGWA MOUNTAIN LIMBLESS SKINK
Melanoseps uzungwensis

IDENTIFICATION: A large limbless skink known only from the Udzungwa Mountains. This species has a high midbody scale count of 26–28 and 156–168 ventrals. Size up to 20.2cm total length. The colour above is variable, may be grey or black, or pinkish white; below, with brown lines on each scale row.

HABITAT AND DISTRIBUTION: Evergreen hill forest. A Tanzanian endemic, this species is known only from the forests of the Udzungwa Mountains. It occurs in sympatry with Loveridge's Limbless Skink *Melanoseps loveridgei*, which is smaller (16cm), and has a lower midbody scale count of 18. Both species are known from forest habitat, the latter has been found also on lawns and open areas on a tea estate at the edge of forest. Conservation Status: This species may be threatened by forest clearance for agriculture and other purposes at least in the

Melanoseps uzungwensis

Uzungwa Scarp Nature Reserve, otherwise not threatened within the Udzungwa Mountains National Park.

NATURAL HISTORY: Unknown. Presumably lives in soft soil or leaf litter, individuals were found under logs during the day, diurnal, probably live bearing, eats invertebrates.

Melanoseps uzungwensis Udzungwa Mountain Limbless Skink. Michele Menegon, Udzungwa Mountains.

SLENDER SKINKS *PROSCELOTES*

A genus of small, slim, smooth-bodied burrowing skinks with short, five-toed limbs and small eyes. The scales are smooth and shiny. Three species are known, all occurring within restricted habitats in moist areas of south-eastern Africa. One species occurs in our area; it is endemic to Tanzania.

USAMBARA FIVE-TOED FOSSORIAL SKINK *Proscelotes eggeli*

IDENTIFICATION: A burrowing, Tanzanian endemic skink. Body cylindrical, limbs very short, each bearing five digits. Unregenerated tail slightly over half the total length. There are 22–26 rows of smooth scales at midbody. Largest female about 20.5cm, largest male 20.6cm, hatchlings 7cm. Above, iridescent grey or brown, each scale with a central lighter or darker spot. Below, creamy-yellow on chin and throat, which contrasts sharply with the rest of the under surface, which is a bright salmon pink. Each scale on the throat has a large basal spot. This is absent on those which run down the centre of the body from the forelimb to hindlimbs, but present on the flanks and tail.

Proscelotes eggeli

HABITAT AND DISTRIBUTION: Endemic to the forests of the Usambara Mountains, Tanzania, known localities include Lutindi forest reserve and Kwai, western Usambaras. Conservation Status: Not assessed by IUCN and found in several of the Eastern Arc forests; some of these are protected under Forest Reserve status. The species should be able to survive as long as the existing natural forest cover remains.

Proscelotes eggeli Usambara Five-toed Fossorial Skink. Stephen Spawls, preserved specimen.

NATURAL HISTORY: Poorly known. Lives under the surface, but probably diurnal, found in forest leaf mould, litter and under and in rotten logs. Presumably lays eggs but clutch details unknown. Diet: insects and spiders.

DWARF BURROWING SKINKS *SCELOTES*

An African genus of small burrowing skinks, with smooth-scaled, slender bodies. The snout is pointed, which is a burrowing aid. They live in soft or sandy soil or leaf litter. They show clear evolutionary progress towards the loss of limbs, some species have four, some two and some no limbs. Some lay eggs, some give live birth. The species has radiated widely in southern Africa, with over 20 species occurring there in a wide range of habitats. A single species with four legs is found outside of southern Africa, it is endemic to Tanzania.

ULUGURU FOSSORIAL SKINK *Scelotes uluguruensis*

IDENTIFICATION: A small fossorial skink, the general impression is of an elongate, round-bodied, thin lizard, with reduced limbs, each bearing five digits. The unregenerated tail is slightly longer than the body length. There are 24 midbody scale rows. Largest male about 17cm, female 15.8cm. General appearance is a spotted grey skink with a reddish tail. Grey, reddish or yellow brown dorsally, the head scales have dark edging. The upper labials are black, the others are dusky. Flanks and forelimbs creamy white, each scale with a black spot at the apex, these join on the forearm and the hind limbs making them look black. Translucent white below, the blood vessels visible. The underside of tail is opaque white with a double row of dusky spots laterally. Male and female colour identical.

Scelotes uluguruensis

HABITAT AND DISTRIBUTION: Endemic to the forests of the Udzungwa, Uluguru, Nguru and East Usambara Mountains, Tanzania, localities include Amani nature reserve, Bagilo and Vituri, at altitudes between 800–1,800m. Conservation Status: Not assessed by IUCN; found in Amani Nature Reserve and in other forest reserves; as long as existing natural forest cover persists, should not be threatened.

Scelotes uluguruensis Uluguru Fossorial Skink. David Gower, Uluguru Mountains.

NATURAL HISTORY: Poorly known. Burrowing, found under logs in leaf mould and soil, in forest and at the forest edge in recently-cleared land. Presumably lays eggs but some species give live birth; four ova were discovered in each of two Uluguru females in September. Feeds on insects, isopods, and spiders. The specific name refers to the Uluguru Mountains, the type locality of the species.

SAND SKINKS *SCOLECOSEPS*

A genus of limbless skinks similar to the limbless skinks of the genus *Melanoseps*, but differing in that the nostril is pierced within a very large rostral scale and connected to the posterior of the scale by a straight groove. This is a useful aid to identification. The rostral is bordered by a pair of internal scales, not a single wide scale. Four species are known, two occur in Tanzania, one in Mozambique, a fourth undescribed species has recently been recorded from the Kenyan coast.

KEY TO THE EAST AFRICAN MEMBERS OF THE GENUS *SCOLECOSEPS*

1a	Distinctly striped, on Kenyan coast...... *Scolecoseps sp.*, Gede Sand Skink [p. 178]		
1b	Not striped, on Tanzanian coast......2		
2a	Rostral scale large, one third the length of the head, three supralabial scales beneath the eye.....*Scolecoseps acontias*, Legless Sand Skink [p. 177]		
2b	Rostral scale small, one fifth the length of the head, a single supralabial scale beneath the eye.....*Scolecoseps litipoensis*, Litipo Sand Skink [p. 178]		

LEGLESS SAND SKINK *Scolecoseps acontias*

IDENTIFICATION: A legless skink, known from two Tanzanian specimens alone. The type specimen (now lost) had a large rostral scale, 33% of the head; five supralabial scales, three of which contact the eye, and 18 midbody scale rows. The body was of uniform thickness, the tail was blunt and non-tapering. Length 15cm. Described as grey-brown, with a whitish chin.

HABITAT AND DISTRIBUTION: Coastal plain of Tanzania. Briefly described by Franz Werner in 1913 from a single specimen, now lost, from Dar es Salaam. A single juvenile specimen, collected in October, from Kilwa has been tentatively assigned to this species. Conservation Status: Data Deficient.

Scolecoseps acontias

NATURAL HISTORY: Unknown. Presumably a burrowing species, living in soil or sand.

LITIPO SAND SKINK *Scolecoseps litipoensis*

IDENTIFICATION: A legless skink, known from a single Tanzanian specimen. The type specimen had a large rostral scale, 20% of the head; five supralabial scales, only one of which contacts the eye, two supraciliary and four subocular scales. It has 18 midbody scale rows, 125 ventrals and 50 subcaudals. The body was of uniform thickness, the tail was blunt and non-tapering. Total length 16.8cm, tail 3.3cm. Uniform black with a white patch on the chin.

HABITAT AND DISTRIBUTION: The single specimen was found in a pitfall trap in the Litipo forest, at the base of a steep hillside just east of Lake Lutamba at 180m altitude: Lindi District and Region, Tanzania (10°02'S, 39°29'E). Conservation Status: Data Deficient.

Scolecoseps litipoensis

NATURAL HISTORY: Unknown. Presumably a burrowing species, living in soil or sand.

GEDE SAND SKINK *Scolecoseps sp.*

IDENTIFICATION: A legless skink, from the Kenyan coast. Like other *Scolecoseps*, has the usual large rostral, roughly 25% of the head. It has about 18 midbody scale rows. The body is of uniform thickness, the tail is blunt and tapers abruptly. Total length 16.8cm, tail 3.3cm. Dull pink, each dorsal scale with a large anteriorly-placed brown spot; these spots combine to give the impression the body is finely striped with dotted brown lines. The rostral is unpigmented. The tails of the females are vivid pinkish-white; those of the males are darker and slightly longer.

HABITAT AND DISTRIBUTION: North Kenyan coastal plain. Known from handful of specimens from the north Kenya coast, at Watamu, Gede and the eastern edge of the Arabuko-Sokoke Forest, just above sea level.

Scolecoseps sp.

Scolecoseps sp from Gede Gede Sand Skink. Bio-Ken archive, Gede.

NATURAL HISTORY: Unknown. Presumably a burrowing species, living in soil or sand. A Gede specimen was found in a 2.5m deep pit, another in soil in a Watamu plot, and one in sandy soil in the forest.

Ichnotropis bivittata
Angolan Rough-scaled Lizard
David Lloyd-Jones

SUB-ORDER SAURIA: LIZARDS
FAMILY LACERTIDAE
LACERTIDS, OR 'TYPICAL' LIZARDS

A family of fairly small, slender diurnal lizards, usually with long tails and well-developed limbs, many are striped. About 320 species, in about 40 genera, are known; they are confined to Eurasia and Africa. The East African forms are largely terrestrial and fast moving (some are arboreal), and some are modified for this lifestyle by having fringes or keels on their toes, and often show leg-waving or leg-tucking behaviour, presumably to ensure rapid muscular response. All African species lay eggs, and some are very short lived, hatching, breeding and dying in a single year, so that the entire population exists only as eggs for part of the year. Communal egg deposition is known in the genus *Adolfus*. Twenty-three species, in eleven genera, are known from East Africa.

KEY TO THE EAST AFRICAN GENERA OF *LACERTIDS*

1a	A vertebral series of enlarged scales down the middle of the back......2		
1b	No vertebral series of enlarged scales down the middle of the back, dorsal scales roughly homogenous......3		
2a	Tail strongly depressed and fringed laterally, in forest or woodland......*Holaspis*, gliding lizards [p. 188]		
2b	Tail cylindrical, not depressed or fringed, in dry country......*Philochortus rudolfensis*, Turkana Shield-backed Ground Lizard [p. 191]		
3a	Ventrals keeled......*Gastropholis*, keel-bellied lizards [p. 186]		
3b	Ventrals smooth......4		
4a	Subdigital lamellae smooth or tubercular......5		
4b	Subdigital lamellae keeled......7		
5a	Nostril bordered by two or three nasals only, in savanna......*Nucras*, scrub lizards [p. 192]		
5b	Nostril bordered by two or three nasals and the first labial, or separated from the latter by a narrow rim, largely arboreal and/or in high grassland, woodland or forest......6		

6a	Seven to ten femoral pores......*Congolacerta vauereselli*, Sparse-scaled Forest Lizard [p. 185]	9a	Ventral scales in six longitudinal rows......*Heliobolus*, true sand lizards [p. 197]
6b	10-21 femoral pores......*Adolfus*, forest and alpine-meadow lizards [p. 181]	9b	Ventral scales in 8-10 longitudinal rows......*Pseuderemias*, false sand lizards [p. 200]
7a	Nostril bordered by 3-5 nasals and the first labial, or narrowly separated from the latter......*Latastia*, long-tailed lizards [p. 194]	10a	Frontonasal single, subocular bordering lip, 34-40 midbody scales......*Ichnotropis*, rough-scaled lizards [p. 201]
7b	Nostril bordered by 2-4 nasals, well separated from the first labial......8	10b	Frontonasal longitudinally divided, subocular not reaching lip, 46-58 midbody scales......*Meroles squamulosus*, Mozambique Rough-scaled Lizard [p. 203]
8a	Collar present, head shields smooth, slightly rugose or pitted......9		
8b	Collar absent, head shields keeled or striated......10		

FOREST AND ALPINE-MEADOW LIZARDS
ADOLFUS

A genus of four species, all occur in East Africa, largely at medium to high altitude; they are good climbers on standing and fallen timber, rocky walls, holes and crevices but tend to hunt on the ground.

KEY TO EAST AFRICAN MEMBERS OF THE GENUS *ADOLFUS*

1a	Vertebral scales distinctly larger than those on the flanks; outermost ventral scale rows incomplete and faintly keeled ....*Adolfus africanus*, Multi-scaled Forest Lizard [p. 181]	3b	Vertebral scale row weakly to moderately keeled; rostral and frontonasal scale usually not in contact; maximum number of temporal scales high, 5-12; number of femoral pores on either thigh 10-14 4
1b	Vertebral scales about the same size as the scales on the flanks; outermost ventral scale rows incomplete and smooth......2	4a	More than 50 dorsal scales in a longitudinal row; number of temporal scales 8-12; 5-6 collar scales; 17-20 lamellae under fourth toe; side of the head and flanks rufous or brown; two black-edged dorsolateral stripes......*Adolfus alleni*, Mt Kenya Alpine-meadow Lizard [p. 182]
2a	Temporal scales small and numerous; granular scales beneath the collar present; dorsal scales in 37-48 rows around midbody .... *Adolfus jacksoni*, Jackson's Forest Lizard [p. 182]		
2b	Temporal scales large and few in number; granular scales beneath the collar absent; dorsal scales in 18-24 rows around midbody......3	4b	Less than 50 dorsal scales in a longitudinal row; number of temporal scales 6-9; four collar scales; 15 lamellae under fourth toe; side of the head and flanks black, each scale with a fine white edge or tip; two black-edged dorsolateral stripes absent......*Adolfus masavaensis*, Western Alpine-meadow Lizard, Mt Elgon population [p. 184]
3a	Vertebral and lateral scales keeled; vertebral scales strongly keeled; rostral and frontonasal scale usu¬ally in contact; number of temporal scales low, 3-6; number of femoral pores on either thigh 8-12......*Adolfus masavaensis*, Western Alpine-meadow Lizard, Aberdare population [p. 184]		

MULTI-SCALED FOREST LIZARD Adolfus africanus

IDENTIFICATION: A slender forest lizard with a rather knobbly appearance due to the strongly keeled back scales. The snout is pointed, eye fairly large. Slim, with a long tail, approximately twice the body length. The body scales are rhombic and with strong diagonal keels which converge towards the mid-line of the back; the mid-dorsal scale rows are distinctly larger than on flanks, in 18–24 in transverse row at mid-body. The ventral scales are in six longitudinal rows. There are 12–17 femoral pores under each thigh. The entire top of the head is metallic, copper-bronze in colour. A continuous mid-dorsal band, of the same colour in Albertine Rift specimens, brownish in Kenyan specimens, extends the length of the body and tail, within this band are a series of random black spots; a dark lateral band is bordered below by yellow spots and lines. The tail is brown, with faint lighter bands. The underside is immaculate lime green.

Adolfus africanus

Adolfus africanus Multi-scaled Forest Lizard. Left: Philipp Wagner, Kakamega Forest. Right: Harald Hinkel, Rwanda, underside.

HABITAT AND DISTRIBUTION: A true Guineo-Congolean primary forest dweller, usually associated with clearings where sunlight penetrates within forest, at middle elevations (from 580m to about 1,200m). In our area, known from Nyungwe Forest, western Rwanda, Kakamega Forest, Kenya, Ugandan records include Mabira, Budongo, Kibale, and Mpanga forests, and Bwindi Impenetrable National Park. The type locality is Entebbe, near Kampala, but whether the species still exists there is not known. Also found in the Imatong Mountains of South Sudan; sporadic records elsewhere include the Congo side of the Albertine Rift, north-western Zambia and Cameroon. Might be in the Congo basin but unrecorded. Conservation Status: appears to be a true inhabitant of primary forest, thus vulnerable to forest loss.

NATURAL HISTORY: Diurnal and arboreal, basks in dappled sunlight on fallen tree limbs, trunks and exposed roots within a few metres of the ground in clearings within the forest. Rarely on vertical tree trunks more than 3m above ground suggesting that this species is primarily an inhabitant of undergrowth. It appears to prefer lower rather than higher montane forest. Presumably lays eggs, but no breeding details known. In the Bwindi Impenetrable Forest, it was sympatric with Jackson's Forest Lizard *Adolfus jacksoni* and the Sparse-scaled Forest Lizard *Congolacerta vaueresellii*.

MT KENYA ALPINE-MEADOW LIZARD Adolfus alleni

IDENTIFICATION: A fairly slim, long-tailed lizard from montane moorland. It has a short head and pointed snout. The limbs are short and powerful. The tail is long, and stout at the base, tapering smoothly, two-thirds of the total length. The dorsal scales are noticeably pointed and keeled, in 18–23 rows at midbody. Maximum size about 19cm, average 12–16cm, hatchling size unknown but presumably similar to *Adolfus masavaensis*, i.e. 5–7cm. Ground colour brown or olive, with a broad or fine dark vertebral stripe; there are two black-edged dorsolateral stripes, usually brown or yellow-brown. The flanks may be rufous, pinkish or warm brown, with a diffuse light stripe low on the flanks. Underside orange.

Adolfus alleni

HABITAT AND DISTRIBUTION: A high altitude species, occurring between 2,500–4,500m altitude, usually in montane moorland above the tree line in the heather and *Hagenia-Hypericum* zones. A Kenyan endemic, as presently defined known only from Mt Kenya. Conservation Status: Near Threatened, but as for the previous species, its habitat is within a National Park

NATURAL HISTORY: Diurnal and terrestrial, living in tussock grass and the open patches between; at night gets deep into vegetation or under ground cover to avoid being frozen. Emerges to bask as the morning warms up and then hunts. Fast moving, usually active between about 9.30am and 5pm, but will retire earlier if it clouds over. Presumably lays eggs, but details unknown. Diet: invertebrates.

Adolfus alleni Mt Kenya Alpine-meadow Lizard. Mike McLaren, Mt Kenya.

JACKSON'S FOREST LIZARD Adolfus jacksoni

IDENTIFICATION: A medium-sized, fairly robust forest and woodland lizard. The head is short, snout rounded. Body cylindrical, tail stout and fairly long, two-thirds of total length. The shape varies; those from forests around and near Mt Kenya seem slimmer than those from elsewhere. The dorsal scales are small and granular, smooth or feebly keeled, in 37–48 rows at midbody. Maximum size about 25cm, average 18–24 cm, hatchling size unknown. The colour varies a lot, but usually there is a broad dorsal stripe, which may be olive, dull or vivid green, and a brown, dark brown, red-brown or rufous dorsolateral band, which breaks up into a row (sometimes several rows) of spots low on the flanks. In some specimens this side-stripe is absent; the colour of the spots can vary from white to yellow or vivid blue. The chin and throat are usually white, often spotted black, the belly yellow or light green, becoming orange towards the vent.

HABITAT AND DISTRIBUTION: Riverine woodland, woodland, forest and forest edge, from 450 to over 3,000m altitude. Few records from Tanzania, save the forests around Mts Kilimanjaro and Meru and the extreme north-west. Widely distributed in southern and western Kenya, southern Uganda, western Uganda and northern Burundi, wherever there are mid-altitude forests, woodland, hilltop or riverine forest; notable localities include the Taita and Chyulu Hills, forests around Nairobi, around the Aberdares, Mt Kenya, the Nyambene, Mathews and Cherang'any Range, gallery forest in the Maasai Mara, Mt Elgon, the forest of the western Lake Victoria shore in Uganda and all along the Albertine Rift, including Parc National des Volcans, Nyungwe Forest; elsewhere to the eastern DR Congo.

Adolfus jacksoni

Adolfus jacksoni Jackson's Forest Lizard. Left: Stephen Spawls, Nairobi. Top right: Stephen Spawls, Underside. Bottom: Daniel Hollands, Mt Elgon.

NATURAL HISTORY: Diurnal and largely arboreal, in wooded areas where sunlight penetrates, and when basking they flatten themselves by rib-spreading to a remarkable extent, almost appearing two-dimensional. They will colonise buildings, walls, earth banks and road cuttings, they will live in wooded gardens (e.g. in Karen and Langata, Nairobi) but are not tolerant of agriculture. Individuals were observed daily on the walls of rocky road cuts at 2,356m in Bwindi Impenetrable National Park at temperatures between 20–36°C. In the Parc National des Volcans in Rwanda, these lizards have colonised the buffalo wall surrounding the park; in Nairobi National Park they live on an elevated walkway. At Bwindi, an important predator on this species is the Great Lakes Bush-viper *Atheris nitschei*, many of which were found hidden at the base of the lizard foraging areas; and Jackson's Forest Lizards were found in their stomachs. Three to five eggs are laid; in Bwindi communal nests were found, often in crevices on exposed vertical road cut walls. Seven such sites were found. The largest contained 16 newly-laid eggs and 574 additional older eggs/shells deeper in the crevice. Diet: invertebrates.

WESTERN ALPINE-MEADOW LIZARD   *Adolfus masavaensis*

IDENTIFICATION: A fairly slim, long-tailed lizard from montane moorland. It has a short head and pointed snout. The limbs are short and powerful. The tail is long, and stout at the base, tapering smoothly, two-thirds of the total length. The dorsal scales are noticeably pointed and keeled, in 19–23 rows at midbody. Maximum size about 16cm, average 11–15cm, hatchlings 5–7cm. Ground colour brown or olive, with a broad or fine dark vertebral stripe; there are two black-edged dorsolateral stripes, either lime green, greenish-yellow or red-brown (hard to see in Mt Elgon specimens). The flanks may be rufous, pinkish or warm brown, with a diffuse light stripe low on the side, the underside vivid orange, orange-pink or blue. Some Mt Elgon specimens have extensive dark spotting on the chin and throat. Taxonomic Note: Until recently specimens of this species from the Aberdares and west of the rift valley were called *Adolfus alleni*; that name is now restricted to a Mt Kenya form.

Adolfus masavaensis

HABITAT AND DISTRIBUTION: A high altitude species, occurring between 2,500–4,300m altitude, usually in montane moorland above the tree line in the heather and *Hagenia-Hypericum* zones, although doubtfully reported from the bamboo zone on Mt Kinangop in the Aberdares. An East African endemic, known only from the Aberdares, the Cherang'any Hills and Mt Elgon, might be on the Mau Escarpment. Conservation Status: Near Threatened, but population trend stable. However, most of its habitat lies within National Parks.

Adolfus masavaensis Western Alpine-meadow Lizard. Stephen Spawls, Aberdares.

NATURAL HISTORY: Diurnal and terrestrial, living in tussock grass and the open patches between; at night gets deep into vegetation or under ground cover to avoid being frozen. Emerges to bask as the morning warms up and then hunts. Fast moving, usually active between about 9.30am and 5pm, but will retire earlier if it clouds over. Presumably lays eggs, but details unknown. Diet: invertebrates, known to take beetles and their larvae.

CONGO LIZARDS *CONGOLACERTA*

A genus of two species, erected in 2011 following the discovery of an unusual lacertid in the eastern Congo; molecular analysis indicates that the single East African species shares an ancestor with the new form and was originally wrongly placed in the genus *Adolfus*.

SPARSE-SCALED FOREST LIZARD *Congolacerta vauereselli*

IDENTIFICATION: A medium-sized slender forest lizard. Snout pointed, slim, with a long tail, two-thirds of total length. The dorsal scales are smooth, in 38–50 rows at midbody. Maximum size around 24cm, average 14–20cm, hatchling around 7cm, 5cm of which is tail. There is usually a broad brown, bronze or greenish central dorsal stripe, the flanks are brown, sparsely speckled with two rows of light spots (some specimens unspotted), sometimes a stripe runs from the pale lips back to the forelimb insertion. Underside white, sometimes spotted with black.

Congolacerta vauereselli

HABITAT AND DISTRIBUTION: In clearings in undisturbed forest, at altitudes between 1,000–2,400m. In East Africa, known from forests in northern and western Rwanda and western Uganda, including the Budongo and Kibale Forests, the Rwenzori mountains, the Uganda side of the Virunga volcanoes and Bwindi Impenetrable National Park. Elsewhere, in eastern Congo. The type locality is uncertain, Gustav Tornier described it from 'jungle between Kagera and Congo'; which could be anywhere in north-western Tanzania, Rwanda or the eastern Congo.

NATURAL HISTORY: Little is known; observations in Bwindi Impenetrable National Park, Uganda, indicate that its habits are very similar to *Adolfus africanus*, which it evidently replaces at higher elevations. Hatchlings around 7cm in length were collected in Rwanda in April.

Congolacerta vauereselli Sparse-scaled Forest Lizard. Michele Menegon, Rwanda.

KEEL-BELLIED LIZARDS *GASTROPHOLIS*

A central and East African genus of diurnal, fairly large, slim, secretive, long-tailed lizards. Unusually for a lacertid, the belly scales are distinctly keeled, hence the name. Four species are known, two occur in East Africa, one of which is endemic. A third species, *Gastropholis echinata*, occurs in the Congo very close to the Rwanda border. In the past these lizards have been placed in different genera, in *Lacerta, Bedriaga* and *Tropidopholis*.

KEY TO THE EAST AFRICAN MEMBERS OF THE GENUS *GASTROPHOLIS*

1a Uniform green...... *Gastropholis prasina*, Green Keel-bellied Lizard [p. 186]

1b Striped...... *Gastropholis vittata*, Striped Keel-bellied Lizard [p. 187]

GREEN KEEL-BELLIED LIZARD *Gastropholis prasina*

IDENTIFICATION: A medium-sized, slim, bright-green lizard with a huge prehensile tail. The head is long and narrow, the eye fairly large with a small golden iris. The limbs are long, rear limbs quite stout, digits long and spindly with a hooked claw on each. The tail is very long and smoothly tapering, about 70% of total length. Dorsal scales are smooth, non-overlapping, small and granular, in 28–40 rows at midbody. The ventral scales are keeled, 8–12 rows. Femoral pores 13–15 per side. Maximum size is about 40cm, average 25–35cm, hatchlings 11–12cm. Emerald-green above, yellow-green below, patches of turquoise around the limb-body junction, sometimes fine black speckled lines on the flanks or black speckling on the tail. The tongue is bright red.

Gastropholis prasina

HABITAT AND DISTRIBUTION: Endemic to East Africa, in forests, woodland and thicket of the coastal plain and the Eastern Arc. Few records; known localities include Watamu, Arabuko-Sokoke Forest, Simba Hills, Amani (Usambara Mountains), Tanga, Zaraninge Forest, Bagamoyo, Nguru Mountains; all between sea level and 1,200m. Conservation Status: Probably under threat, as its already small coastal forest habitat is rapidly disappearing, but it appears to be adaptable as it is known from cashew nut plantations near Watamu.

NATURAL HISTORY: Diurnal, arboreal and secretive, living in holes in trees, one such refuge was 12m above ground level. They move about the branches using their prehensile tails as balancing organs, observed to sleep in the branches supported by the tail. Habits poorly known but agonistic behaviour between captive males has been noted, resulting in a tail being bitten off. Mating behaviour included a male biting the female's neck, entwining their tails, the male then encircled the female's pelvic region and appeared to lick her vent. Clutches of five eggs, roughly 1.5cm long, were laid in captivity in September and October at Watamu. Clutches also found in moist tree holes. One clutch took 61 days to hatch, at a temperature of 26–29°C. Diet invertebrates, in captivity known to attack and eat smaller lizards.

Family Lacertidae: Lacertids, or 'typical' lizards

Gastropholis prasina Green Keel-bellied Lizard. Inset: Stephen Spawls, Arabuko-Sokoke Forest. Main: Michele Menegon, coastal Tanzania.

STRIPED KEEL-BELLIED LIZARD *Gastropholis vittata*

IDENTIFICATION: A medium-sized, slim, striped lizard with a huge prehensile tail. The head is long and narrow, the eye fairly large. The digits are long and spindly with a hooked claw on each. The tail is very long and smoothly tapering, about 70% of total length. Dorsal scales are smooth, non-overlapping, small and granular, but becoming keeled low on the flanks and the ventral scales are keeled. Femoral pores present. Maximum size about 35cm, average 20–30 cm. A maroon or brown central dorsal stripe is bordered by two cream or yellow, black-edged stripes. These break up on the tail, which is shades of brown or grey with paired white or yellow dots. In some specimens the nose and front feet are quite yellow. White or cream below. The tongue is bright red.

Gastropholis vittata

HABITAT AND DISTRIBUTION: Forests, woodland, riverine woodland and thicket of the coastal plain south of Mombasa into Tanzania. Few records; known localities include Diani Beach, Shimba Hills, Tanga, Dar es Salaam, Bagamoyo, Liwale and Kilosa; at altitudes between sea level and 600m. Probably occurs all the way along the Tanzanian coast, but no records between Dar es Salaam and Liwale. Elsewhere, known from northern Mozambique.

NATURAL HISTORY: Diurnal and apparently terrestrial, despite its prehensile tail and similarity to the Green Keel-bellied Lizard *Gastropholis prasina*, which is arboreal. Specimens have been collected in pitfall traps and in houses. Male to male combat observed in

Gastropholis vittata Striped Keel-bellied Lizard. Stephen Spawls, Usambara Mountains.

this species in Bagamoyo in November, on the ground. Diet: invertebrates, a Bagamoyo specimen had ants and a solifugid in its stomach. Presumably lays eggs, as other members of the genus do, a juvenile was collected in Liwale in July.

GLIDING LIZARDS *HOLASPIS*

A genus containing two unique, beautiful lacertids, with serrated, blue-spotted tails, astonishingly well-adapted for gliding, the herpetologist Nicholas Arnold of the London Natural History Museum regards this genus as the most highly specialised lacertid lizard. The first joint of their third and fourth fingers are fused, it is suggested this reduces the forces generated upon landing after a glide. Originally classed as a single species with two subspecies, both are now largely treated as full species (but on the basis of informal opinion); one form occurs in west and central Africa, the other in south-central Africa; both are found in our area.

KEY TO THE EAST AFRICAN MEMBERS OF THE GENUS *HOLASPIS*

1a Pale paravertebral stripes confined to the paired rows of enlarged paravertebral scales, in Uganda or extreme north-western Tanzania...... *Holaspis guentheri*, Western Blue-tailed Gliding Lizard [p. 188]
1b Pale paravertebral stripes not confined to the paired rows of enlarged paravertebral scales, extending onto (or even restricted to) adjacent rows of small scales, in coastal Kenya and eastern and southern Tanzania...... *Holaspis laevis*, Eastern Blue-tailed Gliding Lizard [p. 189]

WESTERN BLUE-TAILED GLIDING LIZARD *Holaspis guentheri*

IDENTIFICATION: A small, light-bodied lacertid. The head is long, with a pointed snout. The body and tail are extremely flattened. The undamaged tail is about 55–63% of total length; its wide pointed scales are used as a climbing aid. Midbody scale count 58–84, of which six are ventrals. A single preanal plate is present. 16–24 femoral pores are present on each thigh. Maximum size about 13cm, average 9–11cm, hatchlings are about 2–3cm long. It has a black back with cream stripes along the side of the body, becoming blue posteriorly, and the upper surface of the tail is black with a central row of bright blue spots and yellow lateral scales. The limbs are grey. Below the throat, chest, limbs are cream, the belly is yellow or orange in males, yellow or orange-grey in females, the underside of the tail is black.

HABITAT AND DISTRIBUTION: A mid-altitude forest species. Only two records from our area, one Uganda (Budongo Forest), one Tanzania, Bukoba. Elsewhere, west to Sierra Leone. Conservation Status: probably under no immediate threat, but it does not tolerate deforestation, due to its lifestyle.

Holaspis guentheri

NATURAL HISTORY: Diurnal and largely arboreal, ascending to great heights on tree trunks; Nigerian specimens were seen over 30m up tree trunks, but will descend to ground level, and has been taken in pitfall traps. Remarkably fast moving and agile, it leaps from branch to branch. It can also glide between trees, on a downward trajectory of 45°, by means of some unusual adaptations; it is more flattened than any other lacertid, and this flattening includes the limbs and tail, it has a large surface area/weight ratio, the vertebrae have no neural crests and the free ribs and transverse movement of the scales means it can increase its surface area by 50%; it is arguably the most highly-adapted lacertid. Difficult to see in its forest environment, but can be spotted on lighter-coloured bark as it basks and forages among cracks and crevices in the bark for small arthropods. When basking, it flattens itself to an astonishing degree. Hides under loose bark. Lays two eggs, roughly 0.6×1.2cm, under loose bark or in leaf litter. specimens from north-eastern DR Congo laid eggs in June; they hatched in 3–4 months. Feeds on invertebrates such as insects and spiders.

Holaspis guentheri Western Blue-tailed Gliding Lizard. Eli Greenbaum, eastern Congo.

EASTERN BLUE-TAILED GLIDING LIZARD *Holaspis laevis*

IDENTIFICATION: A small, light-bodied lacertid. Similar to the previous species, but differing as follows: the midbody scale count is 72–96, it has 17–25 femoral pores, maximum size about 13cm. Differs slightly in colour to the western species; in Kenyan and Tanzanian specimens the upper hindlimbs are green with black speckling, Kenyan specimens have a central black stripe, one single broad black dorsolateral stripe and one narrow black lateral stripe, so a total of five black dorsal stripes, Tanzanian ones have seven black stripes. Juveniles are less brightly coloured than the adults, and their underside is black.

HABITAT AND DISTRIBUTION: Largely a closed lowland forest species, but may be found in more open woodland adjacent to forest, in Miombo and coastal forests, and also recorded in exotic tree plantations such as *Eucalyptus*. Occurs at altitudes between sea level and 1,200m. Recently recorded in Kenya, at Kaya Jibana, near Kaloleni, just north of Mombasa, and the south coast forests of Buda, Gongoni, and Longo (this last is within the Shimba Hills National Park); quite widespread in north-eastern, eastern and southern Tanzania in forest, Tanzanian localities include the Usambara, Uluguru and Udzungwa Mountains, Zaraninge Forest Reserve, Liwale and Tunduru. Elsewhere, south to central Mozambique, west to southern Malawi. Conservation Status: Not assessed by IUCN; it is widespread, probably under no immediate threat, does not tolerate deforestation, due to its lifestyle, but in parts of its range copes with exotic plantations.

Holaspis laevis

Holaspis laevis *Eastern Blue-tailed Gliding Lizard. Top: Michele Menegon, northern Mozambique. Bottom: Michele Menegon, Nguru Mountains.*

NATURAL HISTORY: As for the previous species; a diurnal and arboreal species, but will descend to ground level, taken in pitfall traps; one entered a tent. Active on tree trunks and able to glide to escape. Captive specimens from Tanzania laid eggs in November, December, January and April, and are described as laying clutches every 4–6 weeks; eggs incubated at 30°C hatched in 55 days. In southern Tanzania, the herpetologist C.J.P. Ionides noted that villagers often had no idea that this lizard was present in their area, due to its secretive behaviour; likewise no Kenyan specimens were known until very recently.

SHIELD-BACKED GROUND LIZARDS
PHILOCHORTUS

A genus of seven species of slim, fast-moving lacertids; most are striped, with very long tails. In the centre of the back they have 2–6 longitudinal rows of enlarged scales. Their distribution centres on north-eastern Africa; one form reaches Libya and Egypt, another occurs on the Arabian Peninsula. One species occurs in our area. We have chosen to treat it as a full species, *Philochortus rudolfensis*, although some authorities believe it is a subspecies of another, rather different form, *Philochortus intermedius*.

TURKANA SHIELD-BACKED GROUND LIZARD
Philochortus rudolfensis

IDENTIFICATION: A long-tailed, slender lacertid. The head is short but the snout looks sharp from above. Body thin, tail very long, 65–75% of total length. The body scales are smooth and granular, or feebly keeled posteriorly, in 30–32 rows; the two mid-dorsal rows of scales are noticeably enlarged. There are 11–18 femoral pores. Maximum size 18-20cm, average size 15–18cm. The top of the head is uniform yellow-grey, the body distinctly striped with five brown and six yellow stripes, which coalesce beyond the tail base; the final 75% of the tail is vivid orange. The underside is immaculate white.

Philochortus rudolfensis

HABITAT AND DISTRIBUTION: Semi-desert and dry savannah of northern Kenya; known from a handful of very widely scattered localities in sparsely vegetated and in some cases rocky localities at low altitude, less than 800m; the type was from Lokwakangole, on the west shore of Lake Turkana, subsequent specimens recorded from 'The Steps' above Loyengalani, Laisamis, Wajir Bor, Mandera and Hamiye, on the

Philochortus rudolfensis Turkana Shield-backed Ground Lizard. Mike Roberts, The Steps, Loyengalani.

Tana River. It is obviously very widespread but secretive or rare. Probably occurs in South Sudan, Southern Somalia and Ethiopia.

NATURAL HISTORY: Poorly known. Diurnal and terrestrial. *Philochortus* means 'grass loving'. Presumably lays eggs and feeds on invertebrates.

SCRUB LIZARDS *NUCRAS*

A genus of rather blunt-snouted, terrestrial lacertids with fairly long tails and a distinct collar. Most are attractively striped or barred in vivid contrasting colours. Some ten species are known, largely in southern Africa, but two species reach our area.

KEY TO THE EAST AFRICAN MEMBERS OF THE GENUS *NUCRAS*

1a	No granules between supraoculars and supraciliaries, lamellae beneath fourth toe 16–24.....*Nucras boulengeri*, Boulenger's Scrub Lizard [p. 192]	1b	A series of 2–8 granules between supraoculars and supraciliaries, lamellae beneath fourth toe 20–33.....*Nucras ornata*, Ornate Scrub Lizard [p. 193]

BOULENGER'S SCRUB LIZARD *Nucras boulengeri*

IDENTIFICATION: A medium-sized, blunt-snouted lizard, with a short head. Fairly slim, with a long tail, 60–70% of total length. There are 10–12 femoral pores on each thigh. The dorsal scales are smooth, small and juxtaposed, sometimes there is a scattering of slightly larger scales along the spine. Midbody scale count 44–53. Maximum size about 21cm, average 15–18cm. Tremendously variable in colour and the hatchlings and juveniles are different to the adults. Most adults are some shade of brown, with a fine light vertebral stripe, a light dorsolateral stripe and two light lateral stripes; any of these three may be a series of dots or dashes rather than a continuous stripe. The stripes break up on the tail, which is brown. Immaculate white below. Juveniles and hatchlings often have a red tail. The flanks may have broken dark vertical bars or spots, and in some specimens from central and western Tanzania the vertebral and dorsolateral stripes are absent or exist only as scattered spots. Occasional brown, unstriped, black-spotted adults have been recorded in western Kenya.

Nucras boulengeri

HABITAT AND DISTRIBUTION: Wide-ranging, from low to high savannah and grassland, from sea level to over 2,300m, but not in very dry country, where that niche is held by lizards like Speke's Sand Lizard *Heliobolus spekii* and Southern Long-tailed Lizard *Latastia longicaudata*. Widespread in central and southern Kenya, and north-central Tanzania; plus a few sporadic records from southern Tanzania. Elsewhere, to north-eastern Zambia.

NATURAL HISTORY: Diurnal and terrestrial, hunting in open areas, basking in the sun,

Nucras boulengeri Bloulenger's Scrub Lizard. Top: Stephen Spawls, Lake Manyara. Juvenile, Maasai Mara.

often seen on roadsides or degraded open country in southern Kenya. It has been speculated that they may show crepuscular (twilight) activity. Hides in holes, which they may excavate themselves, and under ground cover. Although breeding details for this species are unstudied, other species of this genus lay between 1–14 eggs. Diet: invertebrates; other members of the genus feed largely on termites, grasshoppers and beetles, but the diet included centipedes, spiders and lizard tails; some desert species ate scorpions

ORNATE SCRUB LIZARD *Nucras ornata*

IDENTIFICATION: A medium-sized, blunt-snouted lizard, with a short head; males have a bigger head than females. Fairly slim, with a long tail, 60–70% of total length. There are 11–16 femoral pores on each thigh. The dorsal scales are very small and juxtaposed. Midbody scale count 40–60. Maximum size about 26cm, average 17–24cm. Black or dark brown above, with three fine yellow dorsal stripes, one of which is in the centre of the back; the flanks are heavily speckled yellow, which may form partial crossbars. The underside is white. In juveniles the tail, from the hind limbs, may be red, and the lateral marking restricted to a few yellow dots.

Nucras ornata

HABITAT AND DISTRIBUTION: In our area, known only from Miombo woodland of the Rondo Plateau in extreme south-east Tanzania. Elsewhere, in savanna and grassland to Mozambique, Malawi and Zambia, south to South Africa.

Nucras ornata Ornate Scrub Lizard. Bill Branch, South Africa.

NATURAL HISTORY: Diurnal and terrestrial, hunting in open areas, basking in the sun. Hides in holes and under ground cover. In southern Africa, they appear to remain hidden during the dry season and emerge following rains and the appearance of termites, but habits in Tanzania unknown. They lay eggs, clutch sizes in southern Africa ranged from 4–13 eggs (usually 5–7); their diet there largely consisted of termites, grasshoppers and spiders.

LONG-TAILED SAND LIZARDS *LATASTIA*

An African genus of long-tailed, fast-moving lizards, mostly striped, inhabiting dry savanna and semi-desert. Ten species are known, most in north-east Africa, although two are West African. Three species occur in our area.

KEY TO THE EAST AFRICAN MEMBERS OF THE GENUS *LATASTIA*

1a Dorsal scales usually keeled, in 39–52 midbody rows, only in Tanzania...... *Latastia johnstoni*, Johnston's Long-tailed Lizard [p. 195]
1b Dorsal scales not keeled, in 52–80 midbody rows, in central or northern Tanzania or Kenya......2
2a Centre of the back with distinct discrete blue, turquoise or green spots on a red or rufous background, no obvious mid-dorsal stripes, 6–11 femoral pores on each thigh......*Latastia caeruleopunctata*, Parker's Long-tailed Lizard [p. 194]
2b No discrete blue or green dots in the mid-dorsal area, but distinct, if sometimes diffuse, mid-dorsal stripes present, 7–14 femoral pores on each thigh......*Latastia longicaudata*, Southern Long-tailed Lizard [p. 196]

PARKER'S LONG-TAILED LIZARD/BLUE-SPOTTED LONG-TAILED LIZARD *Latastia caeruleopunctata*

IDENTIFICATION: A beautiful, relatively large, slim, long-tailed lizard from far northern Kenya. The body is slightly depressed, limbs short, toes long and thin. Tail long, 65–75% of total length. The dorsal scales are strongly keeled, in 55–70 rows at midbody. There are 6–11 femoral pores. Maximum size about 33cm, average 20–30cm, hatchling size unknown. Dorsal ground colour usually brick red, sometimes yellowish brown, the back and flanks covered with discrete green, turquoise or blue spots, these are usually more dense on the flanks. Often between the flank spots are fine black vertical bars. The tail is largely green. Uniform white below.

HABITAT AND DISTRIBUTION: Low altitude dry savanna and semi-desert of the Horn of Africa. In Kenya, only recorded from Mandera and east of Moyale, and just across the border at Murri in Ethiopia; a tentative sight record from near Habaswein, so possibly more widespread in north-east Kenya, but secretive.

NATURAL HISTORY: Poorly known. Diurnal and terrestrial, reported to inhabit both open country and dense acacia scrub; when alarmed it seeks cover in holes among tree roots. In northern Somalia specimens were seen to inhabit burrows. Presumably lays eggs. Diet: invertebrates.

Latastia caeruleopunctata

Latastia caerulopunctata Blue-spotted Long-tailed Lizard. Tomas Mazuch, Somalia.

JOHNSTON'S/MALAWI LONG-TAILED LIZARD
Latastia johnstoni

IDENTIFICATION: A slim, long-tailed lizard from Tanzania. The body is slightly depressed, limbs short, toes long and thin. Tail long, 65–75% of total length. The dorsal scales are strongly keeled, in 39–52 rows at midbody. There are 11–16 preanal pores. Maximum size about 21cm, average 15–20cm, hatchling size unknown. Dorsal ground colour usually some shade of brown, olive or grey, with a fine dark vertebral line, and three lighter stripes on each side of this; between the fine stripes there may be dark vertical spots; lighter below. Tail brown or rufous. The snout, chin and underside may be yellow, often there are yellow and green ocelli on the anterior flanks. White or yellow below; juveniles have a red tail and the red wash extends to the hindlimbs and the posterior dorsal surface; this retreats at adulthood.

Latastia johnstoni

HABITAT AND DISTRIBUTION: Widespread in the savannas of central Tanzania, from the southern shores of Lake Victoria down to Lake Rukwa, Usangu Plains and Ruaha, and also in the south-eastern corner between Kilwa, Newala and Liwale; at altitudes between sea level and about 1,200m, although there seem to be no records from the Rufiji drainage. An isolated record from Bukoba. Elsewhere, to northern Zambia, Shaba in the Congo and Malawi.

Latastia johnstonii Johnston's Long-tailed Lizard. Left: Ignas Safari, Usangu Plains. Right: Bill Branch, juvenile, Mozambique.

NATURAL HISTORY: Diurnal, terrestrial and fast moving, active on open ground, dry river beds and around rock outcrops. Shelters in holes. A small clutch of 3–4 eggs, measuring 12×7mm, is laid, in July or August in southern Tanzania. Diet: invertebrates. One was found in the stomach of a Bibron's Burrowing Asp *Atractaspis bibronii*.

SOUTHERN LONG-TAILED LIZARD *Latastia longicaudata*

IDENTIFICATION: A relatively big, long-tailed lacertid, very fast moving. Snout pointed, body slim but muscular, with stocky rear limbs. The tail is very long, from 60–75% of the total length. There are 7–14 femoral pores. The body scales are small, smooth (sometimes faintly keeled) and granular, and may be slightly shiny, in 52–80 rows at midbody. The ventrals scales are rectangular, in six rows (sometimes eight) and curiously broken up between the forelimbs to form an irregular cluster. Maximum size about 40cm, average 25–35cm, hatchling size unknown. Very variable in colour, but the general pattern is of fine, irregular dark crossbars, interrupted by seven fine or broad longitudinal lines; immaculate white below. Specimens from central Tanzania are often vivid rufous or orange, and the lateral lines consist of blue or green ocelli; southern Kenyan specimens may be dull, yellowy grey; those from extreme north-west Kenya and Moroto in Uganda had faded dorsal markings and looked uniform sandy-grey or dull rufous. Hatchlings have been described as 'reddish', but are usually distinctly striped black and yellow, with a broad brown dorsal stripe, a lateral stripe of yellow spots, and the rear half of the tail is blue.

Latastia longicaudata

Latastia longicaudata Southern Long-tailed Lizard. Left: Stephen Spawls, Dodoma. Inset: Stephen Spawls, Olorgesaile. Right: Robert Sindaco, Lake Turkana.

HABITAT AND DISTRIBUTION: Coastal thicket, moist and dry savanna and semi-desert, from sea level to about 1,300m (possibly higher, there is a curious record from Subukia). Occurs from west of Iringa and the Ruaha National Park, north in a narrow strip to south of Lake Eyasi, then spreads north-east and north-west into Kenya. Widespread in east and northern Kenya and a strip along the southern border. So far Moroto is the only Uganda record, but may well occur more widely in the east and north-east. Elsewhere, west to Senegal across the Sahel, north to Ethiopia and the Yemen.

NATURAL HISTORY: Diurnal, terrestrial and very fast moving, probably the fastest East African lizard, and it is alert, wary and hard to approach. Active during the hottest hours; it waits under bushes, dashing out onto open ground to seize prey, but will venture into thickets and cover, even foraging in leaf litter. Shelters in holes. Clutches of 3–12 eggs laid by captive specimens (Kenyan specimens usually 3–5) and these take from 8–12 weeks to hatch at temperatures between 24–28°C. The eggs hatch to coincide with the start of the rainy season. The presence of both an old corpus luteum and fresh yolk deposition in a Latakwen specimen indicates multiple clutching in a single year. Diet: invertebrates, including scorpions, a Tanzanian specimen ate a Speke's Sand Lizard *Heliobolus spekii*. One was found in the stomach of a Link-marked Sand Snake *Psammophis biseriatus*.

TRUE SAND LIZARDS *HELIOBOLUS*

An African genus of four small terrestrial lacertids, living largely in dry savanna and semi-desert; one species occurs in southern Africa, the other three are found in our area.

KEY TO THE EAST AFRICAN SPECIES OF *HELIOBOLUS*
NB: The colour aspects of this key will not work with juveniles, all of which are striped black and yellow, but locality may give clues.

1a Longitudinally striped, with broken black crossbars, frontal not in contact with the supraoculars, head shields flat and strongly but finely striated, widespread......*Heliobolus spekii*, Speke's Sand Lizard [p. 199]
1b Longitudinally striped, without broken black crossbars, frontal in contact with the supraoculars, head shields smooth, not striated 2

2a Flanks of adult with distinctive large green or yellow ocelli, midbody scale rows 52–64, in northern Uganda......*Heliobolus nitidus*, Glittering Sand Lizard [p. 198]
2b Flanks of adult without ocelli, midbody scale rows 40–42, rare, not in Uganda......*Heliobolus neumanni*, Neumann's Sand Lizard [p. 197]

NEUMANN'S SAND LIZARD *Heliobolus neumanni*

IDENTIFICATION: A small, rare lacertid. Snout pointed, body slim, tail relatively long, 60–70% of total length. Body scales are small, strongly keeled but not pointed, not overlapping and arranged in 40–42 transverse rows at midbody. Maximum size 14.5cm, average 9–14cm, hatchling size unknown. Two fine yellow longitudinal lines in the centre of the back, between them brown, and on each side two rows of yellow dots or elongate dashes, on a dark brown or black background. There is usually a largish pale spot on the flanks just above the insertion of the forelimbs. The posterior body and tail are suffused by a red wash; the underside is white.

HABITAT AND DISTRIBUTION: Known only from a handful of localities between sea level and 900m altitude, in savanna; in Kenya only known from the Tana delta, in Tanzania known from Dar es Salaam, the Pugu Forest (just outside Dar es Salaam) and from Kafukola, Lake Rukwa; the only other records of this species are from the southern rift valley in Ethiopia.

Heliobolus neumanni

Heliobolus neumanni Neumann's Sand Lizard. Kim Howelli, Pugu Forest.

NATURAL HISTORY: Virtually nothing is known of this oddly-distributed species. Presumably similar to other members of the genus, i.e. diurnal and terrestrial. The Pugu Forest specimen was taken on white kaolin sand.

GLITTERING SAND LIZARD *Heliobolus nitidus*

IDENTIFICATION: A beautiful lacertid with flank ocelli. Snout pointed, body slim, tail relatively long, 65–75% of total length. Body scales are small, strongly keeled but not pointed, not overlapping and arranged in 52–64 transverse rows at midbody. Maximum size 23cm, average 14–22cm, hatchling size unknown. Adults have four yellow dorsal stripes on a dark background, the central two formed by two coalescing stripes on the neck, the flanks distinctly marked with vivid green and yellow ocelli, interspersed with black and white speckling, the tail is rufous, with a diffuse dark lateral line. Yellowish-white below. Juveniles have a red or orange tail, the ocelli are small and poorly defined, and there is a broad brown mid-dorsal stripe.

Heliobolus nitidus

HABITAT AND DISTRIBUTION: Moist and dry savannah. Only recently recorded from our area, at Murchison Falls in northern Uganda, but known from Garamba in northern Congo, westwards through the Guinea and Sudan savanna to Guinea and Mali.

NATURAL HISTORY: Diurnal and terrestrial, fast-moving and wary, it is a difficult species to observe. Only a single specimen was collected in two years in northern Ghana. Shelters in holes; lays eggs, eats small arthropods.

Heliobolus nitidus Glittering Sand Lizard. Left: Davide dal Lago, Murchison Falls. Right: Stephen Spawls, Northern Ghana, juvenile.

SPEKE'S SAND LIZARD *Heliobolus spekii*

IDENTIFICATION: A small, relatively common dry country lacertid with a pointed snout. It is slender, the tail is relatively long, 65–70% of total length. The scales at midbody are small, rhombic and strongly keeled, usually arranged in 63–71 transverse rows; the ventrals are in six straight rows. Maximum size about 18cm, average 10–16cm, hatchlings about 6–7cm. The back and flanks are usually rufous brown with diffuse fine black crossbars, these are interrupted by five or six fine light yellow longitudinal stripes, in southern Kenyan and Tanzanian specimens there may be only three or four stripes, the lowest stripe on each flank breaking up into yellow or white spots. White below. The juveniles have no barring; they are black, with fine yellow stripes and a red or orange tail.

Heliobolus spekii

HABITAT AND DISTRIBUTION: Coastal thicket and light woodland, moist and dry savanna and semi-desert, from sea level to about 1,500m altitude, although most common in dry savanna at low altitude, below 1,000m. Occurs from west of Iringa, northwards through north-central and north-east Tanzania, most of eastern and northern Kenya and the coast, a handful of records from eastern Uganda, but also from Nimule, on the South Sudan/Uganda border, so probably in north-central Uganda. Elsewhere, to South Sudan, eastern Ethiopia and Somalia.

NATURAL HISTORY: Terrestrial and diurnal, shuttling between shade and sun to avoid lethal high temperatures; often common in suitable habitat, where it can be seen waiting in ambush in the shade of a bush. It is fast moving, and hard to catch. Shelters in holes and occasionally under ground cover. *H. spekii* is an egg-layer, probably laying 2–6 eggs per clutch. Diet: invertebrates; and preyed upon heavily by snakes, and small raptors.

Heliobolus spekii Speke's Sand Lizard. Stephen Spawls, Southern Ethiopia. Inset: Stephen Spawls, Hatchling.

FALSE SAND LIZARDS PSEUDEREMIAS

A genus of small active terrestrial lacertids, similar to *Heliobolus*, seven species are known, all found in north-east Africa, three confined to Somalia. Two occur in our area.

KEY TO EAST AFRICAN MEMBERS OF THE GENUS *PSEUDEREMIAS*

1a	Upper head scales strongly striated, subocular scale usually entering the lip, 13–17 femoral pores......*Pseuderemias striata*, Peters' Sand Lizard [p. 200]	1b	Upper head scales pitted but not striated, subocular scale excluded from the lip, 17–22 femoral pores......*Pseuderemias smithi*, Smith's Sand Lizard [p. 201]

PETERS' SAND LIZARD *Pseuderemias striata*

IDENTIFICATION: A small lacertid with a pointed snout and slightly bulbous nostrils, giving the snout a crocodile-like appearance. Body slim, tail long, 65–75% of total length. The body scales are small, juxtaposed, keeled and arranged in 53–67 rows around midbody. Maximum length about 160mm. Adults were described as 'cream to pale buff above with 7 brown or black streaks as wide as or wider than the spaces between them; the lower parts are white'. The juvenile coloration includes 4 white streaks separated by black; the belly is black or blackish. A northern Somali subspecies is striped olive and brown, tail brown, whitish below, and the Somali juveniles are conspicuously striped black and yellow, white below.

Pseuderemias striata

HABITAT AND DISTRIBUTION: In our area, known only from two specimens collected in 1892 by William Astor Chanler and Lieutenant Ludwig von Höhnel on 'Tana River, coast to Hamiye'; Hamiye is roughly on the equator, at 39°E, so might occur anywhere along the lower Tana; elsewhere known from eastern Ethiopia and Somalia, either in the riverine forest or surrounding dry savanna.

NATURAL HISTORY: Poorly known. Diurnal and terrestrial. Shelters in holes in Somalia.

Pseuderemias striata Peters' Sand Lizard. Left: Tomas Mazuch, Somaliland. Right: Tomas Mazuch, Somaliland.

SMITH'S SAND LIZARD Pseuderemias smithii

IDENTIFICATION: A small lacertid with a pointed snout and slightly bulbous nostrils, giving the snout a crocodile-like appearance. Body slim, tail long, 65–75% of total length. The body scales are smooth, granular and juxtaposed in 68–82 rows across the middle of the body; ventral scales are usually in eight, rarely ten, straight, longitudinal and 26–30 transverse series. Has six light longitudinal stripes, the two in the centre of the back are close. Ground colour rich rufous brown to dark brown or grey. Back unmarked between the central dorsal lines, longitudinal lines of heavy pale spots between the other stripes. White below. The juvenile colour is described as being darker than the adults.

Pseuderemias smithii

HABITAT AND DISTRIBUTION: Dry savanna and semi-desert of northern Kenya, between 200–900m altitude. It occurs in a broad crescent from North Horr down to Samburu national reserve and the Kalama Conservancy, then east and north to Wajir and Wajir Bor, and also recorded from Mandera. An imprecise record from 'Hamiye, Tana River'. Known localities also include Garba Tula, Archer's Post, Lerata, Ilaut, Serolipi and the Koroli Desert. Elsewhere, occurs in Somalia and southern Ethiopia.

Pseuderemias smithii Smith's Sand Lizard. Left: Robert Sindaco, Lake Turkana. Right: Tomas Mazuch, Archers Post.

NATURAL HISTORY: Diurnal and terrestrial, very fast moving and alert. Little is known of the natural history of this species; it is somewhat smaller than Speke's Sand Lizard *Heliobolus spekii* but very similar to it in habits. There have been no examinations of stomach contents, but the extremely narrow, pointed snout may indicate something significant about its feeding habits. Presumably lays eggs but no clutch details known.

ROUGH-SCALED LIZARDS *ICHNOTROPIS*

A southern African genus of terrestrial lacertids, There are seven species distributed in various savanna and Miombo woodland habitats in south and central Africa; two species occur in our area. The Mozambique Rough-scaled lizard, originally in this genus, *Ichnotropis squamulosa*, was recently moved to another genus, *Meroles*, as a result of molecular analysis.

KEY TO EAST AFRICAN MEMBERS OF THE GENUS *ICHNOTROPIS*

1a Dorsal head shields strongly keeled, prefrontal separated from the supraciliaries by small scales......*Ichnotropis bivittata*, Angolan Rough-scaled Lizard [p. 202]

1b Dorsal head shields feebly ridged, prefrontal in contact with the supraciliaries ......*Ichnotropis tanganicana*, Tanzanian Rough-scaled Lizard [p. 203]

ANGOLAN ROUGH-SCALED LIZARD *Ichnotropis bivittata*

IDENTIFICATION: A medium-sized lacertid with a pointed snout. The body is slim, the tail long, 65–70% of total length. The dorsal and lateral scales are enlarged, pointed, overlapping and strongly keeled in 34–40 rows at midbody. Maximum size about 24cm, average 15–20cm. Hatchling size unknown. Colour quite variable; adult females and non-breeding males have a broad brown or grey-brown dorsal stripe, bordered with a yellow dorsolateral and lateral stripe on each side, the flanks are dark but heavily suffused with paler spots, in some specimens there is a reddish stripe or spotting low on the flanks. The underside is unmarked white. The breeding males have a broad stripe, which commences on the top of the head, as vivid bronze or rufous, becoming brown or grey posteriorly, the flanks are black, a chrome-yellow lip stripe breaks up beyond the forelimbs into blue dots and the lower flanks are vivid red.

Ichnotropis bivittata

HABITAT AND DISTRIBUTION: Savanna, grassland and open woodland at medium to low altitude, between 600–1,700m altitude. In East Africa known only from southern Tanzania, at Ipeni, Udzungwa (south of Mufindi), and near Lukwati, north of Lake Rukwa, but it is also known from Mbala, in Zambia, just across the border at the south end of Lake Tanganyika, so may be more widespread in south-west Tanzania. Elsewhere, to Malawi, Shaba Province, Congo, and Angola.

NATURAL HISTORY: Diurnal and terrestrial; a fast-moving, wary species, active during the hottest hours of the day. Shelters in holes and under rocks. The Lukwati specimen was

Ichnotropis bivittata Angolan Rough-scaled Lizard. Left: Bill Branch, female, Zambia. Right: David Lloyd-Jones, Lake Rukwa, male.

caught in a pitfall trap in grassland bordering *Brachystegia* woodland, and showed curious leg-tucking behaviour, elevating its body and folding its legs into its sides. Presumably lays eggs, but no breeding details known. Diet: invertebrates.

TANZANIAN ROUGH-SCALED LIZARD *Ichnotropis tanganicana*

IDENTIFICATION: Known only from the type specimen, which had a snout-vent length of 3.8cm and a broken tail; it is probably a sub-adult. The head shields are weakly striated. The dorsal and lateral scales are enlarged, pointed, overlapping and strongly keeled in 36 rows at midbody; the ventral scales are smooth and in eight longitudinal and 25 transverse series. It was originally described as 'bronzy olive above...a few small transverse blackish spots in three longitudinal series on the nape and two on the body; a black streak from the nostril to the eye, and another on the edge of the mouth; a white, black-edged streak from below the eye, through the ear, to above the axil; white, black-edged ocellar spots on the posterior part of the back, on the hind limbs, and on the tail; lower parts white.'

Ichnotropis tanganicana

Ichnotropis tanganicana Tanzanian Rough-scaled Lizard. Stephen Spawls, preserved type.

HABITAT AND DISTRIBUTION: The single, type specimen was described from the 'East Coast of Lake Tanganyika' and was presented to the British Museum by a Mr H.W. Nutt. Conservation Status: Data Deficient.

NATURAL HISTORY: Nothing known, never knowingly seen alive.

DESERT AND SAVANNA LIZARDS *MEROLES*

A south-western African genus of eight species of slim brown dry-country lacertids; molecular analysis recently revealed that the Mozambique Rough-scaled Lizard, originally placed in the genus *Ichnotropis*, is nested within this genus.

MOZAMBIQUE ROUGH-SCALED LIZARD *Meroles squamulosus*

IDENTIFICATION: A medium-sized lacertid, in south-east Tanzania. It has a small head and a fairly stout body. The tail is long, 60–70% of total length. The scales of the head are moderately rough and bear rather large, discrete keels. Dorsal scales small, rough, strongly-keeled, overlapping, rather bluntly pointed and arranged in 40–58 rows at midbody (usually 40–50 in Tanzanian animals). Maximum size about 23cm, average 16–20cm, hatchlings about 7cm. Colour quite variable, shades of grey or brown, Tanzanian specimens are usually brownish, with a diffuse broad light dorsolateral line, which may extend onto the tail or fade. The dorsal surface between these lines is marked with ocellate black and white spots. The flanks are brown or rufous with scattered diffuse light spots; the tail may be brown or rufous, with or without spotting. Juveniles are much paler in colour than adults, and the dorsolateral stripe is pinkish-buff. The underside is white (females) to grey (males).

Meroles squamulosus

HABITAT AND DISTRIBUTION: Savanna and woodland of Tanzania, at altitudes between sea level and 900m, although further south it occurs in semi and near desert. There is a cluster of records from south-east Tanzania (including Newala, Liwale, Kitaya and Kilwa), also known from Kitungulu, just east of the southern end of Lake Tanganyika, and a single record from Unyanganyi, just north-west of Kondoa. Elsewhere, south to Namibia and Angola.

NATURAL HISTORY: An active, diurnal predator in open flat clearings. Several individuals may share a branching burrow system, which is frequently dug at the base of a tree. Females lay 8–12 eggs; gravid females with ten eggs, roughly 8×12mm, found in May–July in southern Tanzania. In southern Africa, strangely, this lizard is an annual species; sexual adulthood is reached after only 4 or 5 months, and adults die after laying one or two clutches of eggs; at times during the dry season the entire population exists only as eggs; it is not known if this applies to Tanzanian specimens. Diet: invertebrates, recorded feeding on termites, beetles and grasshoppers.

Meroles squamulosus Mozambique Rough-scaled Lizard. Andre Zoersel, Zambia.

Platysaurus maculatus
Spotted Flat lizard
Michele Menegon

SUB-ORDER SAURIA: LIZARDS
FAMILY CORDYLIDAE
GIRDLED LIZARDS AND THEIR RELATIVES

An African family of lizards. For a long time, it contained only four genera, but a recent molecular analysis has split the genus *Cordylus* into seven genera, one of which is named for the dragon in Tolkien's *The Lord of the Rings, Smaug*. Three genera of Cordylids occur in our area; they are the grass lizards, *Chamaesaura*, the 'typical' girdled lizards, *Cordylus* and the flat lizards, *Platysaurus*. The Cordylids are the only totally African lizard family. No fossils are known. All the lizards in this family have a short, pointed tongue, which is sometimes notched, coated with long papillae. The body scales lack osteoderms (apart from girdled lizards). The tail scales may be very spiny. The body is usually flattened (save in grass lizards), an adaptation for living in a narrow recess. All except the flat lizards give live birth. They largely hunt from ambush, waiting in a suitable crevice until prey goes by.

KEY TO THE EAST AFRICAN CORDYLID GENERA

1a: Snake-like in appearance, legs absent or merely tiny buds......*Chamaesaura*, grass lizards [p. 206]
1b: Not snake-like in appearance, limbs well-developed......2

2a: Vividly coloured, body very flattened, no spines on the tail, scales on the back tiny and granular......*Platysaurus maculatus*, Spotted Flat Lizard [p. 208]
2b: Usually brown or grey, body not flattened, tail with spines, back scales large......*Cordylus*, girdled lizards [p. 209]

SNAKE OR GRASS LIZARDS *CHAMAESAURA*

A south-east African genus of unusual, effectively limbless, lizards that glide through the grass like snakes; hence sometimes called 'grass swimmers'. They have long heads, large eyes and prominent brow ridges. Their limbs are rudimentary, consisting of little more than fleshy spikes or buds. They have very long tails, 4–5 times the body length. Their scales are elongate, strongly keeled and overlapping, forming regular long and transverse rows. Active by day, they live in grassy or bushy areas, sliding rapidly through the grass, at rest they use their tiny limb as stabilisers. They can shed their tails, but are unstable without them. Their pattern of stripes on a sombre background disguises them well in grass, and also serves to confuse attacking predators as to the speed with which they are moving. Grasshoppers appear to be their main food; hunted actively during the warm hours of the day. Three to 12 young are produced by live birth. They are apparently at risk from grass fires, with population crashes recorded in burnt areas. Five species are known, two of which occur in East Africa, in grassland areas; the East African forms have been recently elevated from subspecies to full species.

KEY TO THE EAST AFRICAN MEMBERS OF THE GENUS *CHAMAESAURA*

1a	A distinct forelimb, in northern Tanzania or Kenya......*Chamaesaura tenuior*, East African Highland Grass Lizard [p. 207]	1b	Forelimb minute, in southern Tanzania ......*Chamaesaura miopropus*, Zambian Grass Lizard [p. 206]

ZAMBIAN SNAKE LIZARD/ZAMBIAN GRASS LIZARD
Chamaesaura miopropus

IDENTIFICATION: A slim, apparently legless lizard with a very long tail, the forelimbs are reduced to minute buds, the single-toed hindlimb looks like a spike, with a short claw. It has a flattened head, a pointed snout, an obvious eye and an amiable expression. It resembles a snake at first glance. The tail is 70% of total length and tapers gently to the tip. Body scales keeled, 22 rows at midbody, 38–40 rows lengthwise. There are one or three femoral pores on each side of the vent. Maximum size about 60cm, average 30–50 cm, hatchlings size unknown. Brown, with two darker dorsolateral stripes, these may break up into a series of elongate blackish spots; the flanks are pinky brown, yellow-brown or straw-coloured, the belly is white.

Chamaesaura miopropus

HABITAT AND DISTRIBUTION: Montane grassland and grassy savanna, from 1,500–2,500m altitude. Occurs in south-western Tanzania, from Ugalla and Katavi to the Mbizi Mountains south of Lake Rukwa to the Tukuyu volcanoes (Mbeya, Rungwe, Kipengere and

Family Cordylidae: Girdled lizards and their relatives

Chamaesaura miopropus Zambian Grass Lizard. Michele Menegon, Udzungwa Mountains.

Livingstone Mountains), north to the southern Udzungwa Range. Elsewhere, to Zambia.

NATURAL HISTORY: Poorly known, probably similar to other grass lizards. Diurnal and terrestrial, although they will climb up in grass tufts. They glide through the grass like a snake, looking for prey. If threatened they will dive into a nearby clump or slide rapidly away through the grass. They give live birth, details unknown. Their diet is presumably arthropods.

EAST AFRICAN HIGHLAND GRASS LIZARD
Chamaesaura tenuior

IDENTIFICATION: A slim, apparently legless lizard with a very long tail. It has a flattened head, a pointed snout and an obvious eye. It looks like a snake at first glance. The minute stick-like limbs have one or two tiny clawed digits. The tail is 75–80% of total length and tapers gently to the tip. Body scales keeled, 24–30 rows at midbody (usually 24–26 in East African animals), 36–42 rows lengthwise. There are one or two femoral pores on each side. Maximum size about 62cm, average 30–50cm, Rwandan hatchlings were 17–18 cm. The head is dark brown above, straw-coloured below, the body is striped, but the colour and width of the stripes vary. In some Kenyan specimens, along the centre of the back is a light brown or straw-coloured stripe, bordered by two darker brown dorsolateral stripes, there are two lengthwise straw-coloured flank stripes. The belly is uniform pale brown or yellow or with a number of fine darker longitudinal lines. Some specimens appear uniform yellow or light brown in colour; in some the stripes are very pale.

Chamaesaura tenuior

Chamaesaura tenuior East African Highland Grass Lizard. Miroslava Jirku, Loita Hills.

HABITAT AND DISTRIBUTION: High grassland of the East African plateau, from about 1,200m altitude up to 3,000m. Found from eastern Rwanda north-east through south-west Uganda, also occurs on the Sesse Islands and northern lakeshore. It is sporadically distributed through high central Kenya, mostly on hills (Cherang'any, Kakamega area, Mau, central rift escarpment,

Aberdares, Lebetero Hills, Loita), also in the Chyulu Hills in Tsavo. In Tanzania known from the southern Loita Hills, Crater Highlands and the Usambaras.

NATURAL HISTORY: Diurnal and terrestrial, although they will climb up in grass tufts. In their highland habitat they shelter in the grass tussocks or under piles of vegetation. As the morning starts to warm up they climb up into the tufts to bask, they will then hunt, gliding through the grass looking for prey. If threatened they will dive into a nearby clump or slide rapidly away through the grass. They give live birth to 2–9 young. Their diet includes grasshoppers and other arthropods.

FLAT LIZARDS *PLATYSAURUS*

A genus of spectacular, diurnal, rock lizards, the largest species are over 30cm total length. Their very flat bodies enable them to hide in narrow rock cracks. Males are usually brightly coloured, females are duller, often striped. They live in colonies under a dominant male who confronts rival males, circling and standing high to expose his vivid chest colours; he may also flaunt his colours in courtship. Females lay their two eggs, often communally, in damp leaf litter in rock cracks. They eat arthropods, some eat plant material. They are long-lived, one specimen lasted 14 years in captivity. They live on rocks, without exception, and thus, in the ancient, dissected landscape of the southern third of Africa, where rock outcroppings have become isolated, so have their flat lizard colonies. Small changes accumulate, there is no genetic exchange with other colonies, leading to speciation. Sixteen species are known, some with several races; one form has nine subspecies. A single species just reaches our area.

SPOTTED FLAT LIZARD *Platysaurus maculatus*

IDENTIFICATION: A small, vividly coloured flat lizard from southern Tanzania. The body is flattened, the snout pointed, the tail is long, 60–70% of total length. The dorsal scales are tiny and granular; ventral scales relatively broad, the tail has distinctive whorls of scales and can be shed and regenerated. Males are larger than females, maximum male about 20cm, largest female 17cm. The colour is variable, adult males usually have a brown or rufous head, the back is greenish or turquoise, shading to yellow and brown, with a dorsal stripe, which often broadens at the tail base; the tail is orangey-pink; the limbs blotched brown. The throat and chest are yellow or orange, usually with conspicuous black patches on the chin and just forward of the front limb insertion;

Platysaurus maculatus

the rear underside is black. Females are dark brown or black, with five light stripes, which may be sharp edged or dots, or a mixture of both. The central stripe becomes green near the rear limbs and persists onto the tail; sometimes there is reddish wash on the head and neck. Below, white becoming yellow at the sides, some individuals have a vivid orange

Platysaurus maculatus Spotted Flat Lizard. Left: Michele Menegon, northern Mozambique, male. Right: Michele Menegon female.

chin and chest, with heavy black infusions towards the tail. Juvenile males have the same colour as females.

HABITAT AND DISTRIBUTION: Restricted to suitable rocky outcrops, granite, gneiss and sandstone that weather to produce the narrow fissures which they require as refuges. In East Africa, known only from south-eastern Tanzania at Masasi, 450m altitude, elsewhere to northern Mozambique. Conservation Status: not assessed. Appears not to be threatened by human activities but is restricted by its specialised habitat requirements. The Tanzanian populations at Masasi are the northernmost so far reported.

NATURAL HISTORY: Diurnal and rock dwelling, spending the day shuttling between sun and shade. Females lay two large, elongate eggs; several females may lay in a suitable communal depository. They have an interesting social life, being gregarious save in the breeding season, when adult males mark out and defend a territory, during which time they threaten and drive off rival males. They feed largely on small invertebrates, some species eat plant material.

GIRDLED LIZARDS *CORDYLUS*

An African genus of diurnal, secretive, small to medium-sized spiny, squat brown lizards, with regular, girdle-like scale rows. Their external ears are sometimes hard to see, concealed by spines. The body scales have osteoderms and are heavily keeled, the tail is spiky, but it can be shed and regenerated. Males have femoral pores. In defence they jam themselves in crevices by inflating their body and using an unusual hinge structure to shorten and thicken the skull. They eat invertebrates, some eat plant material. They give live birth to a few large young. Some species are territorial, in colonies under a large dominant male. They may use chemical cues in their social behaviour, hence the glandular pores. Captive specimens have lived 30 years and they make charming and confiding pets. They are at risk from collecting for the pet trade in Tanzania. Like *Platysaurus*, they have radiated in the rocky southern African landscape, with some 22 species in the group (and 19 that were in this genus have been moved to fresh genera). Three species of *Cordylus* occur in East Africa, two are endemic. Girdled lizards are secretive and hard to find, and the enigmatic presence of a species in southern Ethiopia indicates that undiscovered species are probably living in the rocky hills of Kenya.

KEY TO EAST AFRICAN MEMBERS OF THE GENUS *CORDYLUS*

In using this key, the specimen's locality (if known) can be useful; one species lives on the coast and inland in parts of eastern Tanzania, another lives in the medium altitude Maasai country straddling the central Kenya-Tanzania border, the third is known only from a couple of localities on hills in southern Tanzania.

1a	Large (usually more than 7cm snout-vent length), nasals separated or in short contact, on the coastal plain or south-eastern Tanzania ......*Cordylus tropidosternum*, Tropical Girdled Lizard [p. 211]	2a	Body depressed, lateral scales rectangular, head relatively elongate, in northern central Tanzania or south-central Kenya, ...*Cordylus beraduccii*, Maasai Girdled Lizard [p. 210]
1b	Small (usually less than 7cm snout-vent length), nasals in broad contact, not on the coastal plain......2	2b	Body not depressed, lateral scales subtriangular, head short and deep, in south-central Tanzania......*Cordylus ukingensis*, Ukinga Girdled Lizard [p. 212]

MAASAI GIRDLED LIZARD *Cordylus beraduccii*

IDENTIFICATION: A small, stout, speckled yellow-brown spiky lizard. The head is triangular and flattened; the nose sharp, the eye is medium sized, pupil round. The body is quite depressed and broad. The limbs are short and stout. The tail is short and slender, about 50% of total length, tapering sharply, and has regular whorls of very spiny scales. Dorsal scales are heavily keeled. There are 4–6 femoral pores. Maximum size reported as about 16–17cm, but this is unusually large, average 10–14cm, hatchling size in captivity 6–7cm. Light to medium brown, sometimes uniform, some dorsal scales have black keels, there is a broad vague dorsolateral bar of lighter blotches, the tail is mottled brown and tan. Upper labials, chin and throat light yellow-brown, underside cream or yellow. The hatchlings are vivid brownish black, with a dorsolateral band of yellow and dark brown blotches, the tail banded yellow and black, chin and throat yellow or tan.

Cordylus beraduccii

HABITAT AND DISTRIBUTION: A rock-dwelling, secretive species, usually on outcroppings and rocky hills, but also on trees, in mid-altitude savanna and grassland, at 1,500–2,000m altitude. An East African endemic, known from outcrops below the western flank of the Ngong Hills and from the rocky hills and sheet rock outcroppings around Kajiado, in Kenya; thence into northern Tanzania, to Ngorongoro and the central Serengeti. Some unusually large specimens from Mpwapwa region, near Dodoma (not shown on map).

NATURAL HISTORY: Poorly known. Diurnal, lives on rocks and trees, hiding in fissures, cracks and under bark. Emerges to bask in the morning sunshine and will then hunt. Gives live birth, clutch details unknown, a hatchling was captured

Cordylus berraduccii Maasai Girdled Lizard. Stephen Spawls, Ngong Hills.

west of the Ngong Hills in September; a captive specimen gave birth in August. Diet unknown but presumably small invertebrates.

TROPICAL GIRDLED LIZARD Cordylus tropidosternum

IDENTIFICATION: A medium-sized, stout, brown spiky lizard. The head is triangular and flattened, the snout sharp, the eye is medium sized, pupil round, with a brownish-yellow iris (hard to see), the head scales are granular, except for the supraoculars. The scales on the collar may be very spiny. Body slightly depressed, limbs strong, fourth toe on the hindlimbs very long. The tail is short and slender, about 40% of total length, tapering sharply, and has regular whorls of very spiny scales. Dorsal scales are large and strongly keeled, in 21–28 rows at midbody, ventral scales 11–14 rows. There are 6–9 femoral pores in East African specimens. Maximum size about 20cm, average 14–17cm, hatchlings 6–8cm. Usually brown in colour, with light flecks, fading to cream on the belly. There is often a short dark horizontal stripe from the nose to just behind the shoulder. Lips, chin and throat are cream.

Cordylus tropidosternum

HABITAT AND DISTRIBUTION: Coastal forest, woodland and savanna, from sea level to 1,800m altitude. Occurs from the Arabuko-Sokoke Forest south along the Kenyan coast to the Shimba Hills, there is also a curious record from Makindu (not shown on map). Thence widespread across eastern and southern Tanzania, although records sporadic in the southwest. Elsewhere, south and west to Shaba in the DR Congo, south to Mozambique, Malawi and eastern Zimbabwe.

NATURAL HISTORY: A diurnal, arboreal lizard, living in trees in East Africa, sheltering in holes and under bark but known to live in rock crevices elsewhere. Shy and secretive, usually slow-moving but can run quite fast. If approached, it may freeze or surreptitiously move around the tree. They give live birth, clutches of three recorded in coastal Kenya in

Cordylus tropidosternum Tropical Girdled Lizard. Bio-Ken archive, Arabuko-Sokoke Forest.

March, one to five in Tanzania, a single young born to a female in southern Tanzania in November. Diet mostly insects and other arthropods, observed in Kenya coastal forests prising out moths and spiders from under dry bark.

UKINGA GIRDLED LIZARD Cordylus ukingensis

IDENTIFICATION: A small, stout, mottled brown spiky lizard. The head is triangular and flattened, the nose sharp, the eye is medium sized, pupil round. The body is quite depressed and broad. The limbs are short and stout. The tail is short and slender, about 38–40% of total length, tapering sharply, and has regular whorls of very spiny scales. Dorsal scales are heavily keeled. There is no loreal scale, but a big preocular is present. Total length about 16cm, average 9–14cm. An attractive lizard, the dorsal surface is rufous-brown, mottled with black and a diffuse light vertebral line, the flanks are black with pale spotting and speckling, which increases downwards. The tail is warm brown, the underside yellowy-grey or pale grey.

Cordylus ukingensis

HABITAT AND DISTRIBUTION: Occurs in forest and forest-associated grassland patches. A Tanzanian endemic, described from Tandala in the Kipengere Range, just north-west of the northern tip of Lake Malawi, also found in the southern Udzungwa Mountains, recorded near Mufindi and at Kibengu village between Usokami and Uhafiwa, at 2,000m altitude. Conservation Status: Near Threatened. Little known of its habitat requirements and there is no information on population numbers, but its habitat is exploited for agriculture.

NATURAL HISTORY: Poorly known. It appears to be terrestrial, as a specimen was collected in a hole under a rock in montane grassland. Habits presumably similar to other girdled lizards, i.e. diurnal, gives live birth, diet probably small invertebrates.

Cordylus ukingensis Ukinga Girdled Lizard. Michele Menegon, Udzungwa Mountains.

Gerrhosaurus flavigularis
Yellow-Throated Plated Lizard
Stephen Spawls

SUB-ORDER **SAURIA: LIZARDS**
FAMILY **GERRHOSAURIDAE**
PLATED LIZARDS AND THEIR RELATIVES

These relatively large, tough-scaled lizards were originally regarded as a subfamily of the *Cordylidae*. They are ancient and their presence on both mainland Africa and Madagascar indicates that they evolved before the two separated, in the Cretaceous, 80 to 100 million years ago. There are seven genera in the family, five African and two Madagascan. Three genera occur in East Africa: *Broadleysaurus*, *Gerrhosaurus* and *Tetradactylus*. Most species have long tails, the head has large, symmetrical scales with osteoderms. The body scales are hard and rectangular, also with osteoderms, overlapping, sometimes keeled. Along the flanks there is a curious lateral fold of soft, granular-scaled skin. They have well-developed eyes and obvious ear openings. The femoral glands have clearly visible pores in most species, especially in males.

KEY TO THE EAST AFRICAN GENERA OF THE FAMILY GERRHOSAURIDAE

1a Limbs absent or minute, tail very long, more than twice body length......*Tetradactylus*, Long-tailed seps [p. 218]
1b Limbs present, tail about same length as body or slightly longer......2
2a Belly scales in ten (rarely nine) longitudinal rows, body thickset, stripes if present not continuous......*Broadleysaurus major*, Great Plated Lizard [p. 214]
2b Belly scales in eight longitudinal rows, relatively slim, body stripes usually continuous, fine, black and yellow......*Gerrhosaurus*, plated lizards [p. 215]

GREAT PLATED LIZARD BROADLEYSAURUS

A genus of sub-Saharan Africa containing a single species, the Great Plated Lizard *Broadleysaurus major*. Originally placed within the genus *Gerrhosaurus*, molecular analysis now indicates that this species and another huge southern African form, *Matobosaurus validus*, are well separated from the smaller plated lizards, and are placed in new genera. A 10-million year-old fossil of this species is known from Mfangano Island in Lake Victoria

GREAT PLATED LIZARD *Broadleysaurus major*

IDENTIFICATION: A large, stout, armoured lizard, with a short head and rounded snout; the eye and earhole are large and obvious. The body is thickset and compressed, limbs short and powerful. The tail is about 60% of total length, thick, tapers smoothly and is not easily shed. The scales are hard, rectangular, in 14–18 rows at midbody, there are ten rows of ventral plates. Males and females have 9–17 femoral pores; these are more obvious in the males. At a maximum length about 55cm this is one of our biggest lizards, average size 30–45cm, hatchlings 14–16cm. Colour usually some shade of brown, but adult males may develop pinkish or even blue throats; some adults have a yellow dorsolateral and rufous lateral stripe. Some juveniles resemble the uniformly-coloured adults, others are black with yellow speckling, as they become adult the black gradually fades from the head backwards. Specimens from western Kenya, Uganda, and further west retain the juvenile coloration as adults, but the yellow stripes tend to become long lines or rows of yellow spots. Taxonomic Note: Also called Rough-scaled Plated Lizard, Tawny Plated Lizard and Sudan Plated Lizard in the literature.

Broadleysaurus major

HABITAT AND DISTRIBUTION: Coastal woodland, thicket, moist and dry savanna, often around hills, from sea level to about 1,700m altitude (found on Lukenya Hill but not the Nairobi area). Widely distributed in south-east Kenya, south of the Tana River, few records from the north (save Malka Murri, Laisamis, Isiolo, Samburu, Wamba and the Merille River), widespread in eastern and south-eastern Tanzania, isolated records from Smith Sound and Tatanda; the striped western colour form is known in Kenya from Mt Suswa, Baringo north to Amaler and the Kerio Valley, along the Uganda border from Karita north to Moroto. There is a sight record from Lolgorien, Transmara, Kenya. A handful of Uganda records (Mabira Forest, Lake Albert shore, Kidepo Valley, probably more widespread); elsewhere south to South Africa, west to Ghana.

NATURAL HISTORY: A terrestrial, diurnal lizard, sheltering in holes, termitaria, rock piles and rock crevices. Sometimes found around the base of rocky hills, on the coast it lives in the coral rag, hiding in crevices; in Tsavo often utilises the rock piles used to support signposts. Occasionally climbs, one was found in a tree hole 3m above ground level. Quite tolerant of urbanisation, many coast resorts have a few living in their grounds. The males are territorial; in eastern Kenya territory-holding males develop vivid pink throats.

Broadleysaurus major Great Plated Lizard. All Stephen Spawls. Top: Kilifi. Bottom left: Ghana, western subspecies. Inset left: Sub-adult. Inset right: Juvenile.

They will fight with other encroaching males, biting each other's legs and tail, trying to flip their opponents over. The females lay a small clutch of 2–7 eggs, roughly 1.5–2.5×4–5cm, usually beneath a rock, in a deep, damp hole or a termitarium; a clutch of seven eggs was found in southern Tanzania in June. They are omnivorous, eating arthropods, fruits, flowers and other lizards. They make excellent, confiding pets. In southern Tanzania, one was seized by a Black Kite *Milvus migrans*, another was killed by a pack of Banded Mongooses *Mungos mungo*, one found in the stomach of a Black-necked Spitting Cobra *Naja nigricollis*.

PLATED LIZARDS *GERRHOSAURUS*

An African genus of seven species of diurnal, largely striped, medium-sized lizards, with thick armoured scales and a curious skin fold along the flanks. A recent revision elevated several forms out of synonymy or sub-specific status to full species. Two species are found in East Africa.

KEY TO THE EAST AFRICAN MEMBERS OF THE GENUS *GERRHOSAURUS*

1a Four supraciliary scales, body fairly stout, scales on soles of the feet keeled and spiny......*Gerrhosaurus nigrolineatus*, Black-lined Plated Lizard [p. 217]

1b Five supraciliary scales, body fairly slim, scales on soles of the feet smooth and tubercular......*Gerrhosaurus flavigularis*, Yellow-throated Plated Lizard [p. 216]

YELLOW-THROATED PLATED LIZARD Gerrhosaurus flavigularis

IDENTIFICATION: A medium-sized striped slender lizard. The head is short, snout pointed, neck long. The eye is large, pupil round. The ear opening is triangular and obvious. The body is flattened dorsally, toes thin, some of the rear toes are very elongate. The tail is long, two-thirds of the total length, tapering smoothly, and is easily shed. The scales are strongly keeled, in 22–24 rows at midbody. Males have 11–17 femoral pores, females have none. Maximum length about 40cm, average 25–35cm, hatchlings 10cm. Colour quite variable, usually brown; there is a pair of fine yellow and black dorsolateral stripes, sometimes there is a fine yellow, black-bordered vertebral stripe, the back may be speckled yellow. The flanks may be vertically barred black and cream, black and yellow, they may be uniform or blotched red. The body stripes extend onto the tail if it is not regenerated. In breeding males, the chin and throat may turn vivid pink, red, orange or yellow. The belly may be cream, white, yellow or blue-grey.

Gerrhosaurus flavigularis

HABITAT AND DISTRIBUTION: Woodland, high grassland, moist and dry savanna and coastal bush, from sea level to about 2,000m altitude. Widespread in north central, central and south-eastern Kenya and north-eastern Tanzania, including the coastal strip. Records lacking from dry north-eastern Kenya. In Uganda recorded between Kaabong and Moroto on the north-east border but might be more widespread. In the west, occurs from south-east Rwanda and eastern Burundi down through Ugalla and Katavi to Lake Rukwa. Elsewhere, south to South Africa, north to southern Sudan and Ethiopia.

NATURAL HISTORY: A terrestrial, diurnal lizard, often living in holes, under big rocks, leaf piles and in hollow logs. It is secretive and fast-moving, often mistaken for a striped snake as it rushes through the vegetation. In the breeding season the male's throat becomes suffused with bright colours. The females lay from three to eight eggs, roughly 1.5×2cm, in a moist hole or in leaf litter under a thick bush. No clutch details for East Africa, but hatchlings were collected in the Kerio Valley in July. Diet: a wide range of arthropods, including grasshoppers, termites and millipedes.

Gerrhosaurus flavigularis Yellow-throated Plated Lizard. Stephen Spawls, Right: Diani Beach. Left: Male throat colours, Botswana.

BLACK-LINED PLATED LIZARD Gerrhosaurus nigrolineatus

IDENTIFICATION: A long, striped, fast-moving slender lizard. The head is short, snout fairly rounded, neck long. The eye is quite large, pupil round, iris orange or red-brown. The ear opening is triangular and obvious. The body is sub-cylindrical, slightly flattened dorsally. The toes are thin, the third and fourth toes on the hind feet are very elongate. The tail is long, about 70% of the total length, tapering smoothly, and is easily shed. The body scales are hard, keeled and rectangular, in 22–24 rows at midbody. Both sexes have 15–20 femoral pores on each side. Maximum length about 51cm, average 25–40 cm, hatchlings 14–16 cm. Usually brown above with two prominent yellow and black dorsolateral stripes, flanks irregularly speckled yellow or black, brown or rufous, often the sides of the neck and the anterior flanks are brick red. Underside cream, grey or blue-grey. Taxonomic Note: Some authorities suggest that eastern and southern African populations should be assigned to the species *Gerrhosaurus intermedius*, but in the absence of a rigorous definition of that form, we leave the East African examples in *G. nigrolineatus*.

Gerrhosaurus nigrolineatus

HABITAT AND DISTRIBUTION: Moist and dry savanna, coastal bush and grassland, from sea level to about 1,700m altitude. Widespread in south-eastern Kenya, most of eastern and western Tanzania (records lacking from the centre) and eastern Rwanda. Records sporadic elsewhere in East Africa, include near Kampala, (Uganda), Kaapsorok Valley, Burguret, Laikipia, Meru National Park and Shaffa Dika in Kenya. Elsewhere, south to north-eastern South Africa, west to Congo-Brazzaville, south-west to Angola.

NATURAL HISTORY: A terrestrial, diurnal lizard, often living in holes, including termite hills, mouse burrows, squirrel warrens and similar places, on the coast it uses holes and recesses in the coral rag. Shy and fast moving, so rarely observed except when running. It lays 4–9 eggs, roughly 1.5×2.5cm, usually in a deep moist hole or under a large rock, sometimes in a decomposing log or stump. Incubation times unknown in East Africa but 70–80 days in South Africa. It eats a range of arthropods, including grasshoppers, millipedes, centipedes and ants, has been seen to take snails.

Gerrhosaurus nigrolineatus Black-lined Plated Lizard. Stephen Spawls, Moshi.

SEPS OR PLATED SNAKE-LIZARDS
TETRADACTYLUS

A southern African genus of curious plated lizards, some have normal (albeit tiny) limbs, others have greatly reduced limbs or no forelimbs and tiny rear limbs. The body scales are arranged in straight rows. As with the plated lizards, there is a lateral fold of soft skin. They have a long tail. They are rather snake like, resembling the grass lizards. When startled, they respond with almost a jump, flexing their coils against the ground like the release of a compressed spring. Seven species are known, mostly in South Africa, two species occur in southern Tanzania, one is endemic.

KEY TO THE EAST AFRICAN MEMBERS OF THE GENUS
TETRADACTYLUS

1a	No front limbs, body scales in 65 transverse rows......*Tetradactylus ellenbergeri*, Ellenberger's Long-tailed Seps [p. 218]	1b	Has two front limbs, body scales in 67 transverse rows......*Tetradactylus udzungwensis*, Udzungwa Long-tailed Seps [p. 219]

ELLENBERGER'S LONG-TAILED SEPS *Tetradactylus ellenbergeri*

IDENTIFICATION: A small, snake-like plated lizard, with no front limbs and minute rear limbs, a few millimetres long; these are clawless. The shields of the head have fine striations on them but are not keeled. The body is long and slim, the dorsal scales are striated and have a strong median keel, in 14 longitudinal and 65 transverse rows. There is a strong lateral fold along the body. The tail is very long, roughly 75% of total length. Maximum size about 31cm, average 15–25cm, hatchling size unknown. Slightly bluish olive or brown above, a light (yellow or blue-grey) vertebral line is bordered by black lines, and in some cases a diffuse light dorsolateral stripe. The side of the head and neck are ivory or white with black or very dark brown vertical bars, interrupted by the lateral skin fold, these bars become increasing browner and heavier posteriorly. Below, pale olive or light grey.

Tetradactylus ellenbergeri

Tetradactylus ellenbergeri Ellenberger's Long-tailed Seps. Michele Menegon

HABITAT AND DISTRIBUTION: Moist savanna in southern Tanzania, three records only, (Tatanda, Songea and Gendawaki Valley near Bomalang'ombe Village in the southern Udzungwa), between 1,000–2,100m altitude; elsewhere in eastern

Angola, Shaba Province of the DR Congo and the western and northern portions of Zambia. Reported from southern Rwanda (not shown on map) but without supporting specimens. Conservation Status: within the East African region it is known only from southern Tanzania, but it is relatively widely distributed in woodland elsewhere, it is unlikely to be threatened.

NATURAL HISTORY: Not well known, presumably diurnal and terrestrial. Found in grassland and wooded grassland, in which its movements are extremely rapid and snake-like. Diet arthropods. The members of this genus lay eggs; two Tanzanian females were gravid in February and October; each with two eggs, roughly 1×0.5cm.

UDZUNGWA LONG-TAILED SEPS *Tetradactylus udzungwensis*

IDENTIFICATION: A small, snake-like seps, with four small limbs, having three fingers and two reduced toes. Maximum recorded length is about 170mm. Body elongated and cylindrical, dorsal scales in 14 longitudinal and 67 transverse rows, striated with a strong median keel. There are 12 evident longitudinal keel rows on the body and tail and a strong lateral fold along the flanks. Four femoral pores are present on each hind limb, and the first subcaudal scale bears a claw-like scale with its tip pointed in the cranial direction. Head brown, irregularly spotted with dark-brown dots; in lateral view the head is white with vertical black bars in correspondence with the posterior scale edge. Upper parts of body and tail uniformly matt olive-brown. Belly uniformly cream or grey. Males bear a well-developed claw-like process on the first subcaudal.

Tetradactylus udzungwensis

HABITAT AND DISTRIBUTION: Known from swampy grasslands in the Gendawaki valley in the southern Udzungwa only. To date very few faunal surveys have been conducted in the grasslands of the Udzungwa Mountains, it is possible that this species will be found in other similar habitats in other parts of the mountain block.

NATURAL HISTORY: The Udzungwa Long-tailed Seps is very shy and when discovered immediately try to escape by fleeing into tall grass. In March they were observed active in the first part of the day and in late afternoon just before sunset. Individuals were found on the margins of the flooded area in grass tussocks separated by small areas of open standing water. At the only known locality this species is sympatric with other two specialised grass-dwelling reptiles: the Ellenberger's Long-tailed Seps *Tetradactylus ellenbergeri* and the Zambian Grass Lizard, *Chamaesaura miopropus*.

Tetradactylus udzungwensis Udzungwa Long-tailed Seps. Michele Menegon, Udzungwa Mountains.

Acanthocercus atricollis
Blue-headed Tree Agama
Stephen Spawls,
juvenile, Maasai Mara.

SUB-ORDER **SAURIA: LIZARDS**
FAMILY **AGAMIDAE**
AGAMAS

A family of Old World lizards, with nearly 60 genera and close to 500 species, found in Australia, Asia, southern Europe and Africa. Most African species (and all those from our area) belong in the subfamily Agaminae. Agamas have large heads with prominent eyes and thin necks, their bodies are depressed and they have short powerful limbs and a long tapering tail that can only be shed with difficulty. They often have a well-developed pineal eye, manifest as a small depression on top of the head. Some species are highly visible and some are tolerant of urbanisation. Several species are territorial, living in structured colonies under a dominant male, who drives away rivals. Others appear to live in pairs, or sometimes a male and two or three females. The males of several species are vividly coloured, and bask in prominent places; they can also change colour to a limited extent and the colour displayed may be indicative of their position in the hierarchy. They feed largely on ants and termites. All the African agamas lay eggs. As with crocodiles, the sex of the young is determined by the incubation temperature of the eggs, and evidence from some species indicates that females may be able to retain developing eggs within the body over a long time, until suitable weather conditions for laying arrive (primarily a heavy storm). At present, the indigenous East African agamas are assigned to two genera, *Acanthocercus* (two species) and *Agama* (fourteen species). A small population of an introduced species, the Indian Garden Lizard, *Calotes versicolor*, exists around Mombasa Railway Station. Agamas are sometimes called 'cokkamondas' or 'koggelmanders' in parts of East Africa (the names originated in South Africa, and translate loosely as 'little vermin man'). In some areas, they are incorrectly thought to be venomous. Their nearest relatives are the chameleons, with which they form a monophyletic clade.

KEY TO THE EAST AFRICAN GENERA OF AGAMAS

1b	Adpressed hindlimb very long and spindly, extends forward well past earhole, dorsal spines on male extend past hindlimbs, only on Mombasa Island.....*Calotes versicolor*, Indian Garden Lizard [p. 241]	2a	Interparietal (occipital) scale no larger than adjoining head scales, dorsal scales heterogeneous, with a broad vertebral band of enlarged spinose scales......*Acanthocercus*, large agamas [p. 221]
1a	Adpressed hindlimb not very long and spindly, does not extend past earhole, dorsal spines if present only on anterior body, not confined to Mombasa Island.....2	2b	Interparietal (occipital) scale larger than adjoining head scales, dorsal scales homogeneous or enlarged, forming more or less regular longitudinal rows......*Agama*, other agamas [p. 224]

LARGE OR TREE AND ROCK AGAMAS
ACANTHOCERCUS

Big agamas lacking an occipital scale. The males have large, heavy heads. There are about 8 species in the genus, in Africa and the Middle East. Two species occur in our area. There is some taxonomic debate about both species, and a third species, the Black-necked Tree Agama, *Acanthocercus cyanogaster*, is now considered to be restricted to Eritrea and Ethiopia; individuals of that species recorded from our area are now assigned to *Acanthocercus atricollis*.

KEY TO EAST AFRICAN MEMBERS OF THE GENUS *ACANTHOCERCUS*

1a	Flanks covered with uniformly smooth scales, in dry country of northern Kenya only......*Acanthocercus annectans*, Eritrean Rock Agama [p. 221]	1b	Flanks with some conspicuously enlarged, keeled or spinose scales, widespread in East Africa......*Acanthocercus atricollis*, Blue-headed Tree Agama [p. 222]

ERITREAN ROCK AGAMA *Acanthocercus annectans*

IDENTIFICATION: A huge, elegant agama, vividly coloured, in far northern Kenya. The head is triangular, neck narrow. The eye is fairly large and there is a cluster of small spiky scales above and below the ear opening. The body is slightly flattened, legs strong and muscular. The tail is broad at the base (especially in males), tapering smoothly, roughly 60–70% of total length. The mid-dorsal scales are keeled, but the lateral scales are uniformly small and smooth, an excellent aid to identification. Maximum size about 50cm, average 25–40cm, hatchling size unknown but probably around 6–7cm. Displaying males have a blue and yellow head and throat, the back is blue with yellow ocellate-type spotting, the tail base is yellow, shading to blue or turquoise, the

Acanthocercus annectans

Acanthocercus annectans Eritrean Rock Agama. Left: Tomas Mazuch, Ethiopia, female. Right: Tomas Mazuch, Somaliland, male.

limbs are blue. Some larger males have a broad pale or bluish vertebral stripe, which tapers towards the tail. Non-displaying males may be brown, pink or rufous, with a black neck. The females are greeny-grey, with black speckling and pale dorsal crossbars; juveniles have diffuse brown crossbars on a grey background, and a black neck band.

HABITAT AND DISTRIBUTION: On hill and rock outcroppings in dry rocky savanna and semi-desert. In Kenya found at altitudes of 700–1,500m, but up to 2,200m on the northern plateau of Ethiopia. Within Kenya, known only from Dandu and Malka Murri in the extreme north-east. Elsewhere, to east and northern Ethiopia, northern Somalia and Eritrea. Conservation Status: Least Concern.

NATURAL HISTORY:. Little known. In northern Ethiopia it is diurnal, living on rocky hills, it lives in loosely structured colonies. It is fast and wary, running away at high speed if threatened. Dominant males bask on prominent rocks. Suitable refuges may shelter a number of individuals, and around Axum in northern Ethiopia, interestingly, several large males may be found sheltering in the same refuge. Females lay eggs, no clutch details known. Diet mostly insects and other arthropods.

BLUE-HEADED TREE AGAMA *Acanthocercus atricollis*

IDENTIFICATION: A big, stocky tree agama, the adult males have blue heads. The head of adult males is huge, triangular and broad, with swollen cheeks, with a cluster of spines around the earhole and on the narrow neck. The body is squat, legs strong and muscular. The tail is broad and tapers smoothly, roughly 60% of total length. Dorsal scales small and keeled, but along the spine of the male there are irregular rows of big spiky scales, the tail looks like a fir cone. Scales in 100–136 rows at midbody. Male maximum size about 37cm, females less, average 22–30cm, hatchling size 8cm. Colour variable, but displaying males nearly always have a bright blue or turquoise head and throat and a black collar (this may be hard to see); the body is often speckled green or blue on a dark background. Sometimes there is a broad vertebral stripe, which may be yellow, green or blue. Tanzanian animals often have narrow rufous crossbars; Kenyan coastal animals have electric-blue heads and forelimbs. The tip of the tail is usually blue. Females are less bright, their heads are grey with green speckling and blue patches, the back is grey-brown or grey, with diffuse dark crossbars and pale speckling, gravid females develop rufous dorsal marbling. Juveniles have a line of light lozenges along the spine. Taxonomic Note: Several subspecies have been described, these may prove to be full species

Acanthocercus atricollis

HABITAT AND DISTRIBUTION: Moist savanna, woodland and forest clearings, usually provided there are some large trees. From sea level to 2,400m altitude, although in East Africa it is most common in mid-altitude woodland and savanna, from 1,300–2,000m. Widespread in central and western Kenya (common in wooded suburbs of Nairobi, Nakuru, Kisumu and Kitale), most of Uganda save the northeast, and northeast Rwanda. Widely but sporadically distributed in Tanzania in the south-east and south-west, and east of Lake Victoria, but records lacking from the centre and north-east. Sporadic records from Moyale, Hurri Hils, Marsabit and Maralal in northern Kenya, Mkonumbi, Gede, Mombasa and Voi in south-east Kenya. Elsewhere, south to northern South Africa, west to eastern Congo.

NATURAL HISTORY: Diurnal and arboreal, living on the trunks of big trees, sometimes on rocky and termite hills, in northern Tanzania and the Maasai Mara they sometimes live

Acanthocercus atricollis Blue-headed Tree Agama. Top right: Daniel Hollands, Ngorongoro Crater, displaying male. Top left: Michele Menegon, male, Rwanda. Inset: Stephen Spawls, female, Naivasha.

in holes on treeless plains. Shelter at night in holes, or under bark. Strongly territorial, they live in structured colonies, controlled by a big, dominant male who challenges and fights male intruders. In suburbia they will live on ornamental trees, walls, etc. They quickly become used to humans, males will continue to display even if approached, but in the bush they are more wary and will move surreptitiously to the far side of the tree trunk if approached, using one eye to peer over the barrier. They can run quickly (and noisily) up big tree trunks. If restrained or threatened, they display the orange interior of their mouths, and they bite with enough force to draw blood. Four to fifteen eggs, roughly 1.5×2.5cm are laid. The mother may dig a hole in soft soil or utilise an existing one; in Kisumu the females often excavated 'pseudo nests' as well as true nest holes, on average seven eggs were laid, at a depth of around 8cm, to coincide with the start of the rainy season. In South Africa, the eggs take about three months to hatch. Ants are the favoured food, although a wide range arthropods is taken; one specimen ate a juvenile chameleon; one was observed eating acacia gum. Widely (but incorrectly) believed to be venomous in highland East Africa.

AGAMAS *AGAMA*

An African genus of stocky, short-bodied lizards, with triangular heads, well-developed, strong, clawed limbs, the toes are thin. Males have precloacal pores. The scales are small, hard, overlapping and often spiny. They are diurnal, living on the ground, on rocks, sometimes on trees. Some live in structured colonies, with a dominant male, recognisable by size and vivid colours, subordinate males are tolerated provided they are smaller and have subdued colours. They eat mostly arthropods, especially ants. Just under 50 species are known, fourteen species occur in East Africa, seven of these are endemic. Their taxonomy is not yet totally clarified. Originally, all agamas in East Africa with orange or red heads were assigned to the species *Agama agama*; that species is now confined to Central Africa, and the East African specimens of the complex split into four species; there may be more.

KEY TO EAST AFRICAN MEMBERS OF THE GENUS *AGAMA*

Agama keys are difficult to construct, many species have overlapping scale counts and no unequivocal distinguishing features in the laboratory. However, they can often be identified with certainty in the field by a combination of locality, colour (particularly the male throat colour) and habits and our key below makes use of this, but be aware that even experts find difficulty identifying some individuals. The first ten couplets deal with the larger and vividly coloured agamas, the final section of the key deals with the smaller species, three of which are brownish terrestrial animals.

1a	Large, males growing to more than 30cm, usually on rocks, sometimes trees, males usually with vividly coloured heads......2	2b	Hanang Agama [p. 230] Adult males without a large dark bar at the base of the throat, not confined to central Tanzania......3
1b	Small, males usually less than 30cm, on the ground, males frequently without vividly coloured heads......11	3a	Breeding males with blue or turquoise heads, range south-eastern Tanzania, usually on ground or on trees......*Agama mossambica*, Mozambique Agama [p. 240]
2a	Adult males with a large dark bar at the base of the throat, confined to a small area of central Tanzania between Lake Eyasi and Dodoma......*Agama turuensis*,	3b	Breeding males without blue or turquoise heads, range not just south-

	east Tanzania, usually on rocks......4		
4a	Males with a pink or rusty red head, usually at over 1,100m altitude in west and/or central Kenya and/or north-west Tanzania......5		*Agama finchi*, Brian Finch's Agama/Malaba Rock Agama [p. 229]
4b	Male with bright orange head, not usually in west and central Kenya and north-west Tanzania 7	9b	Tail of male predominantly blue throughout, occasionally with white rings, but never red or yellow, females with two maroon or brick red dorsolateral patches, widely separated, midbody scale rows 67–91......10
5a	Adult males with the head, body and tail orange, occasionally violet, but the same colour on the head as the body and tail......*Agama kaimosae*, Kakamega Agama [p. 231]	10a	Throat of displaying male virtually all orange, if dark marking present only at the base of the throat......*Agama lionotus*, Kenya Red-headed Rock Agama [p. 225]
5b	Adult males with the head, and at least the forepart of the body a different colour to that of the tail......6	10b	Throat of displaying male rich orange in the centre, surrounded by a deep navy blue or near-black border......*Agama dodomae*, Dodoma Rock Agama [p. 227]
6a	Adult male with very broad tail base and rust-coloured head, secretive, in high central Kenya......*Agama caudospinosa*, Elmenteita Rock Agama [p. 233]	11a	Midbody scales rows less than 65, in northern, north-eastern and eastern Kenya and north-east Tanzania......12
6b	Adult male with moderately wide tail base and pink head, not secretive, in western Kenya, north-west Tanzania, eastern Rwanda and Burundi......*Agama mwanzae*, Mwanza Flat-headed Agama [p. 232]	11b	Midbody scale rows more than 70, in Tanzania or western Kenya......13
		12a	Spines around the ear opening partially conceal it, these spines as long as the diameter of the eye opening, midbody scales 54–64, mostly in arid central and northern Kenya, adults usually 18–25cm total length......*Agama ruppelli*, Ruppell's Agama [p. 235]
7a	Midbody scale rows 99–114, range extreme southern Tanzania......*Agama kirkii*, Kirk's Rock Agama [p. 238]		
7b	Midbody scale rows 58–91......8		
8a	Midbody scale rows 58–67, has a fine non-tapering pale vertebral stripe, confined to south-central Kenya, south of Nairobi......*Agama hulbertorum*, Ngong Agama [p. 228]	12b	Spines around the ear opening don't conceal it, these spines about half as long as the diameter of the eye opening, midbody scales 52–57, mostly in extreme north-east Tanzania, arid east and north-east Kenya, adults usually 10–15 cm total length ......*Agama persimilis*, Somali Painted Agama [p. 234]
8b	Midbody scale rows 67–91, pale vertebral stripe either absent or tapers posteriorly, not confined to south-central Kenya, ...9	13a	In eastern arc forests of Tanzania...... *Agama montana*, Montane Rock Agama [p. 239]
9a	Tail of male red, orange or yellow at the base, blue at the end, females with one or two elongate yellow or light orange dorsolateral patches, not widely separated, midbody scale rows 70–79......	13b	In savanna and grassland of east and central Tanzania and south-western Kenya......*Agama armata*, Tropical Spiny Agama [p. 237]

KENYA RED-HEADED ROCK AGAMA *Agama lionotus*

IDENTIFICATION: A big, rock-dwelling lizard. The head is large and triangular, males have a small crest on the nape. The eye is fairly large, the ear opening is obvious. The body is depressed, legs strong and muscular, toes long and thin. The tail is laterally compressed (in some males it has a prominent ridge), tapering smoothly, roughly 60% of total length. In some specimens there is a club-like bulge at the end of the tail, used in a fight. The body scales are homogenous, in 67–91 rows at midbody. Maximum size about 35cm (males, females are smaller), average 20–30cm, hatchling size 8–10cm. Displaying males have a vivid red-orange or orange-yellow head, the throat is orange. The body is blue, often with a pale tapering vertebral stripe and darker ocelli; the tail often ringed with white or pale blue. Males with pale heads are usually non-dominant adults. Non-displaying males brown, or even whitish, with faint darker cross-bars and green/yellow head speckles. Females and juveniles are brown, with green head speckling, vague dark cross bars and a vertebral stripe of irregular, sub-rectangular markings; usually a maroon or rufous patch where the upper limbs touch the body, these patches may coalesce to form a maroon stripe. There is often some local variation. Females from extreme northern Kenya were vivid yellow and black dorsally. Taxonomic Notes: Previously known as *Agama agama lionotus*, a subspecies of *Agama agama* (a species complex with a huge distribution), now elevated to a full species. A subspecies with a dark throat bar, *A. l. elgonis*, occurs in western Kenya from the base of Mt Elgon south to Lake Nakuru, and is also known from Laikipia; it may be worthy of specific status.

Agama lionotus

HABITAT AND DISTRIBUTION: In coastal thicket and woodland, moist and dry savanna and semi-desert, from sea level to 2,200m altitude, although most common below 1,500m. Largely absent from areas without rock outcrops or big trees, so, although widely distributed, records are sporadic in some areas of east and northern Kenya. Widely distributed in Kenya, save the high south-west, where the Mwanza and Elmenteita Agamas occur. Ugandan records problematic, due to confusion of this form with *Agama agama* and *A. finchi*, only known with certainly from parts of the east and north. Widespread in a broad band from north-east to south-west Tanzania, records lacking from the north-west, and

Agama lionotus Kenya Red-headed Rock Agama. Top left: Stephen Spawls, Mzima Springs, male. Top right: Stephen Spawls, Voi, female. Bottom left: Stephen Spawls, gravid female, Mzima Springs. Bottom right: Bio-Ken archive, male preying on hatchling.

absent from the far south; but the distribution of this species in Tanzania at present poorly understood, due to confusion with *Agama dodomae*. Elsewhere, to southern Ethiopia and South Sudan.

NATURAL HISTORY: A diurnal, rock- or tree-dwelling colonial lizard, will use buildings and walls, where dense colonies may form due to the abundance of suitable sites and refuges. The dominant male (and sometimes subordinate males) bask in a prominent spot; activity is greatest in mid-morning. They hide in rock and tree cracks. If threatened, they run rapidly for cover, they can jump considerable distances (50cm or more). When basking, males bob the head up and down (this is disliked by some Moslems, who believe it is a mockery of their movement during prayer). Intruders in a dominant male's territory are attacked. The occupant bobs his head as he approaches; both sidle sideways and then rush in, biting savagely, and using the tail as a club. The loser flees. A study at Malindi found that if the dominant male was removed from the colony, the subordinate male quickly assumed control and displayed bright colours. The females lay 5–10 eggs in a hole they excavate, usually at the start of the rainy season, but sometimes to coincide with unexpected storms. In gravid females the red patches on the flanks become vivid, and may signal to a male that she is unavailable. In Ethiopia incubation was 50–60 days. Hatchlings and small juveniles are sometimes eaten by large adults, and hence do not share refuges with them. The diet includes a wide range of insects, especially ants and beetles, plant material is sometimes taken; large adults may eat small lizards including their own species. Mongooses have been seen digging up egg clutches; the adults are often taken by small birds of prey and snakes; predation on this species by Red Spitting Cobras *Naja pallida* has been seen in Ethiopia, the snake carefully climbing a wall to catch sleeping agamas.

DODOMA ROCK AGAMA *Agama dodomae*

IDENTIFICATION: A big, rock-dwelling lizard from the Dodoma area in Tanzania. Essentially similar to the previous species, the Kenya Red-headed Rock Agama (and originally regarded as a subspecies of that form), but differing in a number of subtle ways. The body scales are homogenous, in 70–84 rows at midbody. Males are larger than females Maximum male size about 35cm, average 20–32cm, hatchling size 8–10cm. Displaying males have a vivid red-orange or orange-yellow head, the throat is distinctly marked, with an orange centre and a deep blue-black border. The body is blue, often with a pale tapering vertebral strip and darker ocelli; the tail often ringed with white or pale blue. Females and juveniles are brown, with green head speckling, vague dark cross bars and a vertebral stripe of irregular, sub-rectangular markings; the lateral red patches may extend into a diffuse stripe.

Agama dodomae

HABITAT AND DISTRIBUTION: A Tanzanian endemic, found at altitudes between 500–2,400m altitude. It occurs from the area around Ngorongoro south through high central Tanzania to Singida, Dodoma, Ruaha National Park and the Morogoro area, isolated records from Lake Rukwa and the Rovuma River; and a subspecies, *A. d. ufipae*, with a uniformly dark throat, is known from Kipili on the extreme southern Lake Tanganyika shore. Conservation Status: Least Concern.

NATURAL HISTORY: As for the previous species; a diurnal, colonial species, usually on rocks, but also on houses, and on big trees (particularly baobabs); the dominant male basks in a prominent position. Around Dodoma nocturnal activity has been noted; the lizards feeding on insects around a lamp. A female in central Tanzania had eight eggs in her ovaries in February and many hatchlings about 5cm length, were also noted at that time. Diet includes insects, especially termites and ants, and also plant material.

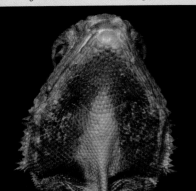

Agama dodomae Dodoma Rock Agama. Left: Bill Branch, male throat colours. Right: Michele Menegon, Dodoma, male.

NGONG AGAMA   *Agama hulbertorum*

IDENTIFICATION: A small rock agama, endemic to southern Kenya; recently recognised as separate from *Agama lionotus*. The head is broad, neck thin, snout short and rounded. Body squat and flattened. The tail is thin and tapering, about 55–65% of the total length. The vertebral scales are feebly keeled, dorsal and lateral scales keeled, dorsal scales are in 58–67 rows at midbody. Males have a single row of 10–11 precloacal pores. Maximum size about 24cm, average 15–20cm, hatchling size unknown but probably 6–7 cm. Displaying males look like a dull *Agama lionotus*, with a red head and blue body, non-displaying males have a brick-red head and back, the limbs and flanks are dull blue, and there is a fine pale, non-tapering vertebral line. The tail is banded blue and white, in some specimens it appears dull grey-white. Throat range, the underside is blue, the underside of the tail is white. Females are rufous-grey, with diffuse dark crossbars and vertebral and dorsolateral lines of light elongate spots.

Agama hulbertorum

HABITAT AND DISTRIBUTION: A Kenyan endemic, known only from dry savanna and rocky hills, at altitudes between 1,200–1,800m. Occurs from the ridge below Corner Baridi on the southern slopes of the Ngong Hills southwards through Olorgesaille and along the eastern wall of the Rift to Elangata Wuas, just west of Kajiado. Probably occurs further south.

NATURAL HISTORY: A diurnal, colonial species; usually on rocks, although observed on buildings at Olorgesaile, where they allowed close approach. The dominant male basks in a prominent position, often right on the top of a boulder. Presumably lays eggs. Diet: invertebrates.

Agama hulbertorum Ngong Agama. Left: Stephen Spawls, Olorgesaile, male. Right: Daniel Hollands, female, Olorgesaile.

MALABA ROCK AGAMA/BRIAN FINCH'S AGAMA
Agama finchi

IDENTIFICATION: A fairly large rock-dwelling agama, discovered in 2001 by the prominent Kenyan ornithologist Brian Finch. The head is broad, neck thin, snout short and rounded. Body squat and flattened. The tail is thin and tapering, about 55–70% of the total length. Dorsal scales keeled, scales at midbody in 70–79 rows. Maximum size about 28 cm, average 18–25 cm, hatchling size unknown. The displaying males have a red, orange or yellow head, often black spotted, and the orange extends down the spine, often to half-way between the pairs of legs, the body is blue-black and the tail is blue-black, orange and lighter blue at the tip. The females are light or dark brown, with green head spots, and a broad, irregular and diffuse lateral stripe of orange, yellow and black (yellow in eastern Congolese specimens). Juveniles resemble females, but the flank stripe is not so prominent. Taxonomic Note: the population in the Murchison Falls area is a subspecies of the nominate form; it is *Agama finchi leucerythrolaema*.

Agama finchi

DISTRIBUTION: Rocky hills in medium-altitude savanna and forest, at altitudes between 600–1,600m. Originally described from rock outcrops in savanna on the Malaba–Busia lroad, on Kenya's western border, south-west of Mt Elgon, it occurs north-west from there in a broad band from south-east Uganda through the centre to the north-west; and also north-east from Lake Kyoga to Moroto. Known localities include Katakwi, Kakoge, Masindi, Teso, Mabira Forest, Wanseko and Murchison Falls; an isolated record from Kidepo. Elsewhere, in south-west Ethiopia, eastern and north-eastern Congo.

NATURAL HISTORY: Poorly known. Diurnal and terrestrial; living on rocky hills and outcrops, but will also utilise tree and buildings. They are often relatively abundant and

Agama finchii Malaba Rock Agama. Top: Daniel Hollands, male, Murchison Falls. Bottom: Daniel Hollands, female, Murchison Falls.

allow close approach. Live structured colonies under a dominant male who basks in a prominent position; up to 14 females have been observed with one male. No clutch details are known, mating was observed in northern Uganda in late June; numerous juveniles also observed at this time. Diet probably insects and other arthropods.

HANANG AGAMA *Agama turuensis*

IDENTIFICATION: A medium-sized rock agama with a distinctive dark throat bar, endemic to north-central Tanzania. The head is broad, neck thin, snout short and rounded. Body squat and flattened. The tail is thin and tapering, about 55–65% of the total length. The vertebral scales are feebly keeled, dorsal and lateral scales keeled, dorsal scales are in 71–85 rows at midbody. Males have a single row of 11–29 precloacal pores. Maximum size about 30cm, average 18–25cm, hatchling size unknown. Displaying males have a brick-red head and dark blue body, the hind limbs and the base of the tail are light blue, the throat is brick-red with a distinct dark U-shape or broad cross bar at the base. The ventral surface is dark blue, lightening posteriorly. Females and juveniles are brown or grey, with rufous lateral patches, diffuse dark crossbars and vertebral and dorsolateral lines of light elongate spots or ocelli.

Agama turuensis

HABITAT AND DISTRIBUTION: A Tanzanian endemic, known only from dry savanna, rocky hills and mountain foothills, at altitudes between 1,300–2,200m. Occurs in central Tanzania, in Kondoa District, from various villages (Unyanganyi, Mangasini, Usandawi), all within 50km to the west and south-west of Kondoa, and also known from the slopes of Mt Hanang, north-northwest of Kondoa. Probably more widespread; reported from the Serengeti (not shown on map).

Family Agamidae: Agamas 231

Agama turuensis Hanang Agama. Left: Patrick Krause, Mt Hanang, female. Right: Patrick Krause, Mt Hanang, male.

NATURAL HISTORY: A diurnal, colonial species; usually on rocks. The dominant male basks in a prominent position. No breeding details known. Diet: invertebrates.

KAKAMEGA AGAMA *Agama kaimosae*

IDENTIFICATION: A large rock-dwelling agama from west of the rift valley. The head is broad, neck thin, snout short and rounded. Body squat and flattened. The tail is very broad at the base, but tapers abruptly (a useful field character), 55–65% of the total length. The dorsal scales are in 76–95 rows at midbody. Males have 10–13 femoral pores. Maximum size about 36cm, average 20–30cm, hatchling size unknown. Colour quite variable; it appears that where it is the only large species of agama, it has a predominantly purple back, but where it occurs in sympatry with *Agama mwanza*e it is more orange. The displaying males have a pink or orange head and back, and blue limbs and an orange or pinkish tail. The throat is purplish with darker purple longitudinal lines or orange with dark orange lines; the chest pinkish-purple. In Serengeti males the orange extends onto the limbs to some extent. The adult females are brown, with poorly defined darker cross bars and yellow scales on the sides of the tail base.

Agama kaimosae

Agama kaimosae Kakamega Agama. Dave Bygott, Serengeti, male.

DISTRIBUTION: An East African endemic. On rocky outcrops in medium to high savanna, at altitudes between 1,100–1,900m. In Kenya, known from the Kakamega area, from Kaimosi, Ngoromosi, Serem, the North and South Nandi Forest, and from the Loita Hills; probably occurs in the intervening country but records lacking; in Tanzania sporadically recorded from various kopjes in the Serengeti, westwards to near Shinyanga. Conservation Status: Least Concern.

NATURAL HISTORY: Poorly known. Diurnal and terrestrial; living on rocky hills and outcrops. Like other large agamas, they live in colonies with a dominant male who displays at a prominent place. They appear to live in sympatry with the Mwanza Flat-headed Agama *Agama mwanzae* in part of their range. They lay eggs, although no clutch details known; diet probably insects and other arthropods.

MWANZA FLAT-HEADED AGAMA *Agama mwanzae*

IDENTIFICATION: A big rock agama, the males have a vivid pink head and back. The head is triangular, ear openings large and obvious, neck narrow. On the neck there are clusters of large, spiky scales, a reduced vertebral crest is present. The body is depressed, the limbs long and muscular, the toes long and sharply clawed. The female tail is long and slim, tapering smoothly, the adult male has a fairly broad tail base (not as broad as in the Elmenteita Rock Agama *Agama caudospinosa*). In both sexes the tail is about 65% of the total length. The dorsal scales are feebly keeled, in 60–90 midbody rows. There are 8–13 femoral pores. Maximum size about 32cm, possibly larger, average 20–30 cm, hatchling size unknown but probably around 8–9cm. The displaying males are attractively coloured, the head, neck, shoulders, throat and chest are pink, with a violet or blue-white vertebral line, the lower back and tail mottled blue-white, the lower front limbs are blue, rear limbs green or blue. Non-displaying males range from grey-brown to blue, with extensive white speckling and ocelli on the flanks, with an electric blue vertebral stripe, tail and limbs; the chin is striped brown. Females and juveniles are brown with irregular dark crossbars and a vertebral stripe of light dashes, underside yellow or cream. Within its range, big males are unmistakeable, in the hand the reduced vertebral crest gives useful clues to identification. Taxonomic Notes: Regarded in the past as a subspecies of the Namibian Flat Agama *Agama planiceps*, now recognised as a full, distinctive species (and not closely related to *A. planiceps*). This species and the Elmenteita Rock Agama have, apparently, mutually exclusive ranges, but possibly overlap in the Loita Hills.

Agama mwanzae

HABITAT AND DISTRIBUTION: Endemic to East Africa. Rock dwelling, living on inselbergs in medium to high savanna and grassland, from 1,000m up to 2,200m. Occurs from the Maasai Mara south-west through the Serengeti, right around the southern Lake Victoria shore, across to eastern Rwanda and Burundi, south into southern Tabora district. Might occur north of the Mara in Kenya, but not documented; a curious small isolated population is recorded from around Mogotio near Nakuru. Conservation Status: Least Concern.

Agama mwanzae Mwanza Flat-heaed Agama. Left: Stephen Spawls, Maasai Mara, female. Right: Stephen Spawls, Displaying male, Seronera.

NATURAL HISTORY: Diurnal, living on boulders, outcrops, rocky cliffs and hills, even on quite small patches of sheet rock, provided there are fissures in which to hide. Sometimes on termite hills. They are quite wary, and although they will live on buildings and walls, they are often reluctant to allow humans to approach closely. Mwanza agamas live in structured colonies, with a dominant male, he basks in an exposed spot, bobbing up and down, and will attack intruders within his territory. Sometimes a whole colony of 20 or more individuals will inhabit a single fissure, and they present a remarkable spectacle when they all emerge to bask in the early morning sun. No breeding details known, presumably they lay eggs either in a rock fissure or in a hole in the soil, hatchling were observed in the southeast Mara in early January. Small juveniles and hatchlings never seem to share the same refuge as adults; it is suggested they are at risk of predation by the adults. Diet: arthropods, fond of ants. A juvenile was eaten by a kingfisher.

ELMENTEITA ROCK AGAMA   *Agama caudospinosa*

IDENTIFICATION: A huge, flat, secretive rock agama. The head is triangular, neck narrow. There are big, fang-like teeth at the front of the upper jaw. Around the earhole, at the angle of the jaw and on the neck there are clusters of large, spiky scales. The body is broad and flattened, the limbs stout and muscular (particularly in males), the toes long and sharply clawed. The tails of adults are very broad and flat at the base, 60% of the total length. The dorsal scales are almost smooth. There is usually a double row of 7–12 femoral pores. Maximum size about 45cm, possibly larger; it is the biggest Kenyan agama. Average 20–35 cm, hatchling size unknown. Colour very variable, in most areas the throat, chest and upper back of displaying males is rufous or orange, the back

Agama caudospinosa

is dark with light diffuse cross bars, the tail strongly banded light and dark. The females are brown with lighter cross bars. In parts of its range (rift valley between Nakuru and the Kedong) the head and the upper limbs of the males are blue and the dorsal surface is blue-black. Often there is a fine curved light horizontal stripe above each shoulder; this is a useful field character. Taxonomic Note: Those from the northern slopes of Mt Kenya (Meru west to Laikipia) and from the western wall of the Kerio Valley, between Iten and Moiben, are of the subspecies *Agama caudospinosa spawlsi*; all others the nominate subspecies.

HABITAT AND DISTRIBUTION: Endemic to Kenya. Lives in grassland and light woodland of the central rift valley and environs, at altitudes of 1,800m and above. Occurs from the Kedong Valley north through Lakes Naivasha, Elmenteita and Nakuru, north-east to Murang'a and Nanyuki, round the northern side of the mountain as far as Meru but no further east, north across the high country to Maralal, up the western side of the rift to Moiben and the trans-Nzoia, west to Kisumu. Conservation Status: Least Concern; a Kenyan endemic with a restricted range, but its favoured habitat; rocky hills and outcrops, means it is probably not under any threat. Habitat protected in Hell's Gate and Lake Nakuru National Parks.

NATURAL HISTORY: A secretive and poorly-known lizard, even though present in some well-populated areas of Kenya. It is diurnal, living on sheet rock, outcrops, rocky cliffs and hills, in and around the central rift valley. It is shy, moving quickly and unobtrusively away

Agama caudospinosa Elmenteita Rock Agama. Top: Colin Tilbury, male, Rumuruti. Left: Stephen Spawls, Nakuru, female. Right: Stephen Spawls, Iten, juvenile.

when approached; it can run extremely fast. Presumably, like other agamas, it is territorial, lays eggs, eat insects and other arthropods. In the Moiben area, quite dense colonies of over 30 adults were found on relatively small sheet rock outcrops of less than 50m². This species and *Agama lionotus* are not usually found in the same area, although on Laikipia both species were found cohabiting rock outcrops around 1,800m altitude. Enemies include the Augur Buzzard *Buteo rufofuscus*.

SOMALI PAINTED AGAMA *Agama persimilis*

IDENTIFICATION: A beautiful little ground-dwelling agama. The head is rounded, with a big cluster of prominent spiky scales on the neck behind the ear opening. The body is squat and depressed. The limbs are long, the hind toes very long. Tail broad at the base, tapering sharply, 60–65% of total length. The body scales are keeled and overlapping, in 52–57 rows at midbody. Maximum size about 16cm, average 10–14cm, hatchling size unknown, probably 5–6cm. It seems that females (average snout-vent length 5.6–6.4cm) are larger than males (average snout-vent length 4.3–5.4cm), which is unusual in agamas. Overall colour rufous or rufous-brown. There is a broad pale or brown bar between the eyes, the snout is pink or brown above, a big pink or red patch on the crown and a fine or broad grey, white, red or brown vertebral line. On either side of this line are paired brown and black or orange blotches, the back is suffused with grey or blue grey towards the back legs. There is a big brown blotch on each anterior flank. The limbs are mottled brown. Underside white.

HABITAT AND DISTRIBUTION: A species of the Somali arid zone, living in dry savanna and semi-desert, at altitudes between 300–1,200m altitude. Occurs from Mkomazi National Park in extreme north-east Tanzania, northwards through Kilibasi, and Rukanga to Voi and the low country surrounding the Taita Hills. Also known from Ukambani, in the Kitui-Ngomeni area, Buffalo Springs, Samburu and the Kalama Conservancy, isolated records from Wajir Bor and Mandera. Probably widespread in the intervening country (all suitable habitat) but no records. Elsewhere, in south and central Somalia and the Ogaden.

Agama persimilis

NATURAL HISTORY: Poorly known. Diurnal and ground-dwelling, in open country, living in holes positioned under a shady, spreading bush. It hunts around the perimeter of the bush, rushing out to snap up prey. If threatened it usually dashes back under the bush, and may dive into its hole, but occasionally it may remain motionless, hoping to avoid detection. A specimen at Voi was active during the middle to late afternoon. Not a colonial species, usually solitary, although it may live in true pairs or occasionally one male with two females; in Samburu National Reserve several pairs lived in holes under bushes in close proximity. Like the Tropical Spiny Agama *Agama armata*, they sometimes ascend into bushes and small trees to bask, and may be seen hanging from a flimsy twig, looking dead. A gravid female was collected in late August in Somalia, no other reproductive details known. Diet: arthropods.

Agama persimilis Somali Painted Agama. Stephen Spawls. Samburu National Reserve. Inset: Samburu National reserve.

RUPPELL'S AGAMA *Agama ruppelli*

IDENTIFICATION: A medium-sized, attractive, dry-country agama. The head is triangular, with a prominent cluster of big spiky scales round the ear opening, partially concealing it. The body is squat and depressed, limbs slender and fairly long. The tail is long and thin, 60–65% of total length. The body scales are keeled, occasionally spiky, in 54–64 rows at midbody. There are 9–13 precloacal pores (lower values in females). Maximum size about

28cm, average 18–25cm, hatchling size unknown. Ground colour brown or grey, some specimens uniform and patternless, others have a faint or distinct pattern. Usually a pale brown bar between the eyes with a chocolate-brown bar in front of and behind this, back of the head grey-brown. There are two broad sub-rectangular blotches on either side of the vertebral scales on the neck, a fine pale vertebral stripe runs the length of the body, with three pairs of big irregular warm brown blotches on either side. On the flanks there is either a brown bar or a series of brown blotches, the limbs are mottled brown and yellow, the tail barred light and dark brown. Yellow or cream below. Some gravid females retain

Agama ruppelli

Agama rueppelli Ruppell's Agama. Inset top: Eduardo Razzetti, Lake Turkana. Main image: Darcy Ogada, Huri Hills, gravid female. Inset bottom: Eduardo Razzetti Lake Turkana, patternless.

the usual adult pattern, others develop a green or turquoise head and vivid orange flank patches.

HABITAT AND DISTRIBUTION: A species of dry savanna and semi-desert, largely below 1,000m altitude. Found from southern Tsavo West and Voi, west to Elangata Wuas, north through Tsavo, Ukambani and Kora, thence north and west to the environs of Lake Turkana and into south-east Sudan. The only Uganda record is from Moroto. Seems to be absent from apparently suitable country in north-east Kenya, but this might be due to undercollecting, as it is widely (if sporadically) distributed in Somalia and the Ogaden.

NATURAL HISTORY: Poorly known. Diurnal and ground-dwelling, although it will climb into shrubs and bushes to bask. Lives in open country, in holes positioned under a shady, spreading bush or in thickets, hunting around the thicket and out in the open, dashing back to cover if threatened. It will also live on sheet rock, provided there are crevices in which it can hide, but not usually on vertical rocks, or anywhere where larger, rock-dwelling agamas live. Population numbers tend to fluctuate with the amount of available cover, for example in Tsavo East and Ukambani (where it is often seen basking or hunting on roads) it is apparently abundant during periods of high rainfall, but populations either crash or are all aestivating during dry periods. Not a colonial species, usually solitary, sometimes in pairs. No breeding details known, presumably lays eggs in a moist hole. Diet insects and other arthropods.

TROPICAL SPINY AGAMA *Agama armata*

IDENTIFICATION: A small, brown or grey, ground-dwelling agama. The head is broad, neck thin, snout short and rounded. Body squat and flattened. The tail is thin and tapering, about 55% of the total length. The scales on the front top of the head overlap towards the snout, those on the back are keeled, there are several longitudinal rows of enlarged spiny scales along the back; this is a good field character. The scales are in 88–105 rows at midbody. Males have a single row of 9–18 precloacal pores. Maximum size about 22cm, average 15–20cm, hatchling size unknown but probably 6–7cm. Colour grey, brown or rufous, depending on the soil/rock colour of its habitat. There are two or three cross-bars on the head, the one between the eyes is V-shaped. There is a pale

Agama armata

vertebral stripe, sometimes with a very fine darker line down the centre, which may fade at the level of the back legs or extend to the tail tip. There are four or five pale short dorsal cross-bars. In breeding males, the cross bars become greenish, the lips and chin vivid green or turquoise. The white chin has fine black vermiculations and a blue dot at the base, the chest speckled rufous, the remainder of the underside white.

HABITAT AND DISTRIBUTION: Mid-altitude savanna of south-west Kenya and the central plateau of Tanzania, at 1,400–2,000m altitude. From the Maasai Mara south through Serengeti, Olduvai Gorge and Shinyanga, east to Dodoma, south to the Usangu Plains and the Lake Rukwa area, west to Ugalla and Katavi. An apparently isolated population recorded from Kajiado, Kenya. Elsewhere, south to Botswana and South Africa.

Agama armata Tropical Spiny Agama. Stephen Spawls. Left: Maasai Mara, grey. Right: Northern Tanzania, rufous.

NATURAL HISTORY: Diurnal and ground-dwelling, on open plains, but may live on small outcrops and sheet rock, provided they are flat, not steeply tilted. Usually lives in a hole, but sometimes under a suitable rock, in a ground crack or a rock fissure. They bask in the open, but are quite wary, usually moving rapidly to cover if approached, although some adults will freeze, relying on their camouflage, which is excellent, to avoid detection. They sometimes ascend into bushes and small trees to bask, and may be seen hanging from a flimsy twig, apparently dead They lay 9–16 eggs, roughly 1×1.5cm, but no further details known in East Africa. Diet: insects and other arthropods, fond of ants, reported to eat plant material in southern Africa.

KIRK'S ROCK AGAMA *Agama kirkii*

IDENTIFICATION: A large agama, recently recorded in extreme south-west Tanzania. A southern African form, it is remarkably similar in appearance to *Agama lionotus*, despite being in another clade. The head is short and slightly bird-like, neck thin, body slightly depressed. The tail is long and thin, 60–70% of total length. The dorsal scales are strongly keeled but not spiky, in 99–114 rows at midbody; a dorsal crest extends from the nape to the tail. Maximum size about 30cm, average 20–28cm, hatchling size 6–7cm. Displaying males have an orange head and blue body, with a paler vertebral stripe, the tail may be blue or turquoise, with pale or even white rings. Breeding females have a turquoise head, the body is yellow-grey, with extensive diffuse rufous flank blotches; the limbs are grey, the tail banded dark brown and yellow-grey. Non-displaying males and females are grey or grey-brown, finely speckled black, with diffuse darker brown or rufous cross bars and a banded tail.

Agama kirkii

HABITAT AND DISTRIBUTION: Lives on rock outcrops in the savanna. In our area, only known from Tatanda, at 1,900m attitude, just north of Mbala in Zambia. Elsewhere south through Zambia and Malawi to Botswana.

NATURAL HISTORY: Diurnal and terrestrial; living on rocky hills and outcrops. Like other large agamas, they live in colonies with a dominant male who displays at a prominent place. In Zimbabwe they are wary and difficult to approach. They lay around ten eggs. Their diet consists largely of ants, plus insects and other arthropods; and in southern Africa they have been observed waiting beside ants trails around fig trees growing on outcrops, eating ants attracted to the fruit.

Agama kirkii Kirk's Rock Agama. Left: Andre Zoersel, northern Zambia, male. Right: Andre Zoersel, northern Zambia, female.

MONTANE ROCK AGAMA *Agama montana*

IDENTIFICATION: A small, lightly-built rock- and ground-dwelling agama, in the Eastern Arc Mountains of Tanzania; originally regarded as a montane sub-species of the Mozambique Agama *Agama mossambica*. The head is broad, neck thin, snout short and rounded. Body squat and flattened. It has thin legs. The tail is long, thin and tapering, 60–70% of total length. Males have a single row of 12–14 precloacal pores. Maximum size about 27cm, average 15–20cm, hatchling size unknown but probably 6–7cm. Colour variable; displaying males have a blue head and a rusty-red throat band and a pale or bluish, broad or narrow vertebral line; the tail is barred. The upper limbs are brown, sometimes the upper limbs have a blue-green wash. The throat is rich blue shading to black, the underside white. The females are brownish, sometimes with a blue or turquoise ear patch, gravid females are noticeably red-brown. Underside pale, often with diffuse black streaking.

Agama montana

HABITAT AND DISTRIBUTION: Usually in forested or deforested areas of the Usambara and Uluguru Mountains, eastern Tanzania, at altitudes between 700–1,400m, but also recorded near Handeni, at Tamota, Kama Mountain, north of the Uluguru Mountains and sporadically recorded from the northern Selous Game Reserve, north of the Rufiji River. Conservation Status: Vulnerable B2ab(iii). Largely dependent on montane forest and woodland, and vulnerable to alteration and destruction of its forest habitat. However, it seems to be able to adapt to agriculture, and will live in fields and gardens.

Agama montana Montane Rock Agama. Left: Walter Jubber, Uluguru Mountains, male. Right: Michele Menegon, Kimboza Forrest.

NATURAL HISTORY: Poorly known. Diurnal, lives on the ground, on earth banks, isolated rocks, rocky outcrops and on trees in the forest, basks in patches of sunlight. In the breeding season the males develop the usual bright colours. Females bury the eggs in pits they have dug themselves in the forest soil. In the Usambaras, clutches of 8–10 eggs, roughly 1×1.5 cm, recorded in October. A recently hatched juvenile of 7.6cm was caught in October. Diet mostly ants, but take other insects.

MOZAMBIQUE AGAMA *Agama mossambica*

IDENTIFICATION: A large elegant-looking agama with a flattened body. The head is short, the eye fairly large and high on the head, snout pointed. The tail is long, 55–70% of total length. A dorsal crest is present and the dorsal and lateral scales are keeled, with a scattering of vague lines of enlarged, spiky scales, in 69–94 midbody rows. There are 12–15 precloacal pores. Maximum size about 31cm (male), largest female 25.9cm, hatchling size unknown. The sides of the head in displaying males is vivid blue or turquoise. In both sexes, the back is grey, brown or rufous, with vague darker or rufous cross bars, which may be broad or narrow, and a vertebral line of light lozenges (sometimes very vague); the tail has fine dark bars, the underside light yellow-grey. Gravid females develop extensive rufous flank patches. In non-displaying males and females, there is often a dark blotch behind the eye, extending in a line to the earhole.

Agama mossambica

HABITAT AND DISTRIBUTION: Woodland and moist savanna, at low altitude, from sea level to about 1,200m. Found in south-east Tanzania, from Tanga south and south west to the Rovuma River, westwards to Ruaha National Park. Elsewhere, to Malawi, Mozambique and eastern Zimbabwe.

NATURAL HISTORY: Diurnal, often seen on the ground in leaf litter but when disturbed races for the nearest tree. Sleeps in suitable tree cracks at night. Presumably territorial, as males display, but details of habits unknown. Females lay eggs towards the end of the short rainy season or at the beginning of the long rains; a clutch of 13 eggs recorded; southern Tanzanian females laid in November and December. Diet: it favours ants, but takes beetles, other insects, millipedes, etc.

Agama mossambica Mozambique Agama. Left: Michele Menegon, male, Iringa. Right: Walter Jubber, Selous Game Reserve, female.

ORIENTAL AGAMAS *CALOTES*

An Asian genus of spindly-limbed arboreal diurnal agama lizards. Around 27 species are known, and one widespread species has been introduced into Kenya, on Mombasa Island.

INDIAN GARDEN LIZARD *Calotes versicolor*

IDENTIFICATION: A large spindly-limbed oriental agama, recently observed on Mombasa Island. The head is broad, neck thin, snout pointed. Body laterally compressed (unlike East African agamas which are flattened). The tail is thin and tapering, about 60–70% of the total length. There is a prominent neck and vertebral crest, a good field character, extending to the hindlimbs, the dorsal scales are smooth, in 39–55 rows at midbody. Maximum size about 45cm, average 25–35cm, hatchlings about 7cm. The colour is astonishingly variable, but male specimens from Mombasa Station had pinkish heads, and the back was light brown with fine diffuse dark barring, the bars encompassing light spots. The females are brown, with diffuse dark broken bars on the dorsal surface; in both sexes the tail is banded.

Calotes versicolor

HABITAT AND DISTRIBUTION: Very widespread in Asia, from India eastwards; in a variety of moist habitats; in our area known only from the railway station on Mombasa Island, where presumably it was inadvertently transported in a cargo. Conservation Status: Not assessed but this is an introduced species, and thus it needs to be monitored, to gauge its effect, if any, on the environment.

NATURAL HISTORY: A diurnal, colonial and largely arboreal species. The dominant male basks in a prominent position. Lays eggs; usually 4–12, but huge clutches of nearly 70 eggs known from India. Diet: invertebrates.

Calotes versicolor Indian Garden Lizard. Stephen Spawls, Maldives, male.

Kinyongia tavetana
Mt Kilimanjaro two-horned Chameleon
Stephen Spawls

SUB-ORDER **SAURIA: LIZARDS**
FAMILY **CHAMAELEONIDAE**
CHAMELEONS

The chameleons are a family of unusual, diurnal arboreal lizards, famous for their ability to change colour. They are celebrated in folklore and superstition, even Aristotle mused on chameleons. Unusually for lizards, the tail cannot be shed and regrown, and is prehensile in all large species. Their eyes, set in small turrets, move independently, their vision is sharp and can be binocular, they have no external ears and poor hearing. Their feet are uniquely adapted for climbing, with opposed bundled toes with sharp claws. Chameleons have a telescopic tongue, longer than their bodies, that can be shot at prey, an adaptation to silent predation on a flimsy branch. They have small, non-overlapping scales without bony plates. The sex organs of the male are located in sheaths at the base of the tail, which is thus broad in adult males.

Most can change colour rapidly, their colour being connected with their emotional/hormonal state, and sensitive to light, shade and temperature; the chromatophores (pigmented light-reflecting cells) in their skins are controlled by the autonomic nervous system, enabling instant response to a stimulus. Their colour usually matches their background; hence the amusing legend that a chameleon placed on a tartan rug will explode with frustration. Slow moving; their active defences are limited to hissing, biting, squirming, jumping from the perch and surreptitiously moving to keep the branch between them and danger (some pygmy chameleons can also vibrate), so their camouflage, by colour and body shape, has to be good. However, if sexually aroused or facing a rival (some are territorial and asocial), vivid, stunning colour patterns may appear, some become black with rage. In cold areas, they will also turn black in the early morning, flatten themselves to increase surface area and align themselves perpendicular to the light, in order to efficiently absorb heat and light from the sun (black is the best absorber of light and heat).

Some have blade-like or annular horns, rival males use these in fights, they may also bite and claw. Chameleons occupy the entire range of East African habitats, from sea-level thicket to montane moorlands. Low-altitude chameleons lay eggs; most high altitude ones give live birth. They are most useful animals, eating insect pests. They are preyed on by many animals, including snakes and birds, particularly hornbills, shrikes and starlings. Predation can cause chameleon populations to peak and crash. Some chameleon species are social, especially the horned species, where males are intolerant of conspecific individuals, and only come together to mate; but the small side-striped chameleons tolerate each other, and will co-exist in close proximity.

There are more than 200 known species of chameleon, in 12 genera (there were 130 species, in four genera when we first published the first edition). Found in Africa, Madagascar and some Indian Ocean islands, a few reach the Middle East and Southern Europe, one occurs in India and Sri Lanka. Chameleons can be loosely split into two groups, pygmy chameleons and large chameleons. There are four genera of pygmy chameleons; the two in Africa are *Rieppeleon* and *Rhampholeon*. Most pygmy chameleons are small brown forest dwellers. Of the eight genera of 'large' chameleons, three occur in East Africa: *Chamaeleo*, *Trioceros* and *Kinyongia*. Large chameleons are bigger, predominantly green and occupy a huge range of habitats. In our first edition, we described 40 chameleon species in two genera, but continuing field work by enthusiasts, a robust new taxonomy based on both morphology and DNA, a greater appreciation of the evolutionary species concept and the fact that the chameleons are fairly easily collected in the field means that species numbers have dramatically increased; some 60 species, in five well-supported genera, are now known to occur in our area, and this number will increase. Tanzania, with 40 species, has the greatest chameleon diversity in Africa, although Kenya is close behind, with 25 species; that diversity is largely due to allopatric speciation creating species with small ranges on isolated forested highlands. Evidence also indicates that the chameleon lineage is over 60 million years old; they originated in Africa as small, possibly ground-dwelling animals, twice dispersed to Madagascar by rafting, and the big flamboyant tree dwellers of the *Kinyongia* genus diversified over 40 million years ago.

In much of sub-Saharan Africa, chameleons are greatly feared, an irrational superstition connected with their strange appearance and behaviour. Some think they are venomous but many associate the presence of a chameleon with bad luck or evil. A book could be filled with African chameleon legends. And yet, to many people, chameleons are the most charming, beautiful and endearing of Africa's reptiles. In the west, they are popular pets, and this has created problems, because they are easily harvested from the wild. At night, many species sleep on the outer edges of vegetation, and are surprisingly visible in torchlight; a collector with a powerful lamp can devastate chameleon populations. In addition, many species have small ranges, often in woodland and forest, which itself is vulnerable to commercial exploitation in land-hungry countries. Thus many chameleon species are of conservation concern and the demand continues, because they are not easy to maintain and breed. Although the pet trade, agriculture and forestry provide a living to many poor people in Africa, there is a need to monitor chameleon populations and protect vulnerable habitats.

We have provided keys to identify our chameleons in the field, and diagrams showing the important points of chameleon anatomy, but in many cases, they can be fairly easily identified by a combination of their locality, size, colour and pattern, head shape, presence or absence of horns, ear flaps and ventral and gular crests. If you see a chameleon in the bush and don't have this book handy, try to note the factors above, and photograph it with your phone.

KEY TO THE EAST AFRICAN CHAMELEON GENERA

As Colin Tilbury, African chameleon expert, has often stated, no key based on external characters alone can be 100% accurate. However, when used with flexibility and with specimens from known locations, these keys should give a reasonably certain identification. There are some 'difficult' species; for example *Rhampholeon spinosum* and *R. viridis*, which have unusually long tails, *Trioceros bitaeniatus*, which is rarely if ever green (and *Rhampholeon viridis* is often green!), *Chamaeleo anchietae*, which is quite small, and *Kinyongia asheorum*, which unlike other members of the genus, has a short 'beard' of seven or fewer enlarged conical scales. Bear these in mind when using this next key.

1a Tail 40% or less of total length, often short and stumpy and rarely or weakly prehensile and rarely concentrically coiled, maximum size never larger than 11cm, overall colour usually predominantly brown or grey, with fine horizontal or oblique stripes......2

1b Tail 50% or more of total length, long and tapering smoothly, strongly prehensile and can be coiled up, maximum size always over 12cm, overall colour usually predominantly green, without fine body stripes......3

2a Soles of feet covered with sharply pointed spinous tubercles, usually lacks a rostro-nasal projection (minute if present), if pattern present usually consists of one or more horizontal stripes......*Rieppeleon*, Rieppel's pygmy chameleons [p. 245]

2b Soles of feet covered with sub-conical to rounded tubercles, usually has a rostro-nasal projection, pattern if present usually consists of one or more oblique stripes......*Rhampholeon*, African pygmy chameleons [p. 248]

3a Have no horns or rostral projection, body scales homogenous and not varying greatly in size, total length at least 25cm or more (except in *Chamaeleo anchietae*), usually in low to medium altitude savanna or woodland......*Chamaeleo*, savanna chameleons [p. 257]

3b Do not have the above combination of characters......4

4a No gular or vertebral crests, (although some males may have a row of vertebral spines, and *Kinyongia asheorum* has a 'beard' of conical scales), often have blade-like horns, never have annulated or bony horns, usually in forest or recently deforested areas......*Kinyongia*, East African montane chameleons [p. 262]

4b Gular and vertebral crests present, usually without horns although a small rostral process may be present, a few species have annulated or bony horns, usually in woodland or savanna, occasionally in forest......*Trioceros*, East African side-striped Chameleons [p. 280]

RIEPPEL'S PYGMY CHAMELEONS *RIEPPELEON*

A genus of three species of little striped chameleons, all three occur in our area. Unusually, all have fairly wide ranges. They also differ from other pygmy chameleons in the shape of certain bones in the skull, and their DNA puts them in a discrete clade that separated from the other African pygmy chameleons over 50 million years ago. Interestingly, several of these pygmy species, of both this genus and *Rhampholeon*, can vibrate violently and do so when held or poked; this presumably is an antipredator device but it is not known exactly how the vibration is produced; possibly by exhaling minute pulses of air.

KEY TO EAST AFRICAN MEMBERS OF THE GENUS *RIEPPELEON*

1a A single small beardlike tuft of tubercles present under the chin......*Rieppeleon brevicaudatus*, Bearded Pygmy Chameleon [p. 246]
1b Chin and throat smooth or with conical tubercles scattered or in divergent rows......2

2a Tail very short, usually less than 20% of total length......*Rieppeleon brachyurus*, Beardless Pygmy Chameleon [p. 245]
2b Tail averaging 25-30% of the total length......*Rieppeleon kerstenii*, Kenya Pygmy Chameleon [p. 247]

BEARDLESS PYGMY CHAMELEON *Rieppeleon brachyurus*

IDENTIFICATION: A small striped brown chameleon, without a rostral process. The casque is flattened. Tail not prehensile, 15–18% of the total length. No gular, ventral or vertebral crest, but has a single thin, slightly raised ridge on the flanks. Maximum size about 6cm, hatchling size 1.5cm. The males are smaller than the females. Colour usually brown or grey, with faint longitudinal stripes of darker brown or grey along the body.

HABITAT AND DISTRIBUTION: Low- to medium-altitude forests, extending into surrounding grassland, but up to 1,200m in the Usambara Mountains. Occurs southwards from the Usambara Mountains along the Tanzanian coast, and inland south of the Rufiji River; known localities include Usambara, Dar es Salaam, Kilwa, Liwale, Lindi and Songea, elsewhere south to northern Mozambique, west to Malawi. Conservation Status: IUCN Least Concern. Although seemingly rare, it has a wide range, so probably under no threat.

Rieppeleon brachyurus

Rieppeleon brachyurus Beardless Pygmy Chameleon. Michele Menegon, Ukaguru Mountains.

NATURAL HISTORY: Poorly known, but spends a lot of its time on the ground, in grass and leaf litter, will climb into small plants, sleeps in low vegetation at night. Said to be attracted to certain savanna trees in the fruiting season, presumably to get flies that come to the fallen fruit. Inflates itself when angry and if picked up, vibrates violently. Lays up to 14 eggs, gravid females taken in January in southern Tanzania. Eats invertebrates.

BEARDED PYGMY CHAMELEON   *Rieppeleon brevicaudatus*

IDENTIFICATION: A little, striped, brown chameleon without a rostral process, but it has a single tiny chin tuft of spiky scales. Casque flattened. The tail is short, longer and broader in males (20–30% of the total length) than in females (15–20% of the total length). No gular or ventral crest, but an undulating ridge or crest with a few raised scales is present along the spine, and there are thin, wavy, raised ridges lengthwise along the flanks. Males are larger than females, maximum size about 10cm, average 5–8cm, hatchling size unknown. Colour quite variable, females often orange-red or rufous, often mono-coloured, males grey or brown, the eye turret skin often distinctly green and the eyelids yellow, the flank stripes various rich shades of brown or chocolate, sometimes yellow brown.

Rieppeleon brevicaudatus

HABITAT AND DISTRIBUTION: Evergreen forest and coastal thicket, from sea level to altitudes of 1,300m. An East African endemic (might extend into extreme north-east Mozambique), occurs from the Shimba Hills in south-east Kenya and the Usambara Mountains, south along the eastern edge of Tanzania, inland to the Uluguru, Nguru and Udzungwa Mountains, south to the Rondo Plateau and Masasi, just north of the Rovuma River. Conservation Status: IUCN Least Concern.

Rieppeleon brevicaudatus Bearded Pygmy Chameleon. Top right: Michele Menegon, Ulugum Mountains. Left: Michele Menegon, specimen imitating a leaf, Mbarika Mountains. Bottom: Michele Menegon.

NATURAL HISTORY: Poorly known. Active on the ground, in leaf litter, moving slowly and often freezing for long periods, climbs into grass and low vegetation, sleeps there at night. Lays up to nine eggs, of length 10–12mm, gravid females in Tanzania were captured in September and October. Diet includes spiders, beetles, grasshoppers and crickets, plus other invertebrates. Vibrates when handled, as an antipredator mechanism, it is suggested this is caused by pulsed exhalations of air.

KENYA PYGMY CHAMELEON *Rieppeleon kerstenii*

IDENTIFICATION: A small brown or grey chameleon, often with longitudinal stripes, without a rostral process. Casque flattened. The tail is short, 25–42% of total length. No true gular, ventral or dorsal crest, but it has little spiky scales and scale clusters under the chin, on the tail, and on the prominent raised eyebrows and face. Maximum size about 10cm, average 6–8cm, hatchling size unknown, females often larger than males. Colour quite variable, may be grey, brown or yellow-brown, sometimes uniform (especially specimens from Kenya's north coast), often there is a series of light and dark horizontal body stripes or lines. A fairly broad dark stripe usually extends through the eye, even on mono-coloured specimens. Taxonomic Note: The subspecies *Rhampholeon kerstenii robecchii* just reaches extreme northern Kenya, it differs in having a more elevated ridge above the eye.

Rieppeleon kerstenii

HABITAT AND DISTRIBUTION: An unusually adaptable pygmy chameleon; found in coastal woodland and thicket, moist and dry savanna and semi-desert, from sea level to

Rieppeleon kerstenii Kenya Pygmy Chameleon. Top left: Michele Menegon, Mkomazi Game Reserve. Top right: Tomas Mazuch, Ethiopia, subspecies robecchii. Bottom: Stephen Spawls, Northern Tanzania.

1,400m altitude. The subspecies *R.k. robecchi* occurs in far northern Kenya, known localities include Malka Murri, Moyale and possibly Mandera, and meets up with the nominate subspecies somewhere between Moyale and the Huri Hills; from there it occurs south past the base of Mt Marsabit to the northern Nyambene Hills, Kitui, Makueni and the eastern Tana River, it occurs all along the coast and into north-eastern Tanzania, inland in Kenya to Taru, Voi and Mudanda Rock, Tanzanian inland records include Mkomazi and Tarangire National Park, also known from Elangata Wuas south of Nairobi, so probably occurs all around Mt Kilimanjaro in suitable low country. Elsewhere, to eastern Ethiopia and northern Somalia. Conservation Status: IUCN Least Concern. With such a huge range, in often dry country, in conservation terms it is under no threat.

NATURAL HISTORY: Lives in bush and grass, hides in thickets. In the Tsavo area, it seems to be crepuscular, active at dawn and dusk, but on the Kenya coast it moves around during the day. Hides in holes and termite hills in dry savanna and semi-desert, in open country fond of thickets and thorn bush. Its camouflage is superb, even in tiny leafless thorn bushes it is virtually impossible to spot. Lays eggs, a relatively large clutch of 9–12 eggs recorded, roughly 1×0.6cm in size. Six gravid females were collected near Mt Kasigau between October and December, one clutch was laid in early November and took just over 50 days to hatch. Eats invertebrates. Vibrates violently, like other pygmy chameleons; a naturalist who picked one up in Malindi dropped it and said he had received an electric shock!

AFRICAN PYGMY CHAMELEONS   *RHAMPHOLEON*

A genus of 19 species of forest-dwelling pygmy chameleons, most have small ranges in hilltop forests in south-east Africa, although two are fairly widely distributed in the forests of west and central Africa. Nine species occur in our area.

KEY TO EAST AFRICAN MEMBERS OF THE GENUS *RHAMPHOLEON*

This key will only really work in the laboratory and requires the use of a binocular microscope or a powerful hand lens. However, most of the chameleons in this genus may be identified by their locality, as they live in mutually exclusive areas, with a few exceptions.

1a	Claws of feet are simple, non-bicuspid…2		255]
1b	Claws of feet strongly bicuspid……4	3b	Accessory plantar spines weak or indistinct, often green, in Pare Mountains (and possibly the West Usambara Mountains)……*Rhampholeon viridis*, Pare Pygmy Chameleon [p. 256]
2a	Rostral process prominent and cushion-like, occurs in both East and West Usambara Mountains……*Rhampholeon spinosus*, Usambara Spiny Pygmy Chameleon [p. 253]		
		4a	Deep dermal pits in the groin/inguinal region……*Rhampholeon beraducci*, Mahenge Pygmy Chameleon [p. 250]
2b	Rostral process short, stubby or indistinct……3	4b	Groin/inguinal pits absent or indistinct……5
3a	Accessory plantar spines well-developed and prominent, usually brown or grey, in the Eastern Usambara Mountains……*Rhampholeon temporalis*, Usambara Pitted Pygmy Chameleon [p.	5a	Axillary dermal pits prominent……6
		5b	Axillary dermal pits absent……8
		6a	Temporal crest weak and indistinct, in

	forest of western East Africa......*Rhampholeon boulengeri*, Boulenger's Pygmy Chameleon [p. 250]		*moyeri*, Udzungwa Pygmy Chameleon [p. 252]
6b	Temporal crest prominent, in southern or eastern Tanzania......7	8a	Tail short in both sexes, less than 25% of total length, in our area in the mountains around the north end of Lake Malawi (Lake Nyasa)......*Rhampholeon nchisiensis*, Pitless/Nchisi Pygmy Chameleon [p. 253]
7a	11–13 tubercles between the bases of the supraoptic peaks, in the Uluguru, Nguru, Ukaguru or Nguu Mountains...... *Rhampholeon uluguruensis*, Uluguru Pygmy Chameleon [p. 255]	8b	Tail averages more than 25% of total length, in our area in the Nguru Mountains ......*Rhampholeon acuminatus*, Nguru Spiny Pygmy Chameleon [p. 249]
7b	15–19 tubercles between the bases of the supraoptic peaks, in Udzungwa or Rubeho Mountains......*Rhampholeon*		

NGURU SPINY PYGMY CHAMELEON *Rhampholeon acuminatus*

IDENTIFICATION: A small brown or green chameleon with short tail and a prominent, laterally compressed rostral process extending several millimetres off the tip of the snout. Maximum total length of about 8.5cm, tail not exceeding one-third of total length and lined by a row of enlarged, spine-like tubercles. Casque is elevated posteriorly and bears a series of acuminate spines along the later edges. Dorsal crest is crenulated, marked by prominent clusters of tubercles in correspondence of each vertebral body. Temporal crest is prominent and fin-like, formed by a ridge of 3–8 enlarged tubercles. No pit in axilla or groin region. Soles of feet are smooth, claws strongly bicuspid. Two transversal darker stripes, extending from the keel across the flanks are always present; patches of blue may be present on the casque and yellow to orange spots on top of the head. Eyeball orange. The relatively longer tail can distinguish males from females. Can be distinguished from all other species of *Rhampholeon* by the combination of prominent, laterally compressed rostral process and acuminate spines on casque and tail.

Rhampholeon acuminatus

HABITAT AND DISTRIBUTION: Endemic to the Afromontane rainforest of the Nguru Mountains, only known between 1,500–1,600m, from a area of about 28 km² within the newly-established Mkingu Nature Reserve. Conservation Status: IUCN Critically Endangered. It is threatened by deforestation and habitat alteration and by collection in the wild for the international pet trade, which is currently unregulated.

Rhampholeon acuminatus Nguru Spiny Pygmy Chameleon. Left: Michele Menegon, Nguru South Forest Reserve, male. Right: Michele Menegon, Nguru South Forest reserve, female.

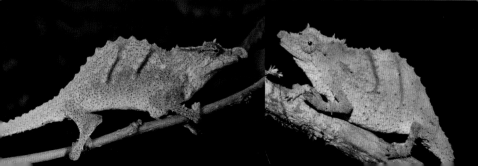

NATURAL HISTORY: Poorly known, active in the leaf litter during the day, roosting on ferns, stems and small branches from 30 up to several metres from the ground. Oviparous, between two and four eggs are laid, with hatching in January.

MAHENGE PYGMY CHAMELEON *Rhampholeon beraduccii*

IDENTIFICATION: The smallest chameleon in Africa with a maximum total length of about 4cm. A tiny brown or yellowish chameleon with short tail, not exceeding one-quarter of total length, and a soft tuberculated rostral process projecting off the tip of the snout. Temporal crest weakly developed, dorsal crest crenulated, each crenulation is formed by a cluster of tubercles. A single deep pit in each axilla and groin are present. Soles of feet are smooth, claws bicuspid. The shape of the body and colour mimics a dead leaf. Two transversal darker stripes, extending from the keel across the flanks, sometimes dotted with back spots are often present. The relatively longer tail can distinguish males from females. Eyeball orange. This species can be distinguished from the other Tanzanian members of the subgenus *Rhinodigitum* (all bearing a rostral process), by the presence of dermal pits both in axilla and groin regions.

Rhampholeon beraduccii

HABITAT AND DISTRIBUTION: Mahenge Mountain only, on the edge of the Kilombero flood plain. Known from a single forest fragment of about 40 km^2. It occurs in submontane rainforest, at about 1,000m. Conservation status: IUCN Vulnerable.

NATURAL HISTORY: Poorly known, during the day is active in the leaf litter, at night they roost on ferns, stems and small branches, usually very close to the forest floor. They lay eggs. Named for the Arusha-based American herpetologist Joe Beraducci and his unstinting, kindly assistance to researchers.

Rhampholeon beraduccii Mahenge Pygmy Chameleon. Left: Michele Menegon Mahenge Mountain, female. Right: Michele Menegon, Mahenge Mountain, male.

BOULENGER'S PYGMY CHAMELEON *Rhampholeon boulengeri*

IDENTIFICATION: A tiny brown or grey chameleon, with a small scaly rostral process on the snout, like a small soft horn, this may be pointed or club-shaped. A prominent spiky crest is present above each eye. Casque flat. Tail not prehensile, although it may be used for gripping, tail length 17–25% of total length, the male usually has a fat and longer tail, the female a thin shorter one. No gular or ventral crest, but there is a scattering of spines on the throat. Dorsal keel weakly crenulated. Has a pit in each axilla (armpit), but none in the groin. Small spines are present at the base of the digits. Maximum size about 8cm, average 5–7 cm, hatchling size unknown. Colour usually shades of brown, rufous or grey, can be so dark brown as to appear black. There are usually two (sometimes one or three) fine, black, oblique lateral stripes on the flanks, heightening the resemblance of this species to a dead leaf.

Rhampholeon boulengeri

HABITAT AND DISTRIBUTION: Forest and woodland, from 1,400–2,400m altitude in East Africa, lower elsewhere. A Central African species and probably a complex which may be split in the future. Sporadic records from extreme north-west Tanzania (Minziro Forest), western Kenya (Kakamega, Cherang'any Hills, North Nandi Forest); Iganga on the northern Victoria shore, fairly widespread along Uganda's south-western border, western Rwanda and Burundi. An isolated record from the Budongo Forest. Elsewhere, to the eastern DR Congo. Conservation Status: IUCN Least Concern.

NATURAL HISTORY: Forest fringes, clearings and along streams, on the ground in leaf litter or in low plants, where its close resemblance to a leaf means it is largely unnoticed. Often remains motionless for a long time, until prey passes by. At night, climbs into low vegetation (usually 30–60cm above ground level, but will go higher). Moves very slowly and carefully, like all chameleons, and will freeze, even with legs raised, for a long time if a predator approaches. When angered, inflate themselves, they are known to sham death, and if picked up or seized will produce a burst of inaudible vibrations. This vibration may also be a factor in combat and territorial displays. They lay from one to three eggs, 12–15 mm long, which may take up to 6 months to hatch; gravid females recorded in September in Kakamega Forest. They eat invertebrates; spiders, grasshoppers, flies, beetles and termites recorded. Several collections of these chameleons contained only females, leading to suggestions that in parts of its range it may be parthenogenic.

Rhampholeon boulengeri Boulenger's Pygmy Chameleon. Left: Michele Menegon, female, Rwanda. Right: Max Dehling, male, Rwanda.

UDZUNGWA PYGMY CHAMELEON Rhampholeon moyeri

IDENTIFICATION: A tiny brown or green chameleon with short tail and a soft tuberculated rostral process extending up to 2mm off the tip of the snout. Maximum total length of about 6.5cm, tail not exceeding one-quarter of total length. Temporal crest is obvious and formed by a ridge of 7–8 enlarged tubercles, dorsal crest crenulated, each crenulation is formed by a cluster of tubercles. A single deep pit is present in each axilla, no pits in the groin region. Soles of feet are smooth, claws strongly bicuspid. The shape of the body and overall colour mimics a dead leaf, usually brown or greenish in colour, sometimes yellow or grey, with a lichen-like pattern. Two transversal darker stripes on the flanks are always present, legs often broadly banded. Eyeball orange. Males can be distinguished by their relatively longer tail. Until recently there has been some confusion with the taxonomy due to its close morphological similarity to *Rhampholeon uluguruensis*.

Rhampholeon moyeri

HABITAT AND DISTRIBUTION: Afromontane forest, from 1,500–2,200m altitude. Recent molecular investigations shows this species occurs in the Udzungwa Mountains only. The population in the Rubeho, Ukaguru, Nguru and Kanga Mountains, as well as the population occurring at lower elevation in the Uzungwa Scarp Nature Reserve, originally assigned to this species, actually belong to genetically very different species, although the map shows the entire complex currently in the process of being described species. Conservation Status: IUCN Least Concern

NATURAL HISTORY: During the day they are active in the leaf litter, where they are very hard to spot, but are easily located at night by torchlight, roosting on ferns, stems and small branches from 30cm up to 3m from the ground. In some localities it is common to see them roosting upside down on thin vines on trees, probably as an anti-predator behaviour. Oviparous, a female was observed excavating a hole 1cm deep, depositing the eggs in it and partially covering them with soil. For the next two days the female remained around the deposition site. When touched some specimens produce a perceptible buzzing vibration.

Rhampholeon moyeri Udzungwa Pygmy Chameleon. Left: Michele Menegon, Rubeho Mountains, female. Right: Michele Menegon, Udzungwa Scarp Nature reserve.

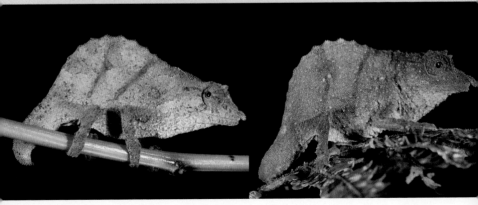

PITLESS/NCHISI PYGMY CHAMELEON
Rhampholeon nchisiensis

IDENTIFICATION: A small brown or grey chameleon, resembling a dead leaf, with a small rostral process. The casque is flat. The tail is short, 18–25% of total length. No gular, ventral or dorsal crest. No pits in the axilla or groin. Maximum size about 8.5cm, average 5–7cm, hatchling size unknown. Colour variable, usually shades of brown or grey, with two to three thin, oblique, dark or black flank stripes, heightening the resemblance to a dead leaf.

Rhampholeon nchisiensis

HABITAT AND DISTRIBUTION: Evergreen forest, at altitudes of 1,800m and above at the north end of Lake Malawi (Lake Nyasa). Known from Mt Rungwe, Poroto, Ukinga Mountains and the Kipengere Range in southern Tanzania. Originally described from the Nchisi Mountains in Malawi. Conservation Status: IUCN Least Concern.

NATURAL HISTORY: Poorly known. Active in leaf litter, climbs into low vegetation and sleeps there at night. Inflates the body, flexes the limbs and brings up the tail when picked up. Lays up to 15 eggs, may lay two clutches per year. Eats invertebrates.

Rhampholeon nchisiensis Pitless Pygmy Chameleon. Left: Michele Menegon, southern Tanzania. Right: Michele Menegon, southern Tanzania.

ROSETTE-NOSED CHAMELEON/USAMBARA SPINY PYGMY CHAMELEON *Rhampholeon spinosus*

IDENTIFICATION: A tiny, bizarre, spiny chameleon. It has a strange, club-like, rosette-shaped, soft, spiky horn on the end of its long snout. The casque is slightly raised to a broad, blunt point posteriorly. The tail is short and prehensile, about two-fifths of the total length. The body scales are heterogeneous, there are enlarged granular scales on the flanks, two rows of soft spines on the flanks and clumps of spines on the limbs and tail. No gular or ventral crest is present, but a distinctive dorsal crest of irregularly-spaced, soft spiny scales extends along the back onto the tail. Maximum size about 9cm, hatchling size unknown, probably 2–3 cm. Predominantly ashy-grey, males sometimes have a curious wash of lime-green on the flanks and tail and light blue on the head, females often have irregular brown blotches on the head and to a smaller extent on the body, and brown bands on the limbs. Taxonomic Note: Originally in the genus *Chamaeleo*, until anatomical and molecular analysis indicated it is a species of *Rhampholeon*.

Rhampholeon spinosus

HABITAT AND DISTRIBUTION: A Tanzanian endemic, found in and at the edges of forest and woodland of both the eastern and western Usambara Mountains at altitudes of 700–1,200m. Thus very vulnerable to destruction of its tiny, restricted habitat, its ability to adapt to cultivation is poor, it is unable to adapt to deforestation, and cannot compete with the blade-horned chameleons in the Usambara Mountains. Conservation Status: Endangered B1ab(ii,iii).

NATURAL HISTORY: Lives at the edge of the forest, in scrubby vegetation, but known to ascent to 2–3m height. Sympatric with several other species of chameleon. When threatened, it carefully shifts to impose the branch between itself and danger. The horn might be used in combat. Lays eggs, clutches of 3–4 recorded, both March and September, suggesting two clutches may be laid per year. Eats invertebrates.

Rhampholeon spinosus Rosette-nosed Chameleon. Top: Michele Menegon, East Usambara Mountains. Bottom left: Stephen Spawls, male by day. Bottom right: Male by night.

USAMBARA PITTED PYGMY CHAMELEON
Rhampholeon temporalis

IDENTIFICATION: A small brown or grey chameleon, resembling a dead leaf, with a little rostral appendage. The casque is low. The tail is short, 27–36% of total length; shorter tails in females. No gular, ventral or dorsal crest. The claws are simple, without a secondary cusp. Maximum size about 8cm, hatchlings 2.6cm. Colour variable, usually shades of brown, grey or pinky-grey, with one, two to three thin, oblique, dark or black flank stripes, heightening the resemblance to a dead leaf.

HABITAT AND DISTRIBUTION: A Tanzanian endemic. Evergreen forests of the East Usambara Mountains and Magrotto Hill, at 900–1,500m altitude. Conservation Status: Endangered B1ab(i-ii)+2ab; at risk from habitat loss.

Rhampholeon temporalis

NATURAL HISTORY: Poorly known. They are active by day in leaf litter in the deep forest but also forest fringes; they climb into low vegetation (10–50 cm above ground level) and sleep there at night. When handled, they show a curious behaviour; inflating their bodies, closing their eyes, flexing their limbs and then lapsing into immobility. They don't vibrate. A captive laid two eggs in November, which hatched four months later at around 22°C. Eats invertebrates.

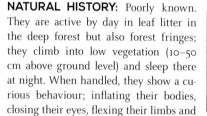

Rhampholeon temporalis Usambara Pitted Pygmy Chameleon. Stephen Spawls, Usambara Mountains.

ULUGURU PYGMY CHAMELEON *Rhampholeon uluguruensis*

IDENTIFICATION: A small brown, grey or green chameleon, recently described from the Uluguru Mountains; it resembles a dead leaf, with a short rostral appendage and prominent canthal process. The casque is flat. The tail is short, 19–22% of total length in females, 23–27% in males. No gular crest but there are large conical tubercles on the chin and a weakly undulating dorsal crest is present. Maximum size about 5cm, hatchling size unknown. Colour variable, usually light to dark brown, sometimes grey or green, mottled with darker blotches of the ground colour, one or two darker, complete or partial, oblique flank stripes are present, heightening the resemblance to a dead leaf. During courtship, one male became green on the flanks and back, with rich light brown markings, the belly became deep brown, the female was a similar colour but with many dark brown spots, the belly heavily speckled with white and yellow.

Rhampholeon uluguruensis

HABITAT AND DISTRIBUTION: In sub-montane to montane, closed-canopy evergreen forest, at altitudes between 1,500–2,000m. An Uluguru landscape endemic, known from

Rhampholeon uluguruensis Uluguru Pygmy Chameleon. Left: Michele Menegon, Uluguru Mountains, female. Right: Michele Menegon, Uluguru Mountains, male.

the Ukaguru Mts and the Mkungwe forest Reserve only. Conservation Status: IUCN Least Concern.

NATURAL HISTORY: Poorly known. Active on the ground by day, in moist open rocky areas of the forest floor, climbs into low vegetation and sleeps there at night. During courtship a male approached a female from behind and began jerking back and forward, as both moved slowly forward. The male grasped the female's hindlimbs and back. The female stopped and turned sideways. The male grasped the female's neck and then climbed onto her back. A clutch of three eggs recorded, 6×10mm, deposited in a hole 3cm deep, which was excavated by the female. After laying, the hole was closed and was perfectly concealed.

PARE PYGMY CHAMELEON *Rhampholeon viridis*

IDENTIFICATION: A small green or brown chameleon with relatively short tail, exceeding 30% of the total length, which is about 9cm. Sometimes a small rostral bulge, projecting slightly off the snout and formed by subconical tubercles, is present. Casque flat, temporal crest is prominent and formed by a ridge of 4–7 enlarged conical tubercles, which continues upward toward the top of the casque. Dorsal crest is crenulated, a cluster of tubercles forms each crenulation, it fades towards the sacrum and it reappears over the tail. Soles of feet smooth, claws simple. Axillary and inguinal dermal pits are present. Two transversal brownish to bright orange stripes, extending from the keel across the flanks are always present. Eyeball orange. The shape of the body and overall colour mimics a dead leaf, sometimes a combination of rust and whitish coloured patches gives the chameleon a lichen-like pattern. The relatively longer tail and the rich emerald green colour, when undisturbed, can distinguish males from females.

Rhampholeon viridis

HABITAT AND DISTRIBUTION: Confined to Afromontane forest, from 1,400–2,070m, in the North and South Pare Mountains of northern Tanzania. It probably does not occur in the West Usambara Mountains, despite the existence of a specimen from these mountains assigned to this taxon. The habitat is under threat; the total size of the forest presently remaining and available for this species is approximately 152 km^2. The South and North Pare mountains have lost all of their submontane forest in the last 50 years, and only montane

Family Chamaeleonidae: Chameleons

Rhampholeon viridis Pare Pygmy Chameleon. Left: Michele Mengon, North Pare Mountains, female. Right: Michele Menegon, North Pare Mountains, female. Inset: Michele Menegon, Pare Mountains, male.

forest fragments remain, which are entirely surrounded by transformed landscapes. The species is also known to be in the pet trade, but the degree of harvest is unknown. Conservation Status: IUCN Endangered.

NATURAL HISTORY: During the day they are active in the leaf litter, roosting on ferns, stems and small branches at night. Sometimes found at forest edges, but not outside forested areas in transformed habitats. Oviparous, a captive specimen laid eight eggs which hatched after 81 days of incubation. As with other species in the genus, when touched some specimens produce a perceptible buzzing vibration.

TYPICAL OR SAVANNA CHAMELEONS
CHAMAELEO

At one time all African 'non-pygmy' chameleons were placed in this genus, but after recent taxonomic changes, only 14 species remain there, including those that occur around the Mediterranean, in the Middle East and India. Four species occur in our area; they are relatively large, hornless, egg-laying savannah dwellers with huge ranges. They can be hard to distinguish, but locality, ear flap details and tail length should give useful clues.

KEY TO THE EAST AFRICAN MEMBERS OF THE GENUS *CHAMAELEO*

1a	Occipital lobes present, relatively large and moveable......*Chamaeleo dilepis*, Flap-necked Chameleon [p. 258]	3a	Male always with tarsal spurs, tail about 50% of total length, dorsal crest usually moderate in size, not confined to the western side of East Africa...... *Chamaeleo gracilis*, Slender Chameleon [p. 260]
1b	Occipital lobes absent, or minute and immoveable......2		
2a	Tail noticeably short, less than 43% of total length, dorsal keel formed by a double row of tubercles......*Chamaeleo anchietae*, Angola Chameleon [p. 258]	3b	Male without tarsal spurs, tail about 40-45 % of total length, dorsal crest usually minute, in western East Africa,......*Chamaeleo laevigatus*, Smooth Chameleon [p. 261]
2b	Tail not always noticeably short, usually 43% or more of total length, dorsal keel formed by a single row of tubercles......3		

ANGOLA CHAMELEON Chamaeleo anchietae

IDENTIFICATION: A fairly large chameleon without horns or rostral appendage, with a short tail. The casque is slightly raised posteriorly. Tail prehensile, 35–43% of the total length. The gular crest extends onto the belly (ventral crest) and consists of white or pale soft triangular scales. A weak dorsal crest is usually (not always) present. Maximum size about 18cm, average 12–15cm, hatchlings 4–5cm. Colour usually green, brown or yellow-brown, sometimes with three or four broad, darker, vertical flank bars, sometimes a line of short white bars or spots along the mid-flanks.

HABITAT AND DISTRIBUTION: Plateau grassland at altitudes between 700–1,800m. In our area a population occurs in eastern Rwanda, Burundi and far north-western Tanzania; a solitary record from the Udzungwa Mountains. Elsewhere, south and west through savanna in the DR Congo to Angola. Conservation Status: IUCN Least Concern.

Chamaeleo anchietae

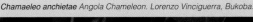
Chamaeleo anchietae Angola Chameleon. Lorenzo Vinciguerra, Bukoba.

NATURAL HISTORY: Active in bush and low savanna trees; it is suggested that the short tail is an adaptive response to life in low bushes rather than trees, where a fall is more dangerous. Strongly territorial, males will not tolerate smaller males in their territory. Lays 10–15 eggs in a burrow which the female excavates. Eats invertebrates.

FLAP-NECKED CHAMELEON Chamaeleo dilepis

IDENTIFICATION: A big savanna chameleon, without horns or a rostral process, but with small to large ear-flaps. The casque is slightly raised at the back. The tail is prehensile, roughly as long as the body. The body scales are heterogeneous, the scales on the ear flaps are often large. A prominent gular crest, consisting of a ridge of pale (often white) spiky scales becomes a prominent ventral crest, a dorsal crest of smaller spiky scales shrinks posteriorly and vanishes on the tail. The interstitial skin in the grooves on the chin may be vivid orange or yellow. Maximum size about 43cm, but most specimens 15–25cm, hatchlings 4–5cm. Colour very variable, mostly shades of green but can be brown, yellowish or grey; spotted black when angry, often a white lateral stripe is present. An orange phase occurs on the coast. Taxonomic Note: Various forms of this chameleon have been described as subspecies or even species, mostly based on flap size; all are just variants of a single polymorphic species.

Chamaeleo dilepis

Chamaeleo dilepis Flap-necked Chameleon. Top: Frank Glaw, Pemba. Bottom left: Stephen Spawls, Ndutu, Tanzania. Bottom right: Royjan Taylor, Tiwi.

HABITAT AND DISTRIBUTION: Coastal forest, woodland, thicket and both moist and dry savanna, from sea level to about 1,500–1,800m altitude. For example, in Kenya it doesn't reach Nairobi but is found at Sultan Hamud and just east of Thika. Widely distributed throughout Tanzania outside of forest and high altitude, enters western Burundi and Rwanda, also in north-east Rwanda. Common in south-east Kenya, along the entire coast and a short way up the Tana River, all around the eastern Lake Victoria basin, and just into eastern Uganda, (and on Lolui Island), also found in the low-lying country east of Mt Kenya. Not in the north, apart from a Mandera record. Elsewhere, south to South Africa, west to Cameroon. Conservation Status: IUCN Least Concern; its huge range, relative abundance and willingness to utilise suburbia means it is under no threat.

NATURAL HISTORY: Lives in bushes, shrubs and trees, but will descend to the ground. Often seen crossing roads, males looking for females may walk eastwards in the morning and westwards in the afternoon. Sleeps on the outer branches of trees at night. If threatened and unable to escape (it can move quite rapidly for a chameleon) it will inflate its body, distend its throat, raise its ear-flaps, hiss and bite furiously; it can also vibrate. In the dry season (and winter further south) they may aestivate, in a hole or some other refuge, such as a tree hole or weaverbird nest. In the mating season, the male's gular groove orange skin goes pale, and the female will then permit him to approach for mating; they mate for several days, with the female becoming increasingly intolerant of the male. Mating observed in southern Tanzania in January and February and in April in the Mara. The females lay on average 20–40 eggs (but up to 73 recorded), 10–15 mm long, in a tunnel dug by the female in damp soil; a Rondo Plateau female laid 38 eggs in February, a Newala female laid 39 eggs in March. The eggs hatch after several months. Diet: mostly invertebrates, occasionally small vertebrates. The distribution of this species and *Chamaeleo gracilis* match closely that of the Boomslang *Dispholidus typus,* which often eats them. Other enemies include predatory birds (shrikes, small birds of prey, coucals) and monitor lizards.

SLENDER OR GRACEFUL CHAMELEON   *Chamaeleo gracilis*

IDENTIFICATION: A big savanna chameleon, without horns, some specimens have tiny ear flaps but East African examples usually do not. Casque slightly raised to a point posteriorly. Tail prehensile, about half the total length. Scales homogenous. A prominent gular crest of white spiny scales is present and becomes a prominent ventral crest, a dorsal crest is also present, shrinking in size towards the tail. The males have tarsal spurs. Maximum size up to 35cm in Ethiopia, but most East African specimens are 18–30 cm in length, hatchlings 5–6 cm. Colour very variable, all shades of green from light to very dark, blue-green, blue grey, brown and rufous. There is usually a prominent light brown or white flank stripe, a white or pale patch on the

Chamaeleo gracilis

Chamaeleo gracilis Slender Chameleon. Main image: Daniel Hollands, Murchison Falls. Inset top: Stephen Spawls, Voi. Inset bottom: Stephen Spawls, Moshi.

shoulder and another pale spot in the middle of the flank, specimens from southern Kenya and northern Tanzania also have a white spot at the angle of the jaw. Angry animals show black spotting. The skin of the gular pouch may be vivid orange (northern Kenyan specimens), dull orangy-yellow (Tsavo/eastern Kenya) or white.

HABITAT AND DISTRIBUTION: Moist and dry savanna, fond of acacia trees; in very dry areas favours riverine thicket. Found from sea level up to 1,600m altitude; in eastern Kenya, where sympatric with *Chamaeleo dilepis*, it appears to be on the plains and *Chamaeleo dilepis* on the hills. Occurs from northern Tanzania into eastern Kenya, thence north and westwards round Mt Kenya across to western Kenya, found virtually throughout Uganda save the south-west. Also in far northern Kenya, from Marsabit and South Horr north to the northern border, east to Mandera. Elsewhere, north to Eritrea, west to Senegal across the Sahel. Conservation Status: IUCN Least Concern.

NATURAL HISTORY: Arboreal but will descend to the ground to move from tree to tree, lives in trees, shrubs, even quite small bushes and thick grass clumps. When angry, hisses furiously, bites and inflates the gular pouch, it can vibrate, and will jump from branches if threatened. Major enemies includes big rear-fanged savanna snakes like the Boomslang *Dispholidus typus*. Lays eggs, clutch sizes of 44 recorded but 10–25 more usual. In West Africa, the eggs are recorded as taking 3–4 months to develop in the females body and up to seven months to hatch, but East African breeding details unknown. Captive males will fight eat other, biting throats. Eats a wide range of invertebrates, including beetles and grasshoppers (and mantids in arid regions).

SMOOTH CHAMELEON *Chamaeleo laevigatus*

IDENTIFICATION: A fairly large savanna chameleon, without horns, rostral processes or ear flaps. The casque is slightly raised at the back. The tail is prehensile, but about two-fifths of the total length, shorter than in most green chameleons. The body scales are homogeneous. A white or pale gular crest of cone-like soft spiny scales is present and extends onto the belly, to the vent and sometimes beyond, below the tail. The dorsal crest is low, almost non-existent. Maximum size about 25cm, average 15–22cm, hatchling size 4–5cm. Colour quite variable, but usually shades of green, greenish brown or yellow-brown, with irregular darker spots and blotches, when the animal is irritated these spots become larger and very dark. In some specimens, on either side of the vertebral ridge is a series of 4–6 large oval saddles, which may be darker green or brown. A white, cream or light grey colour form is also known, with scattered fawn-coloured spots.

Chamaeleo laevigatus

HABITAT AND DISTRIBUTION: Moist savanna, from 1,000–1,500m altitude in East Africa, but down to 300m and possibly lower elsewhere. It occurs north from Lake Rukwa all along the Lake Tanganyika shore, western Burundi and Rwanda, into southern Uganda. Also found along the Albertine Rift, and widespread in north-west Uganda. An isolated record from the Sesse Islands, and a population exists east from the Victoria Nile into Kenya, and north along the Uganda–Kenya border, south to Kakamega and Kisumu. Elsewhere, north to Sudan, west to Chad. Conservation Status: IUCN Least Concern.

Chamaeleo laevigatus Smooth Chameleon. Steve Russell, Moroto.

NATURAL HISTORY: Arboreal, living in bushes, small trees and thicket, sometimes in sedge or reedbeds. Will readily descend to the ground to cross treeless areas, and thus often seen crossing roads in savanna areas, especially in the morning in the rainy season. Lays eggs, clutch sizes of up to 60 eggs known but 15–30 more usual; taking 4–5 months to hatch. A series of specimens from the DR Congo were captured in dry savanna as they apparently do not aestivate and their colour (yellow or grey) rendered them conspicuous. Diet invertebrates, recorded as eating many grasshoppers.

EAST AFRICAN MONTANE CHAMELEONS
KINYONGIA

A genus of some 20 species of green, tree dwelling chameleons, 18 of which occur in East Africa, where they live largely in highland forests; many have very small ranges. The males of many species have large blade-like horns and spectacular head ornamentation, they lack a gular crest and a continuous vertebral crest. The generic name for this clade was coined in 2006, and is a Latinisation of the Swahili name for chameleons, Kinyonga; they split from other chameleons over 40 million years ago. We can presume that an ancestral species was widespread in the forest that covered the region, the subsequent rifting, uplift and increasing aridity isolated the populations in the hill forests and isolation lead to speciation.

KEY TO THE EAST AFRICAN MEMBERS OF THE GENUS *KINYONGIA*

1a	Gular area with no projections, tubercles or cone-like scales......2		
1b	Gular area with one or more cone-like tubercles......*Kinyongia asheorum*, Mt Nyiro Bearded Chameleon [p. 265]		
2a	Rosetting of tubercles present on the mid-flanks or lower flanks......4		
2b	Rosetting of tubercles on flanks absent......3		
3a	Tail smooth, males with helmet like casque and a short blade-like rostral process projecting upward from snout, females with smooth rostrum, Rwenzori area only......*Kinyongia carpenteri*, Carpenter's Chameleon [p. 267]		
3b	Tail sometimes with small cones at the base, males with tall and prominent casque and an ovoid rostral process, projecting forward off the snout, nearly as large as the casque, females with an elevated rostral bump, Rwenzori area only......*Kinyongia xenorhina*, Strange-horned Chameleon [p. 279]		
4a	Dorsal keel smooth or with 6 or less small cones......5		
4b	Dorsal keel smooth or rough, with more than 6 cones over the anterior part of the keel......15		
5a	Male specimen......6		
5b	Female specimen......11		
6a	Rostral process formed by a pair of blade-like firm horns......*Kinyongia tavetana*, Mt Kilimanjaro Two-horned		

6b	Chameleon [p. 273] Rostral process a single projection......7	14a	Temporo-lateral crest slopes upward from its origin......*Kinyongia uthmoelleri*, Hanang Hornless Chameleon [p. 276]
7a	Rostral process firm at the base but soft and mobile toward the end......8		
7b	Rostral process firm along the entire length......9	14b	Temporo-lateral crest straight or slopes marginally downward from its origin......*Kinyongia oxyrhina*, Uluguru One-horned Chameleon [p. 272]
8a	Rostral process consisting of a single thin laterally flattened projection extending up to 12mm off the front of the snout, Udzungwa, Rubeho, Uluguru, Nguru, Nguu, Mahenge......*Kinyongia oxyrhina*,Uluguru One-horned Chameleon [p. 272]	15a	Rostrum smooth in both sexes, snout-vent length less than 80mm and dorsal crest weakly developed......16
		15b	Rostrum ornamented with twinned blade-like horns in males, females either with short horns or rostral bumps or ridges......17
8b	Rostral process short, thick and covered in stout tubercles extending up to 5mm off the front of the snout, Shimba hills in Kenya and East Usambara Mts in Tanzania......*Kinyongia tenuis*, Usambara Soft-horned Chameleon [p. 274]	16a	Parietal crest distinct, inferior orbital tubercles prominent, tail about 60% of total length, Mt Kenya, Aberdares and Nyambene Hills......*Kinyongia excubitor*, Mt Kenya Hornless Chameleon [p. 267]
9a	Rostral process single, sub-triangular, upward pointing, Udzungwa Mts and Magombera forest only......*Kinyongia magomberae*, Magombera Forest Chameleon [p. 269]	16b	Parietal crest distinct, inferior orbital tubercles not prominent, tail about 54-57% of total length, Highlands in the Albertine Rift only......*Kinyongia adolfifriderici*, Ituri Chameleon [p. 264]
9b	Rostral process formed by the fusion of the canthi rostrales, forming a low wall of tubercles protruding off the snout......10	17a	Dorsal crest weakly developed......18
		17b	Anterior part of the dorsal crest well developed and formed by prominent cones......20
10a	Casque with a prominent convex parietal crest......*Kinyongia uthmoelleri*, Hanang Hornless Chameleon [p. 276]	18a	Canthus rostrale rugose and terminates on the lateral side of the rostral horn (in males) or bumps (in females)...... 19
10b	Parietal crest weak to flat, rostral process formed by less than ten scales, Poroto Mts and Mt Rungwe only...... *Kinyongia vanheygeni*, Poroto Single-horned Chameleon [p. 277]		
10c	Parietal crest weak to flat, rostral process formed by ten or more scales, Udzungwa and Livingstone Mts. Only...... *Kinyongia msuyae*, Msuya's Forest Chameleon [p. 271]	18b	Canthus rostrale terminates in the base of the rostral horn and continues as the upper margin of the horn. Males and females with horns. In East Usambara Mountains only...... *Kinyongia matschiei*, Giant East Usambara Blade-horned Chameleon [p. 270]
11a	Rostral process present......12		
11b	Rostral process absent......13	19a	Temporo-lateral crest originates from the mid post-orbital rim, maximum snout-vent length 90mm, Taita Hills only......*Kinyongia boehmei*, Taita Hills Blade-horned Chameleon [p. 265]
12a	Rostral process formed by twinned ridges on snout.......*Kinyongia tavetana*, Mt Kilimanjaro Two-horned Chameleon [p. 273]		
12b	Rostral process short, thick, and covered in stout tubercles extending up to 5 mm off the snout, Shimba hills in Kenya and East Usambara Mts. in Tanzania......*Kinyongia tenuis*, Usambara Soft-horned Chameleon [p. 274]	19b	Temporo-lateral crest originates from the upper post-orbital rim, females usually have no horns but my have short rostral horns, maximum snout-vent length 130mm, Nguru and Nguu Mts only......*Kinyongia fischeri*, Nguru Blade-horned Chameleon [p. 268]
13a	Parietal crest weakly developed, top of the head green and white......*Kinyongia vosseleri*, Vosseler's Blade-horned Chameleon [p. 278]	20a	Supra-orbital ridge of prominent conical tubercles, dorsal crest formed by greater than 20 conical tubercles and extends over half way along keel, horns long, maximum snout-vent length 140mm, West Usambara Mountains only......*Kinyongia multituberculata*,
13b	Parietal crest prominent and well developed......14		

	West Usambara Blade-horned Chameleon [p. 272]
20b	Dorsal crest of well-formed cones not extending beyond one-third of the dorsal keel......21
21a	Dorsal crest formed by up to 15 prominent conical tubercles, base of the tail smooth, horns short, maximum snout-vent length 80mm, Uluguru Mts only......*Kinyongia uluguruensis*, Uluguru Two-horned Chameleon [p. 276]
21b	Dorsal crest extends only up to one/third of the keel, base of the tail with a distinct crest, horns well developed in males, East Usambara Mts only...... *Kinyongia vosseleri* (males), Vosseler's Blade-horned Chameleon [p. 278]

ITURI CHAMELEON *Kinyongia adolfifriderici*

IDENTIFICATION: A small green or brown chameleon, without horns, rostral appendages or ear flaps. Casque slightly raised. Tail prehensile, slightly more than half the total length. No gular or ventral crest, males have a weak dorsal crest on the front third of the back, consisting of small pointed tubercles, this is often (but not always) absent in females. Flank scales homogenous, showing typical rosettes. Maximum size about 16cm, average 9–12cm, hatchling size unknown. Colour usually uniform brown, greyish or green, When stressed, yellow or rufous dots may appear on the head and vague broad vertical bars on the body. Taxonomic notes: A recent molecular analysis indicates this species may be a complex. If valid, Rwandan animals are placed in the species *Kinyongia rugegensis*, Ugandan animals in *Kinyongia tolleyae*.

Kinyongia adolfifriderici

HABITAT AND DISTRIBUTION: Forest and woodland along the Albertine Rift, between 1,000–2,200m altitude. Known from western Rwanda, north-west Burundi and western Uganda to the south end of Lake Albert; elsewhere to north-east DR Congo, at Medje and the Ituri Forest. Conservation Status: IUCN Least Concern.

NATURAL HISTORY: Lives in bushes and small trees in the forest understory, but will also enter plantations, including *Eucalyptus*. Very shy, surreptitiously sidling behind branches to avoid detection. If threatened, they will readily hop off their branch, and on hitting the ground they roll up the tail, flex their limbs and remain immobile, some tried to worm their way under leaves and grass stems. A clutch of three eggs recorded, local information is that the eggs are laid in tree hollows or under damp leaf litter. Eats invertebrates.

Kinyongia adolfifriderici Ituri Chameleon. Left: Max Dehling, Rwanda. Right: Michele Menegon, Rwanda.

MT NYIRO BEARDED CHAMELEON Kinyongia asheorum

IDENTIFICATION: A medium sized, slender chameleon with a smooth tail longer than the snout-vent length, reaching a maximum of 22cm in total length. Casque elevated posteriorly, parietal crest prominent, serrated, composed by laterally compressed tubercles. Dorsal crest moderately developed, temporal crest absent. There is a cluster of long and thin conical tubercles on the throat, usually seven in males and three in females. Instead of the large rostral tubercles, females have only a pair of prominent rostral tubercles and a lower number of gular appendages. A brown or green chameleon, gular cones, posterior part of the upper labials and zone around the commissure of the mouth are usually white. Cranial crest, rostral appendages and sides of the casque are often reddish-brown. The species is easy to distinguish and unique within its genus by having very conspicuous gular and rostral appendages. The species is named in honour of James and Sanda Ashe, for their life-long contribution to African herpetology.

Kinyongia asheorum

HABITAT AND DISTRIBUTION: Confined to Mt Nyiro, in northern Kenya. It inhabits Afromontane forest between 2,000–2,500m altitude. Conservation Status: IUCN Near Threatened.

NATURAL HISTORY: Poorly known. A truly arboreal species, the Samburu inhabitants of Mt Nyiro describe this chameleon as a tall tree dweller, however it has been found on *Dodonaea* bushes, at about 2m from the ground. Probably lays eggs like the other members of the genus.

Kinyongia asheorum Mount Nyiro Bearded Chameleon. Left: Robert Sindaco male Mt Nyiro. Right: Tomas Mazuch female.

TAITA HILLS BLADE-HORNED CHAMELEON
Kinyongia boehmei

IDENTIFICATION: A small blade-horned chameleon, known only from the Taita Hills and nearby ranges. Males have a pair of scaly blade-like horns on the nose, females have a blunt snout, sometimes with a small rostral projection. No ear flaps. The casque is moderately elevated posteriorly along the middle. The tail is prehensile, about half the total length or slightly more. Body scales mostly homogeneous, in rosettes. No gular, ventral or dorsal crest. Maximum size about 19cm, average 14–17cm, hatchling size unknown. Colour very variable, can be green, brown, rufous, with blue or yellow blotching. In females the top of the head is nearly always rufous. The male has a pair of rather spiky blade-like horns, covered with small scales forming a serrated profile, and a slightly raised casque; the female has a small double projection on the nose.

DISTRIBUTION: Found only in the woodland, forest, farms and gardens of the Dawida and Mt Mbololo blocks of the Taita Hills, Mt Sagalla and Mt Kasigau, in south-eastern Kenya, between altitudes of 1,200 to just over 2,000m. Conservation Status: IUCN Near Threatened.

Kinyongia boehmei

NATURAL HISTORY: Diurnal and arboreal. They will ascend high in forest trees, but are equally happy in low bushes. Often almost abundant, within the town of Wundanyi they can be found a few metres apart, sheltering on roadside vegetation and in garden hedges; they like agriculture and suburbia and are more common there than in the true forest. They lay on average eight eggs, but a maximum of 12 recorded. They eat a range of small invertebrates.

Kinyongia boehmei Taita Hills Blade-horned Chameleon. Top: Stephen Spawls, Female. Bottom left and right: Stephen Spawls, male.

CARPENTER'S CHAMELEON Kinyongia carpenteri

IDENTIFICATION: A medium-sized chameleon of the Albertine Rift, the males have an astonishing large casque. Adult males bear a superficial resemblance to *Trioceros hoehnelii*, having a tall blade-like casque and a short forward/upward projecting, vertically-flattened rostral horn. The female has only vestigial cranial ornamentation. Body scales weakly heterogeneous, but not clumped in rosettes. The tail is slightly longer than the body length, 55–60% of total length. A weak dorsal crest is present, no gular or ventral crest. Colour very variable, often blue-green ground colour, the casque is suffused with yellow, and the rostral process deep orange. The tail is banded.

Kinyongia carpenteri

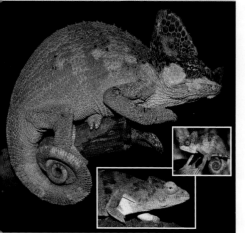

Kinyongia carpenteri Carpenter's Chameleon. Main image: Thomas Price male. Inset left: Colin Tilbury female. Inset right: Colin Tilbury male.

HABITAT AND DISTRIBUTION: Known only from the montane evergreen forest of the Rwenzori Range, on both the Uganda and Congo sides, at altitudes between 1,800–2,500m. Conservation Status: IUCN Near Threatened.

NATURAL HISTORY: Poorly known. Lives in the canopy of large trees. Perches at night between 3–5m above ground, but has also been found in hedges in urban areas. Males probably fight with their large horns. When stressed, they jump off the perch, and a male has been seen rolling into a tight ball as it fell, on hitting the ground it crept under cover and remained motionless. Lays eggs (unusually, as it lives at altitude). Eats invertebrates.

MT KENYA HORNLESS CHAMELEON Kinyongia excubitor

IDENTIFICATION: A medium-sized chameleon from high central Kenya. It is slightly squat. Has no horns, rostral process, or ear flaps. Has a distinctive flattened, slightly raised casque. Tail prehensile, about 54–60% of total length.. Scales largely heterogeneous, with the usual rosettes and a scattering of enlarged scales on the flanks. No gular or ventral crest, but it has a dorsal crest of enlarged, spiky scales, (although this can be almost invisible in some specimens), these spines get shorter towards the tail. Maximum size about 20cm, average 12–18cm, hatchling size unknown but probably 4–5cm. Mostly green, sometimes brown or rufous, when irritated several broad, dark or black vertical bars appear on the body, interspersed with vivid yellow spots. Females mostly uniform green.

Kinyongia excubitor

HABITAT AND DISTRIBUTION: A Kenyan endemic of the eastern central highlands, found in a wide range of woodland between 1,200–2,500m. Originally believed to be confined to the Mt Kenya forest, from Kerugoya anti-clockwise around the mountain to Meru, it has also been recorded from Kimakia on the southern side of the Aberdares, in forest patches in the Nyambene Hills, and also Ngaia Forest, below the Nyambene Hills on the road down to Meru National Park. Probably also in the Imenti Forest north-east of Meru. Conservation Status: IUCN Vulnerable B1ab(i,ii,iii)+2ab(i,ii,iii). Its tiny range, within an area that is rapidly being logged, gives cause for concern, especially as it is not found within any protected areas.

NATURAL HISTORY: Poorly known. Arboreal, living in small to medium forest trees and herbage, does not appear to tolerate agriculture but it is found in Miraa plantations. Some specimens jump from the perch if disturbed. Reproductive biology unknown. Eats invertebrates.

Kinyongia excubitor Mt Kenya Hornless Chameleon. Left: Jan Stipala, southern Aberdares, male. Right: Stephen Spawls, Ngaia Forest, male.

NGURU BLADE-HORNED CHAMELEON *Kinyongia fischeri*

IDENTIFICATION: A big chameleon from the Nguru and Nguu Mountains, Tanzania. The males have two big blade-like scaly horns on the nose. The females are usually hornless. Body scales are heterogeneous, with big granular scales scattered on the flanks in rosettes; the dorsal crest rarely extends more than a third of the way along the back, and consists of a few lanceolate cones (usually less than five in females). The tail is long, about 60–65% of the total length. There is no gular or ventral crest. Maximum size about 35cm, average 20–30cm, and females usually smaller than males. Colour very variable and often spectacular, shades of green, the males often brown, in most adults along each flank is a series of posteriorly expanding relatively fine light lines. Taxonomic Note: The specific name *fischeri* has been used for a number of largely blade-horned chameleons, most of which are now regarded as discrete species. This is the original.

Kinyongia fischeri

HABITAT AND DISTRIBUTION: A Tanzanian endemic. Forest, woodland and farmland

Kinyongia fischeri Nguru Blade-horned Chameleon. Top: Michele Mwenegon, Nguru South Mountains, male. Bottom left: Michele Menegon, Nguru South Mountains, female.

of the Nguru and Nguu Mountains, part of the Eastern Arc, north of Morogoro, at altitudes between 500–1,500m. Conservation Status: Near Threatened.

NATURAL HISTORY: Not well known. Occurs in both the forest and surrounding agricultural land. The males, being horned, are probably territorial and fight. They lay eggs, a clutch of 10 developing ova recorded. Diet invertebrates.

MAGOMBERA FOREST CHAMELEON *Kinyongia magomberae*

IDENTIFICATION: An elongated forest chameleon with low casque, no signs of gular or dorsal crest and indistinct lateral crest. The rostral process of the male is short, sub-triangular and completely ossified; tail is as long as the body or slightly longer. It resembles *Kinyongia tenuis* and, to a lesser extent *Kinyongia oxyrhina* in bearing a single, blade like horn. The other similar species *Kinyonga msuyae*, has a rostral appendage that, on close inspection, is clearly double tipped.

HABITAT AND DISTRIBUTION: Confined to Magombera forest, a remnant of natural habitat in the densely cultivated Kilombero valley, and the forest at low and medium elevation in the Udzungwa Mountains National Park in the area formerly called Mwanihana forest, from 250 to about 1,000m altitude. In the Udzungwa Mountains the species could co-occur with two other *Kinyongia* species: *K. oxyrhina* and *K. msuyae* which at present are known to inhabit higher altitudes and different portions of the mountain block.

Kinyongia magomberae

Kinyongia magomberae Magombera Single-horned Chameleon. Andrew R Marshall, Magombera Forest.

NATURAL HISTORY: Very little is known. The holotype was observed being predated upon by a twig snake *Thelotornis mossambicanus* in Magombera Forest. The snake, disturbed by the presence of the human observer, dropped the already dead chameleon on the forest floor.

GIANT EAST USAMBARA BLADE-HORNED CHAMELEON
Kinyongia matschiei

IDENTIFICATION: A huge chameleon from the Eastern Usambara Mountains, Tanzania. The males have two big blade-like scaly horns on the nose. The females have shorter but distinct horns. Body scales are heterogeneous, with big granular scales scattered on the flanks in rosettes; the dorsal crest consists of a few lanceolate cones, usually 8–10, only extends a short way along the back. The tail is long, about two-thirds of the total length. There is no gular or ventral crest. Maximum size about 43cm, average 25–35cm, and females usually smaller than males. Colour very variable and spectacular, usually shades of green, with turquoise or blue on the head, and most specimens show two fairly broad oblique light lateral bars, forward-pointing as they descend. Taxonomic Note: Elevated in 2008 from a subspecies of *Kinyongia fischeri*. Sometimes called the Giant Monkey-tailed Chameleon.

Kinyongia matschiei

HABITAT AND DISTRIBUTION: A Tanzanian endemic. Forest, woodland and farmland of the East Usambara Mountains, at altitudes between 800–1,500m. Conservation Status: Endangered B1ab(iii)+2ab(iii).

NATURAL HISTORY: Diurnal and arboreal. Occurs in both the forest and surrounding agricultural land, and has been found in low bushes, even on grass. The males probably fight with their horns. When stressed, they flatten this bodies, appearing larger, and will jump from their branches. Copulation lasts 10–15 minutes and a male and female may mate frequently over 7–10 days; the gestation period is 47–55 days. The females lays on

Kinyongia matschiei Giant East Usambara Blade-horned Chameleon. Left: Michele Menegon, East Usambara Moutains, female. Right: Michele Menegon, East Usambara Mountains, male.

average 16–20 eggs (but up to 24 recorded), roughly 17×9mm. They take approximately 11 months to hatch, hatchlings were found in June. Diet: invertebrates, known to take small vertebrates in captivity.

MSUYA'S FOREST CHAMELEON   Kinyongia msuyae

Kinyongia msuyae

IDENTIFICATION: A small, elongated chameleon with a tail longer than the snout-vent length, exceeding 15cm in total length. Casque elevated posteriorly, usually green with broad paler bands and scattered blue scales or aggregation of scales. Males bear a bone-based rostral appendage formed by a converging, scaly elongation of the canthi rostrales, appearing like a double-tipped short horn protruding over the snout by 3–5mm. *K. msuyae* does resemble several species sharing part of its range in size, general body and head shape and by possession of a single, bone-based rostral appendage in males. *K. oxyrhina* and *K. magomberae* lack the double tip at the end of the rostral appendage, while it differs from *K. vanheygeni* in the length of the rostral appendage being longer, formed by more than ten scales and pointing straight forward.

Kinyongia msuyae Msuya's Forest Chameleon. Left: Michele Menegon, Livingstone Mountains. Right: Michele Menegon, Livingstone Mountains.

HABITAT AND DISTRIBUTION: Known only from two small forest fragments in the Livingstone Mountains and Kigogo Forest Reserve in the southern Udzungwa Mountains. An individual tentatively assigned to this species was collected in the Northern part of the Udzungwa. In the Udzungwa Mountains the species could co-occur with two other *Kinyongia* species: *K. oxyrhina* and *K. msuyae*, in the Southern Highlands of Tanzania with *K. vanheygeni,* in both cases these species are known to occur in different portion of the distribution area.

NATURAL HISTORY: Very little is known. Individuals were observed at night from 2–5m high, in bushes or trees in the forest interior.

WEST USAMBARA BLADE-HORNED CHAMELEON
Kinyongia multituberculata

IDENTIFICATION: A fairly large chameleon from the Western Usambara Mountains, Tanzania. The males have two big blade-like scaly horns on the nose. The females have shorter but distinct horns. Body scales are heterogeneous, with big granular scales scattered on the flanks in rosettes; the dorsal crest consists of thirty or more lanceolate cones, extending about halfway or more towards the tail, and in the male the crest reappears on the tail. The tail is long, about 55–65 % of the total length, usually longer in males. There is no gular or ventral crest. Maximum size about 34cm, average 20–30cm, and females usually smaller than males. Colour very variable and spectacular, usually shades of green or brown, with turquoise or blue on the head; the throat is often blue and there is a scattering of light dots on the anterior flanks. Horns usually grey. When stressed, some individuals go purple-grey. Taxonomic Note: Elevated in 2008 from a sub-species of *Kinyongia fischeri*.

Kinyongia multituberculata

HABITAT AND DISTRIBUTION: A Tanzanian endemic. Forest, woodland and farmland of the Western Usambara Mountains, at altitudes between 1,200–2,500m. A small introduced population occurs in the Nairobi suburb of Karen (not shown on map). Conservation Status: Endangered B1ab(iii)+2ab(iii).

NATURAL HISTORY: Diurnal and arboreal. Occurs in both the forest and surrounding agricultural land, and in and around villages and disturbed forest is often common. The males probably fight with their horns. Lays eggs. Diet: invertebrates.

Kinyongia multituberculata West Usambara Blade-horned Chameleon. Left: Philip Shirk, male and female, West Usambara. Right: Stephen Spawls, West Usambara, male.

ULUGURU ONE-HORNED CHAMELEON / EASTERN ARC SHARP-NOSED CHAMELEON *Kinyongia oxyrhina*

IDENTIFICATION: A small chameleon, the male has a distinctive soft blade-like rostral horn, which can be up to 2cm long (possibly longer), the tip is moveable. The female has no horn. No ear flaps. The casque is raised posteriorly. The tail is prehensile, slightly more than half the body length. The body scales are heterogeneous, showing the usual rosetting, and are very large on the horn. There is no gular, ventral or dorsal crest. Maximum size about 17cm, average 10–15cm, hatchling size unknown. Colour very variable, usually green, grey, brown, rufous or intermediate shades, some specimens show a pattern or rust-red and white blotches. The eyeball skin may be bright red. During courtship, the eyeball skin of the female often becomes green and scattered scales on the snout become bright blue, similar bright blue scales appear on the horns and eyeball skin of the male.

Kinyongia oxyrhina

HABITAT AND DISTRIBUTION: A Tanzanian endemic, in forest and woodland, secondary cover and exotic plantations, of the Uluguru, Udzungwa, Nguu, Nguru and Rubeho Mountains, between altitudes of 1,400–1,900m. Thus vulnerable, due to exploitation of its very restricted forest home. Conservation Status: IUCN Near Threatened.

NATURAL HISTORY: Poorly known. Arboreal, in bushes and trees, perching at night between 3–6m above the ground. Usually found at the forest's edge. Males are suspected to be territorial, and fight with their horns. If disturbed in a tree, they will drop to the ground in a tightly-coiled ball, and then twist vigorously, presumably to get into the undergrowth (or dissuade a predator trying to eat them). Lays eggs, a clutch of 13 recorded. Said to be unusually aggressive when handled. Eats invertebrates.

Kinyongia oxyrhina Uluguru One-horned Chameleon. Left: David Gower, Uluguru Mountains, male. Right: Philip Shirk, female, Nguu Mountains.

MT KILIMANJARO TWO-HORNED CHAMELEON
Kinyongia tavetana

IDENTIFICATION: A medium-sized, slim-bodied, green, brown or blue-grey chameleon, males have a pair of scaly blade-like horns on the nose, females have a blunt snout, sometimes with a small rostral projection. No ear flaps. The casque is raised posteriorly along the middle. The tail is prehensile, about half the total length or slightly more. Body scales mostly homogeneous, with a few enlarged scales on the flanks and the flank scales arranged in rosettes. No gular, ventral or dorsal crest. Maximum size about 23cm, average 15–22 cm, hatchling size unknown. Colour mostly shades of green, brown or blue-grey. Both sexes have a broad, irregular side-stripe (white or pale green), males often have three broad oblique green or brown bars high on the sides, with very dark green between. The tail is barred in light and dark green or brown. When stressed, a vivid orange patch appears behind the eye and bars appear on the limbs. The female is light green, the tail barred like the male's, the body dark green or greeny-brown, or with vertical irregular bars of light green on a dark background. the rostral projection and snout often reddish-brown.

Kinyongia tavetana

HABITAT AND DISTRIBUTION: An East African endemic. Forest, woodland and farmland on mountains and hill ranges of south-east Kenya and north-east Tanzania, at altitudes from 1,000m up to 2,200m. Recorded from the Chyulu Hills, North and South Pare Mountains, the south side of Mt Kilimanjaro, Ngurdoto Crater and Mt Meru. Also reported from the thicket forest of the Kibwezi area (but without supporting specimens). The holotype was described from the 'Taveta Forest'; there is no longer any forest there. Range very confined and thus vulnerable to deforestation, although part of its range lies within the Tanzanian National Parks of Arusha and Mt Kilimanjaro. Conservation Status: IUCN Near Threatened.

NATURAL HISTORY: Found in woodland but readily enters well-wooded gardens, hedges and plantations; common in leafy suburbs of Arusha and in surrounding coffee plantations. Its habits are poorly known, but males fight savagely with their horns. Lays eggs, a clutch of nine recorded. Diet: invertebrates.

Kinyongia tavetana Mt Kilimanjaro two-horned Chameleon. Stephen Spawls. Left: Arusha, male. Right: Arusha, female.

USAMBARA SOFT-HORNED CHAMELEON / USAMBARA FLAP-NOSED CHAMELEON *Kinyongia tenuis*

IDENTIFICATION: A small chameleon, males have a small, vertically flattened scaly horn on the snout, females have an even smaller horn or just a few raised scales. No ear flaps. The casque is very slightly raised posteriorly. The tail is prehensile, just over half the total length. There is no gular, ventral or dorsal crest. Body scales mostly homogeneous;

the lateral rosettes of scales align so as to give the impression of dark stripes in some animals. Maximum size about 14.5cm, hatchling size unknown. Colour very variable, may be predominantly green, olive, grey or red-brown. The horn may be brown, blue, green or red, the lips are sometimes blue. Sleeping individuals have been recorded with a rufous head, a dark mantle and broad dark vertical bars on a grey-brown body.

Kinyongia tenuis

HABITAT AND DISTRIBUTION: Woodland and forest, from 100–1,400m altitude. An East African endemic, known only from the forests of the Shimba Hills in south-east Kenya, Magrotto Hill and the Usambara Mountains in north-east Tanzania (found in Amani nature reserve), thus vulnerable to habitat destruction, although some of its range is protected in the Amani nature reserve and Shimba Hills National Park. Conservation Status: Endangered B1ab(ii,iii)+2ab(ii,iii).

NATURAL HISTORY: Little known. Males may fight with the horn. Although specimens have been found in long grass, they usually sleep in low to medium sized trees at the forest edge, at heights up to 6m, several specimens may be in one tree. May jump from the perch when threatened. Lays eggs, clutches of 3–5 recorded; a Usambara specimen contained five eggs in mid-April; clutches laid by captive females in January hatched in April. Eats invertebrates.

Kinyongia tenuis Usambara soft-horned Chameleon. Top: Philip Shirk, Usambara Mountains, female. Bottom left: Philip Shirk, Usambara Mountains, male. Bottom right: Stephen Spawls, Usambara Moutains, male.

ULUGURU TWO-HORNED CHAMELEON
Kinyongia uluguruensis

IDENTIFICATION: A relatively small blade-horned chameleon, endemic to the Uluguru Mountains. The males have two short blade-like horns on the nose. The females have rostral bumps instead of horns. Body scales are heterogeneous, with big granular scales scattered on the flanks and the lateral scales clumped in rosettes, particularly on the lower flanks. The tail is long, about three-fifths of the total length. There is no gular or ventral crest, but a weak dorsal crest of spines that rapidly become short and disappears just beyond the level of the shoulder. Males up to just under 19cm, females 16cm, the hatchling size is unknown. Colour very variable but males are often brown and females dull green; both sexes usually have a pale or white side-stripe.

Kinyongia uluguruensis

HABITAT AND DISTRIBUTION: Medium to high altitude Afromontane forest in the Uluguru Mountains, at altitudes between 1,300–2,200m. Conservation Status: IUCN Least Concern.

NATURAL HISTORY: A rare and secretive species. Said to inhabit the edges of the forest, and at night perch at heights between 1.5–6m above ground level. The adult males are intolerant of each other, and when they meet they become vivid yellow-brown, jerk their heads, flatten the body and then lock horns, thrusting and twisting. The females lay eggs, a clutch of eight recorded. Diet: invertebrates.

Kinyongia uluguruensis Uluguru two-horned Chameleon. Colin Tilbury, Uluguru Mountains.

HANANG HORNLESS CHAMELEON *Kinyongia uthmoelleri*

IDENTIFICATION: A medium-sized, rather thickset chameleon, the canthal crests (the raised lines of scales in front of each eye, at the side of the head) form a sort of shovel or scoop on the snout, this extends a couple of millimetres in front of the mouth. No ear flaps. The casque of the males is strongly raised at the back, to form a heavy looking helmet, with huge granular scales on either side of the median line. In females, the casque is smaller and less prominent. The tail is prehensile and long, three-fifths of the total length. The body scales are heterogeneous, with enlarged scales on the flanks and the casque. No gular, ventral or dorsal crest. Maximum known size 23cm, average 15–20cm, hatchling size

Kinyongia uthmoelleri

Kinyongia uthmoelleri Hanang hornless Chameleon. Left: Michele Menegon, Pare Mountains, male. Right: Stephen Spawls, Ngorongoro Crater, female.

unknown. Usually greenish, adults have a prominent pale (yellow or white) stripe high on the flanks, this stripe may be short or extend all the way to the level of the back legs. Males from the South Pare Mountains show vivid pale or pinkish heads. The body is usually barred with vertical bands of darker and lighter green.

HABITAT AND DISTRIBUTION: A Tanzanian endemic of the high woodland and grassland of north-central Tanzania, at altitudes between 1,700–2,500m altitude. It has a disjunct distribution; recorded from Mt Hanang, on the Singida–Babati road, at 2,300m, on Mt Oldeani, in the Ngorongoro Crater rim forest and, oddly, the South Pare Mountains. Might be found on the craters north-east of Ngorongoro or somewhere around Kilimanjaro and thus link the gaps. Conservation Status: IUCN Least Concern.

NATURAL HISTORY: Little known. Lives in tall trees, where specimens slept at heights between 2–10m. A clutch of ten eggs recorded. Diet: invertebrates.

POROTO SINGLE-HORNED CHAMELEON *Kinyongia vanheygeni*

IDENTIFICATION: A small, elongate chameleon, with casque elevated posteriorly, tail longer than the snout-vent length, about 14cm of maximum length. Usually brown or green with broad paler bands. Males bear a short blunt horn, protruding about 3mm of the snout and formed by two parallel longitudinal rows of three enlarged conical scales each. This species resembles *Kinyongia msuyae*, which has a different range in the Southern Highlands, in size, general body and head shape and by possession of a single, bone-based rostral appendage in males. It differs from it in the length of the rostral appendage being shorter and formed by less than ten scales.

Kinyongia vanheygeni

Kinyongia vanheygeni Poroto single-horned Chameleon. Petr Necas, top: male, bottom: female.

HABITAT AND DISTRIBUTION: Known from the Poroto Mountains and Mt Rungwe in the Southern Highlands of Tanzania, at altitudes between 2,000–2,500m. In that area it is the only chameleon with a single rostral appendage.

NATURAL HISTORY: Very little is known. Individuals were observed at night from 2–3m high in bushes or trees in the forest interior.

VOSSELER'S BLADE-HORNED CHAMELEON
Kinyongia vosseleri

Kinyongia vosseleri

IDENTIFICATION: A large chameleon, exceeding 30cm in total length with two large, laterally flattened, blade-like horns in males. Casque barely elevated posteriorly, parietal crest formed by 4–5 low tubercles, serrated, composed by laterally compressed tubercles. Anterior half of lateral crest with double row of tubercles. No gular crest or ornamentation. Horns are parallel at the base, converging towards their tips, upper edge is serrated, while the lower margin is relatively smooth. Females are hornless. A bright to dark green chameleon, with two broad, yellowish or light green lateral bands. Top of the casque black on the sides, with white markings. Tail also banded. A broad, white stripe on the cheek is distinctive. Horns are paler, often white or light brown-grey. It can be distinguished from all the other two horned chameleons by the absence of horns in females and by having a tail crest and a dorsal crest on the first third of the keel only. Taxonomic Note: Until recently this species was part of a complex that included *K. fischeri*, *K. matschiei*, *K. multituberculata* and *K. uluguruensis*. Recent investigations have clarified their respective taxonomic positions.

HABITAT AND DISTRIBUTION: It occurs in the Afrotemperate forest of the East Usambara Mountains only, up to 1,500m altitude. Individuals have been observed along forest edges and in moderately to highly disturbed habitats like small-scale farms with some tree cover, near forest. They avoid extremely altered habitats, such as tea plantations or small-scale agricultural land. Wild population are in decline and CITES Trade Data indicates that between 1977 and 2011 a total of over 78,000 live individuals were exported from Tanzania for the pet trade. Conservation Status: IUCN Endangered B1ab(i-ii)+2ab(iii).

Kinyongia vosseleri Vosseler's Blade-horned Chameleon.
Top: Philip Shirk, East Usambara Mountains, male.
Bottom: Philip Shirk, East Usambara Mountains, female.

NATURAL HISTORY: Oviparous like the other members of the genus. A female with 12 eggs has been recorded. It can be common where it occurs; sometimes in syntopy with *K. matschiei*, *K. tenuis* and *Trioceros deremensis*.

STRANGE-HORNED CHAMELEON / RWENZORI PLATE-NOSED CHAMELEON Kinyongia xenorhina

IDENTIFICATION: A spectacular chameleon of the Rwenzori, the males have a tall, posteriorly elevated casque and a large, vertically compressed oval rostral projection which is flexible and roughly as large as the casque. Females have a tiny casque and rostral process. No gular or ventral crest. A weak dorsal crest is present. Scales are sub-heterogeneous and there is no rosetting of flank scales. Tail around 60% of total length. Males maximum length 27cm, females to 18cm. Males are various shades of blue-green, with chocolate-coloured scales on top of the head, gular pouch shades of light green. Rostral process green, eyeball skin russet, sometimes with a fine dark bar horizontally across it. There are three dark green saddles along the back. Tail yellow-green and banded. When stressed, big vivid orange blotches appear on the flanks, may also turn reddish-brown. Females shades of green, but may turn dark if stressed and display an orange gular pouch.

Kinyongia xenorhina

HABITAT AND DISTRIBUTION: Uganda and the DR Congo, known only from high woodland and forest of the Rwenzori Range, between 1,200–2,600m altitude. Its restricted range means it is vulnerable to habitat destruction, but its entire distribution is protected within Uganda and the DR Congo. Conservation Status: IUCN Near Threatened.

Kinyongia xenorhina Strange-horned Chameleon. Left: Steve Russell, Rwenzori Range, male. Right: Female Luke Verburgt Rwenzori Inset: Matthijs Kuijpers, female. Bottom right: Colin Tilbury, Rwenzori Range, male.

NATURAL HISTORY: Diurnal and arboreal, found in trees, bushes and low vegetation, although they seem to prefer the creeper-festooned boles of large trees. Recorded as perching at heights between 1–7m, and true pairs have been found sleeping close together. When disturbed they will drop from their perches, and a male frantically writhed on hitting the ground, presumably trying to squirm into the undergrowth. They appear to be active throughout the year. The males might fight with their horns. A courting male approached a female while jerking his head, and showing vivid colours that darkened as he approached; the female remained a drab grey-green and retreated. When two males met in the same bush the smaller male became dark as the other brightened, he then jumped off his perch. A clutch of six eggs, roughly 15×10mm, recorded in January. Diet: invertebrates, including caterpillars.

SIDE-STRIPED CHAMELEONS *TRIOCEROS*

A genus of about 40 species of mostly small (but one huge), tree- and bush-dwelling chameleons, widely distributed throughout tropical Africa, with a cluster of species in the highlands of the countries around the Gulf of Guinea, and most of the remainder in East Africa and Ethiopia. About 25 species occur in our area, where they live largely at medium to high altitude. Many have very small ranges, several have been described in the last few years. The males of some species have annular horns (but never blade-like horns); others lack horns or have small rostral projections, many are small with lateral stripes, the 'side-striped' group. Most species have distinctive gular and dorsal crests. Nearly all give live birth. They are often present in huge numbers and are largely tolerant of other members of the same genus; their populations appear to peak and crash. A molecular analysis indicates a common ancestor more than 35 million years ago, an east-west split 35 million years ago, and an East Africa-Ethiopian split about 32 million years ago. Climatic change and the rising of the East African dome has to some extent fragmented the East Africa populations and lead to the evolution of the various forms, many in small highland refugia, with attached conservation concerns. Some of the species listed here, especially those with more than one isolated population in the Eastern Arc, may well turn out to be species complexes and will be further split.

KEY TO THE EAST AFRICAN MEMBERS OF THE GENUS *TRIOCEROS*

1a Midline gular crest present......2
1b Midline gular crest absent (instead there is either a duel series of cones or a midline field of granules)......14

2a Rostral or pre-orbital ornaments or processes present......3
2b Rostral or pre-orbital ornaments or processes absent......7

3a Casque low, elevated, less than the diameter of the eyeball above a line drawn horizontal to the top of the supra-orbital ridge......4
3b Casque tall, elevated, equal to or greater than the diameter of the eyeball above a line drawn horizontal to the top of the supra-orbital ridge......5

4a A single annulated horn in males, less developed in females, Mt Marsabit only......*Trioceros marsabitensis*, Mt Marsabit Chameleon [p. 297]
4b A single cone-like rostral process in

	males, absent in females, widespread in dry areas of South Sudan, northwest Kenya and north-eastern Uganda......*Trioceros conirostratus*, South Sudanese Unicorn Chameleon [p. 283]		drawn horizontal to the top of the supra-orbital ridge, Northern Tanzanian and central Kenya volcanic highlands......*Trioceros sternfeldi*, Crater Highlands Side-striped Chameleon [p. 304]
5a	Parietal crest present......6	12b	Casque height equal or greater than diameter of the eyeball above a line drawn horizontal to the top of the supra-orbital ridge, Mt Hanang only...... *Trioceros hanangensis*, Mt Hanang Dwarf Chemeleon [p. 288]
5b	Parietal crest absent, males with three annulated horns, Eastern Arc Mountains......*Trioceros deremensis*, Usambara Three-horned Chameleon [p. 285]		
6a	Midline gular crest strongly developed, strongly curved parietal crest widespread in Kenyan highlands......*Trioceros hoehnelii*, Von Höhnel's Chameleon [p. 289]	12c	Casque height less than diameter of the eyeball above a line drawn horizontal to the top of the supra-orbital ridge......*Trioceros bitaeniatus*, Side-striped Chameleon [p. 282]
6b	Midline gular crest weakly developed, strongly curved parietal crest Mt Kulal only......*Trioceros narraioca*, Mt Kulal Stump-nosed Chameleon [p. 299]	13a	Gular crest of conical tubercles, flanks with two rows of large plate-like tubercles. Often paired, or single with a double peak, Mt Kenya only......*Trioceros schubotzi*, Mt Kenya Side-striped Chameleon [p. 303]
6c	Midline gular crest weakly developed, straight parietal crest, Cherang'any Hills and Mtelo Mountains only......*Trioceros nyirit*, Pokot Chameleon [p. 301]		
7a	Background scalation homogeneous or finely heterogeneous......8	13b	Gular crest of conical tubercles, flanks with two rows of large plate-like tubercles. Often paired, or single with a double peak, bright orange gular crest in males, Aberdare Mountains only...... *Trioceros kinangopensis*, Aberdare Mountains Dwarf Chameleon [p. 295]
7b	Background scalation strongly heterogeneous......10		
8a	Head short, less than twice the width, ventral crest indicated by a row of white tubercles, on Mt Nyiro only......*Trioceros ntunte*, Mt Nyiro Montane Chameleon [p. 300]	13c	Gular crest of conical tubercles, flanks with one row of enlarged tubercles, along the upper flank, mountains of the Albertine Rift......*Trioceros rudis*, Rwenzori Side-striped Chameleon [p. 301]
8b	Head greater than twice the length......9		
9a	Occipital lobes present, Eastern Arc Mountains......female of *Trioceros deremensis*, Usambara Three-horned Chameleon [p. 285]	14a	Occipital lobes present......15
		14b	Occipital lobes absent......23
9b	Occipital lobes absent, ventral crest of distinct white tubercles, tail length in males usually greater than snout-vent length, 1–4 prominent black or white pigmented gular grooves usually present......*Trioceros ellioti*, Montane Side-striped Chameleon [p. 286]	15a	Dorsal crest absent, keel smooth, flanks with enlarged tubercles in rows, in the Albertine Rift only......*Trioceros ituriensis*, Ituri Forest Chameleon [p. 291]
		15b	Dorsal crest present......16
		16a	Rostral/pre-orbital ornaments present both in males and females......17
		16b	Rostral/pre-orbital ornaments absent in males and females...... 20
10a	Midline gular crest weakly developed (cone size less than diameter of the eye opening)......11	17a	Dorsal keel fin-like scalloped, large size. Single thick annulated, short rostral horn......*Trioceros melleri*, Giant One-horned Chameleon [p. 298]
10b	Midline gular crest strongly developed (cone size greater than diameter of the eye opening)......13		
		17b	Dorsal keel with acuminate tubercles or conical tubercles......18
11a	Head short, length less twice the width, in our area Rwanda only......*Trioceros schoutedeni*, Schouteden's Montane Dwarf Chameleon [p. 302]		
		18a	Dorsal crest weakly developed, males with three annulated horns, females usually without or with a single rostral horn, Eastern Arc Mountains only...... *Trioceros werneri*, Werner's Three-horned Chameleon [p. 306]
11b	Head length greater than twice the width......12		
12a	Casque height equal or greater than diameter of the eyeball above a line	18b	Dorsal crest strongly developed (tu-

bercles larger than diameter of the eye opening)......19

19a Gular region with two rows of large cones diverging posteriorly, Udzungwa Mountains and Southern Highlands of Tanzania only......*Trioceros tempeli*, Udzungwa Double-bearded Chameleon [p. 305]
19b Gular region smooth, males and females with three horns, Southern Highlands of Tanzania only......*Trioceros fuelleborni*, Poroto Three-horned Chameleon [p. 287]

20a Dorsal crest of large triangular cones, flanks, limbs with large triangular cones, background scalation markedly heterogeneous, Udzungwa Mountains only......*Trioceros laterispinis*, Spiny-flanked Chameleon [p. 296]
20b Dorsal crest weakly developed......21

21a Temporal crest present, background scalation strongly heterogeneous...... *Trioceros incornutus*, Ukinga Hornless Chameleon [p. 290]
21b Temporal crest absent......22

22a Occipital lobes large, flap-like......hornless female *Trioceros werneri*, Werner's Three-horned Chameleon [p. 306]
22b Occipital lobes thin and rudimentary, Udzungwa Mountains, Southern Highlands of Tanzania and northern Malawi......*Trioceros goetzei*, Goetze's Whistling Chameleon [p. 287]

23a Background strongly heterogeneous, males with three horns, females with one, three or none, Kenya and northern Tanzania only......*Trioceros jacksoni*, Jackson's Chameleon [p. 292]
23b Background homogeneous or finely heterogeneous, females usually lack horns, mountains in the Albertine Rift only......*Trioceros johnstoni*, Rwenzori Three-horned Chameleon [p. 293]

SIDE-STRIPED CHAMELEON   *Trioceros bitaeniatus*

IDENTIFICATION: A fairly small brown or grey chameleon, usually with two side stripes, without horns or rostral appendages. The casque is slightly raised posteriorly, the head is narrow. The tail is prehensile, roughly equal to or slightly less than half the total length (females usually 42–48%, males 44–51%). The body scales are heterogeneous, with usually two rows (sometimes only one) of enlarged tubercular scales along the sides. A weak gular crest becomes a weak ventral crest, a prominent dorsal crest of spiky triangular scales is present. Maximum size about 17cm, usually 11–14cm, hatchling size 3–4cm. Colour variable, usually grey, shades of brown, sometimes almost black even when unstressed; always has two prominent pale flank stripes. When

Trioceros bitaeniatus

mono-coloured, the head is sometimes conspicuously darker than the body. May also be blotched black, white and grey, with three or four large pale blotches horizontally in the middle of the flanks. Specimens from around Lake Victoria may be blue-grey, with yellow patches in the body-limb junction.

HABITAT AND DISTRIBUTION: Medium to high altitude open savanna and grassland, woodland and isolated hills, from 1,000–3,000m altitude. Sporadic records from north-eastern Tanzania (Ngorongoro, Arusha, Mt Longido, central Serengeti), widely distributed in central and western Kenya, especially in grassland areas, from Loitokitok and the Chyulu Hills, the plains south and east of Nairobi, north-west through the Rift Valley to as far north as Kabluk. An odd record from Kora. Isolated records from west of Lake Victoria (including Rubondo island and Bukoba in Tanzania). Uganda records sporadic (including Mbarara, Queen Elizabeth National Park, Mabira Forest). A handful of records from volcanic mountains (Marsabit, Mt Nyiro and Mt Kulal) in northern Kenya, where it

Trioceros bitaeniatus Side-striped Chameleon. Stephen Spawls. Left: Gilgil, male. Right: Gilgil, female. Bottom: Gilgil, male.

lives in evergreen forest. Elsewhere, known from the eastern Congo, Somalia and highland Ethiopia. Conservation Status: IUCN Least Concern.

NATURAL HISTORY: Diurnal, rarely found in large trees, usually in small trees, bush, tall grass and sedge, in highland Kenya often on *Leonotis* plants. Also on hedges and ornamental plants, sometimes in large numbers on Maasai manyattas. Capable of withstanding extreme cold. May be abundant in some areas, but populations appear to peak and crash, this may be connected with the fact that this species gives live birth, and possibly also with predation by birds, especially starlings and shrikes. Usually 6–15 young are born, although one specimen gave birth to 25 babies. Two broods of differing sizes have been recorded in the female oviducts. Females have been known to eat their own young. Diet: invertebrates, especially grasshoppers.

SOUTH SUDANESE UNICORN CHAMELEON
Trioceros conirostratus

IDENTIFICATION: A small slender chameleon, rather similar to the Side-striped Chameleon *Trioceros bitaeniatus*. Males have a tiny, long cone-shaped scale on the tip of the snout. It has a short gular crest. Tail slightly less than half total length. The body scales are heterogeneous, with enlarged scales on the flanks, particularly those within the flank stripes. Maximum size about 18cm. Usually some shade of brown in Kenya, with pale side-stripes and orange tints, but green and brown in Uganda, In the middle of each flank, between the upper and the lower stripe, a series of three distinctive oval shaped markings can be present. A member of the *Trioceros bitaeniatus* complex.

DISTRIBUTION: Hill forest and woodland, at altitudes between 1,000–2,300m. Described in 1998 from a single specimen from Lomoriti in the Imatong Mountains in South Sudan, it has now been recorded from a chain of hills along the Kenya–Uganda border, including Mt Kadam, Mt Moroto, Kaabong (all northeast Uganda) and in north-western Kenya on the Mtelo Massif (Sekerr Range) and the Loima Hills, 80km west of Lodwar. Conservation Status: IUCN Least Concern. It seems to be relatively abundant in natural arid areas and modified landscapes as long as shrubs persist.

Trioceros conirostratus

NATURAL HISTORY: Little known. Diurnal and arboreal. The Mtelo specimens were found in shrubs bordering fields in an agricultural landscape on a plateau which was originally covered with *Acacia-Commiphora* woodland. Presumably similar to other members of the genus, i.e. gives live birth, populations peak and crash, eats invertebrates.

Trioceros conirostratus Sudanese unicorn Chameleon. Top left: Colin Tilbury, Mount Mtelo, male. Top right: Colin Tilbury, female, Mount Moroto. Bottom left: Colin Tilbury, Mount Mtelo, male. Bottom Right: Thomas Price, Mount Moroto, male.

USAMBARA THREE-HORNED CHAMELEON
Trioceros deremensis

IDENTIFICATION: A big chameleon, with a bizarre sail-like dorsal ridge, the males have three sharp annular horns, which appear when the juvenile is 10–12cm long. The females have no horns, but have a sharp nose. Small ear flaps. The casque is raised posteriorly to a spike. The tail is prehensile, as long as the body (slightly shorter in females). Body scales mostly homogenous, but there is a sprinkling of large tubercular scales on the flanks. The body appears extremely deep, due to the dorsal ridge, especially when the animal is nervous and sits in a hunched-up position. A prominent gular crest of pale or white spiky scales becomes a prominent ventral crest, but the dorsal crest is poorly developed and may be absent. The interstitial skin in the grooves on the chin is white, grey or pale green. Maximum size of males up to 30cm, but more usually 20–27cm, females slightly less, hatchlings 5–7cm. Colour very variable but usually shades of green, with dark and light swirls and bars of lighter and darker green, very often with yellow streaks and spots, when irritated becomes spotted with black. Hatchlings may be purplish-white.

Trioceros deremensis

HABITAT AND DISTRIBUTION: Forest and high woodland, 1,200–2,300m altitude, usually in clearings at the forest edge, moves into coffee plantations. A Tanzanian endemic, known from the following hill ranges, East Usambara (common), West Usambara (rare), Uluguru, Mkungwe, Nguru and Udzungwa. In conservation terms, vulnerable due to its very small range, where rapid deforestation is taking place, but it seems to adapt fairly readily to coffee plantations, and will live in hedges and ornamental trees. Conservation Status: IUCN Least Concern.

NATURAL HISTORY: Poorly known, but such a large species may utilise a good range of forest habitats, and ascend high in trees. The males probably use the horns for combat. Lays on average 11–32 eggs, maximum clutch size 40. Eats invertebrates, recorded as being fond of beetles and grasshoppers.

Trioceros deremensis Usambara three-horned Chameleon. Top left: Michele Menegon, Nguru Mountains, male. Bottom left: Michele Menegon, Nguru Mountains, female. Right: Michele Menegon, Nguru Mountains, male.

MONTANE SIDE-STRIPED CHAMELEON Trioceros ellioti

IDENTIFICATION: A medium-sized chameleon, without horns or a rostral process. No ear flaps. Casque very slightly raised posteriorly, with one to four distinct longitudinal grooves on the gular pouch, the skin between the scales distinctively black. Tail prehensile, slightly greater than the body length in males, slightly shorter in females. Scales weakly heterogeneous, with enlarged scales lying within the lighter flank stripes. A very prominent gular crest of long, pale spiky scales becomes a prominent ventral crest, and a fairly prominent dorsal crest is present, the spines getting shorter towards the tail. Maximum size about 19cm, average 10–15cm, hatchlings 3–4cm. Colour quite variable, but usually shades of vivid blue, green or brown, often with a broad, prominent, light-coloured (white, orange or yellow) flank stripe. In some specimens, especially those from western Uganda the gular pouch is electric blue.

Trioceros ellioti

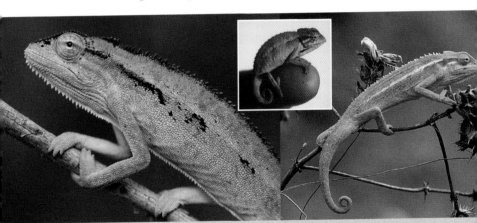

Trioceros ellioti Montane side-striped Chameleon. Top left: Daniel Hollands, Entebbe. Top right: Harald Hinkel, Rwanda. Inset: Stephen Spawls, Kakamega, juvenile.

HABITAT AND DISTRIBUTION: High, moist savanna (above 1,000m, up to 2,800m) and grassland. Occurs in high western Kenya, from Ruma National Park and Kisii north to the Cherang'any Hills, west to Mt Elgon, all along the northern shore of Lake Victoria in Uganda, round to where it just enters Tanzania. Also occurs right along the Albertine Rift in Uganda, from south of Lake Albert southwards into Rwanda and Burundi, where very widespread. Elsewhere, Imatong Mountains in South Sudan, and the western slopes of the Albertine Rift in the DR Congo. Conservation Status: IUCN Least Concern.

NATURAL HISTORY: Usually in low trees, bushes, shrubs and tall grass, fond of reedbeds and sedge in swampy areas. Doesn't ascend high in big trees. Often in hedges and ornamental plants in both rural and urban areas, and will live in plantations and roadside herbage. Often abundant in suitable habitat, and its populations seem to peak and crash, like the side-striped chameleon. Habits poorly known, but can tolerate cold, high-altitude areas, may be able to supercool to avoid freezing. Gives live birth, between 4–12 babies, and the females can retain viable sperm for up to 18 months if not mated. Eats invertebrates.

POROTO THREE-HORNED CHAMELEON Trioceros fuelleborni

IDENTIFICATION: A medium-sized chameleon, males have three short stout horns, females have a single stout nose horn and two tiny horns in front of the eye. Two large ear flaps present. Casque flattened. Tail prehensile, about half the total length. Body scales heterogeneous, with a scattering of large tubercular scales on the flanks and very large scales on the ear flaps. No gular or ventral crest, but a prominent dorsal crest of spiny scales is present, shortening posteriorly and this may disappear on the tail. Maximum size about 22cm, average size 20cm, hatchling size unknown. Colour quite variable, but usually shades of green, the eye turret skin is sometimes yellow or orange, there is a scattering of brown or rufous scales edging the vertebral ridge, and these turn black when the animal is stressed. Males usually have a broad stripe high on the flanks, consisting of three or four touching oval spots.

Trioceros fuelleborni

HABITAT AND DISTRIBUTION: A Tanzanian endemic, found only in the high woodland of the Ngosi Volcano, Poroto Mountains and Kungura Mountains, in south-west Tanzania, at altitudes between 2,000–2,500m. Thus vulnerable in conservation terms, its tolerance to deforestation and exact range are not known. Not protected in any proclaimed areas. Conservation Status: IUCN Least Concern.

Trioceros fuelleborni Poroto three-horned Chameleon. Left: Michele Menegon, Poroto Mountains, male. Right: Michele Menegon, Poroto Mountains, female.

NATURAL HISTORY: Poorly known, but arboreal like other chameleons. Males are probably territorial and fight with their horns. Between 11–24 live young are born, with one litter a year, in the wet season between November and January. Diet: invertebrates; one series contained almost nothing but beetle remains, but this might simply reflect the relative abundance of beetles at the time of collection.

GOETZE'S WHISTLING CHAMELEON Trioceros goetzei

IDENTIFICATION: A small chameleon, without horns or rostral appendage, with small ear flaps. Casque slightly raised at the back, has one or two well-developed distinct grooves on each side of the gular region. Tail prehensile, about half the total length. Body scales homogenous. No gular or ventral crest, but a prominent dorsal crest consisting of soft spiky scales extends onto the tail. Maximum size about 21cm, average 14–19cm, hatchling size 4–5cm. Colour very variable, but usually brown, green or grey, sometimes purplish-grey, with three prominent pale or white stripes on the head, the upper and lower extend down the body, the upper stripe may extend onto the tail. Taxonomic Note: Two subspecies occur in Tanzania, *Trioceros goetzei goetzei*, with obvious gular grooves, and *T. g. nyikae*, where the gular grooves are not so obvious.

Trioceros goetzei

HABITAT AND DISTRIBUTION: Lives on the fringes of high woodland and in well-wooded savanna and high grassland of montane plateaux, at altitudes from 1,500–3,000m. *Trioceros goetzei goetzei* is found in the Udzungwa, Ubena and Ukinga Mountains, *T. g. nyikae* in the Poroto Mountains, Ngosi Volcano, Tukuyu and Rungwe Montains. Conservation Status: IUCN Least Concern.

NATURAL HISTORY: Poorly known. Diurnal and arboreal. Lives in grass, bushes, shrubs and low trees, often in marshy areas where it frequents reeds, sedge and waterside vegetation, avoiding grass fires. Gives live birth, clutch sizes 6–10, in December and January; newborns were found in February and March. Eats invertebrates. The common name stems from the ability of some individuals, when seized, to produce a high-pitched hiss followed by an inhalational wheeze, which sounds like a whistle or squeak, but not all individuals do this.

Trioceros goetzei Goetz's Chameleon. Left: Michele Menegon, Kitulo Plateau National Park. Right: Michele Menegon, Kitulo Plateau National Park.

MT HANANG DWARF CHAMELEON *Trioceros hanangensis*

IDENTIFICATION: A small sized, stocky chameleon, with a maximum recorded total length of about 14cm, tail as long or slightly longer than the snout-vent length. The casque is elevated posteriorly with a conspicuous parietal ridge and no occipital lobes. Scalation is heterogeneous, with two enlarged denticulate scales in two rows on the flanks. The gular crest is visible and consists of elongated conical scales, ventral crest obvious in large individuals only. Essentially green, with lateral yellow broad bands along the row of enlarged scales. Head often light blue. Males are more rugose than females. It is a member of the *Trioceros bitaeniatus* complex and in size, scalation and body shape it resembles *Trioceros sternfeldi* from the mountains of Northern Tanzania, of which it was considered just a geographically separated population until recently, some experts remain sceptical of the validity of this species.

HABITAT AND DISTRIBUTION: Known to occur on Mt Hanang only, above 2,800m in moorland habitat, in the moister north-eastern side of the mountain. Individuals have been observed during daytime perched on bushes along streams. The chameleon population of the Nou Forest Reserve on the Mbulu plateau, north of Mt Hanang may belongs to this species but its actual taxonomic placement has to be investigated. The species is not found in

Trioceros hanangensis

Trioceros hanangensis Mt Hanang Dwarf Chameleon. Stephen Spawls, Mount Hanang.

the surrounding lowland savannah habitat or in the primary forest habitat. Conservation Status: IUCN Near Threatened.

NATURAL HISTORY: Poorly known, viviparous, 4–12 babies are born after a gestation of 5–6 months.

VON HÖHNEL'S CHAMELEON/HIGH-CASQUED CHAMELEON *Trioceros hoehnelii*

IDENTIFICATION: A small chameleon with a small blunt horn-like lump on the nose, without ear flaps, with a large, raised, helmet-like casque. The tail is prehensile, about half the total length. The body scales are very heterogeneous, with big scales on the casque and big tubercles scattered over the limbs and body, these tubercles may form distinct rows on the flanks, corresponding with the pale side-stripes. There is a beard-like gular crest of long and short scales which extends onto the belly, shortening posteriorly; a similar vertebral crest extends along the back onto the tail. Maximum size 16–18cm (possibly slightly larger in highland Kenya, 23cm), average 10–14cm, hatchlings 4–5cm. Colour very variable, any shade of green, greeny-brown, light to very dark grey, rufous or yellow; often with vivid red flank patches or vertebral stripes. There are usually two pale, prominent flank stripes and irregular, broad, darker bars on the flanks. Juveniles are brown or grey, with prominent yellow or cream stripes. The eyeball skin is often dark green, and the casque yellow, grey or brown.

Trioceros hoehnelii

HABITAT AND DISTRIBUTION: An East African endemic. High savanna and grassland, from 1,500–4,000m altitude. Found from Machakos, the Mua Hills and Nairobi north and north-west through the high country and across the rift valley, occurs around and on top of the Aberdares and Mt Kenya, westwards to the Mau, north to Mt Elgon,

the only records from outside Kenya are the western slopes of Mt Elgon and the highlands around Mbale, just west of the mountain. Absent in places on the floor of the high central rift valley, where *Trioceros bitaeniatus* occurs, more on the slopes. Conservation Status: IUCN Least Concern.

NATURAL HISTORY: Arboreal, shrubs, bushes (even very small spindly ones), small trees, thicket, reedbeds, sedge and in tall or tussock grass. Rarely found more than 2–3m above ground level. Not known to be territorial although rival males will fight, circling, jerking their heads and biting each other's flanks and legs and tail base. Hisses and bites when angry, and will quickly drop from its perch if pursued. Reaches sexual maturity at ten months, gives live birth to 7–18 young in sticky embryonic sacs. Populations peak and crash, like the Side-striped Chameleon *Trioceros bitaeniatus*, in the Limuru area population densities of over 300 animals per hectare have been recorded. Eats invertebrates. Capable of supercooling, a study of this species on the moorlands of the Aberdare Mountains found it survived nights when the temperature was −2°C, and became active and started to feed when the air temperature was 7°C. One was eaten by a Kenya Montane Viper *Montatheris hindii* in August.

Trioceros hoehnelii Von Hoehnel's Chameleon. Top: Jan Stipala, Eldama Ravine. Bottom: Stephen Spawls, Tigoni.

UKINGA HORNLESS CHAMELEON *Trioceros incornutus*

IDENTIFICATION: A medium-sized chameleon, without horns or a rostral appendage, with large ear flaps. The casque is flat. The tail is prehensile, about half the total length. The body scales are heterogeneous, the scales on the ear flaps are large, and there is a sprinkling of larger scales on the flanks. No gular or ventral crest, but a dorsal crest of quite widely spaced, spiky scales is present and extends onto the anterior part of the tail. Maximum size about 19cm, average 12–16cm, hatchling size 4–5cm. Colour quite variable, usually shades of green, brown or yellowy-grey, with a series of large dark blotches along the flanks, an irregular, pale stripe may be present high on the flanks.

Trioceros incornutus

HABITAT AND DISTRIBUTION: Montane forest, woodland and plantations. Originally thought to be a Tanzanian endemic, found at altitudes of around 2,000m in woodland and plantation forest in the Ukinga and Poroto ranges, Ngosi Volcano and Mt Rungwe, north of

Trioceros incornutus Ukinga hornless Chameleon. Michele Menegon, southern Tanzania.

Lake Malawi (Lake Nyasa) in southern Tanzania; now also known from the Nyika Plateau in Malawi. Conservation Status: IUCN Least Concern. However, may be vulnerable in conservation terms, as its small woodland habitat is being rapidly felled.

NATURAL HISTORY: Poorly known. Lives in shrubs, small trees and thicket. Gives birth to live young, clutch sizes of 11–16 recorded. Gravid females collected in February and March (although, oddly, females give birth towards the end of the year). Eats invertebrates.

ITURI FOREST CHAMELEON *Trioceros ituriensis*

IDENTIFICATION: A fairly big, slender chameleon without horns or ear flaps. The casque is raised to a point posteriorly. The tail is prehensile, about half or slightly more than half the total length. The body scales are heterogeneous, with a scattering (sometimes in lines) of large flank scales. No gular, ventral or dorsal crest is present. Females are the larger sex, reaching 25cm, average 18–22cm, males maximum 19cm. Hatchling size unknown. Colour very variable, but usually shades of light green, yellowish or grey-green. The big granular scales on the flanks are often pale or white, and a pale line is present along the centre of the flanks. Taxonomic Note: In 1994 *Chamaeleo (Trioceros) tremperi* was described from Maji Mazuri Station, near Eldama Ravine, on the western wall of the rift valley in Kenya. Extensive searches have failed to locate any more, and the types of *Trioceros tremperi* are identical to *Trioceros ituriensis*, so the Kenyan record is probably incorrect.

Trioceros ituriensis

HABITAT AND DISTRIBUTION: Forest and woodland. The type series was taken in the Ituri Forest in north-east DR Congo, at an altitude of 500–1,000m. Two East African records only, from the Bwamba Forest and Kibale, in Uganda, at an altitude of around 1,500m. Conservation Status: IUCN Least Concern. It is probably widespread and common in forested country in north-eastern DR Congo.

NATURAL HISTORY: Arboreal, in bushes and low trees in forest. No details of the

Ugandan population are known but those from the Ituri Forest were found on small trees and the understory, not up in big trees, and they occasionally descended to the ground. Specimens have been found sleeping 2m up a tree, and also in low bushes; when disturbed they may jump from the perch. Slow-moving and placid, they are said to be reluctant to even hiss. Lays eggs, clutch details unknown. Diet known to include spiders, snails, winged termites, beetles and wasps.

Trioceros ituriensis turi forest Chameleon. Top: Harald Hinkel, male, eastern Congo. Bottom: Colin Tilbury, female, easten Congo.

JACKSON'S CHAMELEON/KIKUYU THREE-HORNED CHAMELEON *Trioceros jacksoni*

IDENTIFICATION: A fairly large chameleon, the males have three annular horns, the one on the snout is usually biggest. The female may have one medium-sized, stout nose horn and two little horns, one in front of each eye, or a single short rostral horn, or sometimes none at all. There are no ear flaps. The casque is slightly raised at the back. The tail is prehensile, about half the total length. The body scales are heterogeneous, particularly large on the head and casque. There is no gular or ventral crest, but a prominent vertebral crest of stout, widely-spaced spiny scales extends from the nape to the base of the tail. Maximum size of males about 36–38cm in one subspecies (but see below), females about 25cm, average size males 15–25cm, hatchlings 4–5cm. Colour very variable but males usually shades of green with yellow, females brown and white or black and white, sometimes dull green. In males, the chin is often yellow and there is a broad stripe of darker green high on the flanks, with a lighter stripe of vague geometric blotches on the flanks. In females, the high stripe on the flanks is brown or black and the mid-flank stripe consists of pale and brown or black blotches. When angry, males become almost black, with green and brown speckling. Taxonomic Note: Three subspecies are described: the nominate subspecies *Trioceros jackson jacksoni*, reaching 27cm, *T. j. xantholopus*, reaching 38cm, with yellow edging to the cranial crest, and *T. j. merumontanus*, rarely larger than 22cm, with a yellow forehead, on Mt Meru in Tanzania.

Trioceros jacksoni

HABITAT AND DISTRIBUTION: An East African endemic. Mid-altitude suburbia, farmland, woodland and forest from 1,600 to over 3,000m. Known from the hills around

Trioceros jacksonii Jackson's Chameleon. Top: Jan Stipala, male, Machakos. Right: Stephen Spawls, female, Nairobi National Park. Inset: Stephen Spawls, male, Nairobi. Left: Stephen Spawls, female, Nairobi.

Machakos, in and around Nairobi, north along the wooded highlands east of the rift valley, to the Kikuyu escarpment forest and the mid to high altitude forest south of the Aberdares, thence east and north through the highlands to Mt Kenya, it occurs at mid-altitude right around the mountain, north to Ol Donyo Lossos, east to the Nyambene Hills. A single record from the north end of the Aberdares, at Shamata gate, 2,800m. There is a curious record from Samburu National Reserve (not shown on map). The subspecies *Trioceros jacksoni merumontanus* occurs on Mt Meru, Tanzania. A population (subspecies uncertain) also occurs on Ol Donyo Orok, near Namanga. Introduced on several Hawaiian islands and in southern California. Conservation Status: IUCN Least Concern. Very tolerant of suburbia and agriculture, as long as hedges and insects remain.

NATURAL HISTORY: Arboreal, will ascend high in forest trees to 10m or more, although equally fond of hedges, thicket, bush and shrubbery. If threatened, they will hop off their branch. Males are territorial and fight with their horns and by biting, trying to force their opponents to retreat or fall off the branch. Males in hedges in Nairobi were rarely closer than 20m apart. A male in Nairobi was seen to follow a female he could not see, presumably by scent. The females give live birth to 7–30 young (the smaller subspecies have smaller broods), captive specimens have had broods of 50. Gestation in captivity ten months. A Tanzanian specimen held 16 eggs in November. They eat invertebrates.

RWENZORI THREE-HORNED CHAMELEON *Trioceros johnstoni*

IDENTIFICATION: A big chameleon, the males have three annular horns, two in front of the eyes and one on the nose, the females are hornless or have three minute horns. No ear flaps. The casque is slightly raised at the back. The tail is prehensile, slightly greater than half the total length. The scales are heterogeneous, with a scattering of larger granular flank scales. No gular, ventral or dorsal crest is present. Maximum size about 30cm, average 15–25cm, hatching size 7–8cm. Colour very variable but usually vivid shades of green with yellow, the upper flanks often yellow-green, with dark greenish blotches, the lower flanks darker green, some specimens have oblique side bars of darker and lighter green, or yellow and turquoise, occasionally there is a series of small scarlet blotches along the vertebral ridge.

Trioceros johnstoni

HABITAT AND DISTRIBUTION: Forest and woodland, from 1,000–2,500m altitude along the Albertine Rift; in the high country of western Uganda, western Rwanda and Burundi, and the adjacent high forest of the DR Congo. Known localities include most of the mid to high altitude forest of the Rwenzori range, environs of Lake Mutanda, the Bwindi Impenetrable Forest, Nyungwe Forest and the Virunga Volcanoes. In conservation terms, it has a restricted habitat, but adapts readily to suburbia and plantations. Conservation Status: IUCN Least Concern.

NATURAL HISTORY: Poorly known. Arboreal, big males will ascend high into trees, but also in woodland and thicket. Perch heights vary from a few centimetres to several metres. Males are territorial and fight fiercely with their horns and by biting. When seized, they can vibrate like the pygmy chameleons. Breeding details; clutches of 4–23 eggs in captivity; eggs were laid at four month intervals and hatch in 3–4 months at 20–22°C. A gravid female was collected in June in the eastern Congo. Diet: invertebrates.

Trioceros johnstoni Ruwenzori three-horned Chameleon. Left: Steve Russell, Kasese, male. Top right: Michele Menegon, Rwanda. Bottom right: Steve Russell, Kasese, female.

ABERDARE MOUNTAINS DWARF CHAMELEON
Trioceros kinangopensis

IDENTIFICATION: A small-sized, stocky chameleon, with a maximum recorded total length of about 12cm. Tail shorter than the snout-vent length in both sexes. Casque is slightly elevated posteriorly with conspicuous lateral and parietal ridges and no occipital lobes. Scalation is heterogeneous, with two prominent rows of enlarged tubercles on each flank. Scales on the temporal region are enlarged and decreasing in size posteriorly till merging with the body scales, snout with sloping profile. Dorsal crest well developed and consisting in a single row of elongated conical scales, gular crest is visible and consisting of shorter conical tubercles, ventral crest formed by shorter tubercles that terminate at the vent. A grey, green or brown chameleon, sometimes turquoise in the lower part of the body, with lateral yellow-green broad bands. Enlarged tubercles on the flanks usually paler in colour. Gular crest is bright orange, ocular area usually grey-turquoise in males and grey with green tubercles in females. Males are more rugose than females. It is a member of the *Trioceros bitaeniatus* complex and it closely resembles *Trioceros schubotzi* from Mt Kenya from which can be distinguished by the sloping profile of the snout, which is more squared in *Trioceros schubotzi*, and the lack of raised canthus rostralis, sometimes forming a short rostral process in *Trioceros schubotzi*.

Trioceros kinangopensis

HABITAT AND DISTRIBUTION: The species appears to be restricted to the Afroalpine zone of the Kinangop peak in the Aberdare Mountains between 3,500–3,600m altitude. It could occur in the northern Aberdare range but its presence needs to be confirmed. Individuals were found in low ericaceous shrubs in a habitat dominated by tussock grass. Conservation Status: IUCN Near Threatened

NATURAL HISTORY: Poorly known, probably viviparous like the other species in the *T. bitaeniatus* complex.

Trioceros kinangopensis Mt Kinangop alpine Chameleon. Left:Jan Stipala, female. Right: Jan Stipala, Mount Kinangop, Aberdares, male.

SPINY-FLANKED CHAMELEON *Trioceros laterispinis*

IDENTIFICATION: A small spiky chameleon, resembling a piece of lichen, without horns or a rostral process, but with fairly large ear-flaps. The casque is slightly raised posteriorly. The tail is prehensile, slightly less than half of the total length. The body scales are heterogeneous, with enlarged flattened scales scattered over the flanks. A spiky gular crest is present, and a prominent dorsal crest consisting of big, widely-spaced stiff blunt scales, looking like rose thorns, extends right down the back and onto the tail. On the flanks, the chin, the legs and the tail there are curious, widely-spaced clusters of blade or thorn-like spines. Maximum size about 14cm, hatchling size unknown. Colour quite variable, it may be greenish, grey-green or grey. On the flanks there is often three darker, vertical, dumbbell shaped bars, interspersed with light blotches that, high on the flanks, appear to form a horizontal black and white bar.

Trioceros laterispinis

HABITAT AND DISTRIBUTION: A Tanzanian endemic, found only in the forest and woodland of the Udzungwa Mountains, and at present known only from the vicinity of Mufindi and Kigogo, at altitudes between 1,200–1,800m. Thus vulnerable (possibly endangered) due to habitat destruction, especially from logging and firewood collection. Conservation Status: Endangered B1ab(iii).

Trioceros laterispinis Spiny-flanked Chameleon. Left: Michele Menegon, Udzungwa Mountains. Top right: Stephen Spawls, Udzungwa Mountains. Bottom right: Michele Menegon, Udzungwa Mountains.

NATURAL HISTORY: Poorly known. Arboreal, perfectly camouflaged against the lichen-covered branches of forest trees. Gives live birth, up to 16 embryos recorded in one female; a gravid female was collected in July. Diet: invertebrates.

MT MARSABIT CHAMELEON *Trioceros marsabitensis*

IDENTIFICATION: A small green chameleon, described in 1991. The males have a single small, stout annulated nose horn, females have a low conical tubercle on the nose. No ear flaps are present. The casque is slightly raised at the back. The tail is prehensile, about half the total length. The body scales are heterogeneous, with a row of enlarged tubercular scales on the flanks. There is a low but distinct gular crest of enlarged scales, becoming a ventral crest and extending backwards to the vent. A dorsal crest of widely-spaced spiky scales extends onto the tail, getting smaller posteriorly. Maximum size about 18–19cm, hatchling size unknown but probably 4–5cm, 4.5cm juveniles have been found. Speckled greenish-brown, a brown line runs through the eyeball skin and the throat is suffused with lime green.

Trioceros marsabitensis

HABITAT AND DISTRIBUTION: A Kenyan endemic, known only from rainforest on the slopes of Mt Marsabit at an altitude of 1,250m, and thus dependent for its survival on the continued existence of the Marsabit forest. Much of the forest lies within the National Park, although the degree of protection there is patchy. Conservation Status: IUCN Near Threatened.

Trioceros marsabitensis Mount Marsabit Chameleon. Left: Colin Tilbury, male, Mt Marsabit. Right: Colin Tilbury, female, Mt Marsabit.

NATURAL HISTORY: Lives in trees and bushes in the forest, recorded as ascending to a height of 6m. Juveniles were found in low bushes. The presence of the rostral horn suggests that males may be territorial and fight. Probably gives live birth, but breeding details unknown.

GIANT ONE-HORNED CHAMELEON Trioceros melleri

IDENTIFICATION: Africa's biggest chameleon, with a single small annular horn, a stiff scaly rostral projection, or an enlarged rostral process. Both sexes have big ear flaps. The casque is slightly raised. The tail is prehensile, slightly over half the total length. The body scales are heterogeneous, very large on the ear flaps and big flat granular scales are scattered all over the body. No gular or ventral crest, but a huge, fin-like undulating dorsal crest extends the length of the back and along most of the tail. Maximum size around 60cm, average 30–50cm, hatchling size 6–7cm. Colour very variable but the usual overall impression is of an animal with narrow yellow and broad darker, vertical bars, usually three or four yellow bars on the body and 6–10 yellow bars on the tail.

Trioceros melleri

The broad bars may be blue-grey, dark green, brown or black. The enlarged flank scales are often pale, giving a spotted appearance, the nose, throat and sides of the head are usually light green, speckled with dark green, there is often a row of 4–5 dark or black spots low on the flanks. Juveniles are greyish-white and black.

HABITAT AND DISTRIBUTION: In Tanzania, well-wooded savanna and woodland (not high forest) at low altitude, sea level to 1,200m, but found at over 1,500m in Malawi. Sporadic records from eastern and southern Tanzania, from Mtai Forest Reserve south, mostly along the coast (occurs near Dar es Salaam) but inland at Morogoro, Kilosa and the Uluguru Mountains, down to the Rufiji River, thence across northern Mozambique to Malawi. Conservation Status: IUCN Least Concern.

NATURAL HISTORY: Arboreal, ascending up to 10m or more in woodland trees, so often present and unseen, but may descend and is seen crossing roads and paths; this may be

Trioceros melleri Giant one-horned Chameleon. Michele Menegon, Pwani, Dar es Salaam.

males looking for females or females looking for a suitable spot to lay eggs. In southern Tanzania, they were observed close to the ground between November and January, and higher in trees at other times. When angered they will inflate the body, raise the ear flaps, hiss and bite. The males are territorial, and will fight with their horn and by biting, often pugnacious during the breeding season. The males jerk their heads at a possibly receptive female, who responds by rocking. This species lays eggs in a hole dug by the female; clutch sizes between 38–91 eggs, roughly 2–2.5cm long, recorded, in December and January in Malawi, and February in Tanzania; mating observed in southern Tanzania in February. Diet: invertebrates but big enough to take small vertebrates, known to eat birds, other chameleons and skinks. The tongue may be up to 42cm long.

MT KULAL STUMP-NOSED CHAMELEON *Trioceros narraioca*

IDENTIFICATION: A medium sized, relatively slender chameleon, with a maximum recorded total length of about 17cm. Tail as long as the snout-vent length in both sexes. On tip of the snout it bears a short, ossified, blunt rostral process that can extend from 1.5mm in females to over 5mm in males off the snout. The casque is tall and markedly elevated posteriorly with strongly curved parietal crest and prominent para-parietal ridges, no occipital lobes or temporal crest. Scalation is heterogeneous, two rows of enlarged tubercles on the flanks, dorsal crest well developed, serrated and consisting in clusters of 3–4 conical tubercles, sometimes extending into the first third of the tail. Gular crest consisting of short conical tubercles, extending along the midline of the belly. A

Trioceros narraioca

Trioceros narraioca Mt Kulal Chameleon. Top: Robert Sindaco, Mt Kulal. Bottom left: Tomas Mazuch, Mt Kulal. Bottom right: Tomas Mazuch, hatchling.

brownish-green chameleon, with a broad, paler band that extend from the shoulder area to the base of the tail. A shorter pale band between axilla and groin is often present. The gular crest is white, ocular area usually turquoise in males. Males tend to be bigger and more rugose than females, with a slightly shorter tail and a more prominent rostral process. It is a member of the *Trioceros bitaeniatus* complex and it closely resembles *Trioceros hoehnelii* and *Trioceros nyirit* from highlands of Kenya, it can be distinguished from the former by shorter gular conical scales, from the latter by the convex profile of the parietal crest.

HABITAT AND DISTRIBUTION: Known only from relict montane forest of Mt Kulal, a volcano on the southern slopes of Lake Turkana, Kenya, at altitudes between 1,700–1800m. It inhabits montane forest, gardens and plantations near the forest margin, where it can be found both on bushes and on trees up to several meters high. It doesn't inhabit the Acacia woodland and savanna at lowers altitude or the montane dry grassland on the top of Mt Kulal. According to local people it prefers trunks and thick branches of trees rich in epiphytic lichens and mosses. Conservation Status: IUCN Near Threatened

NATURAL HISTORY: Very poorly known, probably viviparous like the other species in the *T. bitaeniatus* complex.

MT NYIRO MONTANE CHAMELEON *Trioceros ntunte*

IDENTIFICATION: A small stocky chameleon with a short head, described in 2005. Maximum recorded total length of about 15cm. Tail slightly shorter than body length. It is hornless. There is a gular crest of slightly enlarge conical tubercles. Scalation is heterogeneous, two rows of enlarged tubercles on the flanks, dorsal crest well developed, serrated and consisting in clusters of 3–4 conical tubercles, decreasing in size towards the tail. In colour it is usually green, yellow-brown or brown, with the usual pair of light flank stripes, when stressed it shows vertical dark bars; these may be rich brown. The chin and throat may be blue-white or blue. It is a member of the *Trioceros bitaeniatus* complex.

Trioceros ntunte

Trioceros ntunte Mt Nyiru Chameleon. Roberto Sindaco, Mt Nyiru.

HABITAT AND DISTRIBUTION: Endemic to Kenya, known only from Mt Nyiro, just south of Lake Turkana, where it inhabits low vegetation above the remnants of the evergreen forest, at altitudes between 2,500–2,700m. Conservation Status: IUCN Data Deficient. The mountain it lives on has no formal protection and is exploited for agriculture and stock raising.

NATURAL HISTORY: Very poorly known. Capable of coping with low temperatures. A female had 11 babies in March.

MT MTELO STUMP-NOSED CHAMELEON/POKOT CHAMELEON *Trioceros nyirit*

IDENTIFICATION: A medium sized, relatively slender chameleon, with a maximum recorded total length exceeding 19cm. Tail as long as the snout-vent length in both sexes. On tip of the snout it bears a short, ossified, blunt rostral process that can extend from 1.5mm in females to over 5mm in males. The casque is tall and markedly elevated posteriorly with a straight or weakly curved parietal crest, no occipital lobes or temporal crest. Scalation is heterogeneous, two rows of enlarged tubercles on the flanks, dorsal crest well developed, serrated and consisting in clusters of 3–4 conical tubercles, sometimes extending into the first two thirds of the tail. Gular crest composed by short conical tubercles, extending along the midline of the belly. Usually a bright green chameleon, with a broad, paler band that extend from the shoulder area to the base of the tail. Females more variable in colour than males, green to light brown. Toes and soles of the feet bright yellow in males from Mtelo massif. Gular crest is white, ocular area usually turquoise in males, head red in many individuals. Males tend to be bigger and more rugose than females, with slightly shorter tail and rostral process more prominent.

Trioceros nyirit

Trioceros nyirit Pokot Chameleon. Left: Jan Stipala, Cherang'any Hills, male. Right: Jan Stipala, Mt Mtelo, female.

HABITAT AND DISTRIBUTION: Known from the Mtelo Massif and the Cherang'any Hills in northern Kenya only, where it inhabits montane forest and Afroalpine moorlands, it enters transformed landscapes as long as there are trees and bushes. In the Cherang'any Hills it has been recorded between 2,900–3,150m, while on Mtelo Massif it reaches a lower elevational limit of about 2,300m. At lower elevations in can occur in sympatry with *T. ellioti* and *T. conirostratus*. Conservation Status: IUCN Near Threatened.

NATURAL HISTORY: Very poorly known, probably viviparous like the other species in the *T. bitaeniatus* complex.

RWENZORI SIDE-STRIPED CHAMELEON *Trioceros rudis*

IDENTIFICATION: A small, fat chameleon, without horns, rostral process or ear flaps. The casque is slightly raised posteriorly. The tail is prehensile, about half the total length or slightly more than half. The body scales are heterogeneous, with two distinctive flank lines of larger granular scales. A medium to small gular crest is present, becoming a short ventral crest, which rarely reaches the vent. A medium-sized dorsal crest of spiny scales extends the length of the back and onto the tail. Maximum size about 17cm, average 10–15 cm, hatchlings 4–5cm. Colour usually blotchy green or yellow-green, with one or two distinctive light flank stripes following the line of the big flank scales; these lines may be narrow, or broad, diffuse or with sharp edges, or like a chain of interconnected elongate diamonds. Sometimes broad vertical bars of dark green and greeny-yellow appear on the flanks and tail, sometimes prominent rufous blotches along the vertebral ridge. Occasional specimens are grey or blue-grey.

Trioceros rudis

HABITAT AND DISTRIBUTION: Medium to high altitude grassland and savanna, ranging from 1,000m altitude up to montane moorland at 4,000m in the Rwenzori Range. Essentially it inhabits the high lands of the Albertine Rift Valley, from Bujumbura in Burundi north through Rwanda to the Rwenzori Range in Uganda, common in Bwindi Impenetrable National Park. Elsewhere, known from the eastern slopes of the Rift Valley in the DR Congo and Gilo in the South Sudan; some problematic specimens from Kenya's Mau Escarpment have been assigned to this species. Conservation Status: IUCN Least Concern.

Trioceros rudis Ruwenzori Side-striped Chameleon. Left: Michele Menegon, Rwanda, male. Right: Michele Menegon, Rwanda, female.

NATURAL HISTORY: Arboreal, fond of bushes and thicket, will also live in tall and tussock grass, reedbeds and sedge. Like other members of the side-striped chameleon complex, often present in huge numbers in suitable habitat, especially open grassland with medium-sized, thick bushes. Capable of surviving sub-zero temperatures, in the Rwenzori Range they live above the frost line. Adapts well to cultivation and urbanisation, living in hedges, shrubbery, tall crops and roadside vegetation. Gives live birth, a clutch size of eight recorded. Eats invertebrates.

SCHOUTEDEN'S MONTANE DWARF CHAMELEON
Trioceros schoutedeni

IDENTIFICATION: A small squat chameleon with a spiny back, recently elevated out of the synonymy of *Trioceros rudis*. Maximum recorded total length of about 13cm. Tail slightly shorter than body length. Head short, it is hornless. There is a very small gular crest of tiny conical tubercles. Scalation is heterogeneous, two rows of enlarged tubercles on the flanks, dorsal crest very well developed, consists of regularly spaced spiky scales with smaller ones between. In colour it is usually green, rufous or brown, with the usual pair of light flank stripes, the upper one usually more visible, when stressed it shows vertical dark bars of triangular lozenges. It is a member of the *Trioceros bitaeniatus* complex.

HABITAT AND DISTRIBUTION: In our area, known from Nyamagabe and Gisenyi, in Rwanda, at altitudes between 1,400–2,100m. Elsewhere in the mountains of eastern Congo, south Kivu. Conservation Status: IUCN Data Deficient. Might be able to cope with high-altitude agriculture.

Trioceros schoutedeni

NATURAL HISTORY: Very poorly known. Probably gives live birth.

Trioceros schoutedeni Schouteden's Montane Dwarf Chameleon. Left: Max Dehling, Rwanda, male. Right: Max Dehling, Rwanda, female.

MT KENYA SIDE-STRIPED CHAMELEON *Trioceros schubotzi*

IDENTIFICATION: A small, fairly squat chameleon, without horns, rostral processes or ear flaps. Casque slightly raised. Tail prehensile, 35–50% of total length. Body scales strongly heterogeneous, with big granular scales of the flanks. A gular crest of spiky scales, both long and short, is present, becoming a ventral crest that extends to the vent. A prominent dorsal crest of long spiny scales extends onto the tail. Maximum size about 15cm, average 9–14cm, hatchling size 3–4cm. Usually shades of green, grey-green or yellow-green (males), or light green or brown (females), with two prominent pale yellow, entire or broken flank stripes; these may appear as a line of large oval dots. When stressed, broad dark vertical bars appear.

Trioceros schubotzi

HABITAT AND DISTRIBUTION: A Kenyan endemic, found in the alpine zone of Mt Kenya, known localities include the Sirimon track and Liki Valley, at altitudes over 3,000m, where it lives in shrubs, small trees, giant heather and tussock grass. Thus its range is very

Trioceros schubotzi Mt Kenya Side-striped Chameleon. Left: Jan Stipala, Mt Kenya, male. Right: Jan Stipala, female, Mt Kenya.

restricted, fortunately all of it lies in the Mt Kenya National Park, although it may be threatened by heather fires. Conservation Status: IUCN Near Threatened.

NATURAL HISTORY: Poorly known. Its habitat is characterised by rapidly fluctuating daytime temperatures, depending upon the weather, at night the temperature falls below zero and it may snow. This chameleon is startlingly tolerant of low temperatures and becomes active at sunrise, with hardly any thermoregulatory behaviour, catching invertebrates that have also become active at dawn. At night, it sleeps deep in the interior of bushes. It is able to supercool, to withstand freezing temperatures. When picked up, they inflate themselves. There appears to be no definite breeding season. Gives live birth, clutches of 7–10 young recorded in oviducts. Feeds on invertebrates, known prey items include flies, butterflies and their larvae, beetles, wasps, grasshoppers and spiders.

CRATER HIGHLANDS SIDE-STRIPED CHAMELEON
Trioceros sternfeldi

IDENTIFICATION: A small, fat chameleon, without horns, rostral process or ear flaps. The casque is slightly raised posteriorly. The tail is prehensile, about half the total length or slightly more than half. The body scales are very heterogeneous, with two distinctive flank lines of larger granular scales. A medium to small gular crest is present, becoming a short ventral crest, which rarely reaches the vent. A medium-sized dorsal crest of spiny scales extends the length of the back and onto the tail. Maximum size about 17cm, average 10–15cm, hatchlings 4–5cm. Colour in Tanzania usually blotchy green, with one or two distinctive flank stripes of light green or yellow, following the line of the big flank scales, these lines may have precise edges or just be a series of linked blotches; when stressed vertical dark sub-diamonds appear. Some specimens have vivid blue eye turrets. Loita Hills specimens are green with heavy yellow speckling, good camouflage in the yellow lichen-covered trees, the Maralal specimens were predominantly brown and green.

Trioceros sternfeldi

HABITAT AND DISTRIBUTION: Medium to high altitude grassland and savanna. It has a very disjunct distribution, suggesting there may be cryptic forms within the group; populations are found in the mountains and crater highlands of northern Tanzania (Kilimanjaro, Mt Meru, Ngorongoro, Embagai), in the Loita Hills in Kenya and on the Karisia and Saanta Hills north and east of Maralal, all at altitudes between 1,600–2,500m. Conservation Status: IUCN Least Concern.

Trioceros sternfeldi Crater Highlands Side-striped Chameleon. Top left: Michele Menegon, northern Tanzania. Top right: Stephen Spawls, Mt Hanang, male. Bottom left: Stephen Spawls, Arusha National Park, female. Bottom right: Jan Stipla, Loita Hills, male.

NATURAL HISTORY: Arboreal, fond of bushes and thicket, often on the fringes of forest in the mist belt. Like other members of the side-striped chameleon complex, often present in huge numbers in suitable habitat, especially open grassland with medium-sized, thick bushes. Capable of surviving sub-zero temperatures. Adapts well to cultivation and urbanisation, living in hedges, shrubbery, tall crops and roadside vegetation. Gives live birth, clutch sizes 4–12, after a gestation period of 5–6 months. Eats invertebrates.

UDZUNGWA DOUBLE-BEARDED CHAMELEON
Trioceros tempeli

IDENTIFICATION: A medium-sized chameleon, may resemble a lichen-covered branch, with a small, hornlike lump on the snout and ear flaps. The casque is slightly raised posteriorly. The tail is prehensile, about half the total length. The body scales are heterogeneous, with big scales scattered over the flanks. There is a prominent, double gular crest, consisting of soft white spiky scales, these two chin crests diverge towards the throat. Oddly, there is no ventral crest, but a prominent dorsal crest of widely-spaced spiny scales extends along the back and most of the way along the tail. Maximum size about 23cm, average size 12–18cm, hatchling size unknown. Colour quite variable, usually pale green or brown, sometimes with fine black striations between the scales on the flanks. A broad pale-coloured stripe is usually present, high on the flanks, and the vertebral ridge is edged with reddish-brown. Low on the flanks there is a scattering of light-coloured scales.

Trioceros tempeli

HABITAT AND DISTRIBUTION: A Tanzanian endemic, inhabits farmland, suburbia, woodland and forest, at altitudes from 1,500–2,400m. Known from southern Tanzania, from the Udzungwa Mountains, south of Iringa, to Mufindi and the Kipengere Range (Ubena and Ukinga Mountains), north-east of the north end of Lake Malawi (Lake Nyasa). Conservation Status: IUCN Least Concern.

NATURAL HISTORY: Relatively common where they occur, in gardens, hedges, trees and secondary growth in forests and along the forest margins. At night, their perch heights vary from 1–5m. Gives birth to live young, clutch size from 15–28. Juveniles have been observed in January, February and July, which indicates that two litters a year may be produced. Eats invertebrates.

Trioceros tempeli Werner's three-horned chameleon. Left: Michele Menegon, Udzungwa Mountains, female. Right: Michele Menegon, Udzungwa Mountains, male.

WERNER'S THREE-HORNED CHAMELEON *Trioceros werneri*

IDENTIFICATION: A medium-sized green or brown chameleon, males with three annular horns on the snout, females with a single annular horn, both sexes have a huge single occipital lobe that overlies the flat casque and the neck like a short cape. The tail is prehensile, about half the total length. The body scales are heterogeneous, with big granular scales on the flanks. Maximum size about 23cm, average size 16–22cm, hatchling size unknown. Colour quite variable, usually green with browns and yellows, males have an irregular pale longitudinal stripe high on the flanks and oval dark green blotches along the vertebral ridge. The ear flaps are sometimes barred and there is often a broad light lie behind the eye. The females may be green or brown, sometimes with yellow skin vermiculations between the scales, there may be irregular rufous blotches or stripes on either side of the vertebral ridge. Taxonomic Note: molecular analysis indicates that, as presently defined, *Trioceros werneri* is a complex of at least four and maybe six species, and likely to be split in the future.

Trioceros werneri

HABITAT AND DISTRIBUTION: A Tanzanian endemic. Closed forest in the Uluguru, Udzungwa, Ukaguru, Nguru and Rubeho Mountains of eastern and south-eastern Tanzania, from 1,400m–2,200m altitude. Conservation Status: IUCN Least Concern.

NATURAL HISTORY: Poorly known. Found high in forest trees, where they perch high up, but they have also adapted to suburban gardens. Males territorial, they fight with their horns, and males often show old injuries and scars, presumably as a result of fighting. Gives live birth to 15–20 young; females at Mufindi in July were gravid and hatchlings were found, an Uluguru female held 17 eggs in November. One litter per year is produced, after the start of the rainy season. Diet: invertebrates.

Trioceros werneri Werner's Three-horned Chameleon. Top: Michele Menegon. Bottom: Michele Menegon.

Varanis albigularus
White-throated savanna monitor
Michele Menegon

SUB-ORDER **SAURIA: LIZARDS**
FAMILY **VARANIDAE**
MONITOR LIZARDS

A lizard family and genus found in Australia, Asia and Africa; many are very large. The world's heaviest lizard is a monitor, the south-east Asian Komodo Dragon *Varanus komodoensis*, it reaches 160kg and 3.1m length; the Asian Water Monitor lizard *Varanus salvator* is the longest lizard in the world and grows to 3.5m long; East Africa's biggest lizard is the Nile Monitor *Varanus niloticus*, reaching 2.5m or more. A fossil monitor of nearly 6m length is known from the Pleistocene of southern Asia and Australia. Much taxonomic work has been done on *Varanus* in recent years; since we first wrote this book the number of known species has increased from 45 to around 80; most of the new additions described are from south-east Asia,

Monitors are carnivores, eating a wide range of prey, the Komodo Dragon is said to have killed and eaten humans. Monitors have powerful jaws and long retractile forked tongues, an obvious ear opening and a long flexible neck. Their limbs are well-developed, with large claws. The tail cannot be shed but is long and powerful; used as a climbing aid, in swimming and as a defensive weapon. The tough and leathery skin, covered with bead-like scales is exploited in some countries, leading to their persecution; in some countries they are eaten; in rural East Africa monitors (particularly the White-throated Savanna Monitor *Varanus albigularis*) may be killed by smallholders as they raid chicken runs. In Egypt long-distance lorry drivers may have a stuffed monitor mounted on the radiator grille, to bring luck. All monitors lay eggs. Male combat has been observed in some species. Australia has the greatest diversity of monitor lizards (known there as 'goannas'), with around 30 species. At present only four species are known from Africa. Some years ago the African Nile Monitor was split into two species, *Varanus niloticus* and *V. ornatus*, ('Forest Monitor'); recent molecular studies indicate that it is a single but variable species, within which there are at least three evolutionary lineages, so more splits may be on the cards.

KEY TO EAST AFRICAN MEMBERS OF THE GENUS *VARANUS*

1a Relatively slim, mostly yellow and black, nostril round or oval, snout pointed, usually near water...... *Varanus niloticus*, Nile Monitor [p. 309]
1b Heavily built, colour a mixture of greys, brown, black and dirty white, nostril an oblique slit, snout rounded, usually in dry savanna......2
2a Midbody scale rows 110–167, scales small and granular on the neck,, widespread in Kenya, eastern Tanzania, also in eastern Uganda......*Varanus albigularis*, White-throated Savanna Monitor [p. 311]
2b Midbody scale rows 75–100, scales large and cobblestone-like on the neck, in north-west Uganda......*Varanus exanthematicus*, Western Savanna Monitor [p. 313]

NILE MONITOR/WATER MONITOR *Varanus niloticus*

IDENTIFICATION: A big, dark green and yellow lizard, usually near water sources. It has a pointed snout, the eye is large, pupil round, iris yellow and black, the eyelids are yellow. The tongue is long and forked. The neck is very long, the body is long, relatively slim and cylindrical, with well developed muscular limbs, with strong claws. The tail cannot be shed, it is tough and whip-like, laterally depressed, triangular in section, with a hard vertebral ridge, about 60% of the total length. The skin is tough and leathery, the scales are small and button-like, in 128–183 rows at midbody (usually 137–165). Maximum size about 2.5m, possibly larger, there are reliable anecdotal reports of specimens close on 3m, average 1.5–2.2m, hatchlings 25–33cm. Ground colour black, brown, dark green or grey-green, spotted to a greater or lesser extent with yellow. There are 4–11 dorsal, yellow crossbars, which may be solid or ocelli. The limbs are spotted yellow on black, the flanks vertically barred or blotched green/black and yellow, the tail has 9–18 vertical yellow bars on a dark background. The belly is dirty yellow or cream, with black or dark blue crossbars, vermiculations or blotches. The juveniles are vividly coloured green or greeny-black and yellow. Big adults can become very dull, especially if foraging away from water and in the process of sloughing, but the distinctive pointed head shape means they can be distinguished from savanna monitors. Taxonomic Note: In 1997 examples of this species from the central African forests, with (*inter alia*) fewer crossbars were split into a separate species, *Varanus ornatus*, Forest Monitor; recent molecular work indicates this was unjustified, although further work might split the taxon.

Varanus niloticus

HABITAT AND DISTRIBUTION: Usually near fresh water sources, from sea level to about 1,600m in our area, but rarely higher; for example found in Nairobi National Park but not at Lake Naivasha. Widespread throughout East Africa and recorded from almost all suitable water sources, absent only from high altitude areas, the Rift Valley soda lakes and the dry north and east of Kenya, although records sporadic for central Tanzania. Elsewhere, north to Egypt, west to southern Mauritania, south to South Africa; it has the widest distribution of any African lizard. Conservation Status: Not assessed by IUCN, but hugely exploited in parts of West Africa for their skins, which make a durable leather and for local medicine and magic. They may be under threat because of this; however the proliferation of dams in parts of Africa has provided extra habitat for Nile monitors, they thrive in such places.

NATURAL HISTORY: A diurnal, versatile lizard, it is active in the water and on the ground, it readily climbs trees and rocks; juveniles are more arboreal than adults. It is fast moving, running with a distinct serpentine motion. It swims superbly, limbs tucked in, using its blade-like tail, and can stay submerged for 20 minutes or more. When inactive, it basks or rests on waterside vegetation, trees, logs and rocks, often in a prominent position. In southern Africa it may hibernate in big rock cracks, such behaviour is not recorded in East Africa. Nile monitors are usually wary and, if approached will run away or jump into water, often from a considerable height, swimming away below the surface to a refuge such as a reedbed. They often live in waterside burrows. However, in some areas they have become used to humans and will permit close approach. Small juveniles are more cautious than the adults but will not enter deep or fast flowing water, if approached they prefer to scramble away around the banks or dive into vegetation, if pursued into water they will frantically swim a short distance and then try to hide. Nile Monitors forage on land and in the water, they have a very varied diet, they are fond of freshwater crabs and mussels, which they crush with their rounded, peg-like teeth, but will take any suitable invertebrate, including slugs, spiders and water beetles, they will also take small vertebrates including frogs, fish, lizards, bird and birds eggs; they have been seen 'herding' fish in shallow water. They are notorious raiders of unattended crocodile nests and are also known to open and raid sea turtle and freshwater terrapin nests. The juveniles have sharper teeth than the adults and eat mostly insects and frogs. Nile Monitors have an excellent sense of smell and will hunt out and eat carrion; they are also known to raid chicken runs. They lay eggs, often in an active termite nest. The female claws her way in and deposits the eggs, the hole is then resealed by the termites and the eggs are kept warm and moist inside the hill while they develop. Incubation times are unknown in East Africa, but in South Africa in the wild the eggs take a year to incubate, 4–6 months in captivity. Between 20–60 eggs, roughly 3×5cm, are laid. In East Africa hatchlings have been observed in November in southern Tanzania, January in southern Kenya and July in northern Kenya. Like other monitors, if cornered they will inflate the throat and hiss loudly, raise themselves up high, stiff-legged and lash with their tails, if seized they will bite savagely and scratch with their claws.

Varanus niloticus Nile monitor. Left: Stephen Spawls Mzima Springs. Right: Stephen Spawls, Arusha, hatchling.

WHITE-THROATED SAVANNA MONITOR Varanus albigularis

IDENTIFICATION: A big, heavily-built, grey or brown monitor lizard. The head is large and square, the snout bulbous, with an obvious big nostril, the eye is small, pupil round, earhole large. The tongue is forked, long and blue. The body is broad, somewhat triangular in section, with well developed stout and muscular limbs, with strong claws. The tail is slightly longer than the body and cannot be shed, it is tough and whip-like, laterally depressed (although in big adults it may be flattened horizontally across the base), triangular in section, with a hard ridged top. The scales are coarse, small, non-overlapping and beadlike, in 110–167 rows at midbody (usually 132–150 in East Africa). Two precloacal pores are present. Maximum size about 1.6m, average 1–1.4m, hatchlings 23–26cm. Colour grey or brown, quite variable and the colour and pattern are often obscured by dirt and unsloughed skin patches, they often have ticks as well, particularly round the head orifices. Clean specimens look grey or black above, with 3–8 white crossbars, rows of white spots or ocelli. The top of the head is dark, a broad dark bar on the neck spreads out onto the back, the side of the head, neck and flanks are lighter. The tail is barred grey or brown on yellow or cream. The underside is paler with fine black vermiculations. Juveniles have a conspicuous black chin. Some juveniles from central and southern Tanzania are attractively coloured, the side of the head behind the eyes is grey-blue, the back blue-grey with bright yellow ocelli, on the flanks there are blue-grey stripes on a yellow background and the tail is vividly banded yellow and blue.

Varanus albigularis

HABITAT AND DISTRIBUTION: Dry and moist savanna, coastal thicket and woodland and semi-desert, from sea level to about 1,500m altitude. Probably quite widespread in Kenya and Tanzania, but there are very few museum records, largely due to the problems of preserving such big lizards. Abundant and easily observed in most of eastern and southern Kenya, especially in Tsavo, Ukambani, Tharaka and the coast, records from the north more sporadic but known from the area between Wajir, Buna and Malka Murri. Occurs from Baringo and the Kerio River down to Lake Turkana. The only records from Uganda are from the east and north-east; at Amudat, Moroto and the Kidepo National Park, but probably more widespread. Widespread throughout most of Tanzania save the northwest. No records for Rwanda and Burundi. Elsewhere, south to South Africa.

NATURAL HISTORY: Diurnal (there are a few reports of crepuscular and nocturnal activity), mostly terrestrial, but it climbs well if clumsily, ascending big trees and rock outcrops, often to a considerable height. When inactive, it will hide in thickets, in holes (especially aardvark and porcupine burrows), in rock fissures, in hanging beehives, tree holes and cracks and abandoned termite hills. In the dry season it may aestivate, hiding in a recess, but specimens have been found passing the dry season in such unsuitable places as out on the side branch of a big tree. In some areas they are tolerant of humans, basking until closely approached, in other areas they are more nervous, running to hide at the first sign of danger. If cornered, they will hiss very loudly and ominously and stand up high, stiff-legged. They can lash powerfully and accurately with their tails, they also have a very hard bite (although the teeth are peg-like and blunt), they will hang on with bulldog-like tenacity to anything they seize. Cornered specimens have been known to leap at an aggressor. Research on this monitor in Namibia indicates that the males may have home ranges up to

Varanus albigularis White-throated savanna monitor. Top: Stephen Spawls, Tsavo East. Bottom: Michele Menegon, southern Tanzania.

25km², females have smaller ranges, 8–10 km², and during the rainy season they walk 3–6 km per day looking for food. They lay eggs, in southern Africa clutches of 8–51 eggs, roughly 3.5×6cm, are laid in a hole dug by the female, sometimes she may use a termite hill hole or a hollow tree, in Kenya rock crevices are sometimes used. Incubation time about four months in southern Africa. Hatchlings were collected in October and November in southern Tanzania, and in July in coastal Kenya. The diet includes a wide range of vertebrate and invertebrates (basically, any animal they can overpower), including small mammals, birds, snakes, other lizards, eggs, tortoises, insects, millipedes, other arthropods and even carrion. They catch and kill snakes by seizing them and violently lashing the snake from side to side while clawing at the body and crushing it in their jaws. They seem to have an excellent sense of smell, and will excavate buried nests and holes containing living prey, especially mammals. Their enemies include large birds of prey (especially the martial eagle), ratels and big snakes like the Brown Spitting Cobra *Naja ashei*; mongooses take a heavy toll of the eggs. Savanna Monitors can be abundant in some areas, especially those with suitable refuges; one kopje in the western Tharaka Plain had around 12 Savanna Monitors living permanently on it; but in some areas they are scarce. They have lived 11 years in captivity.

WESTERN SAVANNA MONITOR Varanus exanthematicus

IDENTIFICATION: A relatively small, grey monitor lizard. The head is thick, the snout short and bulbous, with an obvious big nostril, the eye is fairly large, pupil round, iris orange, earhole large. The tongue is long and forked, pink with a grey tip. The body is broad and short, somewhat triangular in section, with well developed stout and muscular limbs, with strong claws. The tail is tough and whip-like, laterally depressed (although in big adults it may be flattened horizontally across the base), triangular in section, with a hard ridged top. The tail cannot be shed, it is slightly less than half the total length. The scales are coarse, non-overlapping and beadlike, they are big and cobblestone-like on the neck, in 75–100 rows at midbody. Two preanal pores are present. Maximum size about a metre, average 50–80cm, hatchlings 18–20cm. Ground colour light or dark grey, sometimes brown, there are several light white spots or ocelli across the back, the tail has light and dark vertical bars, the belly is dirty white with dark crossbars; there are often fine vertical bars on the lips. Like the White-throated Savanna Monitor its general appearance changes with how dirty it is, and how near sloughing.

Varanus exanthematicus

HABITAT AND DISTRIBUTION: Dry and moist savanna, of west and central Africa, from sea level to about 1,400m altitude. Recently recorded from Murchison Falls and Moyo, north-west Uganda, and known from South Sudan and at Garamba in the north-east of DR Congo. Elsewhere west to Senegal. Conservation Status: IUCN Least Concern. Heavily

Varanus exanthematicus Western savanna monitor. Top left: Stephen Spawls, captive. Top right: Davide Dal Lago, Murchison Falls. Bottom: Jean-Francois Trape, West Africa.

exploited for its skin and also for the pet trade, despite being a CITES Appendix 1 animal; it is thus under threat in some west African countries. It is a large inhabitant of relatively open country and thus at risk of local extinction if harvesting is not controlled. A curious recent superstition in north-west Uganda is that the blood of these lizards cures AIDS, and has led to them being exploited.

NATURAL HISTORY: Similar to the White-throated Savanna Monitor. It is diurnal, mostly terrestrial, but it climbs well, ascending big trees and rock outcrops. When inactive, it will hide in thickets, in holes, in rock fissures, tree holes and cracks and abandoned termite hills. It appears to be abundant in some areas, rare in others, a prolonged museum expedition to Garamba found only a single specimen during several months of collecting, but a commercial collector in Nigeria was receiving over 200 a month from one village. Like the White-throated Savanna Monitor it is confident in defence; if cornered, it will hiss very loudly and ominously, lash powerfully and accurately with its tail and bite hard, it will hang on with bulldog-like tenacity to anything it seizes. Females lay from 6–20 eggs, roughly 3×5cm, usually in a hole they dig themselves. In west Africa hatchlings appear in March or April, near the start of the rainy season. The diet includes the usual wide range of vertebrate and invertebrates, they will catch and eat any smaller animal they can overpower.

Geocalamus modestus
Mpwapwa Wedge-snouted Worm Lizard
Stephen Spawls

SUB-ORDER **AMPHISBAENIA: WORM LIZARDS**
FAMILY **AMPHISBAENIDAE**
WORM LIZARDS

Probably the least known of East Africa's reptiles, the worm-like amphisbaenians or worm lizards are a group of bizarre squamate reptiles, specialised for burrowing life. Their relationships have been vigorously debated, and they have been placed in their own sub-order, but recent molecular research indicates they form a monophyletic clade of lizards, in a superfamily of lizards, the Lacertoidea. They are old, fossil worm lizards from 65 million years ago in the Palaeocene epoch are known. Their evolutionary history is quite fascinating, their underground life should hinder dispersal but they occur widely in the Americas, Europe and Africa. Molecular evidence indicates, startlingly, that they dispersed after the supercontinents separated, the African species form a monophyletic subgroup that rafted there from North America about 60 million years ago. A fossil species from the Miocene strata of Rusinga Island, *Listromycter leakeyi,* had a skull 36mm long, the largest known skull of any worm lizard. Worm lizards have many unusual features, including a reduced right lung (snakes and legless lizards have the left lung reduced) and a unique middle ear. They are often pallid, purple or pink. The East African species have no ears or limbs (some species have front legs). Some have reduced eyes visible below the skin; others not. Most are small, less than 30cm, although a 75cm form is known from South America. Their smooth, rectangular scales are arranged in external rings or annuli, which superficially resemble those of earthworms. The head of amphisbaenids may be rounded or highly modified into a wedge or keel which assists in its fossorial (burrowing) way of life. Although many species burrow in soft soil or sand, others push through harder ground. The skin of amphisbaenids is only loosely attached to the muscles below; this permits the head and body to move in such a way that the head can be used as a ram when burrowing. Hence their skulls (unlike those of snakes) are relatively short and strong, and some have a hardened blade-like edge to the snout, that can be used like a shovel and a ram, scraping soil from the front of the burrow and compacting it on the sides. They have large strong teeth with which they seize their prey, mostly invertebrates, which they detect by scent and vibration, the victim is ripped to shreds before being dragged into the burrow.

As with other squamates, male amphisbaenids have hemipenes; females lay eggs, but in some species, these may be retained inside the body and females give birth to live young. Amphisbaenians live mostly in savanna or woodland, they are rarely seen above ground, but are sometimes flooded out of burrows after heavy rains. They are also found under logs, or dug up by cultivators. Because of their secretive habits and small size, amphisbaenians are the least known of East African reptiles. Some species are known only from the type specimens collected decades ago. There are over 195 species known, in 19–25 genera (depending on your point of view) and six families; found throughout sub-Saharan Africa and South America, with a few species found in North America and the Arabian peninsula; one species is found in Spain. The largest family, with 11–15 genera and nearly 180 species, is the Amphisbaenidae, nine genera and 65 species occur in Africa. All East African amphisbaenids belong in this family. In our area there are four genera and ten species, nine of which are endemic to East Africa, seven of these are endemic to Tanzania, one species occurs in Kenya and Tanzania. Much of our knowledge of East African worm lizards is due to the efforts of one man, the assiduous herpetological collector C.J.P. Ionides, who was based in south-east Tanzania. Amphisbaenians are very hard to identify to species level, the key below (and the specific keys) can only be successfully used with the aid of a binocular microscope. Anyone finding a worm lizard should take it to a local museum, little is known of their distribution or variation and museum specimens are needed.

KEY TO THE GENERA OF EAST AFRICAN WORM LIZARDS

1a Head rounded or slightly compressed......2
1b Head modified into a vertical keel by strong lateral compression of its anterior aspect......*Ancylocranium*, sharp-snouted worm lizards [p.323]
2a Snout compressed, pectoral shields differentiated, interannular sutures forming anteriorly directed chevrons......*Geocalamus*, wedge-snouted worm lizards [p. 317]
2b Snout rounded, pectoral shields not modified......3
3a Prefrontal and first supralabial distinct......*Loveridgea*, round-snouted worm lizards [p. 318]
3b Prefrontal and first supralabial fused with nasal......*Chirindia*, round-headed worm lizards [p.320]

WEDGE-SNOUTED WORM LIZARDS
GEOCALAMUS

Amphisbaenians with a compressed snout and with well-developed pectoral shields (large scales in the 'chest' area). Three species are known, two are found in East Africa, one extends from south-eastern Kenya into north-eastern and central Tanzania, the other is known from central Tanzania.

KEY TO EAST AFRICAN MEMBERS OF THE GENUS *GEOCALAMUS*

1a Body annuli 195–226, tail annuli 19–24, suture between nasal and first supralabial complete, infralabials usually two, underside of tail past autotomy site never completely pigmented......*Geocalamus acutus*, Voi Wedge-snouted Worm Lizard [p. 317]

1b Body annuli 238–243, tail annuli 26–27, suture between nasal and first supralabial incomplete, infralabials three, underside of tail past autotomy site completely pigmented......*Geocalamus modestus*, Mpwapwa Wedge-snouted Worm Lizard [p. 318]

VOI WEDGE-SNOUTED WORM LIZARD *Geocalamus acutus*

IDENTIFICATION: A medium-sized amphisbaenian with a distinct wedge-like snout from the drier parts of Kenya, ranging south-east to the dry, central portion of Tanzania. 196–226 annuli on body, 38–42 segments (18–20 dorsals and 20–22 ventrals) in a midbody annulus, pectoral segments feebly differentiated, slightly longer than broad, forming an anteriorly-directed angular series; 19–26 annuli on tail; six anals, much divided; four preanal pores. Total length about 28cm. Pink or pinky-brown above, flanks pinkish-purple, the head mottled white/pink, close examination shows the dorsal segments to be pinky-brown with white edging, purply-pink below with some darker mottling on the tail. The grooves between the segments are pink. When preserved the dorsal surface become brown, the underside tan.

Geocalamus acutus

HABITAT AND DISTRIBUTION: Dry savanna at medium to low altitude, from sea level to 1,300 m. An East African endemic, it is found in sandy soils of south-eastern Kenya

Geocalamus acutus Voi Wedge-snouted Worm Lizard. Stephen Spawls, Arabuko-Sokoke Forest.

and north-eastern Tanzania from the Arabuko-Sokoke Forest in Kenya westwards to Mt Lolkisale, in Tanzania at the eastern edge of Tarangire National Park. Isolated records exist for Dodoma.

NATURAL HISTORY: Little known. Terrestrial, burrowing, probably diurnal. Some specimens were dug up by ploughs. Presumably lays eggs, feeds on invertebrates.

MPWAPWA WEDGE-SNOUTED WORM LIZARD
Geocalamus modestus

IDENTIFICATION: A median sized amphisbaenian with 238–241 annuli on body, 34–38 segments (16–18 dorsals and 18–20 ventrals) in a midbody annulus. The two median ventral segments are nearly equilateral; the pectoral segments are only feebly differentiated, slightly longer than broad and forming an anteriorly-directed angular series. 29 annuli on the tail; six anal scales, 3–4 preanal pores. Total length of about 28cm. Colour above uniformly violet to pinkish or grey brown, often becoming pinker towards the head and darker towards the tail with pale pink grooves between the segments. Below, pinkish, white or translucent white.

Geocalamus modestus

HABITAT AND DISTRIBUTION: A Tanzanian endemic, sporadically recorded from the moist central savanna in two areas; around Dodoma (vicinity of Dodoma town, Ikikuyu, Mpwapwa), and further north-west around Ushora in Mkalama; all at altitudes between 800–1,200m. Conservation Status: Not assessed by IUCN. Rarely collected, unlikely to be threatened by human activities. However, its microclimate requirements are not known, and it is possible that human activities such as large scale agriculture or overgrazing might damage it in the long run.

NATURAL HISTORY: Little known. Terrestrial, burrowing, probably diurnal. The species is found in sandy soil and presumably feeds on invertebrates. It falls prey to small carnivores such as mongooses.

ROUND-SNOUTED WORM LIZARDS *LOVERIDGEA*

A genus endemic to Tanzania, of two slender, small-sized amphisbaenians with adult snout-vent length of 15–20cm. The head is long and conical, the snout is compressed and strongly bent. One species is known from southern Tanzania, the other from near Ujiji on the eastern shores of Lake Tanganyika.

KEY TO MEMBERS OF THE GENUS *LOVERIDGEA*

1a	Eye visible beneath a discrete ocular, frontals fused, caudal annuli 20–26, with autotomy site at annulus 8–11,	precloacal pores 4......*Loveridgea ionidesi*, Liwale Round-snouted Worm Lizard [p. 319]

| 1b | Eye invisible, no discrete ocular, a pair of frontals, caudal annuli 19–20, with autotomy site at annulus 5 or 6, precloacal pores 6......*Loveridgea phylofiniens*, | Ujiji Round-snouted Worm Lizard [p. 319] |

LIWALE ROUND-SNOUTED WORM LIZARD *Loveridgea ionidesi*

IDENTIFICATION: A small- to medium-sized round-snouted amphisbaenian endemic to southern and south-eastern Tanzania with an even brown dorsal colour with the centre of each segment slightly more densely pigmented. This species has a discrete ocular shield through which the eye is visible. Body annuli (rings) 232–257; caudal annuli 20–26. Four round precloacal pores. 12–16 dorsal and 14–18 ventral segments to a midbody annulus. The precloacal pores of sexually mature males appear to have a 50% larger diameter than those of females. The largest male was 21cm of which 2.5cm was the tail, largest female 21.2cm, of which 2.25cm was the tail.

Loveridgea ionidesi

Loveridgea ionidesi Liwale Round-snouted Worm Lizard. Stephen Spawls, preserved specimen.

HABITAT AND DISTRIBUTION: Endemic to Tanzania, widely distributed in low altitude moist savanna, below 1,000m, in the south-east. Known localities include Liwale, Nachingwea, Msuega, Kilwa, Songea and Tunduru. Might be in northern Mozambique. Conservation status: IUCN Least Concern; fairly widespread.

NATURAL HISTORY: Burrowing, presumably diurnal. Most specimens have been collected between December and May. A female 13cm long contained two embryos of 4.2 and 4.5cm; the species appears to be viviparous (giving live birth). It betrays its presence by pushing up small heaps of still damp black soil in areas where a river has receded. Predators include various small burrowing snakes, the purple-glossed snakes *Amblyodipsas polylepis* and *Amblyodipsas katangensis*, Butler's Black-and-yellow Burrowing-snake *Chilorhinophis butleri* and Bibron's Burrowing Asp *Atractaspis bibronii*.

UJIJI ROUND-SNOUTED WORM LIZARD
Loveridgea phylofiniens

IDENTIFICATION: A round-snouted amphisbaenian which lacks a discrete ocular scale, the eye in adults is not visible externally. Body annuli (rings) 240–260, caudal annuli 19–20. Six oval precloacal pores. 14–16 dorsal and 16–17 ventral segments to a midbody annulus. A medium sized species, 16.5–19.5cm in length, maximum length 20.3 cm.

HABITAT AND DISTRIBUTION: A Tanzanian endemic, known only from the vicinity of the type locality, Ujiji, Kigoma Region, at 780m altitude, on Lake Tanganyika, Tanzania. Conservation Status: IUCN Data Deficient. Poorly known, might be at risk from agricultural development.

NATURAL HISTORY: Nothing known, not collected in nearly 120 years. Presumably similar to other worm lizards, i.e. burrowing, diurnal, feeds on invertebrates.

Loveridgea phylofiniens

ROUND-HEADED WORM LIZARDS *CHIRINDIA*

An African genus of small, elongate worm-like worm lizards, they are mostly pink or whitish pink. They live in soft or sandy soils, or in leaf litter, in moist savanna or woodland. As the common name suggests, the head lacks any keel or wedge-shaped snout, but is instead rounded, most of the head shields are fused behind the rostral scale into a big, tough shield for burrowing. Some species can lose their tails if seized, although it doesn't appear to be regenerated. Five species are known, all from southern and eastern Africa, four in our area, three of these are endemic to Tanzania, the fourth is more widely distributed in southern Tanzania, Mozambique and Zimbabwe. The generic name is from the high-altitude and species-rich Chirinda Forest in eastern Zimbabwe.

KEY TO THE EAST AFRICAN MEMBERS OF THE GENUS *CHIRINDIA*

1a	Parietals not in lateral contact with post-supralabials, 2-4 segments in the first postgenial row......2		or 3 postgenials in first row, 12 + 12 segments to a midbody annulus, 23 caudal annuli......*Chirindia swynnertoni*, Swynnerton's Round-headed Worm Lizard [p. 322]
1b	Parietals laterally in lateral contact with post-supralabials, three segments in the first postgenial row......3		
		3a	Body annuli 200-256, ten segments in a midbody annulus ventral to the lateral sulci......*Chirindia rondoensis*, Rondo Round-headed Worm Lizard [p. 322]
2a	A discrete ocular, two supralabials, four postgenials in the first row, 14 + 14 to 14 + 16 segments to a midbody annulus, 24-25 caudal annuli, found only near Mpwapwa......*Chirindia mpwapwaensis*, Mpwapwa Round-headed Worm Lizard [p. 321]	3b	Body annuli 238-243, 12 segments in a midbody annulus ventral to the lateral sulci......*Chirindia ewerbecki*, Ewerbeck's Round-headed Worm Lizard [p. 320]
2b	Ocular fused with the nasal-prefrontal-labials, a single supralabial, 2		

EWERBECK'S ROUND-HEADED WORM LIZARD
Chirindia ewerbecki

IDENTIFICATION: A round-headed worm lizard from extreme south-eastern Tanzania. Found in red laterite or sandy soils. 264-280 annuli on body; 25-28 annuli on tail;

22–24 segments in a midbody annulus, the two median ventral segments two times as broad as long; six anals, the outer ones frequently divided; six preanal pores in males, none in female. Largest male 16cm; largest female 17cm. Colour: pink. Taxonomic Note: Two subspecies described: *C. e. ewerbecki,* the Mbanja Round-headed Worm Lizard, with 26–28 caudal annuli, and *C. e. nanguruwensis,* the Newala Round-headed Worm Lizard, with 23–25 caudal annuli.

Chirindia ewerbecki

HABITAT AND DISTRIBUTION: A Tanzanian endemic, from the low-altitude woodland and moist savanna of the extreme south-east; known localities are Mbanja, near Lindi (the subspecies *C. e. ewerbecki* Mbanja Round-headed Worm Lizard) and Newala and Nanguruwe, Mtwara Region, Tanzania (subspecies *C. e. nanguruwensis* Newala Round-headed Worm Lizard). Conservation Status: Not assessed by IUCN and known from only a few localities, the habitat in which they are found would not appear to be under any specific threat.

NATURAL HISTORY: Poorly known. Burrowing, specimens were collected from soil in a cassava plantation or by digging deeply beneath fallen logs. Females may lay single large eggs, 31×2mm. Female with eggs collected in April, just after heavy rains. Diet: probably invertebrates, most *Chirindia* are fond of termites. Enemies include small burrowing snakes.

MPWAPWA ROUND-HEADED WORM LIZARD
Chirindia mpwapwaensis

IDENTIFICATION: A round-headed worm lizard known only from the type series, from Mpwapwa. 269–273 annuli on body, 26 on tail; 30 (14 dorsal and 16 ventral) segments in a midbody annulus; six anals; 5–6 preanal pores. Male, 19.4cm, female, 16.2cm. Life colour unknown, probably pink, the museum specimens have faded.

HABITAT AND DISTRIBUTION: A Tanzanian endemic, collected over 70 years ago and not seen since. The type locality, Mpwapwa, is in moist savanna at around 1,000m altitude in eastern central Tanzania. Conservation Status: IUCN Data Deficient. Nothing is know about the biology of this species. Its habitat may be under some threat from degradation, but it unless it has very specific microclimate requirements which have changed since its original collection, it would seem unlikely that this species is threatened directly by human activities.

Chirindia mpwapwaensis

NATURAL HISTORY: Poorly known. Burrowing, specimens were collected by digging in dry earth beneath a fallen tree lying beside a stream. Females probably lay a single large egg. Diet: probably invertebrates, most *Chirindia* are fond of termites. Enemies include small burrowing snakes. Sympatric with the Mpwapwa wedge-snouted Worm Lizard, *Geocalamus modestus* at Mpwapwa.

RONDO ROUND-HEADED WORM LIZARD
Chirindia rondoensis

IDENTIFICATION: A round-headed worm lizard endemic to south-east Tanzania. 227–247 annuli on body; 23–28 on tail; 20 segments (10 dorsal and 10 ventral) in midbody annulus. 227–247 annuli on body and 24–28 on tail. Six anals, the outer ones frequently divided; six preanal pores in males, none in females. Largest male 14.4cm; largest female 15.2cm. Colour in life, bright pink.

HABITAT AND DISTRIBUTION: Known only from woodland and moist savanna, at low altitude, below 1,000m, of the Makonde and Rondo plateaux, southern Tanzania. Conservation Status: Not assessed by IUCN. Probably dependent on the continued presence of adequate natural forest and the associated soil microclimate for its survival.

Chirindia rondoensis

NATURAL HISTORY: Poorly known, presumably similar to other *Chirindia*. Burrowing, specimens were found in sandy and laterite soil beneath rotting logs or matted vegetation at the edge of the primary forest.

SWYNNERTON'S ROUND-HEADED WORM LIZARD
Chirindia swynnertoni

IDENTIFICATION: An elongate, round-headed worm lizard, closely resembles an earthworm. It has two upper labials, the nasal, first and second upper labials, ocular scale and prefrontal are fused into a single shield. There are 235–265 annuli on body (usually around 245 in Tanzanian specimens), usually 24 (12 dorsal and 12 ventral) segments in a midbody annulus (range 24–28), males have six preanal pores, females none; total length around 15cm. Colour uniform pinky-grey or pink.

HABITAT AND DISTRIBUTION: In East Africa, known from only Mikindani and Tunduru in the low-altitude, moist savanna of south-east Tanzania; also recorded from central and northern

Chirindia swynnertoni

Chirindia swynnertoni Swynnerton's Round-headed Worm Lizard. Luke Verburgt, northern Mozambique.

Mozambique and adjacent Zimbabwe in the Chirinda forest. The type specimen was found in the stomach of a kingfisher. Conservation Status: Not assessed by IUCN. Widespread, unlikely to be threatened.

NATURAL HISTORY: Poorly known. Burrowing, living in loose soil, takes cover under large stones, in and below rotting logs, vegetation heaps, etc. It writhes wildly when handled. A female contained a single large elongate egg, 2.2×0.3cm, in December. Diet: probably invertebrates, most *Chirindia* are fond of termites. Enemies include small burrowing snakes, in Mozambique the Dwarf Wolf Snake *Lycophidion nanum* is a specialist feeder on these worm lizards; this wolf snake isn't found in Tanzania but other wolf snakes are.

SHARP-SNOUTED WORM LIZARDS
ANCYLOCRANIUM

An African genus of worm lizards. Three species are known, one in Somalia, the other two are endemic to savanna and woodland of south-eastern Tanzania. As the name suggests, some of the scales on the head are modified into a distinctive sharp-edged snout, visible from above. They live underground, but virtually nothing is known of their lifestyle, they are known only from the original specimens, collected over 40 years ago.

KEY TO THE TANZANIAN MEMBERS OF THE GENUS
ANCYLOCRANIUM

1a	Head elongate, body annuli more than 250, caudal annuli more than 15...... *Ancylocranium ionidesi*, Ionides' Sharp-snouted Worm Lizard [p. 324]	1b	Head short, body annuli less than 250, caudal annuli less than 15...... *Ancylocranium barkeri*, Barker's sharp-snouted Worm Lizard [p. 323]

BARKER'S SHARP-SNOUTED WORM LIZARD
Ancylocranium barkeri

IDENTIFICATION: A sharp-snouted worm lizard endemic to south-eastern Tanzania. Looks like a worm. It has an elongate body and short head; viewed from above, the snout is distinctly pointed. The nasal scale is fused with the rostral. There are two subspecies, the body annuli range from 209–212 in *Ancylocranium barkeri newalae*, the Newala sharp-snouted Worm Lizard, and to 220 in *Ancylocranium barkeri barkeri*, the Lindi Sharp-snouted Worm Lizard. The tail cannot be shed. A series of seven Newala specimens captured in May and June ranged from 21.9–25.9cm length. Colour of the living animal not known, but probably pinkish-brown.

Ancylocranium barkeri

HABITAT AND DISTRIBUTION: Moist savanna, below 1,000m altitude, in south-eastern Tanzania. *Ancylocranium barkeri barkeri* is known only from the holotype, collected at Mbemkuru, Lindi District, Tanzania. *A. b. newalae,* the Newala Sharp-snouted Worm Lizard, is known from the Makonde Plateau, Tanzania. It may be more widespread than the present records would indicate, or simply overlooked. Conservation Status: IUCN Data Deficient. Unlikely to be under threat, unless patterns of land use are affecting its habitat and microclimate requirements, about which nothing is known.

NATURAL HISTORY: Unknown. Probably similar to other worm lizards, i.e. burrowing, lays a single elongate egg, eats invertebrates, fond of termites. The specific name refers to R. de la Barker, hunter and resident of south-eastern Tanzania; *newalae* refers to the locality of Newala in Tanzania.

IONIDES' SHARP-SNOUTED WORM LIZARD
Ancylocranium ionidesi

IDENTIFICATION: A sharp-snouted worm lizard, which looks like a worm. It has an elongate body and short head, from above the head comes to a distinctive sharp point. Body with 34 segments (18 dorsal and 16 ventral) to midbody annuli. 302–327 annuli on body, 19–23 on tail. Rostral scale enormous, compressed and arched with a sharp cutting edge. The largest known adult male was 229mm, largest female 249mm. Taxonomic Note: Two subspecies known, see below.

Ancylocranium ionidesi

HABITAT AND DISTRIBUTION: Endemic to Tanzania, found in the low altitude moist savanna of the south-east. Known from four specimens from the east coast north of Lindi. Two subspecies are described: *Ancylocranium ionidesi ionidesi* (with body annuli 319–328) is known only from Kilwa District, Tanzania, and *Ancylocranium ionidesi haasi* (body annuli 304–320) from only the type locality at Mtene, Rondo Plateau, Lindi District, Tanzania. Conservation Status: IUCN Data Deficient.

NATURAL HISTORY: Unknown. Probably similar to other worm lizards. The species name honours the collector C.J.P. Ionides.

Ancylocranium ionidesi Ionides' Sharp-snouted Worm Lizard. Stephen Spawls, preserved specimen.

ORDER CROCODYLIA
CROCODILES

Osteolaemus tetraspis
Dwarf Crocodile
Stephen Spawls

ORDER CROCODYLIA

The crocodile is a familiar villain in folk tales old and new, symbol of danger and deceit; the expression 'crocodile tears' is used to describe insincere sympathy, as the crocodile with gently-smiling jaws welcomes in the little fishes in Lewis Carroll's poem. The crocodilians are the last remaining examples of the Archosaurs, the ruling reptiles, including the dinosaurs, that dominated the earth for 150 million years. However, modern crocodiles are more closely related to birds than to any other reptiles. They are unusually advanced reptiles, with a four-chambered heart (most reptiles have a three-chambered one, with a common ventricle), the total separation of the ventricles means that the blood is efficiently oxygenated, meaning they can keep moving a lot longer. They have a third eyelid, the nictitating membrane, which sweeps dirt from the eyeball. They have an advanced limb structure, and can walk with their legs below them, the 'high walk'; they can even gallop, as well as slide on their belly. In the roof of their mouth is a hard palate, they have a longitudinal cloacal aperture and males have a single penis. There are a number of crocodile fossils; crocodile-like reptiles existed well over 100 million years ago, they haven't changed much in the last 65 million years, and they offer us a modern glimpse of the magnificent giant reptiles that once ruled the earth.

There are 25 modern species of crocodilians (the term includes alligators, caimans and gavials), in three families, although the status of a couple of species is debated. They occur in the tropics; a couple of species just reach temperate regions. The crocodiles form a natural clade, which split from the largely American alligator/caiman clade about 50 million years ago. South and Central America have the greatest diversity with 12 species, there are four African species. All crocodiles are aquatic carnivores and most are endangered; several species are on the brink of extinction, despite official world-wide legal protection. They are persecuted for their skins and their largely unjustly deserved reputation as predators on humans and stock. In reality only two species, the Estuarine or Saltwater Crocodile *Crocodylus porosus* of south-east Asia and Australia, and the Nile Crocodile *Crocodylus niloticus,* regularly take humans. Some of the other species have caused a few deaths and severe injuries, usually as a result of ignorance or carelessness by the person concerned in the presence of a big, powerful carnivore.

The size of crocodiles is a matter of much debate and exaggeration. The Estuarine Crocodile is the biggest species (and the heaviest living reptile); a skull in the Natural History Museum in London is 0.93m long, and since the ratio of skull to total length for an adult crocodile is about 1:7, this indicates that the owner of the skull was at least 6.5m long. There are stories of larger animals, but unsupported by hard evidence. Living crocodiles are dwarfed, however, by the extinct giant crocodile *Sarcosuchus* which was over 11m long. Four million years ago a fossil species *Crocodylus thorbjarnarsoni,* 7.6m long, lived in the Turkana Basin.

FAMILY CROCODYLIDAE

Crocodiles are distinguished from alligators by the fact that the fourth mandibular tooth is visible when the mouth is closed; it is concealed in a socket in alligators. There are sixteen species in the family, four species occur in Africa, three in East Africa. Only the Nile Crocodile is widespread, the Slender-snouted Crocodile, within our area, is confined to Lake Tanganyika, the Dwarf Crocodile known only from extreme western Uganda. The fourth African species, the West African Crocodile *Crocodylus suchus* was elevated from the synonymy of the Nile Crocodile on the basis of molecular evidence. It found in West and Central Africa, is a smaller species than the Nile Crocodile and is believed not be as dangerous to humanity as that species, but its status as a valid species is not accepted by some herpetologists.

KEY TO EAST AFRICAN CROCODILES

1a Snout very elongate, at least two and a half times as long as wide at the level of the eyes, found only in Lake Tanganyika and environs......*Mecistops cataphractus*, Slender-snouted Crocodile [p. 327]

1b Snout quite broad, less than two times as long as its width at the level of the eye......2

2a Snout very short and broad, less than 1.5 times longer than the width at the level of the eyes, maximum size 1.5m, in our area only in western Uganda......*Osteolaemus tetraspis*, Dwarf Crocodile [p. 333]

2b Snout not very short and broad, more than 1.5 times longer than the width at the level of the eyes, maximum size more than 4m, widespread......*Crocodylus niloticus*, Nile Crocodile [p. 329]

SLENDER-SNOUTED CROCODILES *MECISTOPS*

A presently monotypic African genus, although the single species may be split in the near future.

SLENDER-SNOUTED CROCODILE *Mecistops cataphractus*

IDENTIFICATION: A fairly big crocodile with a long, slender snout, in East Africa found only in Lake Tanganyika. Maximum size about 3m (possibly slightly larger), average 1.5–2.5m, hatchlings 22–25 cm. The teeth are very prominent, sticking out from the upper and lower jaw even when it is closed. The tip of the long snout is bulbous, with two valved nostrils on top. The skin is covered with regular horny plates, those on the back are keeled and bony, The limbs are strong and muscular, the rear toes have slight webbing. The tail is quite long, 30–40% of total length, with two raised keels posteriorly. Adults are usually shades of uniform grey or brown in Lake Tanganyika, but may have black blotches or

Mecistops cataphractus

Mecistops cataphractus Slender-snouted Crocodile. Stephen Spawls, Malagarasi River, Lake Tanganyika.

spots on the body in other regions, especially in forested areas. The belly and throat is lighter, cream, yellow or orange. Juveniles are vividly marked, grey-green with irregular black blotches and crossbars, they have a short snout. Taxonomic Notes: Recent anatomical and molecular analysis shows this species and the Dwarf Crocodile form a clade separated from the 'true' crocodiles, *Crocodylus*; the generic name *Mecistops* is available for this species, and those from west of the Dahomey gap may prove to be a different species.

HABITAT AND DISTRIBUTION: In East Africa, confined to Lake Tanganyika and some of the large streams entering it on the eastern side. It has been speculated that it is extinct there, but it was always rare, and is still found around the Malagarasi Delta and associated small streams. Elsewhere, west to Senegal, south-west to Angola, also on the island of Bioko; from sea level to about 1,000m altitude. Lives in lakes and rivers, in forest and forest-savanna mosaic, in savanna rivers in some areas. It will utilise dams, and swamps if there are large enough pools. Conservation Status: Critically Endangered A2acde+3cde+4acde and on CITES Appendix 1. Strictly protected by international agreement and widespread over a large area. Nevertheless under threat from bushmeat and skin hunters in west and central Africa, its body parts are also used for local medicine; it is suggested that it is extinct in several countries.

NATURAL HISTORY: A poorly known, shy and secretive crocodile, moving away rapidly from humans. Said to attempt to hide in waterside vegetation when molested, rather than dive as a Nile Crocodile would. In Lake Tanganyika it basks in the daytime in secluded spots; and in that lake it makes up only 1% of the crocodile population. It is a fast and agile swimmer. It may take refuge in waterside holes if available. In the Ivory Coast mating was seen in February, the females constructed nests of vegetable matter on elevated forest stream banks in March or April. The nest was 1–2m wide and 50–80cm high. Clutches of 12–30 eggs were recorded, which were then covered with layer of vegetation 30cm or so thick. At an incubation temperature of 27–34°C, the eggs hatched after 90–100 days. The female guarded the nest and became very sluggish. The juveniles quickly entered the water, which is high in June or July in West Africa, they used backwaters and shallow flooded areas, hiding among floating vegetation, catching frogs, tadpoles, fish and insects. The juveniles have a startlingly loud distress call if seized, which presumably may cause a predator to drop them. The adults eat mostly fish, as do other long-snouted members of this order, but may opportunistically take small mammals, swimming birds, reptiles and amphibians. They do not, however, ambush large animals coming to drink and are thus no danger to humans. A juvenile was seen at night waiting in shallow water for small fish to swim into its open jaws. Like the Nile Crocodile, they often have gastroliths (stones) in their stomachs. A specimen in Nairobi Snake Park, captured as a juvenile in Gombe Stream game reserve, Tanzania, lived 30 years in captivity, and reached a length of 2.4m.

TRUE CROCODILES *CROCODYLUS*

A tropical genus, of some 13 species, two are found in Africa, one in our area. They all look very similar in general appearance, colour, body and tail shape, although the length of the snout varies a bit. They live mostly in water, lay eggs on land, most eat fish although the two big species (Nile and Estuarine Crocodile) will take game and humans.

NILE CROCODILE *Crocodylus niloticus*

IDENTIFICATION: A big, thickset crocodile with a fairly broad snout. Males are larger than females. Maximum size about 5.5m (anecdotal reports of larger specimens exist), average 2–3.5 m, hatchlings are 28–32 cm. Big adults may weigh as much as one tonne. The teeth are prominent, sticking out from both the upper and lower jaw, a big incisor on each side near the front of the lower jaw fits into a groove on the outside of the upper jaw, and acts as a locking mechanism. Crocodile teeth do not interlock, but they can cut; humans have had their arms bitten off.

Crocodylus niloticus

The skin is covered with regular, thick, horny scales, those on the back are like armour plate, keeled and bony. The limbs are short and powerful, with thick claws, the rear toes have partial webbing. The tail is about 40% of total length, with two raised keels posteriorly. Adults usually grey, brown or tan, sometimes greenish. Juveniles vivid yellow or tan with prominent black blotching and speckling. Some adults are bright yellow with black blotches. In Lake Tanganyika it may be distinguished from the Slender-snouted Crocodile by its broad snout.

HABITAT AND DISTRIBUTION: Widespread in suitable water bodies throughout East Africa. Known localities include most suitable rivers and lakes, from sea level to about 1,600m, possibly higher. Some curious records include the rivers of the Songot Range, in north-west Kenya near Lokichoggio, where a dwarf form lives in the hillside streams, isolated water holes in the Kidepo National Park, Uganda, in Umani Springs near Kibwezi, Lake Chala at the foot of Mt Kilimanjaro, and in water holes in and around Dodoma. Absent from the Rift Valley lakes, save Baringo and Turkana. They are quick to utilise pools and pans in suitable country, walking there at night, in Somalia a specimen was found 8km from the nearest river. In Uganda, Nile Crocodiles were absent from Lakes George and Edward, exterminated 7,000 years ago by volcanic activity, but are returning to Lake Edward and the Kazinga channel. Elsewhere, south to eastern South Africa, west to West Africa (although in some areas of West Africa *Crocodylus suchus* occurs, see the introduction). Conservation Status: IUCN Lower Risk/Least Concern. Nile Crocodiles are protected by CITES legislation (Appendix 1, with some provisos). Crocodile farms in Africa provide meat and skin to the trade. Nile Crocodiles are still widespread and present in many protected areas.

NATURAL HISTORY: Nile Crocodiles are the most thoroughly studied East African reptile, much data is available. They may be active at any time throughout the day and night, spending the day between basking, sleeping and hunting, at night they are usually in the water. They select differing spots according to size. Small individuals are usually in shaded, shallow, vegetated backwaters, where they can hide. They don't bask in the open but will utilise patches of sunlight. In South Africa, juveniles have dug riverbank burrows up to 3m long, using their jaws. The juveniles also forage on the banks. Subadults of 1–2m length will bask on open mud and sand banks, usually in groups of two or three, but don't use the best sites, which are favoured by big adults. When basking, they finely regulate their temperature by opening and closing their mouths. Large animals will venture into open water, although studies in Lake Turkana showed that they avoided very deep water, other than crossing deep stretches while travelling. The lake crocodiles also disliked windy beaches and rough water, favouring warm, calm, shallow water. At night Nile Crocodiles will gather in size classes in suitable areas, resting quietly in the water with the head on the surface and the body angled downwards at about 30–40 degrees. Their eyes reflect light and they can be easily spotted from a distance with a spotlight or torch, a technique used by professional crocodile hunters.

Nile crocodiles swim powerfully using the tail, limbs tucked in. They can stay under water for over 45 minutes, there are anecdotal reports of them staying down for two hours, the heart rate of a quiescent submerged crocodile slows to a few beats per minute. Adults are neutrally buoyant and hence comfortable at any depth in the water. Juveniles are less dense and float. Adults swallow stones (presumably to affect buoyancy, but this isn't proven); virtually every adult crocodile examined has had stones in its gut. On land, crocodiles can run at speeds over 10km/h; there are reports of them reaching 25km/h for short bursts. They walk with the legs below them and can slide on their bellies. They will jump from a high bank if disturbed. Large males are territorial, defending a suitable beach or stretch of water and a group of females (usually 6–10), and attacking intruding rival males. Some group behaviour has been noted, including co-operative fish 'herding', a group carefully working a shoal of fish into shallow water.

Nile Crocodiles have excellent sight, hearing (the external ear opening is a slit just behind the eye but inside the ear drum is bigger than the eye) and taste. Juveniles in captivity in South Africa could detect, and refused to eat, meat laced with antibiotics. Crocodiles can smell carrion over long distances. They are also vocal, young ones can yelp loudly and they have a number of communication calls. Adults can hiss, growl and roar loudly, producing an almost bull-like sound, and they may also slap their chins on the water. Mothers signal to their young by vibrating their bodies, producing a low frequency sound, the babies will come towards the source, and a similar vibration by the mother, but presumably of different frequency, warns the babies of a predator, on hearing it they dive.

Nile crocodiles have elaborate courtship rituals, the male rubs the underside of his jaw and throat on the female's neck, this releases a flow from the glands under the lower jaw, the smell of which stimulates the female. A submissive female has been seen to approach a dominant male, raise her head, open her jaw and growl gently. Mating occurs in the water, the male gently bites the female's throat, climbs on her back and twists his tail base to be opposite the female's cloaca. Mating takes only a few minutes. An alpha male in the presence of a suitable female inflates his body, inclines his head to submerge his nostrils and blows vigorously for 5–6 seconds. In Lake Turkana mating mostly occurred in October, laying in November; mating in September and egg laying in October recorded in Lake Baringo; captives in coastal Kenya mated between August and December and hatched between March and April. A suitable nest site, usually on a sunny well-drained sandbank above flood level, will be reused for years by the female. She digs a hole at night, using her hind legs, and lays, depending on her size, between 20–95 (usually 30–50) white, hard-shelled eggs, roughly 7.5×5.5cm, mass about 100g, the whole process taking 2–4 hours. Some females guard the nest site vigorously against predators such as monitor lizards, hyenas and even other crocodiles; at Lake Turkana baboons excavated nests. A vigilant female will leave the nest only to drink, other females may cover the nest and simply leave it, returning just before hatching. Eggs laid around Lake Turkana in November hatched in February, hatchlings were seen in Lake Baringo in December and on the Tana River at Garissa in July; in Tanzania egg laying is recorded in November in the north, August/September at Lake Rukwa and December in Lake Victoria, eggs were laid on the Ruzizi plain, north of Lake Tanganyika, between April and August. The Lake Victoria crocodiles seem to have two breeding seasons, August/September and December/January.

Just prior to hatching, the babies make a high-pitched chirping call inside the egg. On hearing this, the female excavates the nest, takes the young in her mouth, carries them to water, washes and then releases them. She may also guard them after returning to eat the eggshells and membranes. The sex of the hatchling depends on the incubation temperature, in a cold year (26–30°C) the lower temperature produces a clutch of mostly females, in a hot year the higher temperature (31–34°C) produces mostly males. The age of sexual maturity varies; in South Africa it is 12–15 years old, when they are 2–3m long and 70–100kg mass.

Crocodylus niloticus Nile Crocodile. Top left: Steve Russell, Kazinga Channel. Top right and inset: Stephen Spawls, Ethiopia. Bottom: Michele Menegon, Selous Game Reserve.

In Lake Turkana it is reckoned to be six years. Mortality is high among hatchlings, only 1% of them reach maturity. When small, they have many enemies; other crocodiles, big fish, pythons, birds of prey, predatory mammals; but a fully grown adult has no enemies save man and possibly hippos; an adult hippo has been known to bite a crocodile in two.

Crocodiles have a varied diet. Hatchlings eat mostly insects, tadpoles and frogs, but they soon begin to take fish, which for the major part of the diet of all age classes. In Lake Turkana, tilapia were preferred to catfish. They will also take any swimming creature, including water birds and smaller crocodiles. Adults will snatch prey at the water's edge; the list of such prey includes domestic stock, humans, giraffe and even rhino. Nile Crocodiles may wait in ambush at a regular drinking spot, or opportunistically stalk drinking animals from a distance, drifting on the surface or swimming under water, exposing the eyes and nostrils briefly to get a breath and a direction fix as they approach. Their method of prey capture is to either seize the drinking animal by the nose, head or neck and drag it in or, if the animal is not right at the brink, to rush out, often with a tail-propelled thrust and scramble and grab the prey, usually by the leg (although, despite legends, they do not use the tail from within the water like a scythe, to knock the prey down before rushing to grab it). They will also use their tail in the water to steer fish into their mouths, and the Indian Crocodile or Mugger *(Crocodylus palustris)* has been seen to use the tail to stun prey on the land. Crocodiles have been seen to jump from the water to a height equivalent to two-thirds of their length, leaping over a bank, another was seen to rush 8m from the water after prey; thus serving as a caution to anyone sitting near the edge of a crocodile-infested river. Prey is usually held underwater until it drowns, although small prey may be swallowed immediately, or repeatedly crunched in the jaws until it dies. Crocodiles can swallow food underwater, as they have valved nostrils and a gular flap at the back of the mouth. During the annual migration in the Mara and Serengeti, crocodiles take a heavy toll of wildebeest, zebra and other antelope crossing the river. Prey is so plentiful then that the crocodiles ignore old carcasses, preferring fresh meat.

There is a general belief in East Africa that river crocodiles are more dangerous than lake crocodiles. This may be due to the fact that in lakes, where the water is usually clear, the crocodiles can see their preferred prey of fish, but in rivers that are muddy for part of the year, the fish are invisible and the crocodiles switch to taking drinking animals, which can easily be seen. In East Africa, lakes Turkana and Baringo are regarded as safe places to swim, despite the crocodiles there, although fatal attacks are documented for both lakes. The really dangerous rivers in East Africa for crocodile attacks are the Tana, the Galana, the Rufiji and the Rovuma.

Large prey animals are torn to bits, either by shaking or jerking the carcass; in the water they may seize part of the prey and then rotate on their long axis, twisting the bit off. They do not, as legend has it, store prey in underwater caverns until it decomposes. Crocodiles may also rotate quickly when they have seized part of a large living target, including humans. Crocodiles are attracted to splashing and jerky movement in the water, and in murky rivers may attack anything causing such a disturbance. Crocodiles are long-lived and have survived 60 years in captivity; they are estimated to reach 70 years or more. Their growth rates are not well known, some juveniles monitored in Lake Turkana showed no length increase during a year, others grew 30–40cm in that time.

SAFETY AND CROCODILES: Nile Crocodiles (even small ones) are dangerous to human beings. To avoid the risk of crocodile attack, stay at least 5m from the water's edge. Never swim, paddle or wash in natural water sources unless local experts assure you it is safe; never approach the water after dark.

DWARF CROCODILES *OSTEOLAEMUS*

A tropical African genus with a single small species, inhabitant of water bodies in the forest of central and western Africa.

DWARF CROCODILE/BROAD-FRONTED CROCODILE
Osteolaemus tetraspis

IDENTIFICATION: A little, dark, thickset, broad-snouted, short-headed crocodile, in our area confined to western Uganda. The smallest of the crocodiles, maximum length of the eastern subspecies *Osteolaemus tetraspis osborni* is about 1.2m, although the western subspecies *Osteolaemus tetraspis tetraspis* reaches 1.7m. Average 80cm to 1m, males growing larger than females, hatchlings of the eastern form 18–25 cm. The teeth are prominent, visible at the front of the jaws when they are closed. The tip of the short snout is slightly raised; the eyes are big and prominent. The skin is covered with hard scales, these are horny, keeled and rectangular on the back. The limbs are muscular, rear toes webbed. The tail is broad, about 45 % of total length, with a double keel of elongate scales at the back and a single keel at the front. Adults are usually black, brown or rufous-brown above, sometimes with patches of brown mottling, and black and white blotching on the jaws, black and yellow below. Juveniles are yellow-brown, heavily speckled with black. Taxonomic Notes: Two subspecies are recognised, the western one is larger, has a prominent bump at the end of the snout and 11 pairs of anterior tail scales, the eastern form (found in the eastern DR Congo and western Uganda) is smaller, the snout bump is less prominent and it has 12–14 anterior tail scales.

Osteolaemus tetraspis

Osteolaemus tetraspis Dwarf Crocodile. Stephen Spawls, captive.

HABITAT AND DISTRIBUTION: Small rivers, shallow tributaries of big rivers, backwaters, swamps and pools in the central African forest, venturing out into savanna along well-shaded rivers. Not in the mainstream of large rivers, due to its small size. In East Africa, known only from small rivers in the vicinity of Lake George, Semliki Forest Reserve and National Park and Murchison Falls National Park, at altitudes between 600–1000m, also recorded from the Ituri Forest and north-east to Haut-Zaire, DR Congo. The western subspecies occurs from central DR Congo north-west to Senegal. Conservation Status: IUCN Vulnerable A2cd; on CITES Appendix 1.

NATURAL HISTORY: Not well known. It is sometimes active by day but more often nocturnal. At night it moves around on the river banks, and will wander some distance from the water. It rarely basks. Often lives in riverbank holes or deep muddy cracks, sometimes in true pairs. It is timid, fleeing if approached. Angry specimens are described as growling like dogs. Courtship prior to copulation in captive specimens involved splashing, 'drumming' and neck-rubbing. About two weeks before parturition the female builds a nest of vegetation, about 1.5m in diameter, on the banks and lays a clutch of 6–21 eggs (the more mature the female, the larger the clutch), roughly 4×6.5 cm, each weighing about 60g. The female defends the nest vigorously, attacking any potential predator that approaches. The eggs take about four months to hatch. Specimens from the DR Congo ate mostly fish, but other prey items include river crabs, frogs, and insects. Dwarf Crocodiles are probably generalist feeders, taking anything they can catch on land and in water, a specimen from Brazzaville had eaten a water bird. In parts of its range its flesh is eaten, the tail providing excellent steaks; living trussed-up specimens can be seen for sale in many central African riverside markets. The skin, however, has little commercial value as the scales are very bony. A study of market specimens of the western subspecies around Brazzaville found that long thorns had penetrated and lodged in the internal organs of several specimens, it is not known how.

SUBORDER **OPHIDA (SERPENTES)**
SNAKES

Bitis arietans
Puff Adder
Stephen Spawls

SUB-ORDER SERPENTES

Known to everybody, a survey in Great Britain found that they were the most feared animal, snakes are highly specialised, legless squamate reptiles. They are found world-wide, absent only from some oceanic islands, the Antarctic and very high latitudes, although one species, the European Viper *Vipera berus*, is found inside the Arctic circle. They are unique in many ways; they have a very long spine and many ribs (over 400 in one species) that support the rounded body and are used to move with. Snakes have no legs, although a few have vestiges of a pelvic girdle and little spurs, vestigial legs. All lack eyelids, they have a forked tongue that retracts into a sheath, they have no external ears, sharp but non-interlocking teeth and a greatly enlarged internal cavity, with no left lung (this is retained but very reduced in a few species). All are carnivores, swallowing their prey whole, they can't bite bits off. They are ectotherms, getting their heat from outside; consequently their energy budgets are low and they eat sparingly, some exist on ten or more good meals a year. The bones of the lower jaw are not fused at the front, they also have a flexible trachea, an extensible and flexible junction between the lower and upper jaw and an enormously elastic skin, all adaptations allowing them to swallow animals much wider than their heads.

Some have evolved specialised teeth, fangs, designed to transfer venom. The venom is designed to immobilise prey rapidly and to aid digestion, not to keep humanity on its toes. A few snakes, for example spitting cobras, use their venom in defence, particularly against predatory birds. Non-venomous snakes kill their prey by constriction, the tight squeeze means the prey cannot breathe and blood cannot circulate, death is rapid. Other snakes simply swallow their prey alive. Most snakes lay eggs in a warm, damp place. A few species give live birth, in East Africa notably the Sand Boa, Slug-eater, Mole Snake and most vipers. There is no parental care. In East Africa they occupy a range of habitats, from sea level to montane moorlands over 3,000m altitude, but in general the number of species decreases with increasing altitude.

Python regis Royal Python. Barry Hughes, Ghana.

Rhamnophis aethiopissa Large-eyed Green Tree Snake. Steve Russell, Kasese.

The earliest unequivocal snake fossils are from the Jurassic period, and are nearly 170 million years old. Since then, snakes have evolved and radiated. When we first published this book in 2002, there were just over 2,900 known species of snake, in 18 families, and some 203 species from eight families were found in East Africa. Now there are around 4,600 known species of snake, in six superfamilies and 25–27 families (depending upon which authority you follow). In East Africa there are 228 snake species, in four superfamilies and ten families. Fifty East African snake species are dangerous to humans, 48 of these are venomous, the other two are large pythons. Of these 50, 23 species are known to have killed people. Thus, most people are afraid of snakes, which is a natural reaction. A dangerous snake near one's home must be killed or removed. However, the danger of snakebite depends on circumstances. The rural poor in remote regions are at significant risk, the better-off visitor to East Africa's remote areas is at much less risk, and, should they get bitten, at much less risk of dying. However, those whose work or pleasure takes them deep into the East African bush should get to know which snakes are dangerous and which are not. Snakes have an important part in food webs and wholesale destruction of them will upset the ecological balance. In 1985 Thailand exported 1.3 million snakes, mostly to other south-east Asian countries whose people were under the totally erroneous impression that eating parts of snakes is good for the health and gives you long life, in much the same way that in the same region they believe that the meat of an owl is good for the eyes. Thailand's rat population exploded, freed from snake predators, destroying an estimated 400,000 hectares of rice fields. In central Kenya and northern Tanzania, the harmless and beautiful Mole Snake is a valuable controller of rats in the farms of the central rift valley; it is a snake that local farmers should learn to identify and tolerate.

KEY TO THE EAST AFRICAN SNAKE SUPERFAMILIES

Note: the taxonomic category superfamily has found fairly wide use with snake taxonomists, but is little used with lizards. However, within East Africa it is a useful category for subdivision and identification of the various snake groups and we have chosen to use it here.

1a Body worm-like, head round and blunt, tail blunt, eyes only visible as minute dark dots under the head skin, body scales all the same size......Typhlopoidea, blind and worm snakes [p. 339]
1b Body not worm-like, head not round, tail not blunt, eyes well-developed, enlarged belly scales present....2

2a Ventral plates are as broad or almost as broad as the body, no vestigial limbs, midbody scale rows less than 50...... Colubroidea, typical snakes [p. 383]
2b Ventral plates broader than the body scales but much narrower than the body, vestiges of hind limbs present as short claws on either side of the vent, midbody scale rows more than 70......3

3a Small, adults less than 1.2m, subcaudal scales single, no obvious neck...... Booidea, boas [p. 375]
3b Large, adults over 1.2m, subcaudal scales paired, obvious neck......Pythonoidea, pythons [p. 378]

Pseudaspis cana Mole Snake. Stephen Spawls, Njoro.

Rhinotyphlops ataeniatus
Somali Blind Snake
Stephen Spawls

SUB-ORDER **SERPENTES**
SUPERFAMILY **TYPHLOPOIDEA**
BLIND AND WORM SNAKES

A superfamily of around 440 species of distinctive, largely small, harmless burrowing snakes, in five families. They are loosely known as the scolecophidians. They have blunt rounded heads and tails, the tail often ending in a spike that aids locomotion (and is pushed into a restraining hand); the resemblance between head and tail and their ability to slide backwards means they are sometimes called 'two-headed snakes'. The eye is usually primitive (and indeed invisible in some forms). Their bodies are shiny and do not taper like 'ordinary' snakes. Blind and worm snakes are found on all continents save Antarctica, although their greatest diversity is in the tropics. Two families and nearly 50 species occur in East Africa. They are an ancient group; fossil records of the superfamily extend back 60 million years, and molecular taxonomy indicates a split between Typhlopids and Leptotyphlopids over 150 million years ago. Their relationships are vigorously debated, in the last ten years there have been several major taxonomic revisions that affect the scolecophidians to a greater or lesser extent, some species have been placed in three different genera in the last few years, although this is fortunately not of huge significance in a guide such as this.

Identification of scolecophidions can be difficult. Although some species have been well studied and are of distinctive appearance, others are known only from museum specimens. Most are small and look much the same. Certain identification often requires a dead specimen, a binocular microscope, a good key and a detailed description. Our species descriptions here are often brief, and we have tried to list characters that might be useful in the field, seen with a sharp eye and a hand lens, although to be certain you may have to resort to our key and a microscope. Useful field clues for a living animal are the locality, colour and body shape; thick, medium or thin: if the average length to diameter ratio (given for each species) is around 30 it is thick, around 50 is medium, over 70 is thin. Bear in mind that the colour may change over time; specimens on the point of sloughing may appear silver or silvery-grey, and there is some evidence that worm snakes can actively change colour. Further details of the recent major revisions of this fascinating group can be found from the Reptile Database; or by searching papers on the genera by Donald Broadley, Van Wallach and S. Blair Hedges. If you find a scolecophidion, it really is worth getting it to a museum. A subsistence farmer near Embu, Kenya, finding a blind snake while hoeing, made the effort to take it to a National Museum scientist; it turned out to be a new species and the first example of the genus from central Kenya.

KEY TO THE EAST AFRICAN FAMILIES OF THE *TYPHLOPOIDEA*

| 1a | Usually fairly thick, adults often more than 15cm total length and more than 0.5cm total diameter, more than 20 midbody scale rows, teeth only in upper jaw, ocular shield (the scale with the eye in it) does not touch the mouth...... Family Typhlopidae, blind snakes [p. 341] | 1b | Usually small, thin and wormlike, adults less than 15cm total length and less than 0.5cm total diameter, less than 16 midbody scale rows, teeth only in lower jaw, ocular shield (the scale with the eye in it) touches the mouth...... Family Leptotyphlopidae, worm or thread snakes [p. 363] |

Afrotyphlops lineolatus
Lineolate Blind Snake
Stephen Spawls

SUB-ORDER SERPENTES
FAMILY TYPHLOPIDAE
BLIND SNAKES

A family of about 270 species, in four subfamilies and 18 genera; all East African species save one vagrant are in the subfamily Afrotyphlopinae. All have a unique appearance; they have a very short, blunt, usually rounded head, with a big rostral (snout) scale and their tails are also very short, only 1–3% of total length. Their bodies do not taper like those of 'ordinary' snakes; they resemble metal rods with rounded ends. They have primitive, sometimes invisible or near-invisible eyes, and their body scales are all the same size, they do not have the enlarged ventral scales of typical snakes. They largely live below the surface, in holes, soft soils or leaf litter, where they feed on arthropods. They are fond of termites and ants eggs and may enter or live in termitaria, where they hunt for brood chambers with eggs. They have a Gondwanan origin. They are hard to find and there are relatively few museum specimens, which complicates research. Twenty-six species are known to occur in East Africa. Their taxonomy is complicated (it was suggested in the past that they were actually legless lizards), with many recent changes, some ephemeral. In addition, some distribution records and patterns are at present zoogeographically inexplicable, suggesting some species are not properly defined. At present the East African forms are in four genera.

KEY TO THE GENERA OF TYPHLOPIDAE IN EAST AFRICA

1a	Usually pink in life, relatively slender, eye not visible or visible only as a pigmented spot......*Letheobia*, gracile pink blind snakes [p. 355]	1b	Not usually pink, often not slender, eye clearly visible......2

2a A broad yellow or orange stripe down the centre of the rostral, with fine black side stripes......*Rhinotyphlops*, African beaked blind snakes [p. 361]
2b No broad yellow or orange stripe down the centre of the rostral......3
3a Third supralabial overlaps ocular shield; postoculars 1-2; inferior nasal suture contacts second supralabial or preocular...... *Indotyphlops braminus*, Flower-pot Blind Snake [p. 354]
3b Third supralabial not overlapping ocular shield ; postoculars 3-7; inferior nasal suture contacts first supralabial or rostral..... *Afrotyphlops*, African blind snakes [p. 342]

AFRICAN BLIND SNAKES *AFROTYPHLOPS*

A sub-Saharan African genus of about 30 species of largely thick, relatively big blind snakes, often striped, spotted or blotched. Fifteen species are known from East Africa; the majority have very restricted ranges, usually in forest. Their snouts are rounded and their eyes are visible as light-sensitive, black spots beneath the ocular scale. In East Africa there is a known association between African blind snakes, and safari ants (siafu) *Dorylus molestus*; legend has it that the blind snake gets eaten by the ants when times are hard or, conversely, the snake eats their eggs when it gets the chance but is protected from their bites by its close-fitting scales; the truth is not known. Many species eat termites, a large South African species was observed using its head to bore into an active termitarium, and at the same time dribbling fluid, possibly to soften the soil and assist entrance.

KEY TO THE EAST AFRICAN SPECIES OF AFRICAN BLIND SNAKES *AFROTYPHLOPS*
Note: some species occur more than once in this key, on account of variation in scale characters and colour/pattern. You will need a binocular microscope and our figures showing head scalation.

1a Third supralabial overlaps ocular shield, dorsal rostral narrow to moderate (less than half the interocular head width); nostril pierced laterally; postoculars 1-2; inferior nasal suture contacts second supralabial or preocular......2
1b Third supralabial not overlapping ocular shield, dorsal rostral broad (greater than half the interocular head width); nostril pierced ventrally; postoculars 3-7; inferior nasal suture contacts first supralabial or rostral......3
2a Preocular overlaps second supralabial, snout rounded in profile; superior nasal suture present; mid-dorsal scales 400-425; 24 scale rows; body slender (length/width ratio over 50)...... *Afrotyphlops platyrhynchus*, Tanga Blind Snake [p. 351]
2b Second supralabial overlaps preocular, snout wedge-shaped in profile; superior nasal suture absent; mid-dorsal scales 216-302; scale rows usually 22 ; body short and robust (length/width ratio less than 40)......*Afrotyphlops calabresii*, Southern Wedge-snouted Blind Snake [p. 345]
3a Snout sharply angular in profile, with a keratinised horizontal edge in adults; ventral rostral broad (greater than half internarial snout width......4
3b Snout more or less rounded in profile, lacking a keratinised edge; ventral rostral narrow to moderate (less than half internarial snout width)......5
4a Rostral oval in dorsal view, convex in profile; rostral length/width ratio usually less than 1.25; pattern lineolate or blotched, some large adults uniform grey-brown......*Afrotyphlops mucruso*,

4b	Zambezi Giant Blind Snake [p. 348] Rostral cuneiform in dorsal view, depressed in profile; rostral length/width ratio usually more than 1.25; juveniles lineolate, adults uniform grey-brown...... *Afrotyphlops brevis*, Angle-snouted Blind Snake [p. 344]	10b	*lensis*, Angola Blind Snake [p. 343] Preocular not overlapped by second supralabial......11
5a	A subocular scale usually separates preocular from second and third labials......6	11a	Dorsum heavily blotched with black, underside immaculate; midbody scale rows 26–30; supraoculars transverse......*Afrotyphlops congestus*, Blotched Blind Snake [p. 345]
5b	No subocular scale separating preocular from second and third labials......7	11b	Dorsum with lineate or punctate pattern, underside blotched or spotted like dorsum; midbody scale rows 30–34; supraoculars oblique......*Afrotyphlops punctatus*, Spotted Blind Snake [p. 351]
6a	A blue-grey lineolate dorsal pattern, with or without superimposed black blotches; midbody scale rows 26–28; number of dorsal scale rows 404–463; occurs in the Usambara Mountains...... *Afrotyphlops gierrai*, Gierra's Blind Snake [p. 346]	12a	Preocular overlapped by second supralabial......13
		12b	Preocular not overlapped by second supralabial......14
6b	Dorsum uniform black or blotched black and white; midbody scale rows 30–34; number of dorsal scale rows 464–542; occurs in the Udzungwa, Ukaguru, and Uluguru Mountains...... *Afrotyphlops nigrocandidus*, Bicoloured Blind Snake [p. 350]	13a	Midbody scale rows 30–34; supraocular wedged between ocular and preocular; underside pigmented like dorsum......*Afrotyphlops punctatus*, Spotted Blind Snake [p. 351]
		13b	Midbody scale rows 22–26; supraocular wedged between preocular and nasal; underside immaculate......*Afrotyphlops rondoensis*, Rondo Blind Snake [p. 352]
7a	Number of dorsal scale rows less than 310; less than 200mm in total length......*Afrotyphlops nanus*, Kenyan Dwarf Blind Snake [p. 349]	14a	Ocular overlapped by second supralabial......15
7b	Number of dorsal scale rows more than 310; usually more than 200mm in total length......8	14b	Ocular not overlapped by second supralabial......16
8a	Preocular not contacting supralabials; dorsal scales in 28 rows throughout (no reductions caudally); inferior nasal suture arising from between first and second infralabials......*Afrotyphlops kaimosae*, Kakamega Blind Snake [p. 347]	15a	Midbody scale rows 26–32......*Afrotyphlops lineolatus*, Lineolate Blind Snake [p. 348]
		15b	Midbody scale rows 22–24......*Afrotyphlops tanganicanus*, Tanganyika Blind Snake [p. 352]
8b	Preocular contacting supralabials; dorsal scale rows with some reductions caudally; inferior nasal suture arising from first supralabial and/or rostral......9	16a	Dorsum heavily blotched with black; underside immaculate; midbody scale rows 26–28......*Afrotyphlops usambaricus*, Usambara Blotched Blind Snake [p. 353]
		16b	Dorsum, and often underside, usually lineolate or spotted......17
9a	Supraocular with lateral apex between ocular and preocular......10	17a	Midbody scale rows 26–32......*Afrotyphlops lineolatus*, Lineolate Blind Snake [p. 348]
9b	Supraocular with lateral apex between preocular and nasal......12		
10a	Preocular usually overlapped by second supralabial......*Afrotyphlops ango-*	17b	Midbody scale rows 22–26......*Afrotyphlops tanganicanus*, Tanganyika Blind Snake [p. 352]

ANGOLA BLIND SNAKE *Afrotyphlops angolensis*

IDENTIFICATION: A big brown or rufous blind snake, largely from the Albertine Rift and Uganda, but also central Kenya. Head blunt, snout rounded, the rostal scale is oval.

The eye is visible. The tail is short and ends in a spike. Midbody scale rows 24–36; number of dorsal scale rows 350–501; the length/diameter ratio is 26–60. Maximum size about 60cm, averages 30–50cm. Usually brown or grey, may appear uniform or finely spotted, animals from less forested areas are often brown dorsally and pale or rufous-brown below, those from forest may lack pigment.

HABITAT AND DISTRIBUTION: Found in forest as well as savanna and woodland, in our area mostly at medium altitude, between 700–2,000m. Widespread in Rwanda and Burundi, especially at altitudes above 1,500m, occurs along the Albertine Rift and in south-west Uganda, with a handful of disparate records from further east (Geita, Kibondo, Sekenke, Udzungwa in Tanzania, Chogoria, Chuka, Kitui, Transmara, Kora, Nairobi, Nyambene Hills in Kenya). Elsewhere, west to Cameroon, south to Angola.

Afrotyphlops angolensis

Afrotyphlops angolensis Angola Blind Snake. Max Dehling, Butare, Rwanda.

NATURAL HISTORY: A burrowing, nocturnal species, living in holes, loose soil or leaf litter. They have been taken in bucket pitfall traps; they feed on small invertebrates.

ANGLE-SNOUTED BLIND SNAKE *Afrotyphlops brevis*

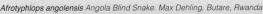

IDENTIFICATION: A big thick brown blind snake from the savanna and dry country of Kenya. Head blunt, snout pointed and quite sharp. The eye is visible. The tail is short and ends in a spike. Midbody scale rows 29–40; number of dorsal scale rows 288–557; the length/diameter ratio is 17–38. Maximum size about 70cm, averages 40–60cm. Juveniles are often striped, but big adults usually (not always) lose the striping, becoming uniform brown.

Afrotyphlops brevis Angle-snouted Blind Snake. Victor Wasonga, Ol Ari Nyiro.

Afrotyphlops brevis

HABITAT AND DISTRIBUTION: Moist and dry savanna and semi-desert, from near sea level to over 2,000 m, although more common at low altitude. Sporadic records from a number of localities in eastern and northern Kenya, include Ngulia Lodge, Chiokarige, Kora, Kitui, Samburu National Reserve, Tambach, Sig-

or, Ol Ari Nyiro and Kakamega. Probably in northern Uganda, as it is known from a cluster of localities in South Sudan, elsewhere to Somalia and Ethiopia.

NATURAL HISTORY: Poorly known, but live in holes and soft soil and sand; they have been collected in pitfall traps and dug out of termitaria, a pair were found poorly concealed at the entrance to a rodent burrow at Chiokarige. On rainy season nights the males emerge and move around on the ground.

SOUTHERN WEDGE-SNOUTED BLIND SNAKE
Afrotyphlops calabresii

IDENTIFICATION: A small wedge-snouted blind snake from the dry country of northern and eastern Kenya. The snout is prominent and distinctly wedge-shaped. The eye is relatively large and visible. The tail is short and ends in a spike. Midbody scale rows 22; number of dorsal scale rows 257–302; the length/diameter ratio is 18–37. Maximum size about 19cm, averages 14–16cm. In life, often looks quite silvery, but is rufous brown, the scales pigmented at the edge so the snake looks speckled. Taxonomic Note: Kenyan specimens were previously known as *Typhlops cuneirostris*.

Afrotyphlops calabresii

HABITAT AND DISTRIBUTION: A snake of dry savanna and semi-desert of the Somali-Maasai region. A handful of records from the northeast (Mandera, El Wak, Wajir) and a single curious record from near Kibwezi, all at altitudes between 200–1,200m. Elsewhere, to Somalia and eastern Ethiopia.

Afrotyphlops calabresii Wedge-snouted Blind Snake. California Academy of Sciences; Wajir.

NATURAL HISTORY: Burrows in sand, although it may emerge at night; most specimens collected under ground cover, such as logs, rocks, brush piles etc. One was dug out 30cm below the surface in sandy soil; at Wajir three were found under a large rock that sheltered an ant colony. At Wajir, one was regurgitated by a North-east African Carpet Viper *Echis pyramidum*.

BLOTCHED BLIND SNAKE *Afrotyphlops congestus*

IDENTIFICATION: A very large thick blotched blind snake from Uganda. Its snout is rounded when viewed from the top, and only obtusely (not sharply) angular when viewed from the side. The rostral scale is broad and truncated posteriorly. The eye is large, and seen behind or below the posterior border of the preocular scale. The mid body scales are in 24–30 rows, number of dorsal scale rows 310–416, the length/diameter ratio is 19–29. Adults average 40–60cm, but may reach 76cm. The colour is white or grey white, with large black blotches as its common name suggests. These may merge so that the dorsum is entirely black and some black blotches may occur laterally, however the center of the ventrum is usually unspotted down the middle.

Afrotyphlops congestus

HABITAT AND DISTRIBUTION: A species of forest and woodland, and most records from our area are from the western and south-western side of Uganda, at altitudes between 700–1,800m. Known localities include the Budongo Forest, Mbarara and Hoima. A single Burundi record, from Ruvubu National Park. A couple of strange records from the Usambara Mountains and south-east Tanzania, (not shown on the map), suggest confusion with a similar-looking species. Elsewhere west to Bioko and Nigeria.

Afrotyphlops congestus Blotched Blind Snake. Max Dehling, Rwanda.

NATURAL HISTORY: Poorly known; a nocturnal burrowing species, living in soil and leaf litter, diet invertebrates. Specimens from Uganda are said to have a particularly offensive smell, produced by cloacal glands, presumably as a defence mechanism. Known predators include file snakes *Gonionotophis*.

GIERRA'S BLIND SNAKE *Afrotyphlops gierrai*

IDENTIFICATION: A big dark blind snake with a rounded snout in the Eastern Arc Forests. The oval rostral shield is narrow ventrally. The eye is visible below the preocular shield. Midbody scale rows 26–28, dorsal scale rows 372–463; length/diameter ratio 32–58. It looks dark overall, but on close examination the dorsal scales are a blue-grey on the middle of the back, the sides and ventrum are white, black or black-blotched. The ventrum is usually a clear white or yellow. Size: averages 25–40cm, the largest individual known is about 47cm total length.

HABITAT AND DISTRIBUTION: A Tanzanian endemic, inhabiting Afromontane forest. Most records from the East Usambara Mountains, at altitudes between 600–1,200m. Also recorded from the Uluguru Mountains. Conservation Status: Endangered B1ab(iii).

Afrotyphlops gierrai

Afrotyphlops gierrae Gierra's Blind Snake. Michele Menegon, Usambara Mountains.

NATURAL HISTORY: Little is known of the biology of this species. It appears to be forest dependent on montane forest of the Usambara Mountains, and like other members of its genus, it lives in loose soil, holes or leaf litter, feeds on invertebrates, and lays eggs.

KAKAMEGA BLIND SNAKE *Afrotyphlops kaimosae*

IDENTIFICATION: Known from a single specimen from Kakamega area, Kenya, collected in 1934 by Arthur Loveridge. It is unique in having an ocular scale which is in contact with the nasal shield below the preocular scale. The snout is rounded and prominent. The rostral is oval dorsally; the eye is visible beneath the ocular scale, adjacent to the preocular border. Midbody scale rows number 28, mid dorsal scales number 390, the length to diameter ratio is 43. Measured 21.5 cm at collection. Loveridge described it as uniformly black, scarcely lighter below, save round the mouth and vent; the scales of the preserved specimen are slightly lighter at their bases. Taxonomic Note: Regarded for a long time as a slightly aberrant specimen of *Afrotyphlops angolensis*.

Afrotyphlops kaimosae

HABITAT AND DISTRIBUTION: Collected at the Friends Africa Mission, Kaimosi Village, just south of Kakamega Forest at 1,670 m altitude. Presumably a 'forest-dependent' species, if it is a valid species, rather than an aberrant example of some more common taxon. Conservation Status: data deficient.

NATURAL HISTORY: Unknown, presumably similar to other members of the genus.

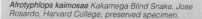

Afrotyphlops kaimosae Kakamega Blind Snake. Jose Rosardo, Harvard College, preserved specimen.

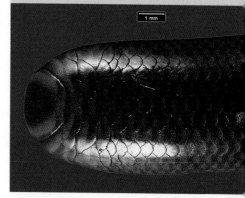

LINEOLATE BLIND SNAKE Afrotyphlops lineolatus

IDENTIFICATION: A big brown or grey, usually spotted blind snake. The snout is prominent and rounded. The rostral shield is broad, 3/4 the width of the head, truncated at the rear. Midbody scale rows 26 to 30; number of dorsal scale rows 323–503, length to diameter ratio is 21–47. Average size is 30–50cm, maximum 64cm. Appears brown, red-brown or grey, on close examination the centre of each scale is seen to be lighter, so the snake appears finely striped with lines of tiny dots; often the pigmentation decreases downwards so the lower flanks look lighter, and the underside may be colourless. In some specimens (especially those from forest or woodland) the light spotting may darken or be absent on all or parts of the body and the snake just looks uniformly dark above; a Ujiji specimen was dark above and blotched below.

Afrotyphlops lineolatus

HABITAT AND DISTRIBUTION: A species tolerant of a wide range of habitats, fairly widespread in forest, woodland and moist savanna, from sea level to 2,600m. Widespread in parts of northern, eastern and western Tanzania, although records lacking from the centre, also known from south-western Kenya, and from the Nairobi area northwards to the eastern side of Mt Kenya. Records from the lakeshore and Albertine Rift in Uganda and south-eastern Burundi; absent from the dry north and east (save records from around Moyale) and also absent from Rwanda. Elsewhere, south and west to Angola, west to Senegal.

Afrotyphlops lineolatus Lineolate Blind Snake. Michele Menegon, Tanzania.

NATURAL HISTORY: Nocturnal and fossorial, living in holes, soft soil and leaf litter. Nairobi specimens were usually found in the rainy season, under ground cover. May emerge on wet nights and move around on the grounds. Diet: invertebrates, often termites. In the Mt Kenya forests, this species has been observed moving along a safari ant (*Dorylus molestus*) migration trail towards the new nest site, it has also been found hidden in the nest when it was opened in a search for the queen, and following an old trail to find a colony

ZAMBEZI GIANT BLIND SNAKE Afrotyphlops mucruso

IDENTIFICATION: A huge, blotched blind snake. The snout is prominent, with an obtusely angled and keratinised edge to the rostrals (this may be absent in younger specimens). The rostral is very large and oval, but tapers posteriorly to a blunt point. The eye is distinct, and seen near the anterior edge of the ocular shield. Midbody scale rows 30–40, number of dorsal scale rows 307–517, length to diameter ratio is 21–58. A big snake, average 50–80cm, but the largest individual known measured 95cm. There are two common colour phases, plain or blotched. Blotched individuals are white with black blotches; juveniles of this phase may appear plain, with the botches faint. The plain form appears grey or blue but each scale is dark-edged, so the snake looks finely striped with lines of dots. Underside usually white, sometimes restricted to a narrow median line. The skin gradually darkens after shedding, sometimes to red-brown, and yellow below,

HABITAT AND DISTRIBUTION: Coastal thicket, moist and dry savanna, mostly at low altitudes, from sea level to around 1,600m. In our area, occurs along the length of the Kenyan coast, although rare in that area, but widespread across central and southern Tanzania. No Uganda, Rwanda or Burundi records. Elsewhere, south and west to Angola and northern South Africa.

Afrotyphlops mucruso

NATURAL HISTORY: Lives deep underground, often in termitaria, and it seeks out the brood chambers of ant nests and eats the eggs and larvae. May emerge at night and has been found under lamps, eating falling termites, on wet nights the males may also emerge and move from one refuge to another, dispersing genetic material. A Tanzanian specimen at Newala was active above ground at 8am in the dry season. Their large size and stout body indicate they may fast for a long time, and eat only two or three times per year. They usually lay 12–40 eggs (broods up to 60 recorded), and these hatch in 5–6 weeks. This snake is widely recognised as 'the two-headed snake' in Tanzania.

Afrotyphlops mucruso Zambezi Giant Blind Snake. Michele Menegon, southern Tanzania.

KENYAN DWARF BLIND SNAKE *Afrotyphlops nanus*

IDENTIFICATION: A small, thick blind snake, known only from two specimens from eastern Kenya. The head is wider than the body and so is the tail. The snout is rounded from above, but wedge-shaped in profile. A large eye is visible beneath an ocular shield. The tail ends in a tiny, thin but distinctive spike. Both specimens have 30 midbody scale rows; the number of dorsal scale rows is 287 and 291, the length to diameter ratio is 23–27. The two specimens are 11.8cm and 12.5cm. The colour in life is unknown; in preservative it is light yellow brown, with around 15 fine pale orange stripes that commence between the eyes, the snout and the underside are yellow. Conservation Status: Data Deficient.

Afrotyphlops nanus

Afrotyphlops nanus Kenya Dwarf Blind Snake. Stephen Spawls, preserved type.

HABITAT AND DISTRIBUTION: At present endemic to Kenya, the two specimens were described in 2009, but collected in 1896, during the construction of the Uganda Railway, at Samburu, Coast Province, 55km west-north-west of Mombasa, 240m altitude, in an area of low savanna with rainfall under 500mm per year.

NATURAL HISTORY: Unknown.

BICOLOURED BLIND SNAKE *Afrotyphlops nigrocandidus*

IDENTIFICATION: A large, distinctive black and white blind snake from the Eastern Arc Mountains in Tanzania. It has a prominent snout. The rostral shield is very broad dorsally (two-thirds the width of the head); it is oval and truncated and barely reaches the level of the eyes which are visible. Midbody scale rows 28–34, number of dorsal scale rows 464–542, length/diameter ratio is 38–52. Average size 30–40cm, the largest specimen recorded has a length of 57.3cm. The general colour impression is of a snake with a black back, white below, although some specimens have white-edged scales, giving a speckled appearance, others are said to have a yellow underside (although this may be an artifact of preservation). In a subadult specimen, the black was confined to a dorsal stripe seven scales wide.

Afrotyphlops nigrocandidus

HABITAT AND DISTRIBUTION: Tanzanian endemic; this species appears to be restricted to montane forests of the southern Eastern Arc Mountains of Tanzania from altitudes between 1,450–1,750m. Known from the Ukaguru, Uluguru and Udzungwa Mountains, probably occurs in the Nguru and Rubeho Mountains.

Afrotyphlops nigrocandidus Bicoloured Blind Snake. Michele Menegon.

NATURAL HISTORY: Little is known of the biology of this species. It appears to be forest dependent on montane forest of the southern Eastern Arc Mountains, and like other members of its genus, to live in loose soil, feed on invertebrates, and lay eggs, follicle studies suggest the average clutch size would be about six.

TANGA BLIND SNAKE *Afrotyphlops platyrhynchus*

IDENTIFICATION: A little-known slender blind snake; four specimens were collected in Tanga in 1910, and it has not been seen since. The snout is prominent and rounded. The rostral scale is oval. The eye is distinct and located under the ocular scale. Midbody scale rows 24, number of dorsal scale rows 400–425, length/diameter ratio is 50–60. The largest specimen is 27.3 cmlength. The colour of living specimens has never been described; the preserved specimens are pale reddish-yellow above, and paler below.

Afrotyphlops platyrhynchus

HABITAT AND DISTRIBUTION: Known from only four specimens taken in the Tanga area of Tanzania, at sea level, presumably in the coastal mosaic.

NATURAL HISTORY: Unknown.

SPOTTED BLIND SNAKE *Afrotyphlops punctatus*

IDENTIFICATION: A thick blind snake from forests and savannas of Uganda. The snout is rounded and prominent, the rostral is broad above. Midbody scale rows 28–32, number of dorsal scale rows 374–465, length/diameter ratio 26–33. Average size 35–55cm, largest specimen was 66cm. There are two main colour phases; one appears brown, or grey, on close examination the centre of each scale is seen to be lighter, so the snake appears finely striped with lines of tiny dots, this breaks up and fades low on the flanks, the underside is reddish or light whitish grey. The other colour form is white, pale pink or grey white, with many irregular black blotches, the underside may be immaculate or have a few blotches. The striped form appears to be more common in Uganda.

Afrotyphlops punctatus

Afrotyphlops punctatus Spotted Blind Snake. Harald Hinkel, Rwanda.

HABITAT AND DISTRIBUTION: Forest and savanna, at low to medium altitude, between about 600–1,400m. Widespread in central and parts of western Uganda, known localities include Lira, Kampala, the Sesse Islands and West Nile. Elsewhere, north to South Sudan, west to Senegal.

NATURAL HISTORY: Poorly known. Like other members of its genus, it lives in holes, leaf litter or loose soil, shelters under ground cover, feeds on invertebrates, and lays eggs. Nigerian specimens preferred slightly acidic, loamy soils, and fed on ants and termites. Specimens collected in the Ghanaian savanna were found on the surface during the rainy season. One specimen was eaten by a toad.

RONDO BLIND SNAKE *Afrotyphlops rondoensis*

IDENTIFICATION: A stocky blind snake of extreme south-east Tanzania. The snout is prominent and rounded. The rostral shield is oval. Midbody scale rows 22–26, number of dorsal scale rows 312–379, length/diameter ratio 30–41. Averages 25–35cm, the largest specimen is 41cm. In life it appears black from a distance, but close examination shows that each scale has a white spot at its apex, giving the impression of a snake striped with lines of dots; the ventral surface is white. In preservative, the spots and the underside turn yellow.

Afrotyphlops rondoensis Rondo Blind Snake. Stephen Spawls, preserved specimen.

Afrotyphlops rondoensis

HABITAT AND DISTRIBUTION: This species is known only from the savanna and woodland of south-eastern Tanzania, at altitudes between 375–620m; known localities include the Rondo Plateau, Newala and Lindi. Conservation Status: Data Deficient.

NATURAL HISTORY: Unknown, presumably similar to other medium-sized blind snakes.

TANGANYIKA / LIWALE BLIND SNAKE *Afrotyphlops tanganicanus*

IDENTIFICATION: A rare endemic of south-eastern Tanzania. The snout is rounded, and is oval and truncated posteriorly. The eye is near the anterior border of the ocular shield. Midbody scale rows 21–24, number of dorsal scale rows 352–425, the length/diameter ratio is 30–44. It reaches a maximum size of around 40cm. In preserved specimens the scales are yellowish towards their middle, but are brown at the edges. The underside is not immaculate but is lightly pigmented; but the life colour is not known.

HABITAT AND DISTRIBUTION: Known only from southeast-

Afrotyphlops tanganicanus

ern Tanzania, from Liwale and Kilwa, at altitudes between sea level and 500m. Conservation Status: Least Concern.

NATURAL HISTORY: Little is known about the biology of this species of very limited distribution. Presumably it burrows in loose soil, lays eggs and feeds on small invertebrates.

USAMBARA BLOTCHED BLIND SNAKE
Afrotyphlops usambaricus

IDENTIFICATION: A big blotched blind snake, found only in the Usambara Mountains. The snout is rounded when viewed from above but from the side is obtusely angular The rostral shield is broad, about 70% of the width of the head; its posterior portion is concave. The rostral extends back to the level of the eyes. The eye is visible through a very large ocular scale. Midbody scale rows 26–28, number of dorsal scale rows 344–389, ratio of body length/width is 27–31. The impression is a thick dark blind snake with light yellow or yellow-pink cross bars, the underside is immaculate, with no dark blotches impinging from the sides. Average size 30–50cm, maximum 60.5cm.

Afrotyphlops usambaricus

HABITAT AND DISTRIBUTION: Endemic to the East Usambara Mountains, Tanzania, currently known only from the Amani area at 750m altitude. It is presumably forest dependent.

NATURAL HISTORY: Because the Amani area is one of a former German botanic garden, cleared from natural forest, it is presumed to be forest dependent, but able to tolerate

Afrotyphlops usambaricus Usambara Blotched Blind Snake. Stephen Spawls, Usambara Mountains.

some disturbance such as possibly low intensity local cultivation, and probably burrows in the loose, relatively moist humus, in which its presumed food, invertebrates, are common. The type was collected in 1935 by Reg Moreau, doyen of African ornithology.

ASIAN BLIND SNAKES *INDOTYPHLOPS*

A genus of around 23 species of thin blind snakes, mostly found in south-east Asia. A single introduced species is known from East Africa, mostly from coastal towns, where it arrived in cargoes, in soil or ballast.

FLOWER POT BLIND SNAKE, BRAHMINY BLIND SNAKE
Indotyphlops braminus

IDENTIFICATION: A small blind snake, introduced on the East African coast. The snout of this species is distinct, it is rounded with a very narrow rostral that is less than one-third of the width of the head, and that does not reach the level of the eye. It has an anterior constriction that is visible dorsally. Midbody scales in 20 rows, mid-dorsal scales 261–368, length/diameter ratio 30–57. Average size 10–16cm, maximum 21cm. It may be grey, dark or light brown or black above, sometimes with a purplish sheen, lighter below. Some specimens have a darker, sharp-edged vertebral stripe. The lower snout chin, and cloacal region and tip of the tail may be pale cream or white. Some specimens that are lighter in colour on the dorsum have a dark apical spot on the anterior part of the dorsal scales. Taxonomic Note: Previously in the genus *Ramphotyphlops*.

Indotyphlops braminus

Indotyphlops braminus Flower-pot Blind Snake. Left: Stephen Spawls, Seychelles. Right: Frank Glaw, Pemba.

HABITAT AND DISTRIBUTION: An introduced species, in East Africa known from a number of coastal towns including Watamu, Mombasa, Dar Es Salaam, Bagamoyo and the Tanzanian Islands. In Tanzania there is an inland record from Liwale, and a couple of unverified inland records (not shown on the map) from Mawaeni, near Dodoma and Nzega, near Tabora. Originally described from India, and now occurs worldwide in tropical areas.

NATURAL HISTORY: A parthenogenic species, all individuals are female, hence it can spread easily (usually in flower pots or bags of soil, hence the common name), as only a single individual is needed to start a population. In East Africa, often found in sand or sandy soil near the sea, sheltering under ground cover. It feeds on the larvae, eggs and pupae of ants and invertebrates. Two to six tiny eggs are laid and are self-fertilising.

GRACILE PINK BLIND SNAKES *LETHEOBIA*

A sub-Saharan African genus, recently revived, of about 30 species of thin, relatively long blind snakes; eight species occur in East Africa. Most are pink in life, and thus sometimes called worm-like blind snakes. Their eyes are primitive, invisible, or visible only as indistinct darker dots beneath the scales. In preservative they turn colourless or tan.

KEY TO THE EAST AFRICAN GRACILE PINK BLIND SNAKES
LETHEOBIA

1a Nasal divided, with nostril pierced laterally; rostral less than half width of head.............. *Letheobia uluguruensis*, Uluguru Gracile Blind Snake [p. 360]
1b Nasal semidivided, with nostril pierced ventrally; rostral more than half width of head 2

2a Midbody scale count 20, in high central Kenya......*Letheobia mbeerensis*, Mbeere Gracile Blind Snake [p. 358]
2b Midbody scale count not 20, not in high central Kenya......3

3a Snout rounded in profile; ocular subequal to subocular 4
3b Snout with angular horizontal edge; ocular usually smaller than subocular......6

4a Scale rows 26-24-24; number of dorsal scale rows fewer than 360; Pemba Island......*Letheobia pembana*, Pemba Gracile Blind Snake [p. 359]
4b Scale rows 24-22-22; number of dorsal scale rows more than 360......5

5a Number of dorsal scale rows fewer than 400; length/diameter ratio 62 or less; coastal Kenya and Tanzania...... *Letheobia swahilica*, Swahili Gracile Blind Snake [p. 359]
5b Number of dorsal scale rows more than 400; length/diameter ratio 62 or more......*Letheobia pallida*, Pallid Gracile Blind Snake [p. 358]

6a Preocular more than half the height of the nasal, ocular separated from the subocular by one or more temporals......7
6b Preocular less than half height of nasal; ocular contacting subocular ...... *Letheobia lumbriciformis*, Worm-like Gracile Blind Snake [p. 357]

7a Midbody scale rows 22......*Letheobia gracilis*, Slender Gracile Blind Snake [p. 356]
7b Midbody scale rows 24......*Letheobia graueri*, Lake Tanganyika Gracile Blind Snake [p. 356]

SLENDER GRACILE BLIND SNAKE Letheobia gracilis

IDENTIFICATION: A long thin pink species from south-western Tanzania and Burundi. It has a prominent, snout with an angular horizontal edge. The eye is not visible. The midbody scale count is 22; there are 629–726 mid-dorsal scales. The length to diameter ratio is 60–107. The largest known specimen was 54cm in length; it averages 30–45 cm. In life it is pink to pinky-purple, in preservative it is colourless.

HABITAT AND DISTRIBUTION: Savanna and woodland, between 600–1,200m altitude. In our area, known from Burundi, and southwestern Tanzania (localities include Tatanda and Kitungulu), elsewhere to northern Zambia and south-western Congo.

Letheobia gracilis

Letheobia gracilis Slender Gracile Blind Snake. Frank Willems, northern Zambia.

NATURAL HISTORY: Poorly known. Has been found in the open, on red soil, and dug up by a plough, a Zambian specimen was attacked and partially swallowed by a Bibron's Burrowing Asp *Atractaspis bibronii*.

LAKE TANGANYIKA GRACILE BLIND SNAKE/GRAUER'S GRACILE BLIND SNAKE Letheobia graueri

IDENTIFICATION: A large gracile blind snake, originally described from the Congo. The snout has an angular horizontal edge, with a very broad rostral that is truncated posteriorly. The frontal is crescent shape. The eye is not visible. There is a nasal suture that arises from the first or second labial. The midbody scale count is 24. There are 454–622 mid-dorsal scales. The length/diameter ratio is 60–89. Maximum length around 40–45cm, average 30–40cm. Very pale pink in life, the tail is almost white and the head looks white; in preservative it appears unpigmented.

HABITAT AND DISTRIBUTION: Found in savanna, forest and woodland, including Miombo woodland; also found in areas of rice cultivation. In our area, recorded from Semliki National Park in Western Uganda and a cluster of records from the northeastern shore of Lake Tanganyika in Burundi (Rumonge) and Tanzania, including Mahale Peninsula, Gombe Stream National Park and Ujiji, at altitudes between 700–1000m. Elsewhere, known from the eastern Congo. Conservation Status: Least Concern.

Letheobia graueri

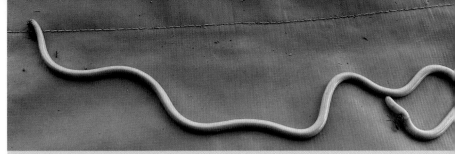

Letheobia graueri Lake Tanganyika Gracile Blind Snake. Craig Doria, western Tanzania.

NATURAL HISTORY: Poorly known, presumably similar to other blind snakes, several specimens have been collected hiding in heaps of vegetable debris.

WORM-LIKE GRACILE BLIND SNAKE/ZANZIBAR GRACILE BLIND SNAKE *Letheobia lumbriciformis*

IDENTIFICATION: An elongate blind snake, resembling a worm in colour and appearance. The snout is angular with a horizontal edge. The rostral shield is very broad and truncated posteriorly. The eye of this species is not visible. The nasal suture arises from the second labial. The midbody scale count is 18, there are 490–607 mid-dorsal scales. The length to diameter ratio is 43–85. The largest specimen is 44.5cm in total length, it averages 25–40cm. Colour: purplish-pink above, blotched with pinky brown, the underside is mottled brown and light grey; in preservative the underside becomes colourless.

Letheobia lumbriciformis

HABITAT AND DISTRIBUTION: An East African endemic, known from coastal savanna, dry coastal forest and farmland, mostly at low altitude, although there is a specimen from 'Usambara' without further details. Occurs from Mkonumbi, on the mainland opposite Lamu, south to Tanga and the Usambara mountains, Kenyan localities include Malindi, Watamu, the Arabuko-Sokoke Forest, Gogoni Forest, Mombasa and Changamwe, Tanzanian records include Amboni, Tanga and the Kilulu Forest reserve, it is also on Zanzibar. Conservation Status: Least Concern.

NATURAL HISTORY: Poorly known. A Gede specimen was on the surface following a rainstorm. Specimens have been found when ploughed up in sisal and cotton plantations and in other cultivation after heavy rain.

Letheobia lumbriciformis Worm-like Gracile Blind Snake. Stephen Spawls, Arabuko-Sokoke Forest.

MBEERE GRACILE BLIND SNAKE   Letheobia mbeerensis

IDENTIFICATION: A recently described elongate pale pink blind snake from Kenya. The snout is rounded and prominent. The rostral scale is very broad and truncated at its rear; the eye is just visible as a tiny black spot. Unusually, it has no subocular scale. On the single known specimen, the midbody scale count is 20, there are 670 mid-dorsal scales. The length to diameter ratio is 62. It is 28cm long. In life, it is pink, the faint pink-brown edging to each dorsal scale gives a faintly speckled appearance, the head appears white.

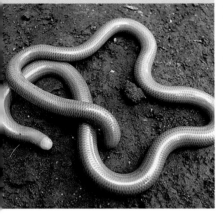

Letheobia mbeerensis Mbeere Gracile Blind Snake. Patrick Malonza, Siakago.

Letheobia mbeerensis

HABITAT AND DISTRIBUTION: Known from a single specimen from agricultural land, in low altitude savanna (1,220m) in the vicinity of Siakago Village, Mbeere District, southwest of Mt Kenya.

NATURAL HISTORY: Poorly known. Presumably similar to other members of the genus, the specimen was found by a farmer tilling the soil.

PALLID GRACILE BLIND SNAKE   Letheobia pallida

IDENTIFICATION: This species, which is known from only a few specimens, is endemic to Unguja, the largest island of Zanzibar. It has a prominent, rounded snout, and the very broad rostral shield is truncated at the rear. The frontal scale is shaped like a crescent. The eye is visible as a pink spot in some specimens. The midbody scale count is 22, there are 418–433 mid-dorsal scales. The length/diameter ratio is 53–62. It is small, the longest recorded individual was 192mm in total length. It is colourless.

Letheobia pallida Pallid Gracile Blind Snake. Stephen Spawls, preserved specimen.

Letheobia pallida

HABITAT AND DISTRIBUTION: Endemic to Zanzibar island, where it lives in coastal mosaic vegetation, at sea level or just above. Conservation Status: Data Deficient.

NATURAL HISTORY: Two of the few specimens known were taken from a well. It presumably is a burrower in soft soil; nothing is known of its biology.

PEMBA GRACILE BLIND SNAKE *Letheobia pembana*

IDENTIFICATION: A recently-described elongate pale pink blind snake from Pemba, a member of the *L. pallida* species complex. The snout is very rounded and prominent. The rostral scale is very broad and truncated at its rear; the eye is just visible as a pale pink spot. On the single formally described specimen, the midbody scale count is 24, there are 353 mid-dorsal scales. The length/diameter ratio is 53. In life it is unpigmented and thus appears various shades of pink, due to the blood and organs visible through the skin. Taxonomic Note: collected more than 100 years ago but only recently recognised as separate from *Letheobia pallida* by its more numerous scale rows (26-24-24 vs 24-22-22) and lower number of mid-dorsal scales (353 vs 376–487).

Letheobia pembana

Letheobia pembana Pemba Gracile Blind Snake. Frank Glaw, Pemba.

HABITAT AND DISTRIBUTION: Endemic to the coastal mosaic of Pemba Island, at sea level or just above. Conservation Status: Data Deficient.

NATURAL HISTORY: Poorly known. Presumably similar to other members of the genus, living underground and feeding on small invertebrates.

SWAHILI GRACILE BLIND SNAKE *Letheobia swahilica*

IDENTIFICATION: A medium-sized, colourless blind snake, endemic to the northern East African coast. The snout is rounded and prominent, with a very broad rostral which is truncated posteriorly. The frontal shield is shaped like a crescent. Usually the eye is not visible but if it is, it may be seen beneath the nasal/ocular sulcus. It is fairly thick, with a length/diameter ratio of 49–62 and 376–392 mid-dorsal scales. The midbody scale count is 22. The largest recorded specimen is 190mm in total length. In life it is vivid pink or whitish pink, in preservative colourless. Taxonomic Note: This species is closely related to *Letheobia pallida* which is an insular, island form known only from Zanzibar.

Letheobia swahilica

HABITAT AND DISTRIBUTION: Found in coastal mosaic vegetation of the East African coast, scattered records include Ngatana, in the Tana Delta in Kenya, from various localties between Kilifi and Shimoni (Takaungu, Mazeras, Vipingo and the Shimba Hills) and the Kwamarimba Forest Reserve, Tanzania, at low altitude. Conservation Status: Least Concern.

Letheobia swahilica Swahili Gracile Blind Snake. Patrick Malonza, Shimba Hills.

NATURAL HISTORY: Apparently tolerant of disturbance, it has been recorded from cultivated areas in Kenya and coastal forest of Tanzania. Little is known of its biology; specimens have been collected under vegetable debris and in agricultural land in valley bottoms.

ULUGURU GRACILE BLIND SNAKE *Letheobia uluguruensis*

IDENTIFICATION: A vivid pink blind snake, endemic to the Uluguru Mountains of Tanzania. The snout is rounded and prominent. The rostral scale is 'moderate' in size and oval in shape. The eye in this species is visible as a darker pink blotch below the head scales. The ocular shield is separated from subocular by two or three postoculars. The nasal suture arises from the second labial scale. The midbody scale count is 22, there are 379–416 mid-dorsal scales. The length/diameter ratio is 52–57. The largest specimen is 24.5cm in total length, it averages 15–22cm. In life it is vivid pink, with the larger organs (especially the liver) looking purple-pink; it fades to colourless in preservative.

Letheobia uluguruensis

HABITAT AND DISTRIBUTION: A Tanzanian endemic, from the ancient forests of the Uluguru Mountains, at altitudes between 760–1000m. The only known specimens were found in farmland near forest; some under the rotting thatch of a collapsed hut at the edge of the forest, others dug out of loose soil in small scale, low intensity agriculture. Conservation Status: Endangered B1ab(iii).

Letheobia uluguruensis Uluguru Gracile Blind Snake. David Gower, Uluguru Mountains.

NATURAL HISTORY: As is the case for other members of its genus, it probably spends most of the time underground, in burrows, where it feeds on small invertebrates. Given its very localised distribution, it may be considered dependent on the forest ecosystem.

AFRICAN BEAKED OR SHARP-SNOUTED BLIND SNAKES *RHINOTYPHLOPS*

An African genus of blind snakes, with a beaked snout and a broad rostral scale. A number of species have been moved in and out of the genus since it was named in 1974; at present seven species remain, two of which are found in our area.

KEY TO EAST AFRICAN MEMBERS OF THE GENUS *RHINOTYPHLOPS*

1a A distinctive yellow stripe runs the length of the body......*Rhinotyphlops unitaeniatus*, Yellow-striped Blind Snake [p. 362]

1b No distinctive yellow stripe runs the length of the body......*Rhinotyphlops ataeniatus*, Somali Blind Snake [p. 361]

SOMALI BLIND SNAKE *Rhinotyphlops ataeniatus*

IDENTIFICATION: A dry country species of the Horn of Africa, just reaching our area. Its head shape and shield pattern are similar to those of the previous species, with which it is in some places sympatric. The midbody scale count is 24, number of dorsal scales 443–531, the length/diameter ratio is 34–72. The back is a uniform dark brown to black; the top of the head bears a short, median orange or yellow stripe. The average size is 25–40cm, the largest specimen recorded is 45.5cm.

HABITAT AND DISTRIBUTION: Dry savanna of the Horn of Africa. In our area known only from Malka Murri, on the northern border with Ethiopia, at an altitude of around 1,000m, elsewhere to Ethiopia and Somalia.

Rhinotyphlops ataeniatus

Rhinotyphlops ataeniatus Somali Blind Snake. Stephen Spawls, Ethiopia.

NATURAL HISTORY: Not well known but presumably similar to other members of the genus. A specimen was found in the stomach of a Lizard Buzzard *Kaupifalco monogrammicus*. A captive Ethiopian specimen found in the sand of a dry watercourse and kept in a sand-filled tank emerged during rainstorms and crawled around on the surface.

YELLOW-STRIPED BLIND SNAKE *Rhinotyphlops unitaeniatus*

IDENTIFICATION: Unmistakeable on account of its colour. The snout is very prominent. It has an acutely angular horizontal keratinized on it rostral, which is also very large, longer than broad. It extends beyond the level of the eyes. The eye is visible beneath the anterior edge of the nasal scale. Midbody scale count 24, dorsal scales vary from 467–586. The length/diameter ratio is 38–77. Average size 25–40cm. The largest specimen measured 43.5cm. The colour pattern is distinctive; the dorsum is dark brown to black, with a bright yellow vertebral stripe 3–5 scales wide that runs from the back of the head to about 1cm before the tip of the tail. The ventrum is slightly lighter than the dorsum.

Rhinotyphlops unitaeniatus

HABITAT AND DISTRIBUTION: A species of dry and moist savanna and coastal thicket, from sea level to around 1,200m altitude. In our area, known from Meru National Park (and possibly the lower slopes of the Nyambene Hills) down the Tana River to the coast, thence south along the coastline to Tanga in Tanzania, inland through Tsavo and Amboseli to Namanga, in Tanzania also known from Mkomazi and Tarangire National Parks. Elsewhere, to Somalia.

NATURAL HISTORY: This species presumably has the same habits as other blind snakes, it lives in burrows in loose soil, in sand or in leaf litter, where it feeds on small invertebrates, and lays eggs. Occasionally active by day, one was found crossing at path at Gede during an afternoon rainstorm and sometimes found in houses along the Kenya coast. A Kinna specimen was found while being attacked by a chicken.

Rhinotyphlops unitaeniatus Yellow-striped Blind Snake. Bio-Ken archive, Watamu.

Leptotyphlops merkeri
Merkers's Worm Snake
Victor Wasonga

SUB-ORDER SERPENTES
FAMILY LEPTOTYPHLOPIDAE
WORM OR THREAD SNAKES

Found in the Americas, Africa and western Asia, indicating a West Gondwanan origin, the worm snakes are a family of about 140 species, in 13 genera and two subfamilies. Sometimes called thread snakes, or bootlaces (which they resemble), the East African species are in the subfamily Leptotyphlopinae, in three genera: *Epacrophis* (two species); *Leptotyphlops* (13 species); and *Myriopholis* (3 species). The family includes the world's smallest snake, *Tetracheilostoma carlae* of Barbados, reaching only 10cm; the second smallest is found in Kenya, the Tana Delta Worm Snake *Leptotyphlops tanae*, at 10.2cm. All worm snakes have a unique appearance; they are tiny, thin snakes with non-tapering bodies, blunt heads and tails. They have no teeth in the upper jaw, a single lung and oviduct, their eyes are reduced to a black light-sensitive cell beneath the head shields (a few species have a well-developed eye). They live in burrows and cracks underground and largely feed on the eggs, larvae and workers of social insects (although some have been found, strangely, up trees in bird bests, feeding on bird fleas). They avoid ant attack by chemical camouflage, coiling up and producing a pheromone. One to seven eggs are laid. Most are dark brown or black, those from semi-desert areas tend to be pink.

KEY TO EAST AFRICAN MEMBERS OF THE SUBFAMILY *LEPTOTYPHLOPINAE*

All three genera of East African worm snakes appear in this key. As with the blind snakes, you will need a dead specimen, a binocular microscope and our figures showing head scalation to identify an animal to a species, but the locality may also provide useful clues.

1a Midbody scale rows 16......*Leptotyphlops parkeri*, Parker's Worm Snake [p. 370]
1b Midbody scale rows 14......2

2a A discrete frontal separates the rostral from the supraocular......3
2b Frontal fused with the rostral, which is consequently in contact with the supraoculars......15

3a Posterior supralabial not reaching level of eye; precloacal shield semilunate; light brown above, lighter below......4
3b Posterior supralabial reaching level of eye; precloacal shield heart-shaped or subtriangular; dark brown to black above and below......9

4a Anterior supralabial much larger than infranasal; a thick thorn-like apical spine on the tail......5
4b Anterior supralabial subequal to infranasal; a small needle-like apical spine on the tail......6

5a Dorsal scale rows more than 200, inland in Kenya......*Epacrophis drewesi*, Drewes' Worm Snake [p. 365]
5b Dorsal scale rows fewer than 200, on Manda or Lamu Island......*Epacrophis boulengeri*, Lamu Worm Snake [p. 365]

6a Snout strongly hooked in profile......*Myriopholis macrorhyncha*, Hook-snouted Worm Snake [p. 374]
6b Snout not hooked, or only feebly hooked in profile......7

7a Subcaudals usually more than 35, range south of the Tanzanian border......*Myriopholis ionidesi*, Ionides' Worm Snake [p. 374]
7b Subcaudals usually fewer than 35, range north of the Tanzanian border......8

8a Dorsal scale rows 278–300, subcaudals 30–32......*Myriopholis braccianii*, Scortecci's Worm Snake [p. 373]
8b Dorsal scale rows 227–260, subcaudals 25–30......*Leptotyphlops tanae*, Tana Delta Worm Snake [p. 373]

9a Eye large, beneath a clear window or dome in the ocular shield; number of dorsal scale rows usually more than 250; subcaudals usually more than 30......10
9b Eye small, below a flat ocular shield; number of dorsal scale rows usually less than 250; subcaudals usually fewer than 30......12

10a Eye beneath a dome in the ocular shield; number of dorsal scale rows 272–322; subcaudals 30–44...... *Leptotyphlops macrops*, Large-eyed Worm Snake [p. 368]
10b Eye beneath a slightly domed window in the ocular shield; number of dorsal scale rows 229–269; subcaudals 28–39........11

11a Number of dorsal scale rows 229–237; range south-east Tanzanian coast to western Kenya......*Leptotyphlops howelli*, Kim Howell's Worm Snake [p. 367]
11b Number of dorsal scale rows 247–269; endemic to Pemba Island......*Leptotyphlops pembae*, Pemba Worm Snake [p. 371]

12a Number of dorsal scale rows usually fewer than 245; range southern Sudan, south through Uganda, Kivu, and Rwanda to south-east Tanzania......13
12b Number of dorsal scale rows usually more than 245; range northern Tanzania (Kilimanjaro), Kenya and Ethiopia......14

13a Number of dorsal scale rows less than 205; cloacal shield narrow; range south-east coast of Tanzania......*Leptotyphlops mbanjensis*, Mbanja Worm Snake [p. 369]
13b Number of dorsal scale rows usually more than 205; cloacal shield broad......*Leptotyphlops emini*, Emin Pasha's Worm Snake [p. 366]

14a A white patch below tail tip; total length/diameter ratio 62–77, only on Mt Marsabit......*Leptotyphlops aethiopicus*, Ethiopian Worm Snake [p. 366]
14b No white patch below tail tip; total length/diameter ratio 44–55, near Mt Kenya......*Leptotyphlops keniensis*, Mt Kenya Worm Snake [p. 367]

15a Pale brown above, head darker, tail often with a black tip; white below...... *Leptotyphlops nigroterminus*, Black-tipped Worm Snake [p. 370]
15b Dark brown to black above and below......16

16a Ten scale rows on tail......*Leptotyphlops pitmani*, Pitman's Worm Snake [p. 372]

16b	Twelve scale rows on tail......17	17b	Range: southern Kenya and eastern Tanzania......*Leptotyphlops merkeri*, Merker's Worm Snake [p. 369]
17a	Range: northern shores of Lake Tanganyika and Ruzizi valley...... *Leptotyphlops latirostris*, Uvira Worm Snake [p. 368]		

LAMU WORM SNAKE *Epacrophis boulengeri*

IDENTIFICATION: A pink worm snake confined to Lamu and Manda Islands in Kenya. The body is cylindrical, and the head and neck are flattened, and those areas are thus distinct from the rest of the body. The tail is short, ending in a spine that is recurved ventrally. Midbody scale rows 14, number of dorsal scale rows 179–192, length/diameter ratio 29–37. The largest specimen was 20.3cm in length. Colour in life recorded as pink, lighter below in preservative the dorsum is light brown, each individual scale is beige or pink, with a brown border.

Epacrophis boulengeri

HABITAT AND DISTRIBUTION: Known only from Manda and Lamu islands, at sea level or just above, on Kenya's north coast. Conservation Status: Data Deficient.

NATURAL HISTORY: Virtually nothing known; presumably similar to other worm snakes. Two specimens were found buried in leaf litter, another in red soil under rotting grass at the edge of cultivation.

DREWES' WORM SNAKE *Epacrophis drewesi*

IDENTIFICATION: A tiny bicoloured worm snake known from a single specimen found near Isiolo, Kenya. The body is cylindrical. The tail is short, ending in a spine that is recurved ventrally. Midbody scale rows 14, number of dorsal scale rows 248, length/diameter ratio 29–37. The only specimen was 14.3cm in length. In preservative, it has a light brown dorsal stripe 5 scales wide, a mid-lateral lightly pigmented row of scales, and the seven rows across the underside are creamy white. There is a pale collar two scales wide, broken on the vertebral line. Taxonomic Note: Named for Robert C. Drewes, curator emeritus at the California Academy of Sciences.

Epacrophis drewesi

HABITAT AND DISTRIBUTION: The single specimen was taken in savanna grassland 10km south of Isiolo, where the Somali arid zone meets the wooded grassland of Mt Kenya, at an altitude of 1,250m. Conservation Status: Date Deficient.

NATURAL HISTORY: Presumably similar to other worm snakes, living underground, feeding on small invertebrates.

Epacrophis drewesi Drewes' Worm Snake. California Academy of Sciences; south of Isiolo.

ETHIOPIAN WORM SNAKE *Leptotyphlops aethiopicus*

IDENTIFICATION: A thin worm snake, in our area known only from Mt Marsabit, with a distinctive white tail-tip. The body is cylindrical, the snout rounded. The tail is short, ending in a spine that is recurved ventrally. Midbody scale rows 14, number of dorsal scale rows 239–261, length/diameter ratio 61–77. Maximum size around 16cm, although average 10–14cm. Colour uniformly dark brown to black (red-brown in preservative) with a distinctive white patch under the tail tip.

Leptotyphlops aethiopicus

Leptotyphlops aethiopicus Ethiopian Worm Snake. Tomas Mazuch, Ethiopia.

HABITAT AND DISTRIBUTION: The only Kenyan specimen was taken on Mt Marsabit at 1,700 m, elsewhere occurs in semi-arid woodland and savanna grassland in Ethiopia, east of the Rift Valley.

NATURAL HISTORY: Presumably similar to other worm snakes.

EMIN PASHA'S WORM SNAKE *Leptotyphlops emini*

IDENTIFICATION: A thin black worm snake from western East Africa. The body is cylindrical, the snout rounded. The tail is short, ending in a spine that is recurved ventrally. Midbody scale rows 14, number of dorsal scale rows 198–244, length/diameter ratio 38–74. Colour uniformly dark brown to black (red-brown in preservative) with white patches on the lips and chin and a white cloacal shield. Reaches a maximum of about 16cm, although most specimens average 8–14cm. Taxonomic Note: In 2007 *Leptotyphlops monticolus* was revived from the synonymy of this species, but was insufficiently diagnosed, so specimens referable to *L. monticolus* from Rwanda are at present assigned to *L. emini*.

HABITAT AND DISTRIBUTION: Savanna and forest-savanna mosaic, at altitudes between sea level and 2,200m, although usually between 700–1,300m in our area. Widely distributed in the western half of Uganda, extreme north-west Tanzania and western Rwanda, sporadic records from south-west Tanzania, west of Lake Rukwa, Newala and also from Loa, just north of the Uganda border in South Sudan. Also known from the eastern Congo and northern Zambia.

Leptotyphlops emini Emin Pasha's Worm Snake. Mike McLaren, Lake Rukwa.

Leptotyphlops emini

NATURAL HISTORY: Lives in holes underground, sometimes emerging at night, or flooded out of its hole during the rainy season. Eats invertebrates, lays eggs.

KIM HOWELL'S WORM SNAKE *Leptotyphlops howelli*

IDENTIFICATION: A thin black worm snake, with a large and distinct eye, known only from two widely-separated localities. The body is cylindrical. The tail is short, ending in a small terminal spine. Midbody scale rows 14, number of dorsal scale rows 229–237, length/diameter ratio 61–67. The two known specimens were 15.8 and 14cm in length. Black, with white patch below the chin. Taxonomic Note: The massive geographical separation between the two individuals suggests their resemblance might be coincidental.

HABITAT AND DISTRIBUTION: Known from two localities over 1,000 km apart; the coastal forest of Mchungu Forest Reserve, north of the Rufiji Delta in coastal Tanzania, at 15m elevation, and gallery forest of the Yala River, western Kenya, at an altitude of around 1,200m. Conservation Status: Data Deficient.

Leptotyphlops howelli

NATURAL HISTORY: Described from preserved animals and never knowingly seen alive; habits presumably similar to other worm snakes. The Muchungu type was collected by participants in the Frontier-Tanzania coastal forest programme.

MT KENYA WORM SNAKE *Leptotyphlops keniensis*

IDENTIFICATION: A recently-described worm snake from central Kenya. The body is cylindrical. The tail is short, ending in a blunt tail cone. Midbody scale rows 14, number of dorsal scale rows 239–265, length/diameter ratio 47–70. The largest specimen was just under 19cm long. Life colour unknown but the preserved specimens are brown, all scales edged with white, and the upper lip, the anterior tip of the chin and the cloacal shield are white.

HABITAT AND DISTRIBUTION: High grassland, mid- and low-altitude savanna in the vicinity of Mt Kenya, at altitudes between 1,000–2,000m. Known from Nyeri, Naro Moru and Isiolo, and, without precise details, from 'Kilimanjaro'. Conservation Status: Data Deficient.

NATURAL HISTORY: Described from preserved animals and never knowingly seen alive; habits presumably similar to other worm snakes.

Leptotyphlops keniensis

UVIRA WORM SNAKE *Leptotyphlops latirostris*

IDENTIFICATION: A species of the eastern Congo, named for the Congolese town of Uvira, recorded on Burundi's Lake Tanganyika shore. The body is cylindrical. The tail is short, ending in a small terminal spine. Midbody scale rows 14, number of dorsal scale rows 206–212, length/diameter ratio 43–58. The largest specimen was 15.5cm long. Life colour unknown but the preserved specimens are golden brown, the underside lighter, and the scales there edged with white, chin and throat white

HABITAT AND DISTRIBUTION: In our area known only from the Burundi shore of Lake Tanganyika, at around 800m altitude, an area of savanna but heavily cultivated; Burundi localities include Bujumbura and Rumonge. Elsewhere, in the eastern Congo.

Leptotyphlops latirostris

NATURAL HISTORY: Unknown, presumably similar to other worm snakes.

LARGE-EYED WORM SNAKE *Leptotyphlops macrops*

IDENTIFICATION: A thin black worm snake, with a large and distinct eye, from coastal East Africa. The body is cylindrical. The tail is short, ending in a small terminal spine. Midbody scale rows 14, number of dorsal scale rows 272–322, length/diameter ratio 53–73. Reaches a length of 29cm, but most specimens 18–24cm. Colour black or silvery grey.

Leptotyphlops macrops Large-eyed/goggle-eyed Worm Snake. Florian Finke, Shimba Hills.

HABITAT AND DISTRIBUTION: Forest and thicket of the coastal strip, from Gede and Watamu southwards to Mkwaja Forest in north-east coastal Tanzania, at low altitude, 0–100m. A problematic spec-

Leptotyphlops macrops

imen in Berlin Museum from 'Kikuyu' is not shown on the map. Conservation Status: Least Concern.

NATURAL HISTORY: Found in leaf litter and soft soil. Habits presumably similar to other worm snakes.

MBANJA WORM SNAKE   *Leptotyphlops mbanjensis*

IDENTIFICATION: A recently-described worm snake from coastal south-east Tanzania. The body is cylindrical. The tail is short, and thick, ending in a blunt terminal cone. Midbody scale rows 14, number of dorsal scale rows 185–197, length/diameter ratio 38–54. Reaches a length of 13.5cm. Uniformly dark brown to blackish-brown above and below; the lower edge of supralabials and first pair of infralabials are dirty white, posterior border of cloacal shield and adjacent scales dirty white.

Leptotyphlops mbanjensis

HABITAT AND DISTRIBUTION: The habitat was originally described as 'Coastal orchard forest with high grass', at 90–130m altitude. Known only from south-eastern Tanzania, localities include Mbanja, Mikindani and Tendaguru; near the Rondo Plateau and site of dinosaur excavations. Conservation Status: Data Deficient.

NATURAL HISTORY: Found in leaf litter and soft soil, specimens were located by uprooting grass. Habits presumably similar to other worm snakes.

MERKER'S WORM SNAKE   *Leptotyphlops merkeri*

IDENTIFICATION: A widespread black worm snake. The body is cylindrical. The tail is short, and thick, ending in a terminal spine. Midbody scale rows 14, number of dorsal scale rows 201–304, length/diameter ratio 40–86. Reaches a length of 23cm, but averages 12–20cm. Uniformly dark brown to blackish-brown or silvery grey above and below; occasionally some white patches on the chin and throat. Taxonomic Note: Originally regarded as a subspecies of *Leptotyphlops scutifrons*, a southern African species, but recent research shows this is a paraphyletic group. *L. scutifrons* itself might occur in our area.

Leptotyphlops merkeri

HABITAT AND DISTRIBUTION: Widespread and often common in the savanna of south-central Kenya and almost throughout eastern Tanzania, from sea level to 1,600m. Known localities include Nairobi, Naivasha, Arusha, Dodoma and Dar es Salaam. Elsewhere, may occur in northern Mozambique. Conservation Status: Least Concern.

NATURAL HISTORY: Nocturnal and burrowing, found in leaf litter and soft soil, or in rotting logs, in the Nairobi area often found under ground cover, especially after heavy

Leptotyphlops merkeri Merker's Worm Snake. Victor Wasonga, Wasini Island.

rain. May enter houses and has been found under carpets. Feeds largely on the eggs and pupae of ants, and the workers. Often eaten by other snakes, especially garter snakes, *Elapsoidea*. Usually lays two eggs, but up to three recorded. In central Kenya there is a curious legend about this snake, that it can spring into the nostrils of an unsuspecting observer and strangle them.

BLACK-TIPPED WORM SNAKE Leptotyphlops nigroterminus

IDENTIFICATION: A recently-described worm snake, light brown in colour. The body is cylindrical. The tail tapers slightly to a small terminal spine. Midbody scale rows 14, number of dorsal scale rows 228–300, higher values in females. The length/diameter ratio is 46–84. Reaches a length of about 19cm. In life appears silvery-brown, close examination indicates the dorsal scales have a brown or rufous brown centre and a silvery-grey edge, giving the snake a striped appearance, paler below, some individuals have a black tail tip, but the extent of this pigmentation varies, from obvious to a virtually invisible black patch on the tail tip.

Leptotyphlops nigroterminus

Leptotyphlops nigroterminus Black-tipped Worm Snake. Bill Branch, Kleins Camp.

HABITAT AND DISTRIBUTION: Moist savanna, at medium altitude, between 700–1,800m altitude. Two disjunct populations are known; from the northern Serengeti and Maasai Mara, and south-western Tanzania, from Tabora south and west through Katavi to northern Lake Rukwa and the southern Lake Tanganyika shore.

NATURAL HISTORY: Found in leaf litter and soft soil, shelters beneath ground cover. Habits presumably similar to other worm snakes.

PARKER'S WORM SNAKE Leptotyphlops parkeri

IDENTIFICATION: A rare pink worm snake, distinguished from all other worm snakes in our area by having 16 midbody scale rows. The body is cylindrical. The tail tapers slightly to a small blunt terminal cone. Midbody scale rows 16. The length/diameter ratio is 47–67.

The two measured specimens were 11.3cm and 16cm. In life appears transparent pink, the snout is slightly brownish, and the tiny eyes are obvious; preserved specimens are yellow-brown, lighter below.

HABITAT AND DISTRIBUTION: This distinctive little snake is known from only three specimens, from dry savanna. The type was from Degeh Bur, eastern Ethiopia, a second specimen was collected in Tsavo West, (probably from Kilaguni Lodge, at 850m altitude), a third specimen was recently collected at Sheikh Hussein, south-eastern Ethiopia, by Tomas Mazuch.

Leptotyphlops parkeri

NATURAL HISTORY: Unknown, presumably similar to other worm snakes. The Kilaguni snake was active on the ground at night in December, the rainy season.

Leptotyphlops parkeri Parker's Worm Snake. Left:Tomas Mazuch, Ethiopia. Right:Tomas Mazuch, Ethiopia.

PEMBA WORM SNAKE *Leptotyphlops pembae*

IDENTIFICATION: A distinctive black worm snake with a big eye, from the island of Pemba. The body is cylindrical. The tail tapers slightly to a small spike. Midbody scale rows 14, number of dorsal scale rows 247–269. The length/diameter ratio is 45–74. The largest specimen measured was 22cm, the smallest 12.4cm. In life (and preservative) it appears shiny black, and the eyes are relatively large and obvious; the underside of the neck, chin and throat are blotched grey-white to a greater or lesser extent.

HABITAT AND DISTRIBUTION: Endemic to the island of Pemba on the north Tanzanian coast, at sea level or just above, and appears to occur throughout the island, in both cultivation and remnant areas of natural forest. Conservation Status: Least Concern.

Leptotyphlops pembae

Leptotyphlops pembae Pemba Worm Snake. Frank Glaw, Pemba.

NATURAL HISTORY: Poorly known presumably similar to other worm snakes.

PITMAN'S WORM SNAKE *Leptotyphlops pitmani*

IDENTIFICATION: A distinctive black worm snake, recently described, from the northwest of our region. The body is cylindrical. The tail tapers slightly to a small spike. Midbody scale rows 14, number of dorsal scale rows 217–272. The length/diameter ratio is 40–86. The largest specimen measured was 18.5cm, In life it appears shiny black, and the eyes are relatively large, the underside has some scattered pale blotches; in preservative it appears brown.

HABITAT AND DISTRIBUTION: In savanna and savanna-forest mosaic, at altitudes between 700–1,800m altitude. Found in

Leptotyphlops pitmani

Leptotyphlops pitmani Pitman's Worm Snake. Bill Branch, Kilimanjaro International Airport.

western Kenya, around the Lake Victoria basin, much of Uganda, save the north and extreme south-west, north-west Rwanda and extreme north-west Tanzania; known localities include Kabarnet, Kericho, Kisumu, Sergoit, Jinja, Entebbe, Gulu and Bukoba; sporadic records further to the east include Ukerewe Island in Lake Victoria and Kilimanjaro International Airport.

NATURAL HISTORY: Poorly known, presumably similar to other worm snakes, living in holes, occasionally emerging during rainstorms.

TANA DELTA WORM SNAKE *Leptotyphlops tanae*

IDENTIFICATION: A distinctive, tiny, pink worm snake, recently described, in our area only known from the Tana River delta. The body is cylindrical. The tail tapers slightly to a thick spine. Midbody scale rows 14, number of dorsal scale rows 227–260. The length/diameter ratio is 46–87. The largest specimen measured was 10.2cm, the smallest 4.8cm. In life it appears pink, preserved specimens have turned light brown to tan, paler below.

HABITAT AND DISTRIBUTION: In East Africa only recorded from the Tana River Delta, at Ngatana, an abandoned village, near Wema, just northeast of Garsen, at 30m above sea level. Might be in northern Kenya as it is found in the Omo Delta in Ethiopia, also known from Somalia.

Leptotyphlops tanae

NATURAL HISTORY: Poorly known presumably similar to other worm snakes. Specimens were dug from black cotton soil, less than 15cm below the surface, and also found in termitaria.

SCORTECCI'S WORM SNAKE *Myriopholis braccianii*

IDENTIFICATION: A distinctive, tiny, pink worm snake, in our area only known from just east of the Tana River delta. The body is cylindrical. The tail tapers slightly to a thick spine. Midbody scale rows 14, number of dorsal scale rows 277–305. The length/diameter ratio is 63–111. The largest specimen measured was 14.4cm. In life it appears pink, preserved specimens have turned light brown to tan, pale yellow below.

HABITAT AND DISTRIBUTION: Low altitude savanna. In East Africa recorded from just east of the Tana River Delta, at Peccatoni, an abandoned village and lake 15km east of Witu. Might be in northern Kenya as it is found in southern Ethiopia and South Sudan, also known from Somalia.

Myriopholis braccianii

NATURAL HISTORY: Poorly known, presumably similar to other worm snakes.

IONIDES' WORM SNAKE Myriopholis ionidesi

IDENTIFICATION: A recently-described species, never knowingly seen alive. The body is cylindrical. The tail tapers to a thorn-like terminal spine. The eye is a tiny black dot. Midbody scale rows 14, number of dorsal scale rows 265–306. The length/diameter ratio is 60–75. The largest specimen measured was 15cm. Usually pink in life, but occasionally brown, preserved specimens are grey or tan, paler below, dorsal scales light-edged.

Myriopholis ionidesi

HABITAT AND DISTRIBUTION: Known from Miombo woodland at low altitude. In our area, known from Dar es Salaam and a cluster of localities in south-east coastal Tanzania (Liwale, Mbanja, Newala) and also northern Mozambique and Malawi.

NATURAL HISTORY: Poorly known, presumably similar to other worm snakes.

HOOK-SNOUTED WORM SNAKE Myriopholis macrorhyncha

IDENTIFICATION: A long thin pink worm snake with a distinctive hooked snout, visible in the hand, and a huge range. The body is cylindrical. The tail tapers to a conical terminal spine. The eye is a clearly visible black dot. Midbody scale rows 14, number of dorsal scale rows 315–492. The length/diameter ratio is 49–133. The largest East African specimen, from Porr Volcano, Lake Turkana, was 16cm, but larger specimens, up to 22cm, known from Egypt. Usually pink, pinky-white, purplish pink or red-brown in life, the scales sometimes light-edged to give a reticulate appearance.

Myriopholis macrorhyncha

HABITAT AND DISTRIBUTION: Dry savanna and semi-desert (and desert elsewhere), from sea level to around 1,300m altitude. Sporadically recorded from East Africa; from the low country west of Mt Meru, north and east around Kilimanjaro in Tanzania, around the mountain and north to Olorgesaile. It is known also from Sankuri, near Garissa on the Tana, from Wajir and a cluster of records around Lake Turkana. Elsewhere, north through northeast Africa to the Middle East, Turkey and Pakistan.

Myriopholis macrorhyncha Hook-snouted Worm Snake. Roberto Sindaco, Lake Turkana.

NATURAL HISTORY: A nocturnal fossorial species, although it may emerge in the twilight, living in cracks and holes and also in soft sand, can be found under rocks in the daytime. Feeds on ant larvae and other small invertebrates.

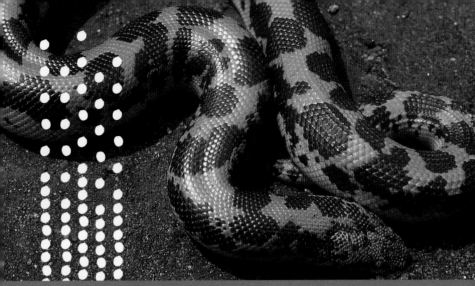

Eryx colubrinus
Kenya Sand Boa
Stephen Spawls

SUB-ORDER **SERPENTES: SNAKES**
SUPERFAMILY **BOOIDAE**
BOAS

A superfamily of around 60 live-bearing boas, it includes some of the world's largest snakes (the anacondas) but also some very small species. The subfamily Erycinae comprises the small, live-bearing sand boas of the genus *Eryx*.

SAND BOAS *ERYX*

A genus of 12 small, harmless specialised boas. They have blunt heads and tails and narrow ventral scales and small claws on either side of the vent, like pythons. They occur in the northern half of Africa, south-western Europe and the Middle East, extending east to India. They are adapted to arid environments and spend most of their lives below the surface, either in burrows or buried in sand or sandy soil. Most have a dorsal pattern of irregular darker blotches on a light background. They are greatly feared in part of their range, possibly due to confusion with desert vipers, some of which have a vaguely similar pattern. The generic name, *Eryx*, is an abbreviation of *Ereyxis*, meaning 'vomiting' in Greek, in reference to the snake's defence mechanism if seized. Five species occur in Africa, one of which is found in East Africa, the Kenya Sand Boa *Eryx colubrinus*.

KENYA SAND BOA *Eryx colubrinus*

IDENTIFICATION: A small blotched snake with a bullet-shaped head. The tiny eyes, with a yellow iris and slim vertical pupil are set well forward, on the angle between the top and side of the head. The top of the head is covered with small scales; this is unusual in a harmless snake. The enlarged rostral scale is broad with an angular edge, for shoving through sand. There is no obvious neck. The body is short and stout, tail short (less than 10 % of total length) and rounded, but with a pointed tip. The scales on the tail and final fifth of the body are strongly keeled, unlike the other body scales, this may aid locomotion or have a defensive purpose. Midbody scales in 44–59 rows (usually less than 53 in Kenyan snakes), ventrals narrow, 165–205, subcaudals single, 19–28, anal scale entire. Maximum size about 1m, average 30–60cm, hatchlings 15–19cm. Ground colour usually grey, yellow or orange (depending on the local soil colour), with sub-circular brown, black or grey blotches, belly immaculate white, cream or pinkish. A uniform brown, rufous or yellow-brown form exists, originally given the specific (and then sub-specific) name *rufescens*, but it appears to be simply a colour morph.

Eryx colubrinus

HABITAT AND DISTRIBUTION: Near desert, semi-desert and dry savanna, from sea level to about 1,500m altitude. Widely distributed over most of dry north and eastern Kenya, although records sporadic in places. South through Ukambani and Tsavo, west through the low country south of Nairobi to Magadi; Loveridge listed it from Nairobi but this was based on a Sultan Hamud record. A single costal record, from Likoni. Occurs in extreme north-eastern Tanzania, known localities include Mkomazi, Kahe, Kilimanjaro Interna-

Eryx colubrinus Kenya Sand Boa. Stephen Spawls, Kilimanjaro International Airport.

Eryx colubrinus Kenya Sand Boa. Left: Stephen Spawls, Kitui. Inset right: Stephen Spawls, Wajir, rufous phase.

tional Airport, Naberera, Tarangire; isolated records further south from Dodoma and Ruaha National Park. Just reaches Uganda in the northeast, at Kidepo National Park and Kaabong area. Elsewhere, north to southern Egypt, north-east to Somali, west to Niger across the Sahel. Conservation Status: Not assessed by IUCN; CITES Appendix II; but widespread in arid areas and probably under no threat; but popular in the pet trade.

NATURAL HISTORY: A burrowing snake, probably nocturnal but its behaviour is unknown, might hunt by day in burrows. It may emerge to bask, especially in the early morning. Lives in holes or buried in sand, may shelter, partially buried, under logs or rocks, but prefers areas with some vegetation rather than open bare soil or sand. In area where the soil is mostly hard (such as the low eastern foothills of Mt Kenya) it will live in sandy riverbeds, which must put it at risk in flash floods. Hunts from ambush, waiting concealed until prey passes, it must be able to detect the approach of small animals as it will suddenly strike from below the sand with remarkable accuracy. Usually good-natured, when initially handled it emits a foul-smelling fluid from glands near the cloaca and may coil tightly around the fingers of the restraining hand. Occasional individuals bite. It gives live birth, usually from 4–20 young (a long-term captive had 33 young at once). An Ngomeni (Kitui) specimen had eight babies and seven infertile eggs in mid-November; a Voi female was captured in a hole with seven babies in late April. The juveniles eat mostly lizards (specially *Heliobolus* and *Latastia*), large adults take rodents as well, and even birds, indicating how well they are concealed and how rapidly they strike. A Lake Magadi snake was heard to fall and was then seen swallowing a bird during the day, it might have climbed a tree to seize it. Sand boas are widely feared in Kenya on account of their 'dangerous' appearance; their Somali name is Apris, legend has it that if bitten, you take seven steps and die.

Python sebae
Central African Rock Python
Anthony Childs, Watamu.

SUB-ORDER **SERPENTES**
SUPERFAMILY **PYTHONOIDEA**
PYTHONS

A family of just over 40 species, in three families; most are pythons but the group includes two strange sunbeam snakes and a Mexican burrowing python. The family Pythonidae, true pythons, found in Australasia, Asia and Africa, includes some of the world's largest snakes.

PYTHONS *PYTHON*

A genus of ten relatively large constricting Old World snakes, six occur in Asia, four in Africa, three in our area.

KEY TO THE EAST AFRICAN MEMBERS OF THE GENUS *PYTHON*

1a Maximum size less than 2m, sensory pits in first to fourth upper labial scales, head pear-shaped......*Python regius*, Royal Python [p. 382]
1b Maximum size up to 5m or more, sensory pits only in first and second upper labial scales, head not pear-shaped......2

2a Central scales on top of head fragmented and small, pattern often dark or dull, a small dark patch behind and in front of the eye......*Python natalensis*, Southern African Rock Python [p. 379]
2b Central scales on top of head large, pattern often vivid, a large dark patch behind and in front of the eye......*Python sebae*, Central African Rock Python [p. 380]

SOUTHERN AFRICAN ROCK PYTHON   *Python natalensis*

IDENTIFICATION: Unmistakeable, a huge thickset snake (rock pythons are the largest snakes in Africa), with sub-triangular head. Snout rounded, eye fairly large, pupil vertical (hard to see). The top of the head is covered with small to medium smooth scales. There are two heat-sensitive pits at the front of the upper labials and three smaller such pits on the lower labials. The body is stout, cylindrical in juveniles, slightly depressed in big adults. Tail fairly short, 9–10% of total length in females, 12–16% in males. Scales smooth, in 78–99 rows at midbody, there are 260–291 narrow ventrals and 63–84 subcaudals. There are small claw-like spurs on either side of the anal scale (vestigial legs) which are larger in males; legend in East Africa says these are used to block the nostrils of the prey being constricted. Maximum size about 5.5m (anecdotal reports of bigger animals exist), average 2.8–4m (in general females are larger than males), hatchlings 45–60cm. Colour a mixture of brown, tan, yellow and grey blotches. The tail sometimes has a light central stripe. White below, with irregular dark speckling. Big adults may look almost black. Freshly sloughed specimens are iridescent. Taxonomic Notes: In 1999, the

Python natalensis

Python natalensis Southern African Rock Python. Top: Michele Menegon, Selous Game Reserve. Bottom left: David Moyer, Iringa. Bottom right: Stephen Spawls, Kitui.

African Rock Python was split into two full species, *Python sebae* and *P. natalensis*, on the basis of some physical differences (including head scales and head patterns). This has been generally accepted, but intermediates exist; molecular evidence may clarify the situation.

HABITAT AND DISTRIBUTION: Coastal thicket, grassland, moist savanna and woodland, often in the vicinity of water or rocky hills, known from semi-desert in part of their range, also thick woodland and mid-altitude forest. Found from sea level to 2,200m altitude but rarely above 1,800m. They are most abundant near low altitude rivers, lakes and swamps; they like to drink, but may be in quite dry country, e.g. the slopes of Mt Suswa. Occurs from the lower north-eastern and eastern slopes of Mt Kenya, south to Kitui area, south-west through the rift valley and the Mara, still common in Nairobi National Park. South and east of Kitui through Ukambani and Tsavo to Kilibasi and the lower Galana River. Occurs across most of Tanzania and southern Burundi, but not in the Lake Victoria basin, and absent from the Tanzanian coast, except the extreme south. Elsewhere, south to northern Namibia and eastern South Africa. Conservation Status: Not assessed by IUCN. On CITES Appendix II. Hunted for their skins and meat, and being large do not co-exist with humans, but still common in many national parks and around large water bodies, for example Lake Naivasha.

NATURAL HISTORY: As for the Central African Rock Python.

CENTRAL AFRICAN ROCK PYTHON *Python sebae*

IDENTIFICATION: A huge, thickset snake. Most details as for the Southern African Rock Python but differs as follows; top of head covered with medium to large scales, scales in 76–99 rows at midbody, 265–283 narrow ventrals, 67–76 subcaudal scales. Maximum size about 6.5m, although there are anecdotal reports of larger specimens up to 9.8m. This form is more brightly coloured than the Southern African Rock Python, usually black, yellow and warm brown, with big, light brown, yellow-edged blotches on the back and flanks. There is a dark arrowhead on the crown, a yellow stripe runs through the eye and in front of and behind the eye is a broad dark patch (narrow in the Southern African Rock Python). The underside of the tail is white, and this is sometimes curled up in the air as a distraction technique, attracting predators such as leopards away from the head.

Python sebae

Python sebae Central African Rock Python. Left: Bio-Ken archive, Galana River. Right: Anthony Childs, Watamu.

HABITAT AND DISTRIBUTION: Similar to the Southern African Rock Python. Coastal thicket, grassland, moist savanna, woodland, forest and along rivers and around water bodies in semi-desert. Occurs from sea level to 2,200m altitude, although most common below 1,500m. Found in coastal Tanzania, from Dar es Salaam, north to Kenya (also on Zanzibar), the Kenya coast, up the Tana River to Kindaruma and Thika area. Also on the Daua River in north-east Kenya, and the Omo. Occurs throughout Uganda (except the north-east), all around the Lake Victoria basin, western Kenya and eastern Rwanda and Burundi. Occurs in the Kilombero Valley and parts of north-central Tanzania, where it overlaps with the southern African rock python, although the exact area of the overlap is unclear, and there are 'hybrid' specimens known. Conservation Status: Not assessed by IUCN. On CITES Appendix II.

NATURAL HISTORY: Usually nocturnal, but will bask and hunt opportunistically during the day. Mostly terrestrial, but juveniles will climb trees. Often aquatic, and adults may spend a lot of time in water, hunting and feeding there (they are excellent swimmers), but emerging to bask. When inactive, shelters in holes (especially warthog and porcupine burrows), in thickets, reedbeds, up a tree, in a rock fissure or under water. Often curls up in a heap of coils, with the head resting on top. Juveniles are active hunters, climbing trees to check nests and holes, prowling about, swimming around looking for prey, big adults tend to hunt from ambush, waiting quietly in a coiled strike position beside a game trail or under a bush until prey passes. Medium-sized juveniles (1.5–2.5m) in Kenya were observed to become active near dusk. Pythons will try to escape if confronted, but if cornered they will strike, and they have very sharp teeth, causing deep wounds and cuts that require stitching. Many captive specimens remain permanently bad tempered and are unsafe to handle (which has, perhaps fortunately, made them unpopular in the pet trade). From 16–100 tennis-ball sized eggs are laid in a thick bush, rock fissure or deep moist hole; an Athi River female laid eggs in October. The female coils around the eggs to protect them. until just before they hatch, leaving only to drink; during this time she elevates her body temperature by a mixture of basking and an unexplained mechanism. Incubation varies from 2–3 months; hatchlings were collected in July near Embu, in August in the Kerio Valley, in May near Arusha and in January near Newala, southern Tanzania. Around Lake Victoria, Central African Rock Pythons are reported to mate and lay eggs throughout the year. There is some evidence that the neonates wait underground until all have hatched, then all emerge at once. Growth rates can be rapid, juveniles reaching 1.4m in a year. Captive specimens have lived over 27 years and fasted over two and half years. They eat a range of prey, from small mammals, birds and frogs up to medium-sized antelope and domestic stock; other recorded prey items include crocodiles, fish, lizards, geese eggs, domestic dogs and poultry. The prey is usually seized from ambush by big adults, but juveniles actively prowl, and in rural East Africa often found after they have entered a chicken run, eaten something large and been unable to get out. Often blamed for killing humans, there are a few documented cases of adults being killed but not eaten and of small children being consumed. Small pythons may be attacked by snake-eating snakes, birds and mammalian carnivores, but big adults have few enemies save leopards, lions, hyenas, crocodiles and humans; in southern Africa some leopards become specialised python eaters. Pythons may be killed for their skin, or their perceived danger to stock. But they do a lot of good, they eat rodents. In some areas they are venerated, a common belief is that if they are killed, rain will not fall; in places in Rwanda the souls of deceased kings were said to live on in pythons.

ROYAL PYTHON Python regius

IDENTIFICATION: A small python with a pear-shaped head, thin neck and stout body. The iris appears dark but is yellow, the pupil is vertical. There are small to medium fragmented scales on the crown. On the front upper labial scales are three or four big, obvious heat-sensitive pits. The body is fairly stout and subtriangular, with a prominent vertebral ridge. The tail is short, 7–10% of total length. Scales smooth, 53–63 rows at midbody, 191–207 narrow ventrals, 28–47 paired subcaudals, anal scale usually entire. Maximum size about 1.5m , possibly larger, average 80cm to 1.2m, hatchlings 35–42 cm. Black or blackish brown, covered with brown or fawn (rarely orange or yellow) sub-circular, light-edged blotches, sometime black spots within these blotches. The head is dark at the front, two light lines run through the top of the eye to the back of the head. The ventral are white in the centre, with black speckling and blotches.

Python regius

Python regius Royal Python.
Barry Hughes, balling behaviour.

HABITAT AND DISTRIBUTION: Grasslands, dry and moist savanna, not in forest (although it will colonise clearings) or very dry country, in Burkina Faso it is absent from regions with less than 600mm annual rainfall. In our area, occurs in north-west Uganda (localities include Murchison Falls, between Gulu and Kitgum, Laufori near Moyo in the far north-west and Hakitenya, north-west of Fort Portal in the Semliki Valley) usually below 1,200m altitude in a high (more than 1,000mm annually) rainfall belt. Elsewhere, north into central Sudan, west to Senegal through the Guinea and Sudan savanna. Conservation Status: IUCN Least Concern. Probably not under threat in Uganda but exploited in West Africa, although the trade is being monitored; it is in demand as a docile and attractive pet, although it can be a finicky feeder.

NATURAL HISTORY: A slow-moving, nocturnal snake, although there are reports of them basking in Ghana. Hunts on the ground and in small trees, looking for prey in tree holes and down burrows. When inactive, it hides in these burrows, tree cracks, hollow tree trunks or rock fissures. May even hide in buildings, under thatched roofs, etc. In West Africa it is mostly active in the wet season, aestivating in a suitable hole in the dry. A good natured snake, if threatened in the open it will curl up into a tight ball with the head totally or partially concealed, hence the alternative names of 'ball python' and, in Sierra Leone, the 'shame snake'. In captivity, this defence mechanism is soon abandoned. A few specimens will bite. Mating occurs in December and January in West Africa, a clutch of around 4–10 eggs (range 1–18) are laid in February to April in a deep moist hole. The eggs are relatively large, roughly 5–6cm wide and 8–9cm long, incubation period is 2–3 months. The diet is mostly ground-dwelling rodents, especially gerbils, they take other rodents and sometimes birds that they can catch in their holes, such as barbets, woodpeckers and parrots. In parts of West Africa this snake is respected and venerated, and may be kept in house, in others it is exploited for its skin and, as mentioned, for the pet trade; some people will also eat Royal Pythons.

Platyceps brevis smithi
Smith's Racer
Stephen Spawls

SUB-ORDER SERPENTES
SUPERFAMILY COLUBROIDEA
ORDINARY SNAKES

A huge, successful family of medium-sized 'ordinary' snakes, widespread on all continents except Antarctica, it includes most snakes save the boas, pythons, blind and worm snakes and a few other specialised animals. It is a monophyletic group; all are descended from a single ancestor. Within this superfamily there are ten families and 14 subfamilies, this has complicated the construction of keys. Hence we have devised two keys; breaking the East African members of the group down in stages, firstly to families and some subfamilies, secondly; a key to the genera within three large families, Colubridae, Lamprophiidae and Natricidae. This is unavoidable; these three families contain such a diversity of snakes that no key (other than a molecular one) will split them. Note that two subfamilies of the Lamprophiidae, the Aparallactinae and Atractaspidinae, split off separately in the first key and thus do not reappear in the second key. However, we have grouped the genera into families for the individual descriptions, and each genus retains a specific key where needed.

KEY TO THE EAST AFRICAN FAMILIES AND SOME SUBFAMILIES OF THE *COLUBROIDEA*

1a One or more pairs of enlarged caniculate or tubular poison fangs present in the front of the upper jaw......2
1b No enlarged caniculate or tubular poison fangs present in the front of the upper jaw, fangs when present usually grooved and set below the eye......4
2a Poison fangs large, moveable, folded back in a sheath when not in use......3
2b Poison fangs relatively small and immobile, not folded back when not in use......Family Elapidae, cobras, mambas and allies [p. 542]

3a Eye very small, dark-coloured, no loreal scale, pupil round......Subfamily Atractaspidinae, burrowing asps [p. 471]
3b Eye relatively large, pupil usually vertically elliptic, (but round in *Causus*, night adders), loreal scale present...... Family Viperidae, vipers [p. 567]
4a No loreal scale present, grooved rear fangs nearly always present......Subfamily Aparallactinae, African burrowing snakes [p. 450]
4b Loreal scale present, grooved rear fangs present or absent......Families Colubridae/Lamprophiidae/Natricidae, colubrids, African snakes/house snakes/water snakes (excluding the Aparallactinae and Atractaspidinae). [p. 387]

KEY TO THE GENERA IN THE FAMILIES *LAMPROPHIIDAE*, *COLUBRIDAE* AND *NATRICIDAE* (excluding the Aparallactinae and Atractaspidinae).

Over 110 species of East African snake, in over 30 genera, belong in these three families, and this key should successfully enable the identification of each genus. Most of these snakes are harmless or, if they have poison fangs (the 'rear-fanged snakes') have relatively weak venom, but there are a few dangerous species in the group, in particular the Boomslang *Dispholidus typus* and three species of vine snake (*Thelotornis*). Be aware also that some large rear-fanged snakes, in particular the tree snakes of the genus *Toxicodryas* and the bigger sand snakes *Psammophis* have been shown to have a fairly toxic venom.

1a No enlarged grooved poison fangs in the upper jaw......2
1b One or more pairs of enlarged grooved poison fangs in the upper jaw......20

2a Dorsal scales smooth (sometimes feebly keeled in *Thrasops*, black tree snakes)......3
2b Dorsal scales keeled......16

3a Nostril pierced in a divided or semidivided nasal......4
3b Nostril pierced in an entire nasal......14

4a Cloacal shield entire, dorsum never bright green 5
4b Cloacal shield usually divided, if entire then dorsum usually bright green in life......7

5a Midbody scale rows 19-33 6
5b Midbody scale rows usually 15-17 (19-21 only in *Prosymna pitmani*)......33

6a Dorsal scales in 19-25 rows at midbody, ventral 152-178, pupil round...... *Lycodonomorphus*, water snakes [p. 389]
6b Dorsal scales in 23-37 rows at midbody, ventral 186-237, pupil vertical...... *Boaedon*, house snakes [p. 391]

7a Snout with a sharp horizontal edge, labials excluded from the eye by suboculars, parietals fragmented......*Scaphiophis*, hook-nosed snakes [p. 533]
7b Snout without a sharp horizontal edge, one, two or three labials enter orbit, parietal not fragmented......8

8a Internasal shield entering nostril, midbody scale rows 21-31......9
8b Internasal shield not entering nostril, midbody scale rows do not exceed 21......10

9a Snout pointed with vertical sides, midbody scales 25-31, body stout, adults more than 90cm......*Pseudaspis cana*, Mole Snake [p. 412]
9b Snout rounded, midbody scale rows 21-25, body slim, adults usually less than 80cm......*Platyceps*, racers [p. 478]

10a Dorsal scales in 13-15 rows at midbody, reducing to 11 posteriorly, usually bright green or yellowy-green in life......15
10b Dorsal scales in 15-21 rows at midbody, if 15 then not reduced posteriorly, not bright green in life......11

11a Body laterally compressed, lateral scales oblique, subcaudals 130-155, adults black......*Thrasops*, black tree snakes [p. 500]
11b Body subcylindrical, lateral scales not oblique, subcaudal usually less than 130......12

12a Ventrals not exceeding 160, dorsal scales in 15-19 rows at midbody, mandibular teeth subequal or smallest posteriorly......13
12b Ventral exceeding 160, dorsal scales in 19-21 rows at midbody, mandibular teeth largest anteriorly...... *Meizodon*, smooth/crowned/semi-ornate snakes [p. 480]

13a Two anterior temporals, subcaudals usually more than 85......*Grayia*, water snakes [p. 536]
13b One anterior temporal, subcaudals

	usually less than 85......*Natriciteres*, marsh-snakes [p. 538]
14a	Pupil round, loreal shield (usually) absent......*Duberria lutrix*, Slug-eater [p. 418]
14b	Pupil vertically elliptic to subelliptic, loreal shield present......32
15a	Body laterally compressed, scales in the vertebral row enlarged, lateral scales narrow, oblique, a pair of large occipital shields.....*Rhamnophis aethiopissa*, Large-eyed Green Tree Snake [p. 502]
15b	Body subcylindrical, scales in the vertebral row not enlarged, lateral scales not narrow or oblique, no enlarged occipital shields......*Philothamnus*, green snakes
16a	Body longitudinally striped in red and black......*Bothropthalmus lineatus*, Red and Black Striped Snake [p. 388]
16b	Body not longitudinally striped in red and black......17
17a	Dorsal scales in 21–27 midbody rows, loreal absent, nostril moderate and pierced in a semi-divided nasal shield, teeth few and rudimentary......*Dasypeltis*, egg-eaters [p. 513]
17b	Dorsal scales in 15–21 rows at midbody, loreal present, nostril large and pierced between two nasal shields, teeth numerous and distinct......18
18a	Body vivid green in life, with or without black lines......*Hapsidophrys*, forest green snakes [p. 497]
18b	Body not green in life......19
19a	Vertebral scales enlarged and bicarinate, eye pupil small and elliptical or subcircular......*Gonionotophis*, file snakes [p. 405]
19b	Vertebral scales not enlarged and with a single weak keel, eye pupil large and rounded or pear-shaped......*Xyelodontophis uluguruensis*, Uluguru Dagger-tooth Snake [p. 509]
20a	Pupil vertically elliptic, head much broader than neck......21
20b	Pupil round or horizontal, head hardly broader than neck......25
21a	Loreal entering orbit, marbled in red-brown and white or yellow above......*Dipsadoboa*, tree snakes (part) [p. 527]
21b	Loreal excluded from orbit by preocular, not marbled red-brown and white or yellow......22
22a	Ventrals 141–183......*Crotaphopeltis*, herald, yellow-flanked and white lipped snakes [p. 523]
22b	Ventrals 195–270......23
23a	Subcaudals more than 100......*Toxicodryas* (Boiga), broad-headed tree-snakes [p. 510]
23b	Subcaudals less than 100......24
24a	Cloacal shield usually entire, a single anterior temporal......*Dipsadoboa*, tree-snakes (part) [p. 527]
24b	Cloacal shield usually divided, two anterior temporals......*Telescopus*, large-eyed and tiger snakes [p. 520]
25a	Pupil horizontal, key-hole or dumb-bell shaped......*Thelotornis*, vine snakes [p. 505]
25b	Pupil round.....26
26a	Dorsal scales keeled...... *Dispholidus typus*, Boomslang [p. 503]
26b	Dorsal scales smooth......27
27a	Small, usually dark in colour, never with pale stripes, tail less than 17% of total length...... *Buhoma*, forest-snakes [p.414]
27b	Small to large, often light in colour with pale stripes, tail usually more than 17% of total length......28
28a	Rostral large, usually projecting, snout hooked in profile and body unstriped...... *Rhamphiophis*, beaked-snakes [p. 446]
28b	Rostral usually normal and snout rounded, if hooked then body striped......29
29a	Nostril pierced in an entire nasal shield......*Hemirhagerrhis*, bark snakes [p. 426]
29b	Nostril pierced between two or more shields......30
30a	Maxillary teeth interrupted below anterior border of eye by two much enlarged, fang-like teeth......*Psammophis*, sand snakes (part) [p. 429]
30b	Maxillary teeth subequal in size and continuous up to the interspace separating the from the posterior pair of enlarged fangs......31
31a	Nostril pierced between two nasal shields only, tail long with more than 80 subcaudals......*Psammophis lineatus*, Striped Olympic/Sand Snake [p. 433]
31b	Nostril pierced between two nasals and an internasal shield, tail moderate, less than 80 subcaudals......*Psammophylax*, skaapstekers [p. 442]
32a	Eight upper labials,......*Lycophidion*, wolf snakes [p. 395]
32b	Seven upper labials......*Chaemalycus*

fasciatus, Ituri Banded-snake [p. 404]

33a Snout strongly depressed, projecting and with an angular horizontal edge...... *Prosymna*, shovel-snouts [p. 419]

33b Snout rounded......*Hormonotus modestus*, Yellow Forest-snake [p. 404]

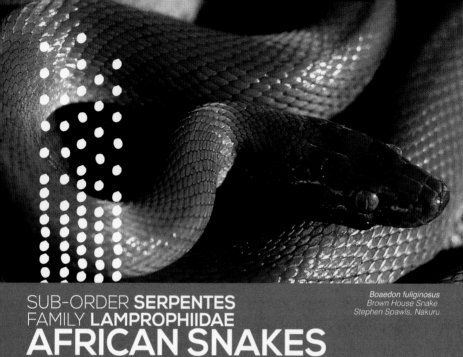

Boaedon fuliginosus
Brown House Snake.
Stephen Spawls, Nakuru

SUB-ORDER SERPENTES
FAMILY LAMPROPHIIDAE
AFRICAN SNAKES
(HOUSE SNAKES AND ALLIES)

A family of African and Madagascan snakes, with seven subfamilies and over 50 genera; over 30 are mainland African, and representatives of all seven subfamilies occur in East Africa. The only members of the group that are dangerous are the front-fanged burrowing asps *Atractaspis*; some of the others are rear-fanged but their venom does not cause any serious problems to humans. Our keys have not split the family into subfamilies, as the key above distinguishes between all the genera in the family; however each subfamily receives a brief description in the appropriate place in the following text.

SUBFAMILY LAMPROPHIINAE

A subfamily of 12 genera, and over 70 species, almost all restricted to Africa (a few species have rafted to the Arabian Peninsula); seven genera and 25 species occur in East Africa. All are harmless, without fangs.

RED AND BLACK STRIPED SNAKES
BOTHROPTHALMUS

An African genus of two species, one from our area, a spectacular harmless, striped, forest-dwelling snake.

RED AND BLACK STRIPED SNAKE Bothropthalmus lineatus

IDENTIFICATION: A medium-sized, vividly striped, harmless colubrid snake. The head is slightly distinct from the neck and a narrow groove runs from the nostril to the eye. The iris is yellow, orange or brown, the pupil is round but slightly elongate vertically. The body is cylindrical, tail fairly long, 19–24% of total length in males, 16–20% in females. The body scales are keeled, in 23 rows at midbody, ventrals 181–210, (usually 190 or above in Ugandan snakes), subcaudals paired, 61–83 rows, anal scale entire. Maximum size about 1.3m, but this is unusually big, average 40–90cm, hatchling size unknown, probably about 20cm. Colour distinctive; there are five fine, longitudinal red stripes, about one scale wide, on a black background. Juveniles have a white or yellow head, with two or three irregular black dashes on top and on behind the eye, the lower lip scales are black-edged on the top margin, chin white. Adults have a brown or fawn head, but retain the black marking. Red below, each ventral scale edged darker red-brown.

Bothropthalmus lineatus

HABITAT AND DISTRIBUTION: Forest and forest islands, from 700–2,300m altitude. Sporadic records from the northern Lake Victoria shore in Uganda, in forest patches, and along the Albertine Rift, from Semliki along the base of the Rwenzoris, down to Rukungiri and Kisoro, Ruhengeri; and a somewhat dubious record from Gisenyi in north-west Rwanda. Also known from the Budongo Forest. Probably more widespread in Uganda. Elsewhere, south-west to northern Angola, west to Guinea.

NATURAL HISTORY: A terrestrial, nocturnal snake, living in forest, often near water. Hides in holes, leaf litter and under or in rotting logs. Might be fairly aquatic; Congo and Ghanaian specimens have been found swimming in streams. Prowls around the forest floor at night. Congo specimens are largely placid, rarely biting, although said to be occasionally irascible elsewhere. The bright colours are probably aposematic (i.e. warning colours), another forest snake not found in our area (*Polemon acanthius*, Red-striped Snake-eater) has very similar coloration, neither are dangerous. They lay eggs, clutches of five eggs, roughly

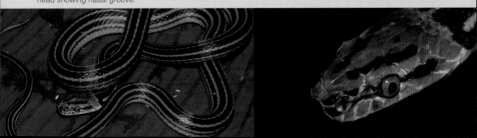

Bothrophthalmus lineatus Red and Black Striped Snake. Left: Konrad Mebert, eastern Congo. Right: Konrad Mebert, head showing nasal groove.

4×2cm, were recorded in two specimens from the DR Congo. However, two Nigerian specimens had 23 and 29 developing eggs in their oviducts. Known prey items include small forest mammals such as shrews and mice.

WATER SNAKES *LYCODONOMORPHUS*

A genus of eight species of small- to medium-sized, harmless, predominantly brown water snakes, found in central and southern Africa. They eat fish and amphibians; some species seem almost entirely aquatic. They have unusual, small circular or elliptic pupils. Two species occur in our area, one of which is endemic to Lake Tanganyika.

KEY TO EAST AFRICAN MEMBERS OF THE GENUS *LYCODONOMORPHUS*

1a Midbody scales in 19 rows, Lake Tanganyika only......*Lycodonomorphus bicolor*, Lake Tanganyika Water Snake [p. 389]

1b Midbody scales in 23 to 25 rows, not in Lake Tanganyika......*Lycodonomorphus whytei*, Whyte's Water Snake [p. 390]

LAKE TANGANYIKA WATER SNAKE *Lycodonomorphus bicolor*

IDENTIFICATION: A medium-sized water snake, in Lake Tanganyika only. Head quite short, eye small, pupil small and round, a useful field character. Body cylindrical, tail fairly short, 15–20% of total length. The scales are smooth, in 23–25 midbody rows, subcaudals paired. Maximum size about 70cm, average 40–60cm, hatchling size unknown. Grey or brown above, sometimes yellow-brown, lower dorsal scales white, cream or yellow, often yellow on the lips, ventrals cream or yellow.

HABITAT AND DISTRIBUTION: Endemic to Lake Tanganyika, at 770m altitude, might venture into the larger rivers flowing into it. Conservation Status: IUCN Least Concern.

Lycodonomorphus bicolor

Lycodonomorphus bicolor Lake Tanganyika Water Snake. Michele Menegon, Lake Tanganyika.

NATURAL HISTORY: Largely aquatic and nocturnal, a study of these snakes in Lake Tanganyika found they were mostly active at night, from dusk onwards; during the day they rested under submerged rocks in the shallows, occasionally raising their heads to breathe. They were affected by the lunar cycle, most activity occurred around the new moon, on moonlit nights activity was reduced. The population density was estimated to lie between 9,000–38,000 snakes per km^2 in suitable habitat. There seemed to be no clear breeding season, gravid females were caught during all months of the study, October to March; most females contained four eggs, with a range of 4–8. One specimen was attacked by a crab.

WHYTE'S WATER SNAKE *Lycodonomorphus whytii*

IDENTIFICATION: A medium-sized brownish water snake. Head fairly short, eye medium-sized, pupil curiously elliptic. The tongue is red with a dark tip. Body cylindrical, tail fairly short, around 12–15% of total length. Body scales smooth, in 19 rows at midbody, 157–173 ventrals, subcaudals 37–47 (females, male data unknown). Most southern African specimens have scales with two apical pits, but these are apparently absent in the Tanzanian specimens. Maximum size 71cm, average 35–55cm, hatchling size unknown. Olive-brown, olive-green, yellow-brown or black, belly orange-yellow with dark spots towards the rear, there is dark median stripe on the underside of the tail. The chin is cream, some specimens have dark spots on the mental and first two pairs of upper labial scales.

Lycodonomorphus whytii

HABITAT AND DISTRIBUTION: Low- to medium-altitude moist savanna and woodland; have been found on flood plains in the vicinity of swamps, rivers and lakes, but also in forest clearings. A handful of records from southern Tanzania, including the vicinity of Mt Rungwe, Livingstone Mountains, Lulanda in the southern Udzungwa and Songea, at altitudes between 1,000–1,900m, but might be more widespread in the vicinity of Lake Malawi. Elsewhere, south to Mozambique and northern South Africa. Conservation Status: IUCN Least Concern

NATURAL HISTORY: Poorly known, might be nocturnal or diurnal. They are often found in the vicinity of water, taking cover in waterside holes, under rocks, logs, etc. They lay eggs, but no clutch details known. Diet: amphibians (one Songea specimen had eaten a large frog, probably an *Amietia*); and possibly fish. The original specimen, interestingly, was named after a Mr Alex Whyte, but still spelt *whytii* by Boulenger.

Lycodonomorphus whytii Whyte's Water Snake. Left: Michele Menegon, Livingstone Mountains. Right: Michele Menegon, head close-up.

HOUSE SNAKES *BOAEDON*

A genus of eight medium-sized (60cm–1.3m) harmless African colubrid snakes. House snake taxonomy is complicated; they have been variously placed, *inter alia*, in the genera *Boaedon* and *Lamprophis*, now their DNA indicates that eight house snake species should be in *Boaedon*, seven in *Lamprophis*. Three species of *Boaedon* occur in our area (one recently described). The huge range and wide colour/pattern variation shown by *Boaedon fuliginosus* has, over the years, prompted several attempts to split what is, almost certainly, a species complex. None have met with unequivocal acceptance. However, a recent revision has described five new species of *Boaedon* from central and West Africa, and some of these may well extend into East Africa. The paper, by Jean-Francois Trape and Oleg Mediannikov, also limits the name *Boaedon fuliginosus* to the uniformly dark West African snakes that have no visible head stripes. East African museum specimens will now need to be re-examined to establish their status. For the time being, pending an East African review, we have retained the name *Boaedon fuliginosus* for all East African house snake with paired subcaudals. However, it is worth noting that specimens fitting the description of one of the new forms, *Boaedon subflavus*, a yellowish form that sometimes does not have head stripes, can be found in East Africa. In addition, specimens fitting the description of the new species *Boaedon paralineatus*, with an obvious line extending back from the head some distance down the flanks, do occur in dry areas of East Africa.

KEY TO EAST AFRICAN MEMBERS OF THE GENUS *BOAEDON*

1a	Subcaudals single......2		Olive House Snake [p. 393]
1b	Subcaudals paired......*Boaedon fuliginosus*, Brown House Snake [p. 391]	2b	Ventral 200–226, three upper labials enter the orbit......*Boaedon radfordi*, Radford's House Snake [p. 394]
2a	Ventrals 185–242, two upper labials enter the orbit......*Boaedon olivaceus*,		

BROWN HOUSE SNAKE *Boaedon fuliginosus*

IDENTIFICATION: Probably the most common and adaptable snake in East Africa; a medium-sized, fairly slim, harmless colubrid. The head is sub-triangular, like the head of a rock python. The eye is medium-sized (prominent in some arid country specimens, leading to suggestions they are a separate species), pupil vertical, iris brown or yellow. Body cylindrical, the tail is 10–15% of total length in females, 15–18% in males. The scales are smooth, in 23–35 midbody rows, ventrals 181–240 (usually more than 201 in East African snakes), subcaudals paired, 42–74, anal scale undivided. Maximum size about 1.2m, average 40–70cm, hatchlings 17–25cm. Colour very variable, the most common colour is some

Boaedon fuliginosus

Boaedon fuliginosus Brown House Snake. Top left: Stephen Spawls, Dodoma. Top right: Dennis Rodder, Pemba. Middle left: Stephen Spawls, Bagamoyo. Middle right: Stephen Spawls, Watamu. Bottom: Stephen Spawls, Kilimanjaro International Airport.

shade of brown, and there is usually a pair of pale lines on each side of the head, one through the eye, the other along the cheek, these lines may extend along the body to a greater or lesser extent. Large adults are often darker than juveniles. Other colour variants include olive grey, light yellow, orange, grey and black. Often several colour forms may occur in the same area. Some coastal specimens, especially juveniles, are attractively spotted pink and red-brown. The head stripes and patterns often fade with age. Belly white or cream.

HABITAT AND DISTRIBUTION: Found in a range of habitats; semi-desert, dry and moist savanna, woodland and forest, from sea level to about 2,400m, possibly higher, although rare above 2,000m. In our area, known from everywhere except montane areas over 2,500m, parts of high central Burundi and Rwanda and most of northern and north-eastern Kenya, although there is a population between Buna, Sololo and Moyale, also known from South Horr and the eastern side of Lake Turkana. Often abundant and tolerant of urban sprawl and intensively farmed land, it is found in all the suburbs of Bujumbura, Kigali, Dar es Salaam, Nairobi and Kampala. The distribution of the East African species elsewhere has yet to be clarified.

NATURAL HISTORY: Nocturnal and terrestrial, emerging at dusk (sometimes late afternoon) to hunt. By day shelters in holes or under ground cover (log, rocks, debris, etc). Quick to bite with needle-sharp teeth if picked up or molested, but tames readily. Clutches of 2–16 eggs, roughly 2×3cm, are laid, incubation times of 2–3 months recorded, depending on the altitude, they may lay more than one clutch a year, hatchlings usually emerge near the start of the rainy season. Young snakes eat mostly lizards (as do adults in dry areas) but adults eat a lot of rodents, they do tremendous good controlling the rat population. Brown House Snakes in East Africa are the farmer's friend and everyone should be able to identify and appreciate them. Other prey items include birds, frogs and snakes. Although often uncommon in undisturbed habitat, this species, due to its wide choice of prey and secretive nocturnal habits, is efficient at exploiting and thriving in peri-urban habitats where other snakes cannot survive. For example, in the farms and *Acacia* woodland on the southern shores of Lake Naivasha in the 1960s, Brown House Snakes were extremely rare; following the urban sprawl that has engulfed the area they are now the most common snake there.

OLIVE HOUSE SNAKE *Boaedon olivaceus*

IDENTIFICATION: A medium-sized harmless colubrid with a red eye. The head is sub-triangular, slightly distinct from the head, the eye is medium sized, pupil vertical, iris distinctly red, orange or rufous brown. Body cylindrical, although big adults can get quite stout and depressed, with broad soft heads. The tail is 11–14% of total length in females, 15–19% in males. The scales are smooth, in 25–31 midbody rows (usually 27 in Ugandan and Rwandan specimens), ventrals 185–242 (usually 192–223 in Uganda), subcaudals single, 38–63, anal scale undivided. Maximum size about 90cm, average 40–70cm, hatchling size unknown, probably 16–20cm. Glossy brown, blackish or grey above, yellow or cream below, the dark flank colour encroaching on the ventral scales, so they look speckled at the edges.

Boaedon olivaceus

Boaedon olivaceus Olive House Snake. Left: Michele Menegon, Rwanda. Right: Kate Jackson, eastern Congo.

HABITAT AND DISTRIBUTION: Forest, riverine forest, forest islands in the savanna and recently deforested areas, from 600 to over 2,000m altitude, elsewhere down to sea level. Sporadic records from western Kenya (Chemilil, North and South Nandi Forests, Kakamega), eastern and central Uganda (Budongo and Mabira Forests, Masaka), widely distributed along Uganda's western border from the south end of Lake Albert south to Kisoro and into western Rwanda (Cyamudongo and Nyungwe National Park). Elsewhere, west to Guinea.

NATURAL HISTORY: Not well known. Nocturnal and terrestrial, emerging at dusk to hunt, when inactive hides in holes or under ground cover (log, rocks, debris, etc). Said to be often near water (one of its common names is 'Gaboon water snake'), but a collection from Mt Karissimbi were in mid-altitude forest. Diet includes rodents, rodent hair found in the gut of a Ugandan specimen, juveniles probably take amphibians and lizards. A captive group maintained at Nairobi Snake Park fed on white mice and were very good natured, much more reluctant to bite than Brown House Snakes. They lay eggs, but no clutch details known.

RADFORD'S HOUSE SNAKE *Boaedon radfordi*

IDENTIFICATION: A recently-described, medium-sized harmless colubrid, closely related to the Olive House Snake, from which it may be distinguished by having two upper labial scales enter the eye (usually three in the Olive House Snake). The head is sub-triangular, slightly distinct from the head, the eye is medium sized, pupil vertical, iris distinctly red or orange. Body cylindrical. The tail is about 12–16% of total length. The scales are smooth, in 27–31 midbody rows, ventrals 200–219, subcaudals single, 37–56, anal scale undivided. Maximum size about 80cm, hatchling size unknown. Glossy olive-grey, grey or brown, yellowish below, the dark flank colour encroaching on the ventral scales, so they look speckled at the edges.

Boaedon radfordi

HABITAT AND DISTRIBUTION: Plateau and forest, from 640m to over 2,300m altitude. Described from a handful of localities in the Congo, along the Albertine Rift, between the north end of Lake Tanganyika and the north end of Lake Albert; in our area recorded from Bwindi Impenetrable National Park.

NATURAL HISTORY: No details known, but probably fairly similar to the Olive House Snake.

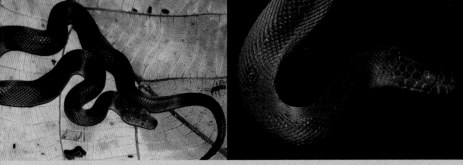

Boaedon radfordi Radford's House Snake. Left: Konrad Mebert, eastern Congo. Right: Konrad Mebert, eastern Congo.

WOLF SNAKES *LYCOPHIDION*

An African genus of small harmless, secretive, flat-headed colubrid snakes, ground dwelling or burrowing, most are speckled grey or brown. All have flattened or cylindrical bodies, smooth scales in 15–17 rows at midbody and the usual nine large scales on top of the head. The eye is small, with an unusual pupil that is sub-circular and distinctive to this genus, the iris is often silvery, shading to orange around the pupil. Wolf snakes have big recurved teeth (hence the common name), which they use to catch sleeping lizards. They rarely try to bite when handled. They lay eggs. Often found in the rainy season, when they have been flooded out of their holes. About 20 species are known; there may be more, the taxonomy is undergoing investigation. Ten species occur in East Africa, three are endemic. They have a superficial resemblance to snakes of several other genera, notably the shovel-snouts (*Prosymna*; harmless), purple-glossed snakes (*Amblyodipsas*, rear-fanged) and burrowing asps (*Atractaspis*, dangerous), do not pick up a supposed wolf snake unless you are totally certain of its identity.

KEY TO THE EAST AFRICAN MEMBERS OF THE GENUS *LYCOPHIDION*

1a	Banded or spotted red and grey or brown, four or more apical pits per scale......*Lycophidion laterale*, Western Forest Wolf Snake [p. 398]	4a	Dorsal scale rows not reduced before the vent......5
1b	Not banded or spotted red and grey or brown, three or less apical pits per scale......2	4b	Dorsal scale rows reduced to 15 before the vent......6
2a	Usually a white collar or blotch on neck, in dry savanna or semi-desert......*Lycophidion taylori*, Taylor's Wolf Snake [p. 402]	5a	Postnasal not in contact with the first labial, subcaudals 41–53 in males, 32–46 in females, dorsal scales with pale stippling, band around snout narrow......*Lycophidion ornatum*, Forest Wolf Snake [p. 401]
2b	Not usually a white collar or neck blotch, not necessarily in arid country......3	5b	Postnasal in contact with the first labial, subcaudals 30–31 in males, 23–24 in females, dorsal scales with a large white apical spot, band around snout broad and orange......*Lycophidion uzungwense*, Red-snouted Wolf Snake [p. 403]
3a	Maximum number of dorsal scale rows 15......*Lycophidion meleagre*, Speckled Wolf Snake [p. 399]		
3b	Maximum number of dorsal scale rows 17......4	6a	Nostril pierced in the middle of a single

or semi-divided nasal, no postnasal, subcaudals 28-31 in males, 19-25 in females......*Lycophidion acutirostre*, Mozambique Wolf Snake [p. 396]

6b   Nostril pierced near the posterior border of a single nasal, which is followed by a much smaller postnasal, subcaudals 31-58 in males, 21-55 in females......7

7a   Ventrals 153-165 in males, 162-173 in females, throat dark like rest of belly, a broad pale band around the snout, usually no other head markings, dorsal scales usually with pale stippling in the apical region......*Lycophidion depressirostre*, Flat-snouted Wolf Snake [p. 398]

7b   Ventrals 170-211 in males, 178-221 in females, throat pale, ventrum dark, pale band around the snout narrow or absent, if absent top of head usually with pale vermiculation, dorsal scales usually with an undivided white apical spot or border.......8

8a   Dorsal scales with white stippling, which may be interrupted, leaving dark crossbands or paired spots......*Lycophidion multimaculatum*, Blotched Wolf Snake [p. 400]

8b   Dorsal scales not stippled, usually bordered with white at the apex.....9

9a   Top of head pale with dark spots, looks dark brown or black, only on Pemba...... *Lycophidion pembanum*, Pemba Wolf Snake [p. 401]

9b   Top of head dark with or without pale spots, not on Pemba Island......*Lycophidion capense*, Cape Wolf Snake [p. 396]

MOZAMBIQUE WOLF SNAKE *Lycophidion acutirostre*

IDENTIFICATION: A small wolf snake, in our area only in southern Tanzania. Head flat, eye small, pupil vertical. The snout is relatively pointed when viewed from the side; hence the specific name. Body slightly depressed, tail fairly short. The scales are smooth, with a single apical pit, in 17 midbody rows, ventrals 132-156 in males, 139-161 in females; subcaudals paired, 28-31 in males, 19-25 in females. Maximum size 31cm, average 20-30cm, hatchling size unknown. Colour: blackish-brown. There is a white snout band, this breaks up behind the eye. The dorsal scales are lightly stippled with white, the lower two or three flank scale rows are tipped white. The underside is grey, with white bars on the outer, free end of each ventral scale.

Lycophidion acutirostre

HABITAT AND DISTRIBUTION: Plantations and woodland at low altitude in southern Tanzania. There is a single Tanzanian record; Liwale, elsewhere to northern Mozambique and southern Malawi. Conservation Status: IUCN Least Concern.

NATURAL HISTORY: Poorly known. Nocturnal and terrestrial, emerging at dusk to hunt, when inactive hides in leaf litter, holes or under ground cover (in rotting logs, under rocks, debris, etc). Diet probably similar to other wolf snakes, i.e. lizards. They lay eggs.

CAPE WOLF SNAKE *Lycophidion capense*

IDENTIFICATION: A small, harmless dark snake. The head is very flat, slightly distinct from the neck, the eye is tiny so the vertical pupil and silvery-grey iris are hard to see. Body cylindrical, tail fairly short, 8-15% of total length (longer in males). The scales are smooth, with a single apical pit, in 17 midbody rows, ventrals 164-221, (higher counts in females) subcaudals paired, 26–48 (higher counts males). Maximum size about 60cm (big specimens usually female), average 20–45cm, hatchling size 14–16cm Looks grey, brown or purplish-brown in the field, occasionally black. Close up the scales are black or brown with fine white edging. Head sometimes black on top and big adults may get very dark. Juveniles have a pale belly but the centre of each ventral darkens as they grow, so big adults have dark, light-edged belly scales.

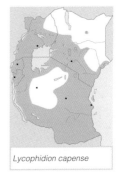

Lycophidion capense

HABITAT AND DISTRIBUTION: High grassland, moist and fairly dry savanna, usually absent from areas with less than 250mm annual rainfall. Found virtually throughout East Africa, from sea level to about 2,400m; can still be found in the suburbs of Kigali and Nairobi. Apparently absent from large areas of dry northern and eastern Kenya (save Marsabit) and records lacking for central Tanzania. Elsewhere, north to South Sudan, west to the Central African Republic, south to eastern South Africa.

NATURAL HISTORY: A slow-moving, inoffensive, nocturnal ground-dwelling snake. Spends much time investigating suitable refuges in its hunt for sleeping lizards; when inactive hides in holes, under ground cover (logs, rocks, piles of vegetable debris, building junk etc.). Most often seen in the rainy season, especially if flooded out of its refuge by rising water. Hardly ever tries to bite if restrained, but it will jerk and twist its body when held and if held by the head, its vigorous twisting can cause the long teeth to puncture the skin. It lays 3–8 eggs, roughly 1×2cm. A Nyambene Hills specimen laid six eggs in September; a hatchling was taken in May in southern Tanzania; incubation time in Botswana was 50–65 days; hatchlings emerged in the middle of the rainy season. Diet largely smooth-bodied skinks (small *Trachylepis, Leptosiaphos, Lygosoma*) which it finds in their refuges and seizes with its long teeth, dragging them out to eat. May feed on other small snakes including worm snakes. A 50cm Kakamega specimen had eaten a 13cm Montane Side-striped Chameleon *Trioceros ellioti*.

Lycophidion capense Cape Wolf Snake. Top left: Stephen Spawls, Ngulia Lodge. Top right: Stephen Spawls, Mau Escarpment. Inset: Harald Hinkel, Akagera National Park.

FLAT-SNOUTED WOLF SNAKE Lycophidion depressirostre

IDENTIFICATION: A small, dark grey or brown, flat-headed snake. The eye is small. In good light, the pupil actually looks round, the iris is silvery-grey but the borders of the pupil are often orange. Body cylindrical, tail fairly short, 8–14% of total length (longer in males). The scales are smooth, with a single apical pit, in 17 midbody rows, ventrals 153–180, subcaudals paired, 26–40 (higher counts males). Maximum size 48cm (big specimens usually female), average 20–35 cm, hatchling size unknown, probably 15–17 cm. May appear almost black, dark grey or grey-brown, but head usually uniform, not stippled or spotted, dorsal scales have tiny light speckles and white edging. Occasionally a light snout band is present. Belly dark grey, sometimes lighter on chin and outer edges of the ventrals.

Lycophidion depressirostre

HABITAT AND DISTRIBUTION: Usually in semi-desert, dry and moist savanna, a few records from more lush habitats, at altitudes between sea level and 1,500m. The only Uganda record is from Kampala. In Kenya it is found from Kakamega north-east out of the savanna into the arid north and east, south through Ukambani and Tsavo and along the coast, widespread in eastern and southern Tanzania. Elsewhere, north to southern Sudan and southern Somalia, might be in Mozambique.

NATURAL HISTORY: Poorly known. Nocturnal and terrestrial, emerging at dusk to hunt, when inactive hides in holes or under ground cover (log, rocks, debris, etc). Diet probably similar to other wolf snakes, i.e. lizards. They lay eggs, but no clutch details known.

Lycophidion depressirostre Flat-snouted Wolf Snake. Left: Anthony Childs, Meru National Park. Right: Anthony Childs, showing flat head.

WESTERN FOREST WOLF SNAKE Lycophidion laterale

IDENTIFICATION: A vividly-banded wolf-snake, in our area only in western Uganda. Head flat, eye small, pupil vertical. Body cylindrical, tail fairly short, 9–14% of total length (longer in males). The scales are smooth, with four apical pits, in 17 midbody rows, ventrals 170–203, subcaudals paired, 27–45 (usually more than 34 in eastern animals, higher counts males). Maximum size 48cm (big specimens usually female), average 20–35cm, hatchling size unknown. Colour very variable; west-central African animals usually banded vivid orange and grey with a broad pale snout band and a dark stripe behind the eye. Snakes from central Congo are grey, each scale with fine white stippling, and a dull pink snout band, northern Congo animals are light brown, with two rows of dorsolateral brown, light-edged spots, the head with a light snout band. Belly scales brown, black, or black with pale edges; the colour of the Ugandan snake is unknown.

Lycophidion laterale

HABITAT AND DISTRIBUTION: Forest. In East Africa, known only from the Bwamba Forest; east of the Semliki River in Bundibugyo, western Uganda, at altitudes around 700m. Elsewhere, west to Senegal, in forest.

NATURAL HISTORY: Poorly known. Nocturnal and terrestrial, emerging at dusk to hunt lizards on the forest floor, when inactive hides in leaf litter, holes or under ground cover (in rotting logs, under rocks, debris, etc). Diet probably similar to other wolf snakes, i.e. lizards. They lay eggs, a Ghanaian specimen contained three oviductal eggs, ready to be laid, 1.2×3.3cm.

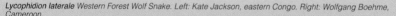

Lycophidion laterale Western Forest Wolf Snake. Left: Kate Jackson, eastern Congo. Right: Wolfgang Boehme, Cameroon.

SPECKLED WOLF SNAKE *Lycophidion meleagre*

IDENTIFICATION: A small, dark flat-headed wolf-snake with a pale snout. Eye fairly obvious, iris yellow. Head slightly distinct from the neck. Body cylindrical, tail fairly short, 8–14% of total length (longer in males). The scales are smooth, in 15 midbody rows (most wolf snakes have 17), ventrals 147–164, subcaudals paired, 22–34 (higher counts males). Maximum size 35cm (big specimens usually female), average 20–30cm, hatchling size unknown. Colour of Tanzanian specimens dark grey, with a white spot on the apex of each scale, tail uniform black. The snout is yellow, cream or pinkish

HABITAT AND DISTRIBUTION: In our area occurs in coastal Kenya and parts of the

Eastern Arc Mountains and closely associated forests in Tanzania, from sea level to 1,600m altitude. Tanzanian records include the eastern Usambaras, Magrotto Hill, Mgambo forest reserve and the Uluguru Mountains; in Kenya known from the Shimba Hills and Arabuko-Sokoke Forest. Elsewhere, south-eastern DR Congo and northern Angola.

NATURAL HISTORY: Poorly known, presumably similar to other wolf snakes, i.e. nocturnal and terrestrial, emerging at dusk to hunt, when inactive hides in holes or under ground cover (log, rocks, debris, etc). Diet probably similar to other wolf snakes, i.e. lizards, an Uluguru female had eaten a Kilimanjaro Five-toed Skink *Leptosiaphos kilimensis*, another had eaten an unidentified snake. They lay eggs, but few clutch details known, a gravid female from Amani was collected in mid-October.

Lycophidion meleagre Speckled Wolf Snake. Natural History Museum Denmark, Usambara Mountains.

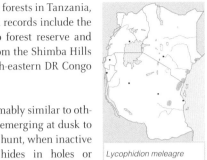
Lycophidion meleagre

BLOTCHED WOLF SNAKE *Lycophidion multimaculatum*

IDENTIFICATION: A small, grey, black-barred wolf-snake. Eye fairly obvious, iris grey. Head slightly distinct from the neck. Body cylindrical, tail fairly short, 8–14% of total length (longer in males). The scales are smooth, in 17 midbody rows, ventrals 153–188, subcaudals paired, 22–38 (higher counts males). Maximum size 53cm (big specimens usually female), average 25–35cm, hatchling size unknown. Colour grey, with dark crossbars, under close examination the grey scales have black and white stippling. Taxonomic Note: Until recently regarded as a subspecies of the Cape Wolf Snake *Lycophidion capense*.

Lycophidion multimaculatum

HABITAT AND DISTRIBUTION: A savanna species. In our area known only from extreme southwest Tanzania, at the northern end of Lake Rukwa, Tatanda and Sumbawanga, at altitudes between 700–1,900m. Elsewhere southwest to northern Namibia and Angola, westwards to Cameroon, at altitudes between sea level and 2,500m.

Lycophidion multimaculatum Blotched Wolf Snake. Philipp Wagner, Zambia.

NATURAL HISTORY: Poorly known, presumably similar to other wolf snakes, i.e. nocturnal and terrestrial, emerging at dusk to hunt, when inactive hides in holes or under ground cover (log, rocks, debris, etc). Diet similar to other wolf snakes, i.e. lizards. Lays eggs.

FOREST WOLF SNAKE Lycophidion ornatum

IDENTIFICATION: A small, attractive forest dwelling wolf-snake. Head flat, eye small but obvious, iris silvery, pupil edged with orange. Body cylindrical, tail longer than in most wolf snakes, 12–18% of total length (longer in males). The scales are smooth, 17 midbody rows, ventrals 175–212, subcaudals paired, 32–53 (usually 33–44 in Uganda, higher counts males). Maximum size 59cm (a Nyambene female), average 25–50cm, hatchlings 14–15cm. Brown, red-brown, grey, pinkish or pinky-grey, sometimes uniform, but there is usually a broad white snout band, extending along the side of the head through the eye, behind the eye the band splits into two, one onto the temporals, the other onto the upper labials. Some specimens have a double row of dark spots along the back. Belly scales black or grey, often with pale edging, so underside looks finely barred.

Lycophidion ornatum

Lycophidion ornatum Forest Wolf Snake. Michele Menegon, Rwanda.

HABITAT AND DISTRIBUTION: Forest and deforested areas, from 700–2,700m altitude. Kenya records include Karen, Ruiru, Karura and the Ngong Forest on the outskirts of Nairobi, mid-altitude forests of south and eastern Mt Kenya and the Nyambene Hills, Mt Elgon, North and South Nandi Forest, Kakamega. Found in forests of the northern Lake Victoria shore in Uganda. Widely distributed along the Albertine Rift, from the environs of Lake Albert south through the high country to western Rwanda and Burundi, extending to Kigoma in Tanzania. An isolated record from Ngorongoro Crater rim, but probably widespread in the crater highlands. Elsewhere, south-west through the DR Congo to northern Angola, and in the Imatong Mountains in Sudan. Conservation Status: IUCN Least Concern.

NATURAL HISTORY: Poorly known. Nocturnal and terrestrial, emerging to hunt on the forest floor at dusk, when inactive hides in leaf litter, in holes or rotting log, under ground cover, etc. Clutches of 2–6 eggs recorded. Gravid females collected in Uganda in January and February, a Kerugoya female laid five eggs, 2–2.6cm × 0.6–1cm in January. Diet largely small, smooth-bodied forest lizards, which it finds by investigating holes and in the leaf litter, but a Congo snake climbed a tree and seized a sleeping Johnston's Chameleon, *Trioceros johnstoni*.

PEMBA WOLF SNAKE Lycophidion pembanum

IDENTIFICATION: A small, dark brown to black wolf-snake, on Pemba only. Head quite broad, eye small, pupil vertical, but this is hard to see. Body cylindrical, tail fairly short. The nostril is pierced at the posterior edge of the nasal, bordered by a postnasal that is in contact with the first upper labial. The scales are smooth, with a single apical pit, in 17 midbody rows, ventrals 170 in males, 178–181 in females, subcaudals paired, 37–40 in females, 46–47 in males. Biggest female 44.4cm, biggest male 33.1cm, average and hatchling size unknown. Appears dark brown to black, the head is pale with dark median spots in each large shield. The dorsal scales have pale edges, the throat is pale.

Lycophidion pembanum

HABITAT AND DISTRIBUTION: Endemic to Pemba Island, at altitudes just above sea level. Known localities there include Gando, Mtambile, Wete, Ziwani and Ngesi Forest. Found amongst leaf litter in clove plantations. Conservation Status: IUCN Data Deficient, its range is restricted and possibly dependent upon the existence of forest, but able to survive in clove plantation and probably in cultivation under moist and shady conditions.

NATURAL HISTORY: Poorly known. Nocturnal and terrestrial, emerging at dusk to hunt, when inactive hides in leaf litter. Diet probably similar to other wolf snakes, i.e. lizards. They lay eggs, but no clutch details known.

TAYLOR'S WOLF SNAKE *Lycophidion taylori*

IDENTIFICATION: A rare, recently-described, arid-country wolfsnake, usually with a distinctive white collar. The eye is medium-sized, pupil vertical, iris brown or red-brown. Head flat, tongue pale or pinkish. Body cylindrical, tail short, 9–13% of total length (longer in males). The scales are smooth, with a single apical pit, in 17 midbody rows, ventrals 161–184 (higher counts females) subcaudals paired, 26–36 (higher counts males). Maximum size 51cm (big specimens usually female), average 25–40cm, hatchling size unknown, a juvenile was 23cm. Grey or grey-brown above, each scale with either a white dot or stipple, head grey with white stippling. Usually a white collar or blotch on the neck, this may be bordered by black patches or bands, occasional individuals have white body markings. Belly scales uniform brown or reddish-brown, sometimes white edged, chin stippled white.

Lycophidion taylori

Lycophidion taylori Taylor's Wolf Snake. Top: Lorenzo Vinciguerra, south of Mt Kilimanjaro. Bottom: Royjan Taylor, Tsavo East.

HABITAT AND DISTRIBUTION: Arid savanna and semi-desert at medium to low altitude. Records sporadic in our area, known only from dry country south-west of Mt Kilimanjaro in Tanzania, from Sala

Gate in Tsavo East, Lodwar and Kakuma in extreme north-west Kenya, but probably occurs in the intervening areas. Also known from Chad, Senegal, northern and eastern Ethiopia, southern coastal Somali and Djibouti; a rare inhabitant of arid central-north Africa.

NATURAL HISTORY: Described in 1993, from museum specimens, so little known. Nocturnal, has been found active at night and under ground cover by day. Habits probably similar to the Flat-snouted Wolf Snake *Lycophidion depressirostre*.

RED-SNOUTED WOLF SNAKE *Lycophidion uzungwense*

IDENTIFICATION: A medium-sized wolf-snake with a most distinctive scarlet V on the snout, extending to the sides of the head. The head is flattened, neck distinct, eye is small, pupil vertical, this is hard to see. Body cylindrical, tail fairly short. Nostril pierced in an undivided nasal. No postnasal present. The dorsal scales are smooth, with a single apical pit, in 17 midbody rows, subcaudals paired, 23–24 in females, 30–31 in males. Biggest females 60.5cm, biggest male 29.1cm. Shiny black, each dorsal scale white-tipped (appears grey from a distance), with a scarlet or orange V or band around the snout and along the sides of the head. Ventrals black.

Lycophidion uzungwense

HABITAT AND DISTRIBUTION: A Tanzanian Eastern Arc endemic, restricted to the forest and plantations of the southern Udzungwa Mountains, at altitudes around 1,500–2,000m. Known localities include Dabaga, Kigogo forest reserve, Udzungwa Scarp Nature Reserve, Bomalang'ombe (Gendawaki Valley) and the Mufindi tea estates. Conservation Status: Probably not threatened as long as the major natural forest block within its range are conserved.

Lycophidion uzungwense Red-snouted Wolf Snake. Michele Menegon, Udzungwa Mountains.

NATURAL HISTORY: Poorly known. Found in forest but also collected on a lawn. Habits presumably similar to other wolf snakes, i.e. nocturnal and terrestrial, emerging at dusk to hunt, when inactive hiding in holes, under ground cover or in leaf litter Diet probably similar to other wolf snakes, i.e. lizards. They lay eggs, but no clutch details known.

BANDED SNAKES *CHAMAELYCUS*

A genus of four small, harmless ground-dwelling snakes of the central African forest. Scales smooth, in 17 rows at midbody. Little is known of their habits. One species just reaches our area, in western Uganda.

ITURI BANDED SNAKE *Chamaelycus fasciatus*

IDENTIFICATION: A little ground-dwelling snake, with a slightly flattened head. Body cylindrical, tail 12–15% of total length. Scales smooth, 17 midbody rows, ventrals 164–198 (usually 175–184 in specimens from the DR Congo), subcaudals paired, 30–56 (usually 41–47 in specimens from the DR Congo). Maximum size 34cm, average 20–30cm, hatchling size unknown. Brown or purple-brown, usually with narrow dark, white-edged crossbars; at close quarters the dorsal scales are finely stippled white and iridescent. Each ventral scale is light grey, edged with darker grey, so the underside looks finely banded.

Chamaelycus fasciatus

HABITAT AND DISTRIBUTION: Forest, from Senegal eastwards to the eastern DR Congo. One record from our area, the Bwamba Forest, east of the Semliki River in Bundibugyo, western Uganda, at altitude around 700m. Known from the north shore of Lake Albert on the Congo side.

Chamaelycus fasciatus Ituri Banded Snake. Jean-Francois Trape, West Africa.

NATURAL HISTORY: Virtually nothing known. Presumably nocturnal and terrestrial, or burrowing. Probably lays eggs. Reported to eat "insects and reptile eggs", but one DR Congo specimen had eaten a Black-lined Plated Lizard *Gerrhosaurus nigrolineatus*.

FOREST SNAKES *HORMONOTUS*

An African genus with a single harmless forest species.

YELLOW FOREST SNAKE *Hormonotus modestus*

IDENTIFICATION: A ground-dwelling snake, with a distinctive pear-shaped head. The eyes are big and prominent, with a pale iris and vertical pupil. The snout is square, body cylindrical, tail long, 19–23% of total length, longest tail in males. Scales smooth, 15 midbody rows, ventrals 220–244, subcaudals paired, 78–103. Maximum size 85cm, average 40–70cm, hatchling size unknown. Colour quite variable, either dull yellow, various shades of brown, yellow-brown or grey-brown. The scales on top of the head are often white-edged, giving a most distinctive reticulated appearance; this is a valuable aid to identification. Each lip scale usually has a dark spot. Belly yellow or cream.

HABITAT AND DISTRIBUTION: Forest or recently deforested country, at medium altitude in East Africa, 750–1,300m, but down to sea level elsewhere. Only handful of widespread East African records exist; Kakamega, Khayega and the Tororo-Webuye Falls road in Kenya; Entebbe, Bwamba and Budongo Forest in Uganda, Mahale Mountains National Park in Tanzania, probably occurs in suitable intervening areas. Known also from Bunyakiri, west of Lake Kivu on the Congo side, so probably in northern Rwanda and south-west Uganda. Elsewhere, south-west to northern Angola, west to Guinea.

Hormonotus modestus

Hormonotus modestus Yellow Forest snake. Bill Branch, Gabon.

NATURAL HISTORY: Little known. Nocturnal, both terrestrial and arboreal, hunting on the forest floor and in trees at night, a Kivu specimen was found 2m up in a bush in a fallow field, a Gabon specimen was 1.8m up a tree. Hides by day in leaf litter or under ground cover. Presumably lays egg, no details known. Diet said to be rodents, a DR Congo specimen had eaten a Tropical House Gecko *Hemidactylus mabouia*. It appears to be rare, with few specimens known, in a collection of over 200 snakes from a cocoa farm in Ghana there were only two examples of this species.

FILE SNAKES *GONIONOTOPHIS*

An African genus of fifteen species, seven of which occur in our area. Originally in two genera, *Mehelya* and *Gonionotophis*, recent research indicates both genera have a common single ancestor; the name *Gonionotophis* has priority. In addition, research shows that in East Africa, the Cape File Snake (originally *Mehelya capensis*) consists of three discrete species: *Gonionotophis capensis*; *G. chanleri*; and *G. savorgnani*. File snakes are harmless, dark-coloured, slow moving, secretive nocturnal snakes; they feed on a range of prey, favouring amphibians and reptiles; including highly venomous snakes. The name refers to the curious triangular body shape, resembling a three-cornered file. The vertebral scales are enlarged, with two or more keels; the flanks scales are small and the interstitial skin visible (especially in the larger forms), consequently they often harbour ticks. There are several superstitions about file snakes, due to their slow movement and sinister appearance. The Cape File Snake *Gonionotophis capensis* has a white vertebral ridge and some believe that anyone seeing this will be blinded. In eastern Kenya, it is believed that if a file snake enters a house or hut then bad luck will affect the owner, nevertheless the snake must not be molested; in western Kenya magicians may keep them for use with rituals. In southern Africa, file snakes have been called 'saw cobras', and legend says that as they rush past you, they saw your legs off with their sharp bodies.

KEY TO EAST AFRICAN MEMBERS OF THE GENUS *GONIONOTOPHIS*

1a Scales in 21 rows at midbody......*Gonionotophis brussauxi*, Mocquard's File Snake [p. 406]
1b Scales in 19 or fewer rows at midbody......2

2a Tail very long, more than 18% of total length......3
2b Tail shorter, less than 18% of total length......4

3a Large, up to 1.4m, in western East Africa, ventral scales more than 234......*Gonionotophis poensis*, Forest File Snake [p. 410]
3b Small, less than 70cm, eastern East Africa, ventral scales less than 185...... *Gonionotophis nyassae*, Black File Snake [p. 409]

4a Scales very feebly keeled, top of head smooth, small, less than 80cm, western Uganda only......*Gonionotophis stenophthalmus*, Small-eyed File Snake [p. 411]
4b Scales strongly keeled, top of head lumpy, large, up to 1.6m, widespread......5

5a Vertebral scale row often quadricarinate; venter usually dark brown to black *Gonionotophis chanleri*, Chanler's File Snake [p. 408]
5b Vertebral scale row bicarinate; underside white, ends of each ventral dark 6

6a Vertebral scale row white; ventrals 215–221 in males, 213–224 in females, *Gonionotophis capensis*, Cape File Snake [p. 407]
6b Vertebral scale row dark; ventrals 220–235 in males, 225–239 in females *Gonionotophis savorgnani*, Congo File Snake [p. 409]

MOCQUARD'S FILE SNAKE *Gonionotophis brussauxi*

IDENTIFICATION: A small, dark snake. The head is broad, the eye dark but quite prominent, the pupil vertical. Body subtriangular in section, tail quite long, 25–28% of total length. Body scales keeled, the vertebral row are enlarged with double keels. Midbody scale count 21, ventrals 167–192, subcaudals paired, 73–99, anal entire. Maximum size about 45cm, average 30–40cm, hatchling size unknown. Dark brown, grey or light grey above, whitish or very pale grey below; the interstitial skin is white and visible between the scales, the upper lip scales light pink or white. Taxonomic Note: Ugandan specimens belong to the eastern subspecies, *G.b. prigoginei*, which has high ventral (182–192) and subcaudal (94–99) counts.

Gonionotophis brussauxi

Gonionotophis brussauxi Mocquard's File Snake. Kate Jackson, eastern Congo.

HABITAT AND DISTRIBUTION: Forest and recently deforested areas. A single Ugandan record from the Budongo Forest, at around 1,000m altitude. Known from the eastern rift escarpment in the DR Congo near the Rwanda and Burundi borders, so might be in those countries and south-west Uganda. Elsewhere, south-west to Angola, west to Guinea-Bissau.

NATURAL HISTORY: A slow-moving, nocturnal, terrestrial snake. Often found near water sources such as rivers, swamps etc. Shelters by day in holes, under vegetation piles, in leaf litter, root tangles etc. Lays eggs, no clutch details known. Diet: small terrestrial amphibians.

CAPE FILE SNAKE *Gonionotophis capensis*

IDENTIFICATION: A fairly big, slow-moving, snake, grey with a white vertebral stripe. The head is broad, with curiously lumpy scales, the eye is medium-sized and dark, the tongue pink. Body distinctly triangular in section, with an enlarged rough vertebral ridge; the tail fairly long, 11–14% of total length, slightly longer in males than females. Females have longer and wider heads and thicker bodies than males of a similar size. The tail is often mutilated and body scales and strips of skin missing, due to attacks by predators. The scales are small and heavily keeled, in 15 rows at midbody, ventrals in our area 203–241, subcaudals paired, 40–61 rows, higher counts usually males. Maximum size about 1.6m, average 70cm–1.3m, hatchlings 40–45cm. Grey, black or dark brown with a white vertebral stripe. The interstitial skin is pale grey or white, the belly is whitish but the outer ends of the ventrals are usually dark.

Gonionotophis capensis

Gonionotophis capensis Cape File Snake. Michele Menegon, Northern Mozambique.

HABITAT AND DISTRIBUTION: Occupies a wide range of habitats including semi-desert; in our area occurs mostly savanna and woodland, at low altitude, rarely above 1,300m. Known from coastal, eastern and southern Tanzania, known localities include Zanzibar Island, Dar es Salaam, Lindi and Songea, elsewhere south to South Africa, west to Angola. Conservation Status: IUCN Least Concern.

NATURAL HISTORY: A slow-moving, nocturnal, largely terrestrial snake, but it will climb into trees and reedbeds in pursuit of prey; one climbed a tree to catch a Boomslang. It emerges after dark (although a juvenile in Botswana was active in the late afternoon of a winter's day) and crawls around, hunting for its reptilian prey. They hide in holes, hollow logs, under ground cover, often in disused termite hills. They hunt by smell and rarely bite if handled, but one specimen bit a handler who had just put down a wolf snake. The teeth are blunt but the jaws are very powerful, with a crushing action, the snake seizing its victim and working its way up to the head, prior to swallowing. File snake are good-natured, but restless in the hand, they also defend themselves by discharging a foul-smelling greenish paste from glands near the cloaca. Cape file snakes lay 5–13 eggs, 3–5×1.5–2.5cm in size. Incubation time in southern Africa is about three months. They eat a wide variety of ectothermic prey; snakes (including highly venomous species, to the venom of which they are immune), amphibians, lizards such as geckoes, skinks and agamas, and possibly smaller mammals; one snake had feathers in its gut. They are not common snakes, possibly due to their secretive lifestyle, and there are few specimens in collections, although their slow rate of movement means they are quite often found as road casualties.

CHANLER'S FILE SNAKE *Gonionotophis chanleri*

IDENTIFICATION: A fairly big, slow-moving, dark snake. The head is broad and flat, the eye medium-sized and dark. The tongue is pink with a white tip. Body triangular in section, with an enlarged rough vertebral ridge; the tail fairly long, 10–15% of total length, slightly longer in males than females. The tail is often mutilated and body scales and strips of skin missing, due to attacks by predators. The scales are small and heavily keeled, in 15 rows at midbody, ventrals in our area 215–243, subcaudals paired, 44–62 rows, higher counts usually males. Maximum size about 1.5m, average 70cm–1.3m, hatchling size unknown. Grey, black, purplish grey or dark brown, the interstitial skin is usually paler than the scales. The ventral scales are grey, with white edging.

Gonionotophis chanleri

HABITAT AND DISTRIBUTION: Occupies a wide range of habitats including semi-desert, dry and moist savanna and woodland, but not forest, from sea level to around 1,800m. In East Africa, widespread but rare in southern and eastern Kenya (localities include Watamu, Voi, Kiboko, Murang'a, Nyambene Hills), isolated records from the Tana Delta, and Sotik in western Kenya. Enters Tanzania from Tsavo to the southern side of Mt Kilimanjaro. Also occurs north from Lake Rukwa in Tanzania right up the Lake Tanganyika shore through Burundi and Rwanda to extreme south-west Uganda. Elsewhere, north to Ethiopia and Somalia.

NATURAL HISTORY: As for the previous species; nocturnal, slow-moving and secretive.

Gonionotophis chanleri Chanler's File Snake. Left: Stephen Spawls, Arusha. Right: Harald Hinkel, Rwanda.

A specimen from Murang'a was collected while trying to eat a Rhombic Night Adder *Causus rhombeatus*, captive specimens in Burundi ate almost any snake offered to them, as well as chameleons. The name honours William Astor Chanler, charismatic American explorer in northern Kenya in the 1890s, who collected the first specimen on the mainland next to Manda Island.

CONGO FILE SNAKE/ DE BRAZZA'S FILE SNAKE
Gonionotophis savorgnani

IDENTIFICATION: General appearance similar to the two previous species. Scales keeled, in 15 rows at midbody. Ventrals 220–239, anal entire, subcaudals 45–62. Maximum size about 1.6m, average 90cm–1.3m. Females are larger than males. Dark brown, purplish brown, various shades of grey or black above, the dorsal scales often white-tipped, giving a speckled effect. Ventral scales white, sometimes with the long edges dark, giving a barred appearance, and outer edges dark.

Gonionotophis savorgnani Congo File Snake. Steve Russell, Kasese.

Gonionotophis savorgnani

HABITAT AND DISTRIBUTION: Savanna, woodland and forest in our area, between 1,000–2,200m altitude; often found in marshy areas in Uganda. Occurs from Chemilil and Kisumu in Kenya west across most of central, south-west and north-west Uganda. Elsewhere north to South Sudan, west to Cameroon.

NATURAL HISTORY: As for the previous species; nocturnal, slow moving and secretive. Ugandan specimens recorded eating snakes, lizards and amphibians.

DWARF FILE SNAKE / BLACK FILE SNAKE
Gonionotophis nyassae

IDENTIFICATION: A slim, small dark file snake. Head not as short as other file snakes and smooth on top, eye small and dark, the vertical pupil is hard to see. The tongue is pink and white. Body sub-triangular in cross-section, with a pronounced vertebral ridge. The tail is fairly long, 18–27% of the total length, usually longer in males. Scales keeled, in 15 midbody rows, the vertebral row enlarged and hexagonal, with a strong or weak double keel. Ventrals 164–184, anal entire, subcaudals 51–79. Maximum size about 65cm (large specimens are usually females), average 35–55cm, hatchlings 20–22 cm. Black or grey in colour, the skin between the scales pink, white or grey, some scales may have fine white dots. A Zanzibar snake had a dorsolateral stripe of fine white dots. The belly may be white, cream, black or grey with pale edging.

Gonionotophis nyassae

HABITAT AND DISTRIBUTION: Low savanna, coastal thicket and woodland, from sea level to about 1,200m. It is sporadically distributed along the East African coastal plain, from southern Somalia and the Boni Forest right down to the Rovuma River, also on Zanzibar. Extends up the Tana River to Garissa, inland to Voi, also in the Usambara Mountains, and across the south-east corner of Tanzania, known from Handeni, Liwale and Tunduru. There is a record in the literature from 'Rwanda-Burundi' (not shown on map). Elsewhere, south to Botswana and north-east South Africa. Conservation Status: IUCN Least Concern

Gonionotophis nyassae Black File Snake. Stephen Spawls, Voi.

NATURAL HISTORY: Poorly known. Nocturnal, terrestrial and secretive. Specimens in South Africa were active at night in the rainy season. The Voi specimen was hiding in the collapsed debris of a baobab tree. Also hides under rocks, in logs, down holes and other suitable refuges. Slow-moving and inoffensive, although reported to produce a noxious cloacal secretion when handled. Between 2–6 eggs are laid. Hatchlings 21–23cm collected in April in Liwale, southern Tanzania. Diet: small lizards, particularly skinks, occasionally snakes and amphibians.

FOREST FILE SNAKE *Gonionotophis poensis*

IDENTIFICATION: A long-tailed harmless forest file snake; so slim it looks like a tree snake. The head is long and looks rectangular, the snout is noticeably square but quite pointed in profile. The eye is relatively large, the neck thin. Body sub-triangular in cross-section (not as pronounced as other large file snakes) and slim, with the usual enlarged row of vertebral scales. The tail is very long for a file snake, 20–25% of total length, generally longer in males. The dorsal scales have a single prominent keel, vertebral scales have two keels. Scales in 15 midbody rows, ventrals 236–262, anal entire, subcaudals 75–124, males counts usually higher. Maximum size about 1.4m, average 80cm–1.2m, hatchling size unknown. Colour olive-grey or grey, the exposed skin between the scales is light grey or blue-grey, belly lighter, light grey to yellowish or white, darker at the outer edges of the ventrals.

HABITAT AND DISTRIBUTION: Forest and forest islands, from medium (700m) to high (2,200m) altitude, elsewhere down to sea level. In our area, occurs from the northern Victoria lakeshore to the east side of Lake Kyoga; also known from Bwindi Impenetrable National Park and Budongo Forest. Recorded from Cyamudongo, western Rwanda; Cibitoke and Bujumbura in Burundi, (and from the eastern DR Congo south-west of Lake Kivu) Elsewhere, west through the forest to Guinea, south-west to Angola.

Gonionotophis poensis

Gonionotophis poensis Forest File Snake. Konrad Mebert, eastern Congo.

NATURAL HISTORY: Poorly known. Slow-moving, nocturnal, terrestrial and secretive. Its relatively long tail suggests that it might be arboreal to some extent. By day hides in holes, leaf litter etc. In Ghana, a female contained eight ova, roughly 2.5×1cm, in May (start of the rainy season), no other breeding details known. Recorded prey items include a mouse in the Cyamudongo animal, Red-headed Rock Agamas (*Agama lionotus*) and Speckle-lipped Skinks (*Trachylepis maculilabris*).

SMALL-EYED FILE SNAKE *Gonionotophis stenophthalmus*

IDENTIFICATION: A small dark file snake, not noticeably triangular in cross-section, with smooth scales from western Uganda. The head is flat, the eye small and dark, the nostril is curiously large and round. The body is weakly sub-triangular in shape (unlike most file snakes, which are clearly triangular), tail 12–17% of total length, usually longer in males. The scales are smooth except for the vertebral row, which is weakly and sporadically keeled. Midbody scale count 15, ventrals 192–228, subcaudals paired, 47–60, anal entire. Maximum size about 76cm, average 40–70cm, hatchling size unknown. Colour glossy black or dark grey; the scales are iridescent. The belly is ivory white, cream or light yellow, darkening at the outer edges of the ventrals.

Gonionotophis stenophthalmus

HABITAT AND DISTRIBUTION: Forest and recently deforested areas, in Uganda from about 600–1,800m altitude, elsewhere down to sea level. Known from three localities in western Uganda: Kibale forest, Fort Portal, and Ntandi in the Semliki National Park, also recorded from near Kabare, south-west of Lake Kivu, and just west of central Lake Tanganyika, both in the DR Congo, so may well be in Rwanda and Burundi. Elsewhere, west to Guinea.

Gonionotophis stenophthalmus Small-eyed File Snake. Bill Branch, Gabon.

NATURAL HISTORY: A nocturnal, terrestrial file snake. Poorly known, might well spend much time burrowing or pushing through leaf litter. One Uganda specimen was found at night on the forest floor. Presumably lays eggs, no details known. Diet said to be snakes, including other file snakes and an Olive Marsh Snake *Natriciteres olivacea*, but little data available, a specimen from the DR Congo had eaten a Western Forest Limbless Skink *Feylinia currori*.

SUBFAMILY **PSEUDASPIDINAE**

A tiny subfamily, as presently defined, of two monotypic genera and two species, both African. The Mole Snake is harmless, although the other member of the subfamily, the Western Keeled Snake *Pythonodipsas carinata*, of the Namib Desert, is rear-fanged. The relationships of the African rear-fanged genus *Buhoma*, forest snakes, are poorly understood but they have been placed by some authors in this subfamily, so we treat them here also.

MOLE SNAKES *PSEUDASPIS*

A monotypic genus containing a large, harmless, south-east African snake, unusual in that it gives live birth.

MOLE SNAKE *Pseudaspis cana*

IDENTIFICATION: A big, stout snake with a small, short head. The eye is small, pupil round, snout blunt and rounded. Neck slightly thinner than the head. The body is cylindrical, thick and muscular. The tail is 16–22% of the total length in males, 13–16% in females. Males have enormous, elongate hemipenes. The scales are smooth (except in some South African specimens), in 25–31 midbody rows (usually 27), ventrals 175–218 (high counts usually females), subcaudals 43–57 in females, 55–70 in males. Maximum size about 1.8m (anecdotal reports of larger animals exist, once from Mount Longonot was reported to be 'at least 2.4m'). Average 1–1.3m, hatchlings 20–25 cm. These snakes show a remarkable colour shift as they grow. Hatchlings are brown, with irregular black cross-bars or Y-shapes, with a double row of irregular white spots on the flanks, the belly is light with dark speckling. Juveniles lose the cross bars and develop a series of dark spinal blotches or an irregular dark zigzag line. The adults are usually uniform, although some retain the juvenile pattern to a greater or lesser extent. Colour of adults very variable, depending on the region, they may be any shade of brown, olive, red-brown, orange or grey. Those from central Kenya are usually light, pinkish or orange-brown, each dorsal scale black-tipped, pinkish below with a dark blotch on the outer edge of every second or third scale. Nanyuki specimens are orange above, northern Tanzanian animals dark brown, olive or grey-green, some northern Tanzanian animals retain dark spots as adults. Southern Tanzanian animals are brown, often with yellow spotting. The Rwandan and Burundi animals were juveniles, their adult colour is unknown but a suspected adult in Rwanda appeared grey.

Pseudaspis cana

HABITAT AND DISTRIBUTION: Grassland, moist and dry savanna, from 1,200 to about 2,600m altitude in East Africa, elsewhere down to sea level. East African distribution curiously disjunct, although being so secretive they may be more widespread and overlooked. In central Kenya, occurs from Eldoret south and east, down across the Rift Valley to Lakes Nakuru, Elmenteita and Naivasha. Also known from the Nyeri-Nanyuki area, this might be connected with the Rift Valley population through Nyahururu. There is an old record from Murang'a and an enigmatic record from Kitui. Reappears in the high country around the south-western side of Mt Kilimanjaro, there is an apparently isolated population in southern Rwanda and Burundi, also found in the southern highlands of Tanzania, southwest from the Udzungwa Mountains around the top of Lake Malawi. Elsewhere, south to South Africa.

Pseudaspis cana Mole Snake. Top left: Stephen Spawls, Njoro. Top right: Stephen Spawls, Ol Donyo Sambu. Bottom left: Stephen Spawls, hatchling, Arusha. Bottom right: Stephen Spawls, juvenile, Botswana.

NATURAL HISTORY: A secretive terrestrial snake, spending much of its time hunting its rodent prey in burrows, when inactive during the day may hide just under the soil surface, also shelters in holes and warrens. One specimen was observed in a tree. Its short pointed head and thick powerful body mean it is well adapted for burrowing. If threatened and unable to escape it will hiss loudly, inflate the body, open the mouth and make huge lunging strikes, it looks very dangerous and is often needlessly killed. However, Mole Snakes are highly beneficial, they occur in farming areas and eat large quantities of rodents, especially mole rats. In Nakuru, they were regularly killed on the golf course by golfers who then complained about the damage mole rats did to the greens. Mole Snakes were once common in the Kenyan Rift Valley, from Mount Longonot north to Nakuru and in the grasslands of the western wall, but the intense farming there has resulted in a steep decline in their numbers, save on farms with enlightened owners. Also recorded eating eggs, juvenile Mole Snakes eat mostly lizards; captive specimens may eat each other. In the mating season, males will fight, savaging each other with their teeth. They give live birth, usually 18–50 young, but one captive female produced 95 offspring. Breeding season unknown in East Africa, at Lake Naivasha hatchlings have been taken in late April, May and November. Mole Snakes make excellent pets, taming well, in South Africa they have been used to train young dogs to avoid snakes, a few bites from a big adult teaching the puppy to leave snakes alone.

FOREST SNAKES *BUHOMA*

An African genus of small, terrestrial, dark-coloured, secretive, rear-fanged forest-dwelling snakes; their affinities are uncertain. Originally placed in *Geodipsas*, a Madagascan genus now synonymised with *Compsophis*, recent work indicates the African representatives warrant their own genus; *Buhoma* is the name by which the Pale-headed Forest Snake is known in the Musigati Region of Burundi. There are three species in the genus; all occur in our area, two are Tanzanian endemics. Although rear-fanged and thus venomous, they are small and pose no threat to humans.

KEY TO MEMBERS OF THE GENUS *BUHOMA*

1a	Has a white or yellow nuchal collar or neck bar; in our area in western Uganda only......*Buhoma depressiceps*, Pale-headed Forest Snake [p. 414]		
1b	No white nuchal collar, in Tanzania......2		
2a	Ventrals 122–133, range Usambara Mountains, Magrotto Hill and Uluguru Mountains......*Buhoma vauerocegae*, Usambara Forest Snake [p. 416]		
2b	Ventrals 145–154, range Uluguru and Udzungwa Mountains......*Buhoma procterae*, Uluguru Forest Snake [p. 415]		

PALE-HEADED FOREST SNAKE *Buhoma depressiceps*

IDENTIFICATION: A small dark forest snake with a short head and fairly small eyes. Two subspecies are known, the following notes apply only to the eastern subspecies, *Buhoma de-*

pressiceps marlieri. Head fairly small, eye pupil round. Body cylindrical, tail short, 14–17% of total length. Scales smooth, in 17 rows at midbody, 150–163 ventrals, 35–43 subcaudals. Maximum size about 44cm (big specimens usually females), average 20–35cm, hatchling size unknown. Colour brown, sometimes with a series of fine dark longitudinal stripes. The head is sometimes pale, but more usually brown, with a distinct white or yellow collar (in juveniles, this may just be a light bar); this is a good field character.

HABITAT AND DISTRIBUTION: Forest, at medium to high altitude, from about 1,000 up to 2,200m. In our area, recorded sporadically along the Albertine Rift; Ugandan records include Kibale, Kalinzo Forest and Bwindi Impenetrable Forest, in Rwanda recorded from Cyamudongo and southern Nyungwe Forests, in Burundi from just north of Musigati. Elsewhere, to eastern DR Congo.

Buhoma depressiceps

NATURAL HISTORY: Little known. Terrestrial, probably diurnal, hunting on the forest floor, takes cover in leaf litter, in holes or under ground cover. It is described as gentle and unaggressive, but when handled produces an unpleasant cloacal discharge. Lays eggs but no further details known. Diet said to include frogs.

Buhoma depressiceps Pale-headed Forest Snake. Left: Michele Menegon, Rwanda, juvenile. Right: Max Dehling, Rwanda, adult.

ULUGURU FOREST SNAKE *Buhoma procterae*

IDENTIFICATION: A small brown or grey, striped or spotted snake. The head is short, eye quite large, pupil round, the iris is orange or red. Body cylindrical, tail fairly short, about 15% of total length. The dorsal scales are smooth, in 17 rows at midbody, ventrals 140–154, subcaudals 33–50, high counts males. Maximum size about 52cm (big specimens usually females), average 25–35cm, hatchling size unknown but a juvenile with an umbilical scar was 14cm. Colour quite variable: brown or grey, with either a dark vertebral stripe or a double row of dark spots, which may coalesce to form crossbars, along either side of the vertebral scale row. Upper labials barred dark and light, ventrals cream, brown or grey.

HABITAT AND DISTRIBUTION: A Tanzanian endemic, confined to the woodland and forest of the Uluguru, Rubeho and Udzungwa Mountains. Conservation Status: IUCN Vulnerable B1ab(iii)+2ab(iii); range restricted, and its forest habitat is threatened, so needs monitoring.

NATURAL HISTORY: A gentle inoffensive little snake, apparently diurnal and terrestrial, although one was found at dusk, located by the screams of a frog it had seized. Found under stones, vegetable debris and other suitable ground cover, in the Udzungwa Scarp Nature Reserve three individuals were found inside a single fallen log; it appears to be relatively common in its favoured habitat. Clutches of between 12–22 eggs recorded, 1.8×1cm, in September. Diet: amphibians, one specimen had eaten a Uluguru Three-fingered Frog *Hoplophryne uluguruensis*.

Buhoma procterae

Buhoma procterae Uluguru Forest Snake. Top left: Michele Menegon, striped, Uluguru Mountains. Bottom left: Michele Menegon, Uluguru Mountains, uniform. Right: Michele Menegon, Uluguru Mountains, barred.

USAMBARA FOREST-SNAKE *Buhoma vauerocegae*

IDENTIFICATION: A little brown forest snake with a fine dark vertebral stripe. The head is small, eye quite large, pupil round. The body is cylindrical, tail short, 15–18% of total length. The dorsal scales are smooth, in 17 rows at midbody; ventrals 122–133, subcaudals paired, 35–48, high counts males. Maximum size about 41cm, large individuals usually females, average 20–35cm, hatchling size unknown but a small female with an umbilical scar was 11.5cm. Colour shades of brown or grey brown, with a fine dark vertebral line extending from the nape onto the tail. There may be a dark V-shaped mark on the crown, tapering to a line and one or two dark crossbars on the nape. Lip scales white with dark edging. Ventrals paler, sometimes brown, in some specimens the ventrals darken towards the tail, in some there is a faint dark median stripe.

HABITAT AND DISTRIBUTION: A Tanzanian endemic species, found only in the forest of the Usambara, Nguru and Uluguru Mountains.

Buhoma vauerocegae

NATURAL HISTORY: Apparently diurnal and terrestrial. Specimens recorded as active on the forest floor, others were found under loose tree bark and fallen timber. They are gentle and unaggressive, but when picked up produced an unpleasant cloacal discharge. Oviductal eggs found in females in September and October, one female had eight eggs measuring roughly 0.6cm. Diet: amphibians.

Buhoma vauerocegae Usambara Forest Snake. Thomas Hakannson, Usambara Mountains.

SUBFAMILY PSEUDOXYRHOPHIINAE

A subfamily of about 90 species and 20 genera, mostly from Madagascar, but a few from other African off-shore islands. Two genera occur on mainland Africa. One species, the Slug Eater *Duberria lutrix*, occurs in our area.

SLUG-EATING SNAKES *DUBERRIA*

A south-east African genus of small, harmless slug-eating snakes; they have short blunt heads, large eyes, the body is cylindrical and the tail very short. They give live birth. Four species in the genus are known, one reaches our area.

SLUG EATER *Duberria lutrix*

IDENTIFICATION: A stout, harmless little brown snake with short head, often in mid to high altitude grassland. The eye is large and dark with a round pupil, but this is hard to see. The body is cylindrical, tail short, 14–22% of total length (males have longer tails). Scales smooth, 15 midbody rows, ventrals 107–149 (counts usually low in East African specimens), subcaudals paired, 17–39 rows, cloacal scale entire. Maximum size about 45cm (big specimens usually females), but East African specimens average 20–30cm, hatchlings 8–11cm. Usually uniform brown (Kenya, north Tanzania, Albertine Rift) or grey (southern Tanzania) above, often a fine black vertebral line is present; the flanks may darken or lighten downwards (Ethiopian specimens have a black side-stripe). Head scales (especially lip scales) often black edged. Ventral scales very variable, may be dark (sometimes with light centres), blue-grey or pale, if pale if may have dark outer edges. All-black (melanistic) specimens are known from the Gregory Rift Valley in Kenya. Taxonomic Note: Several subspecies have been described; molecular work indicates they may be worthy of specific status.

Duberria lutrix

HABITAT AND DISTRIBUTION: High grassland and moist savanna, in East Africa from about 1,000m up to 2,600m, possibly higher. Widespread in the high country of Rwanda, Burundi, south-west Uganda and extreme north-west Tanzania. Another population is found in high western and central Kenya, south to Nairobi, west through the Maasai Mara National Reserve and the Serengeti plains, south-east to the Crater Highlands, Mt Meru and around Kilimanjaro, to Loitokitok on the Kenya side. A population also occurs in southern Tanzania from the Udzungwa mountains south-west to the north end of Lake Malawi (Lake Nyasa), and it is known from Magrotto Hill and elsewhere in the eastern Usambara Mountains. Elsewhere, south to South Africa, also found in the Ethiopian highlands. Conservation Status: IUCN Least Concern.

Duberria lutrix Slug Eater. Left: Bill Branch, Serengeti. Right: Michele Menegon, Livingstone Mountains (subspecies *shirana*).

NATURAL HISTORY: Terrestrial, diurnal and secretive, foraging in the grass and vegetation, when inactive hides in grass tufts, under vegetation heaps, logs, rocks, in soil cracks or down holes. Gentle and good-natured, it never tries to bite, if handled it may squirm and defecate. In South Africa, it may roll up in a flat spiral when threatened (its Afrikaans name translates as 'tobacco roll') but this behaviour has not yet been seen in East Africa. Slug Eaters give live birth, up to 12 young recorded (22 in South Africa), the clutch size is related to the size of the mother. These snakes are beneficial to farmers, eating only slugs and thin-shelled snails, which they find by following the slime trail. Slugs are simply swallowed, the snails are pulled out of their shells first, after eating the snake wipes the slime off on the ground or a plant stalk. In suitable moist areas, they may be very common; on an irrigated strawberry farm in Ethiopia Slug Eaters made up over 90% of the snake population, in the surrounding non-irrigated grassland they made up less than 5%.

SUBFAMILY PROSYMNINAE

An African subfamily containing only the small unusual snakes of the genus *Prosymna*, shovel-snouts; the subfamily is a sister group, oddly, to the sand snake clade.

SHOVEL-SNOUTS *PROSYMNA*

An African genus of small, mostly 20–30cm burrowing snakes, none larger than 45cm, most are grey or brown, often with speckling, although some species are spectacularly coloured and striped or banded. Most have smooth scales, in 15–21 rows at midbody. They have a sharp-edged, broad rostral scale for pushing through soil or sand, short heads and a relatively large eye. They are rarely encountered except after rainstorms, when they may move about on the surface or are flooded out of their holes. Sixteen species are known, from sub-Saharan Africa, their centre of distribution seems to be in the savannas of southern and eastern Africa, although three species reach West Africa. Seven species are known in our area, two are Tanzanian endemics. Several species have a curious defence mechanism, they snap the body into flat coils, looking like the flat spring of an old watch, if prodded they will jerk violently and coil and uncoil. They lay eggs. For many years, their dietary habits were unknown, and much speculated upon. Don Broadley discovered in the 1980s that they feed virtually exclusively on reptile eggs, although C.J.P. Ionides recorded reptile eggshells in the stomach of a *Prosymna pitmani* in 1953 but did not publish it.

KEY TO EAST AFRICAN MEMBERS OF THE GENUS *PROSYMNA*

1a Upper labials 7, ventrals black, only in Tanzanian coastal forests, Usambara Mountains and Shimba Hills....... *Prosymna semifasciata*, Banded Shovel-snout [p. 424]
1b Upper labials not usually seven, ventrals not black, not confined to Tanzanian coastal forests, Usambara Mountains and Shimba Hills......2
2a Dorsal scales in 19–21 rows......*Prosymna pitmani*, Pitman's Shovel-snout [p. 423]
2b Dorsal scales in 15–17 rows......3
3a Upper labials 5......*Prosymna greigerti*, Southern Sahel Speckled Shovel-snout [p. 421]
3b Upper labials 6......4

4a Dorsal scales with paired apical pits......*Prosymna ruspolii*, Prince Ruspoli's Shovel-snout [p. 423]
4b Dorsal scales with a single apical pit......5
5a Head mostly scarlet, body barred black and red, only in Uluguru Mountains...... *Prosymna ornatissima*, Ornate Shovel-snout [p. 422]
5b Head not mostly scarlet, body not scarlet and black......6
6a Rostral very angular in profile, body lacks white dots......*Prosymna ambigua*, Angolan Shovel-snout [p. 420]
6b Rostral rounded in profile, back often with white dots......*Prosymna stuhlmanni*, East African Shovel-snout [p. 425]

ANGOLAN SHOVEL-SNOUT/MIOMBO SHOVEL-SNOUT
Prosymna ambigua

IDENTIFICATION: A small grey shovel-snout from the northwest of our area. Rostral sharply pointed. Body cylindrical and stocky. Tail 12–15% of total length. Dorsal scales smooth, in midbody row of 17. Ventral scales 124–141 in males, 138–157 in females. Subcaudals paired, 25–34 in males, 15–21 in females. Maximum length about 36cm, average 20–30cm. Hatchling size unknown. Usually appears grey, although it can vary from dark purple-brown to blue gray, dorsal scales often dark-edged. Chin white, or with some brown. Ventrals with a blackish half-moon marking proximally, each ventral has its free edge coloured white. Subcaudal scales are patterned similarly to the ventrals. Taxonomic Note: The details here apply to the subspecies *Prosymna ambigua bocagii*.

Prosymna ambigua

HABITAT AND DISTRIBUTION: Found in moist woodland and savanna-forest mosaic around the equatorial lowland forest block. Fairly widespread in northern Uganda, known localities include Kidepo, Gulu, Lira and Serere. A population also occurs in western Kenya (Homa Bay, Lolgorien). Isolated records from north-west Tanzania include Seronera,

Prosymna ambigua Angolan Shovel Snout: Bill Branch, Serengeti.

Nzima, and Bukoba, a single Burundi record, Nyanza-Lac. Elsewhere, west to Cameroon, the nominate subspecies occurs south of the Congo forest. Conservation Status: IUCN Least Concern.

NATURAL HISTORY: Poorly known but probably similar to other members of the genus, i.e. nocturnal, burrowing spending its time in holes or buried in soft soil or sand, most active in the rainy season, eats snake and lizard eggs, lays a clutch of small eggs. It shows the 'watch-spring' behaviour described in the generic introduction.

SOUTHERN SAHEL SPECKLED SHOVEL-SNOUT
Prosymna greigerti

IDENTIFICATION: A small grey snake from the north-west of our area. Snout pointed, the eye is large and dark. Body cylindrical and stocky. Tail 8–12% of total length, longer in males. Scales smooth, in 17 rows at midbody. Ventrals males 144–159, females 159–190, Subcaudals paired, sometimes with a pit near the lateral free edge, 28–41 subcaudals in males, 19–26 in females. Largest male is 30cm, the largest female 34.6cm, average 20–30cm. From a distance, this snake appears speckled grey. The back is dark grey or purple; each dorsal scale has a white edge. The lower edges of the rostral and supralabials are white as is the ventrum; the pale underside sometimes extending up onto the flower flanks

Prosymna greigerti

Prosymna greigerti Southern Sahel Speckled Shovel Snout. Left: Stephen Spawls, Ghana. Right: Stephen Spawls; showing 'watch-spring' behaviour.

HABITAT AND DISTRIBUTION: Moist savanna and woodland, at low to medium altitude. Not yet recorded within our area, but may be expected in northern Uganda as it occurs just across the border in South Sudan and the north-east DR Congo. Elsewhere, west to Senegal.

NATURAL HISTORY: Poorly known but probably similar to other members of the genus, i.e. nocturnal, burrowing spending its time in holes or buried in soft soil or sand, most active in the rainy season, eats snake and lizard eggs. West African specimens laid clutches of 2–4 eggs in early March. This snake shows the 'watch-spring' behaviour described in the generic introduction.

ORNATE SHOVEL-SNOUT Prosymna ornatissima

IDENTIFICATION: An unmistakeable shovel snout with a unique red and black coloration and endemic to Tanzania. Body cylindrical, tail short. Midbody scale rows 15, ventrals 128–131 in males, 148 in the only known female, subcaudals 37–41 in males, 27 in the female. Largest male, 28.4cm (23cm body, 5.4cm tail), female holotype, 28.6cm total length, 25.2cm snout-vent length. Hatchling size unknown. The colour is distinctive, it is vividly banded black and red; the head is red with a black arrowhead on the neck and a black patch under the eye. The throat is pink, ventrals black.

Prosymna ornatissima

HABITAT AND DISTRIBUTION: Endemic to the type locality and vicinity, Uluguru Mountains, Tanzania. Known from cultivation at the edge of rain forest, early specimens from near two villages, Nyange and Vituri, and recently rediscovered in pitfall traps. Conservation Status: Critically Endangered B1ab(i,iii).

NATURAL HISTORY: Poorly known, but probably similar to other members of the genus, i.e. nocturnal, burrowing spending its time in holes or buried in soft soil or sand, most active in the rainy season, eats snake and lizard eggs, lays a clutch of small eggs. It is not known if it shows the 'watch-spring' behaviour described in the generic introduction.

Prosymna ornatissima Ornate Shovel Snout. Left: Michele Menegon, Uluguru Mountains. Right: Michele Menegon, Uluguru Mountains.

PITMAN'S SHOVEL-SNOUT Prosymna pitmani

IDENTIFICATION: A grey shovel-snout from southern Tanzania and Malawi. Body cylindrical, tail short, 7–11% of total length (longer in males). Dorsal scales smooth, with single apical pits, in 19–21 rows at midbody. Ventrals in males 139–149, in females, 155–163. Subcaudals paired, most with a pit near the lateral free edge; 21–28 subcaudals in males, 17–22 in females. Maximum size around 31cm, females larger than males, average 20–25cm, hatchling size unknown. It appears grey (sometimes brown) and speckled, each scale with a pale spot just anterior to the apical pit. Pale vermiculations and spots on the head. Ventrum white.

Prosymna pitmani

Prosymna pitmani Pitman's Shovel Snout. Walter Jubber, Selous Game Reserve.

HABITAT AND DISTRIBUTION: Found in moist savanna, in Miombo woodland and the coastal forest/savanna mosaic, at low altitude, sea level to around 1,000m. Said to favour valleys, near water. Occurs in south-east Tanzania, from the northern Selous Game Reserve south to Kilwa and Liwale. Elsewhere, a single record from Mpatamanga Gorge in Malawi. Conservation Status: IUCN Least Concern.

NATURAL HISTORY: Poorly known but probably similar to other members of the genus, i.e. nocturnal, burrowing spending its time in holes or buried in soft soil or sand, most active in the rainy season. A Liwale female had reptile egg shells in her stomach. A clutch of four small eggs, roughly 2.3×0.8cm were laid in June. It is not known if it shows the 'watch-spring' behaviour described in the generic introduction.

PRINCE RUSPOLI'S SHOVEL-SNOUT Prosymna ruspolii

IDENTIFICATION: A small snake from the dry, arid areas of northern Tanzania, Kenya and Somalia. Body cylindrical, tail short, 10–16% of total length (longer in males). Scales smooth, in 17 midbody rows. Ventrals 145–152 in males and 163–170 in females, subcaudals 35–36 in males, 23–26 in females. About five mid-dorsal scale rows usually have single apical pits, but the pits in the lateral rows are paired. Largest male about 24cm, largest female 30cm. Looks grey from a distance, close up the grey dorsal scales are bordered with white. A few specimens (predominantly those from north-west Kenya) have a black head and neck or just a black collar. White below, sometimes a brown patch on the throat extends back to the sixth ventral. The tongue is vivid pink with a white tip.

Prosymna ruspolii

Prosymna ruspolii Prince Ruspoli's Shovel Snout. Left: Stephen Spawls, Watamu. Right: Tomas Mazuch, Ethiopia, eating gecko egg.

HABITAT AND DISTRIBUTION: Dry *Acacia/Commiphora* woodland, subdesert steppe and coastal thicket, at altitudes between sea level and 1,500m. In East Africa the subspecies *Prosymna ruspolii keniensis* occurs in northern Kenya between Lake Baringo and Lokori, eastwards to Archer's Post, in Tanzania from the Simanjiro Plain round the eastern side of Mt Kilimanjaro, north and west to Sultan Hamud, eastwards through Tsavo to Watamu. An isolated record from Kakuma in northwest Kenya. it is probably more widespread throughout the area but undercollected. Elsewhere, the subspecies *Prosymna ruspolii ruspolii* is found in Ethiopia and Somalia.

NATURAL HISTORY: Poorly known but probably similar to other members of the genus, i.e. nocturnal, burrowing spending its time in holes or buried in soft soil or sand, most active in the rainy season, eats snake and lizard eggs. Clutches of 3–4 small eggs, 2.8×0.7cm recorded. It shows the 'watch-spring' behaviour described in the generic introduction. The Kakuma specimen was found on a dry hillside under a coil of barbed wire.

BANDED SHOVEL-SNOUT *Prosymna semifasciata*

IDENTIFICATION: A small grey snake with distinctive black and white spotting. The body is cylindrical, tail short, about 10–15% of total length. Midbody scale count 17; ventrals 42–45; cloacal scale entire; subcaudals paired; 43 in male, 28 in female. Maximum size about 26cm, averages 20–25cm. Grey or blue-grey, the back marked with numerous elongate black spots which may form dorsolateral rows, these spots edged with white or turquoise spots. The ventral scales are black with white edging and grey stippling laterally. A melanistic (all black) specimen is recorded from Saadani.

Prosymna semifasciata Banded Shovel Snout. Quentin Luke, Shimba Hills.

Prosymna semifasciata

HABITAT AND DISTRIBUTION: Coastal forest and woodland, from sea level to 230m altitude. In Tanzania known from the north coast; localities include Kwamgumi Forest and Saadani National Park, also recorded from the Shimba Hills forest in south-east coastal Kenya.

NATURAL HISTORY: Apparently endemic to lowland coastal forest. Activity patterns poorly understood, most shovel-snouts come out at night but the Shimba Hills specimen was moving by day on the forest floor. Habits presumably similar to other shovel-snouts.

EAST AFRICAN SHOVEL-SNOUT Prosymna stuhlmanni

IDENTIFICATION: A small snake with a shovel-shaped snout. The head is slightly distinct from the neck, eye fairly large, pupil round. The tongue is pink with a white tip. The body is cylindrical, tail fairly short, 11–15% of total length. Scales smooth, in 17 rows at midbody; ventrals 124–153 in males, 138–164 in females, subcaudals 27–39 in males, 17–30 in females. Maximum size about 30cm, average 18–25cm, hatchling size unknown. Colour quite variable, described as red-brown, blue-grey or black, each scale with a pale centre, but most East African specimens are light to dark grey, the scales are shiny. Some specimens have a single or double row of white spots along the centre of the back. The snout is distinctly brownish yellow, the ventrals white, pearly or cream. The tongue is pink with a white tip.

Prosymna stuhlmanni

HABITAT AND DISTRIBUTION: Semi-desert, dry and moist savanna at low to medium altitude, from sea level to about 1,800m. A cluster of records in western Kenya (Ol Ari Nyiro, Kabarnet, Rumuruti). Widespread in south-eastern Kenya, including the northern loop of the Tana River and the coast, and most of eastern and southern Tanzania. Elsewhere, south to South Africa.

NATURAL HISTORY: Poorly known but probably similar to other members of the genus, i.e. nocturnal, burrowing spending its time in holes or buried in soft soil or sand, sometimes under ground cover, most active in the rainy season, eats snake and lizard eggs (one

Prosymna stuhlmanni East African Shovel Snout. Left: Stephen Spawls, Watamu. Right: Lorenzo Vinciguerra, northern Tanzania, showing neck flattening.

was suspected to have eaten a termite). A Voi specimen was under a sisal stump, a specimen from the Tharaka plain was inside a rotten log. Clutches of 3–4 eggs recorded, one was 1.9×0.8cm, the other 3×0.7cm; gravid females collected in January and February in southern Tanzania. This snake doesn't appear to show the 'watch-spring' behaviour described in the generic introduction but it has an unusual threat display, some individuals lifting up the front half of the body and flattening the neck like a cobra, while in this position it flicks the tongue slowly in and out in a deliberate manner, if further teased it tips the head right back, while continuing to flick the tongue.

SUBFAMILY PSAMMOPHIINAE

A subfamily of rear-fanged snakes, with just over 50 species and eight genera. They are mostly African, with a limited penetration into Europe and Asia and a single Madagascan species. Four genera and 21 species occur in our area. Most are longitudinally striped, relatively large, fast moving and diurnal.

BARK SNAKES HEMIRHAGERRHIS

An African genus of slim, little rear-fanged snakes. They are diurnal and partially arboreal, in savanna and semi-desert. Most eat lizards. Their small size and non-aggressive habits mean they are harmless to humans. Recently one of the two known species, the Bark Snake *Hemirhagerrhis nototaenia*, was divided into three species. Thus four species of this genus are now known, three occur in East Africa.

KEY TO EAST AFRICAN MEMBERS OF THE GENUS HEMIRHAGERRHIS

1a	Brown or tan in colour, with either a straight-edged, dark or pale vertebral stripe, or no vertebral stripe at all, ventrals 140–159......*Hemirhagerrhis kelleri*, Striped Bark Snake [p. 427]		
1b	Grey in colour, usually with an irregular black vertebral stripe or line of dots or crossbars, ventrals 152–187...... 2		
2a	Temporal formula usually 1+2, lower labials usually 9, in our area often with distinct cross bars, underside heavily stippled grey-brown, most records from Tanzania......*Hemirhagerrhis nototaenia*, South-eastern Bark Snake [p. 428]		
2b	Temporal formula usually 2+3, lower labials usually 10 or more, in our area usually has a zigzag vertebral stripe, underside pale with grey streaks, most records from Kenya......*Hemirhagerrhis hildebrandtii*, Kenyan Bark Snake [p. 428]		

STRIPED BARK SNAKE *Hemirhagerrhis kelleri*

IDENTIFICATION: A slim brown striped snake with a small eye. The pupil is round or slightly oval vertically, the iris is golden brown but the dark head stripe runs through it. Supraocular slightly raised. Body cylindrical, tail about 21–27% of total length. Scales smooth, in 17 rows at midbody, ventrals 137–159, subcaudals 61–78. Maximum size about 40cm, average 25–35cm, hatchling size unknown. There is usually a dark brown vertebral stripe (sometimes with a light yellow stripe in the centre), a fawn dorsolateral stripe and a pale brown flank stripe. The vertebral stripe is occasionally very pale or absent, such faintly striped individuals may have vertebral and lateral lines of small brown spots. The dorsal scales often have a fine black dot at the posterior end. The head has fine brown and tan reticulations. Ventrals white or cream with three dashed brown or rufous, white-centred stripes.

Hemirhagerrhis kelleri

HABITAT AND DISTRIBUTION: Dry savanna, coastal thicket and semi-desert, from sea level to about 1,500m, but most common below 1,200m. Widespread in north-eastern and eastern Kenya, from Mandera through Tarbaj across to Laisamis, Samburu Game Reserve and Kalama Conservancy, south through Ukambani and Tsavo to Mkomazi National Park and eastwards to the coast, although few records on the actual coastal plain save Lamu, the Tana Delta, Mombasa and Makupa. Also occurs from Olorgesaile and Kajiado south into northern Tanzania, to Olduvai Gorge, thence eastwards round the southern side of Kilimanjaro. No records from north-west Kenya, but a single Uganda record, from Kidepo National Park. Elsewhere, to eastern Ethiopia, Somalia and South Sudan.

Hemirhagerrhis kelleri Striped Bark Snake. Left: Stephen Spawls, northern Tanzania. Right: Stephen Spawls, Aruba Lodge.

NATURAL HISTORY: Poorly known. Diurnal and partially arboreal, climbing into small trees and bushes. Fairly quick moving, at night shelters under ground cover or in holes Lays eggs, a clutch of three recorded. Captive specimens fed readily on lizards, especially day geckoes (*Lygodactylus*). A population in southern Somalia fed almost exclusively on gecko eggs, a Tanzanian snake had eaten a Somali-Maasai Clawed Gecko *Holodactylus africanus*.

KENYAN BARK SNAKE *Hemirhagerrhis hildebrandtii*

IDENTIFICATION: A little grey snake with a zigzag vertebral line. Very similar to the previous species, eyes prominent but small, pupil round iris silvery brown and the facial stripe goes through the eye. Tail 24–34% of total length. Scales smooth, in 17 rows at midbody, ventrals 157–177, subcaudals 79–105, cloacal scale divided. Maximum size about 48cm (possibly larger), average 25–40cm, hatchling size unknown. Colour grey, most specimens have a dark vertebral pattern consisting of broad black subtriangular blotches connected by a dark vertebral stripe, but some specimens also have narrow crossbars; the flanks are mottled grey, sometimes with black blotching. The sides of the head and neck are often lighter than the rest of the body, especially in inland specimens. The tail tip may be orange or rufous. Below light grey or buff, lightly stippled with black or brown.

Hemirhagerrhis hildebrandtii

HABITAT AND DISTRIBUTION: Widespread in semi-desert, dry and moist savanna and coastal thicket, from sea level to about 1,200m altitude. Occurs almost throughout northern and eastern Kenya (although records from the extreme east lacking), and sympatric on the south coast with the South-eastern Bark Snake, notably in the Shimba Hills. Spreads into northern Tanzania, round the base of Kilimanjaro and east to Mkomazi. Elsewhere north to eastern Ethiopia, southern Somalia and South Sudan. Conservation Status: IUCN Least Concern.

Hemirrhagerrhis hildebrandtii Kenyan Bark Snake. Stephen Spawls, Watamu.

NATURAL HISTORY: As for the previous species; fast-moving, diurnal and semi-arboreal. Active during even the hottest hours of the day in northern Kenya. Climbs big trees by working its way up the bark, its camouflage is superb. It may dart out to seize lizards from ambush, often feeds on *Lygodactylus*, dwarf geckoes, but also recorded as preying on the eggs of the Tree Gecko *Hemidactylus platycephalus*, at Kismayu in Somalia. A Tarbaj female laid four eggs in early September.

SOUTH-EASTERN BARK SNAKE *Hemirhagerrhis nototaenia*

IDENTIFICATION: A little grey snake with a zigzag vertebral line and/or cross bars. Eyes prominent but small, pupil round iris yellow, tongue red. The supraocular is raised, giving a 'raised eyebrow' impression. Tail 22–30% of total length. Scales smooth, in 17 rows at midbody, ventrals 151–187, subcaudals 58–90, cloacal scale divided. Maximum size about 44cm, average 25–40cm, hatchling size unknown but an 18cm juvenile was collected in February at Shimoni. Grey, with dark irregular crossbars, and a dark, broad or narrow vertebral stripe or a mixture of the two. The neck is often lighter than the rest of the body. The tail tip may be orange or rufous. Below grey speckled with black.

Hemirhagerrhis nototaenia

HABITAT AND DISTRIBUTION: A snake of dry and moist savanna, woodland and coastal thicket, even mangroves, from sea level to about 1,600m altitude, but most common below 1,200m. A handful of records from coastal Kenya (Mombasa, Diani Beach, Shimoni), widespread in Tanzania, in two apparently disjunct populations: from northern Shinyanga south through Tabora to Lake Rukwa, and south from Lake Manyara across most of eastern Tanzania; the two populations are probably interconnected. No records from Uganda, Rwanda or Burundi. Elsewhere south to Botswana, and a bizarre handful of disjunct records from the central and west African Sahel.

NATURAL HISTORY: Diurnal and semi-arboreal, often found on the rough bark of trees (hence the name), looking for geckoes and their eggs. Non-aggressive and doesn't usually bite if picked up. When approached it often freezes, relying on its camouflage, which is superb on a tree trunk; the bars and vertebral line breaking up the outline. If detected in the open it slides rapidly away. Shelters under bark, ground cover, in tree cracks, etc. It lays from 2–8 elongate eggs, roughly 0.6×2.5cm; gravid females were taken in southern Tanzania in October and May. Diet: mostly lizards, especially tree-dwelling geckoes, gecko eggs, skinks and small agamas also taken, and occasionally amphibians.

Hemirrhagerrhis nototaenia Southeastern Bark Snake. Walter Jubber, Selous Game Reserve.

SAND SNAKES *PSAMMOPHIS*

A genus of diurnal, fast-moving rear-fanged snakes, with long heads and big eyes with round pupils, slim elongate bodies, and long tails; evolved to move fast in lightly vegetated or grassy habitats. Most are striped above and below. Their dorsal scales are smooth, in 11–19 rows at midbody. Many sand snakes are partially arboreal. Lizards are a favoured prey, they hunt them by sight, seizing and chewing until the venom subdues the prey. Unusually, sand snakes have very small, smooth hemipenes, and this may be connected with the size and diurnal habits, enabling them to separate rapidly and flee if disturbed while copulating. Male combat has occasionally been seen in captive animals. Sand snakes have relatively large, grooved fangs and some humans have suffered local symptoms of envenomation such as swelling, haemorrhage, intense itching, local pain and nausea following the bites of some species. If held by the tail, they may lash and struggle so violently that the tail may break off; it is not regenerated. When undisturbed, sand snakes show a curious behaviour known as 'self-rubbing', the snake smearing a secretion from a gland just inside the nose along the belly and lower dorsal scales; the purpose of this is uncertain. About 35 species are known, occurring almost throughout Africa, with a limited penetration into the Middle East and western Asia, in all types of habitat except true desert (but found in oases) and closed forest, although they will enter forest clearings and fringes. Ten species occur in East Africa. Sand snakes are among the region's most visible snakes; if you see a single snake on a safari it is likely to be a sand snake; in some areas of East Africa four or five species of sand snake may be found.

The taxonomy of some sand snake species has caused problems. Recent molecular taxonomic work has shown that snakes of the genus *Dromophis*, olympic snakes, differing from most sand snakes in their tooth arrangements, actually nest within the *Psammophis* clade; thus *Dromophis lineatus*, the Striped Olympic Sand Snake, becomes *Psammophis lineatus*. Other problems concerned the relatively big savanna sand snakes belonging to the so-called Hissing Sand Snake *Psammophis sibilans* complex. Specific names that have been used within East Africa for snakes in this group include *sibilans*, *mossambicus*, *orientalis*, *phillipsii*, *sudanensis*, *subtaeniatus* and *rukwae*. In some areas, members of this group appear to hybridise, and some forms change in colour as they grow. However, molecular work by Rhodes University's Chris Kelly and co-workers indicate that in East Africa, there are four well-supported species; one clade consists of *rukwae*, *sudanensis* and *orientalis*, with *mossambicus* in another clade, and they can usually be identified, as detailed in our key below, although occasional problematic, intermediate individuals exist.

KEY TO EAST AFRICAN MEMBERS OF THE GENUS *PSAMMOPHIS*

1a Transverse dashes on the outer edges of the neck ventrals, no large fang-like teeth present in front of the fangs, which are below the level of the back of the eye......*Psammophis lineatus*, Striped Olympic Sand Snake [p. 433]

1b No transverse dashes on the outer edges of the neck ventrals, enlarged fang-like teeth present below the front margin of the eye, enlarged fangs below the back margin of the eye......2

2a Dorsal scales in 17 rows at midbody......3

2b Dorsal scales in less than 17 rows at midbody......7

3a Upper labials usually 9, with fifth and sixth entering the orbit, ventrals 170–197, subcaudals 143–166, dorsum yellow with three black stripes......*Psammophis punctulatus*, Speckled Sand Snake [p. 438]

3b Upper labials usually 8, with fourth or fourth, fifth and sixth entering the orbit, ventrals 148–183, subcaudals 82–122, dorsum not yellow with three black stripes......4

4a	Usually the first five lower labials in contact with the anterior sublinguals......*Psammophis rukwae*, Lake Rukwa Sand Snake [p. 436]	7a	Dorsal scales in 13–15 rows at midbody.......8
4b	Usually the first four lower labials in contact with the anterior sublinguals......5	7b	Dorsal scales in 11 rows at midbody...... *Psammophis angolensis*, Dwarf Sand Snake [p. 431]
5a	Ventrum uniform or with ill-defined brown or black line or dashes ...... *Psammophis mossambicus*, Olive Sand Snake [p. 434]	8a	Dorsal scales in 13 rows at midbody...... *Psammophis pulcher*, Beautiful Sand Snake [p. 432]
5b	Ventrum usually with a pair of well-defined black lines, yellow between them, ends of ventrals white......6	8b	Dorsal scales in 15 rows at midbody.......9
6a	Pale crossbars on back of the head absent, or extremely faint, body stripes often faint or non-existent......*Psammophis orientalis*, Eastern Stripe-bellied Sand Snake [p. 435]	9a	Two upper labials, usually fifth and sixth, enter orbit, head usually uniform brown, bordered below by a dark stripe through the eye......*Psammophis biseriatus*, Link-marked Sand Snake [p. 439]
6b	Distinct pale crossbars on the back of the head, body stripes usually clear and well-defined......*Psammophis sudanensis*, Northern Stripe-bellied Sand Snake [p. 437]	9b	Three upper labials, usually the fourth, fifth and sixth entering the orbit, head with dark bordered tan blotches on a paler background and a dark-bordered light longitudinal stripe along the junction of intranasal and prefrontals...... *Psammophis tanganicus*, Tanganyika Sand Snake [p. 440]

DWARF SAND SNAKE *Psammophis angolensis*

IDENTIFICATION: A small striped sand snake with fine light crossbars on the head. The eye is large, pupil round, iris red-brown, tongue black. Body cylindrical, tail 27–30% of total length. Scales smooth, in 11 rows at midbody, ventrals 135–156, subcaudals 56–82 (most Tanzanian specimens 56–68), cloacal scale divided. Maximum size about 50cm, average 30–40cm, hatchlings 14–15cm. Distinctly marked; the top of the head is black or deep brown with three fine yellow or white crossbars between the eye and the nape, behind the third is a broad dark collar, followed by one, two or three dark crossbars, which may coalesce. There is a broad dark brown, finely yellow-edged vertebral stripe, flanks light or yellow-brown, sometimes with fine dark hairlines along

Psammophis angolensis

Psammophis angolensis Dwarf Sand Snake. Bill Branch, South Africa.

the lower flanks. The ventrals are white or cream, as are the lips, chin and throat. Can be distinguished with certainty by the combination of locality, size, fine neck crossbars and stripes.

HABITAT AND DISTRIBUTION: Dry and moist savanna and grassland, from sea level to 2,000m altitude, possibly higher. Found almost throughout Tanzania save parts of the north-west and north-east; occurs from the Serengeti Plains southward (but not recorded on any islands), elsewhere south to northern South Africa and Angola, with a disjunct population in highland Ethiopia.

NATURAL HISTORY: Diurnal, terrestrial and secretive, living usually in well-vegetated areas, presumably hides under ground cover or in holes at night. Quick moving, doesn't usually bite if picked up but struggles. They lay from 2–5 elongate eggs, roughly 0.5×1.5cm; females were gravid in September and 15cm hatchlings collected in November and January in southern Tanzania. Diet: lizards, especially skinks and small geckoes, skink eggs and frogs. A rare species, nowhere common, the only other sand snake of similar size is the Beautiful Sand Snake *Psammophis pulcher*.

BEAUTIFUL SAND SNAKE *Psammophis pulcher*

IDENTIFICATION: A small striped snake, with a rounded snout, a big eye and a round pupil. The body is cylindrical, tail long, 36–39% of total length. Scales smooth, in 13 rows at midbody, ventrals 140–147, subcaudals paired, 97–108. Maximum size 43cm. Hatchling size unknown. It is striped; a fine brown, orange or maroon, black-edged vertebral line is bordered by two uniform grey or brown dorsolateral stripes, darkening downwards to becomes a blank flank stripe which extends onto the side of the head but not the lips, which are cream. The lower one and half dorsal scale rows are white. The belly is yellow centrally, bordered by fine red lines, the outer edges of the ventrals white.

Psammophis pulcher

HABITAT AND DISTRIBUTION: One of East Africa's rarest snakes, before 2010 it was known from four specimens. The holotype was collected on the Webi Shebelle River in the western Ogaden, Ethiopia, by Donaldson-Smith in 1894, at an altitude of 500m. A second specimen was found at Voi in 1961, a third at Ngomeni near Mwingi, Ukambani, in 1972 and a fourth specimen is known from 'Somaliland', without further data. However, in August 2011, the safari guide Sean Flatt collected several specimens at Bisanadi National Reserve, east of Meru National Park. It apparently inhabits dry savanna at low altitude, and is probably quite widespread in eastern Kenya, despite its rarity.

Psammophis pulcher Beautiful Sand Snake. Stephen Spawls, Bisanadi National reserve.

NATURAL HISTORY: Poorly known but probably similar to other small *Psammo-*

phis. Flatt noticed that they were active by day, a male and a female were caught outside the same hole on successive days and the captive specimens fed on dwarf geckoes, *Lygodactylus*. Most specimens were seen on the ground. A male introduced into a cage with a female in August wrapped himself around her and rubbed his nose along her flank, she turned and rubbed noses with him, face to face, while they mated.

STRIPED OLYMPIC SAND SNAKE *Psammophis lineatus*

IDENTIFICATION: A medium-sized sand snake, usually in marshy areas. The head is long, pupil round, iris brown. The body is muscular and cylindrical, tail fairly long, about one-third of the total length but often mutilated. The scales are smooth, in 17 rows at midbody, ventrals 138–167, subcaudals paired, 73–105. Maximum size about 1.2m, average 70cm–1m, hatchling size unknown but a Ghanaian juvenile with an umbilical scar was 24cm. Colour essentially brown or olive. Some individuals are uniform above, others striped; juveniles are usually striped. There may be any combination of a broad brown dorsal stripe, a fine yellow vertebral line, a fine yellow dorsolateral stripe and a grey-brown lateral stripe, black-edged on the lower margin. The head is sometimes green. There are usually two or three pale yellow neck crossbars, more common in juveniles. The lip scales are sometimes black edged; the preocular scale is often distinctly yellow, and this can be seen in the field. The belly is yellow, greenish-yellow or cream, with distinct transverse black marks at the outer edges, particularly on the neck ventrals; these dashes are a valuable clue to the identification of this snake.

Psammophis lineatus

Psammophis lineatus Striped Olympic Sand Snake. Top: Stephen Spawls, Ghana. Middle: Jean-Francois Trape, West Africa. Bottom: Caspian Johnson, Ugalla.

HABITAT AND DISTRIBUTION: Moist savannas, especially on flood plains, around lakeshores and marshes and in damp grassland, although it will move some distance away from water. It occurs at altitudes of 500–1,800m in our area, elsewhere down to sea level. East African records are sporadic, in Kenya known from Kisumu and the flats east of there, quite widespread in a broad belt across central and northwest Uganda (and Torit in South Sudan), also known from Lake Nabugabo. Occurs around the northern Lake Tanganyika shore in Burundi, south to Kigoma and Ujiji and the Malagarasi River mouth in Tanzania. Might occur right along the Tanzanian shores of Lake Tanganyika, but unrecorded, known from the north end of Lake Malawi (Lake Nyasa). Elsewhere, south to northern Botswana, west to Senegal.

NATURAL HISTORY: A diurnal, terrestrial snake, fast moving; it swims well and will hunt in the water. If approached it darts away quickly, if seized some specimens seem reluctant to bite but others bite freely. Takes cover in waterside vegetation, in holes or under ground cover at night, a Ghanaian specimen was under a clump of mud. Clutches of 6–9 eggs, roughly 1.5×3cm, recorded. In the wild it feeds largely on amphibians but a captive specimen took a range of lizards.

OLIVE SAND SNAKE/HISSING SAND SNAKE
Psammophis mossambicus

IDENTIFICATION: A big brown snake, common in many areas of East Africa. It has a rounded deep snout (a 'roman nose'), overhanging supraocular and preocular scales (said to have a 'penetrating expression'), a big eye with a round pupil, the iris may be shades of yellow or brown. Body cylindrical and muscular, tail long, 27–33 % of total length. Scales smooth, in 17 rows at midbody, ventrals 151–183, cloacal scale divided, subcaudals paired, 82–110. Maximum size about 1.8m (possibly larger), average 1–1.5m, hatchlings 28–33cm. Quite variable in colour, usually olive brown above, but may be any shade of brown or grey. All the dorsal scales or just the central ones may be black edged, giving a finely striped appearance. The labial scales, chin and sides of the neck may be yellow, orange or dull red, the lips may be black speckled, as may the lower flanks. Belly usually yellow, cream or white (sometimes white anteriorly shading to yellow posteriorly), sometimes with a faint or clear dark line or series of dashes on each side, but never with a pair of clear black lines.

Psammophis mossambicus

HABITAT AND DISTRIBUTION: Widespread in a range of habitats, absent only from dryer semi-desert, near desert, closed forest and montane habitats over 2,500m. Thus it has not been found in eastern or northern Kenya (except at Moyale and the Hurri Hills). Often in waterside vegetation or associated with water sources, and will follow these into dryer areas, for example occurs around the shores of Lake Baringo. Elsewhere, south to eastern South Africa.

NATURAL HISTORY: Diurnal, partially arboreal. Fast moving and alert, if threatened it dashes away and hides. Often mistaken for more dangerous species due to its size and colour. It emerges in the mid-morning to bask and then hunts. If restrained it will bite vigorously. From 8–30 eggs, roughly 1.5×3cm, are laid in a suitable moist hole or damp tree crack, or in leaf litter. In South Africa these take about two months to hatch. Hatchlings collected in November and December in southern Tanzania, and eggs were laid in September there. A wide range of prey recorded, including snakes (one specimen ate a young black mamba), lizards, rodents, frogs and even birds; juveniles eat mostly lizards. Often eaten by birds of prey, caught as they bask on tops of bushes. Some local symptoms of envenomation recorded following bites to snake handlers, including pain, which may be severe, swelling, discoloration, nausea and intense local itching.

Psammophis mossambicus Olive Sand Snake. Top: Stephen Spawls, Baringo. Inset: Stephen Spawls, showing unstriped underside. Bottom: Stephen Spawls, Kilifi.

EASTERN STRIPE-BELLIED SAND SNAKE
Psammophis orientalis

IDENTIFICATION: A fairly big, fast-moving, faintly striped or unicoloured Sand Snake. Head long, snout pointed, supraocular slightly raised, eyes large, pupil round, iris yellow-brown. The body is cylindrical and muscular, tail about one-third of total length. Scales smooth, in 17 rows at midbody, ventrals 148–170, subcaudals 94–116, paired, cloacal scale divided. Maximum size about 1.25m, average 80cm–1.1m, hatchlings 27–32cm. Colour all shades of brown, from light yellow brown to deep brown, unicoloured or with paler dorsolateral stripes. The belly has two broad black or dark brown stripes, with yellow between them and yellow, or white, on the outer edges. The upper labials may be yellow, cream or bluish-green, immaculate or blotched brown. Usually distinguished from other similar sand snakes by the combination of obvious black ventral lines and faint or absent dorsal stripes.

HABITAT AND DISTRIBUTION: Coastal thicket and moist savanna, from sea level to 1,300m altitude. Occurs southwards from Witu along the length of the East African coast, a short way up the Tana River, inland in Tanzania up the Rufiji River and beyond to Dodoma, also inland in the south-east corner of Tanzania. A sight record (not shown on map) from near Tabora, Tanzania, indicates it might occur much further to the west. Elsewhere, inland to Malawi, Mozambique and eastern Zimbabwe.

Psammophis orientalis

NATURAL HISTORY: Fast moving, diurnal, hunting mostly on the ground but will ascent into low trees and bushes, or up on rocks. By night shelters in holes, under ground cover, in hollow logs or sleeping up in a thicket on the branches. It hunts by sight, chasing moving prey items. If restrained it will bite fiercely and thrashes around, rotating vigorously, liable to break its tail if held too far back. Two to ten elongate eggs, roughly 1×3cm are laid, a Shimoni specimen laid nine eggs in mid-February; hatchlings collected in November and December in southern Tanzania. Diet mostly lizards and other snakes but will take rodents, frogs and birds. A common snake on the East African coast, often seen around habitation and in gardens, and tolerant of urbanisation, where rock walls and ornamental trees provide habitat for the lizards it eats.

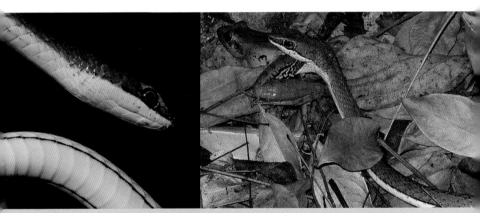

Psammophis orientalis Eastern Stripe-bellied Sand Snake. Left: Stephen Spawls, Watamu. Right: Walter Jubber, Selous Game Reserve.

LAKE RUKWA SAND SNAKE *Psammophis rukwae*

IDENTIFICATION: A large and rather rare sand snake. Head fairly long, eye large, pupil round, iris red-brown. Body slim and cylindrical, tail long, about one-third of total length. Scales smooth, in 17 midbody scale rows, ventrals 148–183, cloacal scale divided, subcaudals 82–122. Maximum size about 1.3m, average 80cm–1.2m, hatchling size unknown. Usually striped; there is fine pale vertebral line astride a broad brown or rufous dorsal stripe, flanks yellow or light brown, ventrals yellow or cream with a very fine black line on each side; sometimes yellow between the lines and white on the outer ventrals; distinguishable from other striped sand snakes of the group by a combination of dorsal stripes (often subdued) and fine (not thick) black ventral lines.

HABITAT AND DISTRIBUTION: Recorded from a number of disparate, widely scattered localities. A few records from western and central Tanzania (including Lake Rukwa), in medium-altitude moist savanna, Kenyan localities include Kaimosi, Kakuma, Lake Baringo and Kakuyuni near Machakos, in Uganda known from Kumi and Rhino Camp, recorded at Torit in the southern Sudan, elsewhere west across the Sahel to Senegal.

NATURAL HISTORY: Not well known. Diurnal and partially arboreal, fast moving, hunts by sight. Lays eggs but no clutch details known. Diet presumably mostly lizards and other snakes.

Psammophis rukwae

Psammophis rukwae Lake Rukwa Sand Snake. Stephen Spawls, Baringo. Inset Stephen Spawls, showing fine belly striping.

NORTHERN STRIPE-BELLIED SAND SNAKE
Psammophis sudanensis

IDENTIFICATION: A fairly big striped sand snake with a long head and a pointed snout. The eye is large, the pupil round, yellow or brown. The body is slim and muscular, tail long, about one-third of total length. Scales smooth, in 17 rows at midbody, ventrals 146–180, subcaudals 82–129, cloacal scale divided. Maximum size about 1.2m, average 70cm–1m, hatchling size 26–28cm. Conspicuously striped, the broad brown vertebral stripe usually has a yellow hairline down its centre. On each side is a fine yellow dorsolateral stripe, often finely edged with black and a broad brown flank stripe. The lowest dorsal scale rows are cream or white, as are the outer edges of the ventrals. There is a clear black line on each side of the ventrals, with yellow between the lines. The head is distinctly marked, with three yellow stripes down the snout, the top of the head is brown with pale cross and angled bars The neck may be barred or dark-blotched. Lips usually light, speckled with brown.

Psammophis sudanensis

Psammophis sudanensis Northern Stripe-bellied Sand Snake. Left: Stephen Spawls, showing black striping dividng yellow and black. Right: Stephen Spawls.

HABITAT AND DISTRIBUTION: Coastal thicket, moist and dry savanna and high grassland, from sea level to 2,700m. North and west from Dar es Salaam and Morogoro through north-central Tanzania to Tabora, east-central Kenya (common in Nairobi National Park), north-west to Nakuru, Gilgil and Mt Longonot (where it is found on the rim of the crater), east of Mt Kenya it extends north to Laikipia, Isiolo, Wamba and the Mathews Range. Also in north-western Kenya and eastern Uganda. Isolated records from Kitgum, and Katwe on Lake Edward, Uganda, Moyale in Kenya. Elsewhere (depending on its status), to Sudan, possibly Egypt, possibly west to Cameroon.

NATURAL HISTORY: Diurnal, swift-moving, partially arboreal. Climbs well, often sheltering in thickets, bushes and hedges. At night it may sleep up in the branches, or in holes, hollow logs or under ground cover. Hunts by sight, chasing its prey, which it seizes and chews, holding the victim in its coils. If threatened, it moves off swiftly, one specimen jumped into a river to escape, juveniles are described at actually jumping off the ground in their attempts to escape. From 4–10 eggs are laid, Nairobi specimens were gravid in January and February, hatchlings collected at Athi River in May. Diet: mostly lizards and other snakes. A favoured prey of harrier or snake eagles.

SPECKLED SAND SNAKE *Psammophis punctulatus*

IDENTIFICATION: A big, thin, striped and speckled snake with a long orange head. The pupil is round, iris golden-yellow. Body slim and cylindrical, tail very long and thin, one third of total length. The scales are smooth, in 17 rows at midbody, ventrals 170–198, subcaudals 130–178, cloacal scale divided. Maximum size about 1.9m, average 1–1.4m, hatchling size unknown. The body is yellow (grey in juveniles) with three black longitudinal stripes, the head is orange or dull red above (very occasionally grey), the lips are white. The nape is usually grey. The flanks and belly are grey or white, heavily speckled black. Taxonomic Notes: East African specimens belong to the subspecies *Psammophis punctulatus trivirgatus*, the typical form, *P. p. punctulatus*, with a single black stripe and a higher subcaudal count occurs in the Sudan and the north-eastern horn of Africa.

Psammophis punctulatus

HABITAT AND DISTRIBUTION: Dry savanna and semi-desert, from sea level to about 1,400m. Sporadic records from northern Tanzania, include Engaruka, just east of Ngorongoro, south-east of Arusha, Same and Mkomazi National Park, but occurs almost through eastern and northern Kenya. Only extends onto the coast at Watamu. Also in the Rift Val-

Psammophis punctulatus Speckled Sand Snake.
Stephen Spawls, Mwingi.

ley south of Nairobi, from Olorgesaille and Magadi east to Ukambani. Elsewhere, on either side of the central Ethiopian massif to the Ogaden and South Sudan.

NATURAL HISTORY: Diurnal and partially arboreal. Probably the fastest moving snake in Africa, its long powerful body enabling it to move at great speed, a useful adaptation for a big diurnal snake in a very hot, sparsely-vegetated habitat. During the day it may hunt actively, and investigate holes, but also waits in ambush in trees and bushes, watching for movement. When prey is spotted, the snake descends swiftly, seizing and holding the victim in its coils, chewing it vigorously until it succumbs. By night, usually shelters in a hole or under ground cover but may also sleep in a tree, out on a branch or in a recess. If restrained it lashes vigorously, gyrating and biting as well. On the Tharaka Plain in June and in the Rift Valley near Olorgesaille in March this species has been observed copulating, the two snakes closely watched by other males that presumably had followed the female's scent trail; a male and female were found in the same hole in July in Voi. Copulation usually occurs in the dry season, the appearance of hatchlings coincides with the rainy season. Between 3–12 eggs are laid, roughly 1×3cm, oviposition occurred in July and August in Wajir, a juvenile with an umbilical scar was captured in July in Tarbaj. Diet includes lizards; known prey species include lacertids such as *Latastia* and *Heliobolus*, also agamas and skinks, and snakes, it is big enough to take small vertebrates as well. A bite from a large specimen caused intense local itching, slight swelling and pain; it is a big snake so bites are best avoided.

LINK-MARKED SAND SNAKE *Psammophis biseriatus*

IDENTIFICATION: A medium-sized, slim grey sand snake. The head is long and pointed, eye large, pupil round, iris yellow-brown. Body thin and cylindrical, the tail is long, 35–40% of total length. The scales are smooth, in 15 rows at midbody, ventrals 138–168, subcaudals 97–134. Maximum size about 1m, average 50–80cm, hatchling size unknown, a juvenile taken in Wajir in July was 32cm. There is a broad olive or brown vertebral stripe, edged with evenly-spaced black subtriangular blotches, the scales on the vertebral row are light yellow-brown; resembling a chain, hence the common name. The flanks and ventrals are light grey with fine black speckling. The upper lips are white, sometimes speckled, bordered above by a dark line; top of head olive or brown, uniform or with irregular but symmetrical dark-edged, light blotching.

Psammophis biseriatus

HABITAT AND DISTRIBUTION: Dry savanna, semi-desert and coastal thicket, from sea level to about 1,300m altitude. Almost throughout northern and eastern Kenya and the coast, extending into northern Tanzania past Lake Natron to Olduvai Gorge and Engaruka,

Psammophis biseriatus Link-marked Sand Snake. Left: Stephen Spawls, Voi. Right: Florian Finke, Diani Beach.

and from Tsavo West into the dry country south of Kilimanjaro. There are also three curiously disjunct southern Tanzanian records, from Tabora, Dodoma and the Ruaha National Park. Elsewhere, north to central Somalia.

NATURAL HISTORY: Diurnal and largely arboreal, living in dry thornbush country where its excellent camouflage makes it hard to see, looks like a dry twig. It spends much time motionless in the branches, waiting for prey to pass, when it advances slowly and carefully before making a final quick rush to seize its victim. At night may sleep in the branches or hide in a hole or under ground cover. It is fast moving but tends to rely on its camouflage, when approached in a bush it may freeze, or move surreptitiously away. If seized it will try to bite, lash wildly and rotate, breaking the tail if held near the tip. Clutches of 2–4 elongate eggs were laid in July and September in north-east Kenya; hatchlings collected in September near Archer's Post. Diet mostly lizards including dry-country geckoes, lacertids, skinks and chameleons. It has a beautiful name in KiKamba, Kyenda Ndeto, meaning 'likes to listen to stories', the implication being that it is so well camouflaged in bushes that it may be resting near a group of people without being seen, and hence hear their stories. A prolonged bite from one of these snakes caused local pain and haemorrhaging, swelling and lymphadenitis.

TANGANYIKA SAND SNAKE *Psammophis tanganicus*

IDENTIFICATION: A medium-sized, slim grey sand snake, often like a brighter version of the previous species. The head is long, the snout is pointed but slightly bulbous, eye large, pupil round, iris yellow-brown. Body thin and cylindrical, the tail is long, 35–40% of total length. The scales are smooth, in 15 rows at midbody, ventrals 146–165, subcaudals 81–114. Maximum size about 1m, average 50–80cm, hatchling size unknown. The colour pattern is variable, the broad vertebral stripe (sometimes with a fine rufous or yellow spinal stripe) is grey, fawn or brown, edged with evenly spaced dark or rufous blotches which may form crossbars, the flanks and ventrals may be any mixture of grey, rufous, fawn or brown, the side of the neck is often heavily mottled orange-brown. The head may be grey or brown above, usually with a clear, but sometimes faded dark double-crescent mark on the nape and a cluster of regular dark blotches on top of the head. The upper lips are white or yellow, immaculate or edged with a mixture of maroon, brown and orange.

Psammophis tanganicus

Psammophis tanganicus Tanganyika Sand Snake. Top: Stephen Spawls, Dodoma. Bottom left: Stephen Spawls, Tarangire National Park. Bottom right: Anthony Childs, Meru National Park.

HABITAT AND DISTRIBUTION: Semi-desert and dry savanna, from sea level to about 1,300m. Sporadically recorded from dry eastern East Africa, from the Dodoma-Singida area north through Tarangire National Park, eastwards to Mkomazi thence into a strip of south-east Kenya, past Voi to the Arabuko-Sokoke Forest. Other Kenyan records include Lake Magadi and Olorgesaille, Nguni, near Mwingi, Meru National Park and Kora and Bisanadi Reserves, Latakwen and the Milgis River, Hurri Hills and the Daua River between Mandera and Ramu. Probably more widespread but overlooked. Elsewhere, north to Eritrea and southern Libya.

NATURAL HISTORY: Poorly known but probably similar to the Link-marked Sand Snake *Psammophis biseriatus*, i.e. diurnal and largely arboreal, living in dry thornbush country, looks like a dry twig. Seems to be more prone to hunting on the ground. A gravid female was captured in December in Ethiopia. A specimen from Olorgesaille was under a rock at midday, one Mandera specimen was crossing the road in the late evening, others resting quietly in bushes during the day. Presumably eats mostly lizards, lays eggs.

SKAAPSTEKERS *PSAMMOPHYLAX*

An African genus of small to medium, usually striped, fast-moving rear-fanged snakes. They are terrestrial and diurnal, inhabiting savanna and grassland. They have a fairly toxic venom, but little is injected when they bite humans, no serious symptoms have been recorded. Unusually, male skaapstekers are larger than females. Six species are known, the numbers boosted after molecular analysis by Rhodes University's taxonomist Chris Kelly found that the Striped Beaked Snake, *Rhamphiophis acutus*, was actually a skaapsteker, and thus it is now called *Psammophylax acutus*. Five species of skaapsteker occur in East Africa. The name skaapsteker means 'sheep-stabber' in Afrikaans, as they were mistakenly believed to kill sheep. There have been moves to change this pejorative name, but we have retained it as it is well known in East Africa.

KEY TO EAST AFRICAN MEMBERS OF THE GENUS *PSAMMOPHYLAX*

1a Snout pointed, ventral scales 155-190......2
1b Snout rounded, ventral scales 139-176......3

2a Distinct black stripe present on outer edges of the ventral scales......*Psammophylax togoensis*, Northern Sharp-nosed Skaapsteker [p. 445]
2b No distinct black stripe present on outer edges of the ventral scales......*Psammophylax acutus*, Southern Sharp-nosed Skaapsteker [p. 445]

3a Usually a single anterior temporal, vertebral stripe absent or edges not parallel if present, ventrals usually grey......*Psammophylax variabilis*, Grey-bellied Skaapsteker [p. 444]
3b Usually two anterior temporals, vertebral stripe usually present, its edges parallel, ventrals not grey......4

4a Occurs north of 7°S......*Psammophylax multisquamis*, Kenyan Striped Skaapsteker [p. 442]
4b Occurs south of 7°S......*Psammophylax tritaeniatus*, Southern Striped Skaapsteker [p. 443]

KENYAN STRIPED SKAAPSTEKER *Psammophylax multisquamis*

IDENTIFICATION: A striped snake. Head quite short, eye medium to small, pupil round. Body cylindrical, tail quite short, 18–20% of total length. Dorsal scales smooth, in 17 rows at midbody, ventrals 160-184, subcaudals 51-66. Maximum size about 1.4m (larger ones always males), average 60-90cm, hatchling size unknown. The central dorsal stripe is fawn, light brown or grey, often with a fine yellow vertebral line, bordered by a darker brown or grey lateral stripe. The ventrals are cream, often with a bright yellow median stripe and a line of orange or rufous dashes on each outer edge.

Psammophylax multisquamis

HABITAT AND DISTRIBUTION: High grassland and moist savanna at medium to high altitude, from about 600 to over 3,300m. Widespread in highland Kenya, especially on open plains and in grassland; north from the Kapiti Plains through Nairobi (where some huge males, up to 1.4m long occur in the grassland to the south of the city), west to the Maasai Mara National Reserve, north through the high country to the northern slopes of Mt Kenya, west to Kakamega, localities include Athi River, Langata, Kijabe, Lake Naivasha, Nakuru, Kinangop Plateau, Molo, Sotik and Gilgil; one was caught at Rutundu at 3,350m altitude on the north-west plateau of Mt Kenya. South from the Mara to Serengeti, east to Mt Kilimanjaro, localities include Loitokitok, Ol Donyo Sambu, Arusha and Moshi, and the Chyulu Hills, some early enigmatic records from Mtito Andei and Voi. Isolated records from Mpwapwa and Tindi in Tanzania and Gabiro in Rwanda. Elsewhere, in highland Ethiopia.

NATURAL HISTORY: Diurnal and terrestrial, shelters under ground cover, in rotting logs,

Psammophylax multisquamis Kenyan Striped Skaapsteker. Stephen Spawls, Naivasha.

in vegetation clumps or in holes. It seems to have a curious pattern of activity (there is speculation that it might be nocturnal to some extent), as it is often under cover during the day. Sometimes basks, and when doing so lies in a strange, kinked fashion. Quite fast moving, slides away if threatened and able to escape, but if cornered or seized, it sometimes shams death, turning its head and neck over, opening its mouth and lolling in a lifeless fashion. It lays eggs, clutches of 4–16 eggs found, and the females coil around the eggs, presumably to protect them. Diet: rodents, lizards and amphibians, but may also take small snakes.

SOUTHERN STRIPED SKAAPSTEKER
Psammophylax tritaeniatus

IDENTIFICATION: A small striped snake. Head quite short, eye medium to small, pupil round. Body cylindrical, tail quite short, 17–23% of total length. Dorsal scales smooth, in 17 rows at midbody, ventrals 139–176, subcaudals 49–69. Maximum size about 90cm, average 50–80cm, hatchling size 22–24cm. A striped snake; a median dark stripe, a grey to olive brown dorsolateral stripe and a dark brown or black flank stripe; the vertebral stripe may be poorly defined, only on the nape or totally absent. The ventrals are white or cream, sometimes with a yellow central band. Similar species: Superficially resembles the sand snakes, but the relatively short tail and short head should identify it.

Psammophylax tritaeniatus

Psammophylax tritaeniatus Southern Striped Skaapsteker. Stephen Spawls, captive.

HABITAT AND DISTRIBUTION: Moist savanna at medium to high altitude, from about 600m to over 2,200m. Occurs in south-west Rwanda (Kwagahunga), southern Tanzania, from the south end of the Udzungwa Range south and east to the border, and from Katavi southwards along the Lake Tanganyika shore, thence into northern Zambia The two populations might be connected. There is an enigmatic record from near Lake Eyasi (not shown on the map). Elsewhere, south to South Africa and Namibia, south west to Shaba in the DR Congo. Conservation Status: IUCN Least Concern.

NATURAL HISTORY: Diurnal and terrestrial, shelters under ground cover, in rotting logs, in vegetation clumps or in holes. Sometimes basks, and when doing so lies in a strange, kinked fashion. Quite fast moving, slides away if threatened and able to escape, but if cornered or seized, it sometimes shams death, turning its heads and neck over, opening its mouth and lolling in a lifeless fashion. It lays eggs, clutches of 5–14 eggs, roughly 1×2cm recorded. Newly hatched juveniles taken in December in southern Tanzania. Diet: mostly rodents, lizards and amphibians; a Tanzanian specimen ate a Black-lined Plated Lizard *Gerrhosaurus nigrolineatus*.

GREY-BELLIED SKAAPSTEKER *Psammophylax variabilis*.

IDENTIFICATION: A medium sized uniform or striped snake. Head quite short, and broad eye medium to small, pupil round. Body cylindrical, tail quite short, 16.5–21.5% of total length. Dorsal scales smooth, in 17 rows at midbody, ventrals 149–167, subcaudals 49–61. Maximum size about a metre, average 50–80cm, hatchling size unknown. Colour grey or olive-brown either uniform or with dark longitudinal stripes. In striped individuals the vertebral scale row is blackish with a pale median hairline, the scales in row 5 are white above and black below in the posterior portion of each scale, a white ventrolateral stripe passes through the lower half of scale row 2 and the upper half of row 1. The lower half of the outer scale row and the entire ventrum are uniform grey. Upper labials and chin whitish, suffused with dark grey.

Psammophylax variabilis

HABITAT AND DISTRIBUTION: Moist savanna at medium to high altitude, also on flood plains. Two discrete populations in our area, known from south-central Rwanda south to north-central Burundi, and from the northern end of Lake Malawi (Lake Nyasa) northeast to the southern portion of the Udzungwa Mountains. Elsewhere, south to northern Botswana.

NATURAL HISTORY: Diurnal and terrestrial, shelters under ground cover, in rotting logs, in vegetation clumps or in holes. Sometimes basks, and when doing so lies in a strange, kinked fashion. Quite fast moving, slides away if threatened and able to escape. It lays eggs. Diet: largely rodents, but also takes lizards and amphibians.

Psammophylax variabilis Grey-bellied Skaapsteker. Michele Menegon, southern Udzungwa Mountains.

SOUTHERN SHARP-NOSED SKAAPSTEKER
Psammophylax acutus

IDENTIFICATION: A small striped snake with a big eye and a sharply pointed snout. Head short and deep, pupil round, iris golden or rufous brown. Body cylindrical, tail 15–21% of total length. Body scales smooth, in 17 rows. Ventrals 155–190, subcaudals 53–72. Maximum size about 1.06m, average 60–80cm, hatchling size unknown. Conspicuously striped, with a fine brown (sometimes yellow-centred) vertebral stripe, on either side of this a fawn stripe, below this there is a brown, black-edged flank stripe, which starts on the snout and runs through the eye. The vertebral stripe splits into a three-pronged trident shape on top of the head. The belly is cream. Taxonomic Notes: This and the following species were originally called the Striped Beaked Snake *Rhamphiophis acutus*, but molecular analysis indicates they are skaapstekers, and the two subspecies have been elevated to full species on differences in colour and geographical separation.

Psammophylax acutus.

Psammophylax acutus Southern Sharp-nosed Skaapsteker. Philipp Wagner, Zambia.

HABITAT AND DISTRIBUTION: In moist savanna and woodland (possibly forest clearings) south of the central African forests, often in marshy areas, floodplains and open country, in our area at mid-altitude, from 400 to about 1,500m, elsewhere from 200–1,800m. Known from Mwaya, at the north end of Lake Malawi (Lake Nyasa), there is a cluster of records from around the south end of Lake Tanganyika, also known from north and west-central Burundi through to north-west Tanzania around Kibondo, and Akagera National Park in Rwanda. Elsewhere, from northern Zambia to Angola.

NATURAL HISTORY: Poorly known. Presumably diurnal. It is terrestrial, spending much time in holes. Clutches of 10–15 eggs recorded. The diet includes rodents, a specimen of the southern subspecies had eaten a worm lizard.

NORTHERN SHARP-NOSED SKAAPSTEKER
Psammophylax togoensis

IDENTIFICATION: A small striped snake, similar to the previous species, from north-west Uganda, with conspicuous black ventral lines. Eye large, snout pointed. Head short and deep, pupil round, iris brown above but the black flank stripe runs through it. Body cylindrical, tail 17–25% of total length. Body scales smooth, in 17 rows. Ventrals 170–188, subcaudals 57–72. Maximum size about 85cm average 50–70cm, hatchling size unknown. Conspicuously striped, with a fine brown vertebral line in the centre of a broad fawn stripe; a broad dark brown flank stripe starts on the snout and runs through the eye, the lower dorsal are yellow or tan. The belly is yellow centrally, bordered on either side by black line, which may be fine or broad but always unbroken. The outer edges of the ventrals are white.

Psammophylax togoensis

HABITAT AND DISTRIBUTION: Moist savanna, from sea level to over 1,600m, although in our area known only from Murchison Falls in north-west Uganda at around 700m altitude; but also known from Garamba National Park in the north-east Congo. Elsewhere west across the savanna to Guinea.

Psammophylax togoensis Northern Sharp-nosed Skaapsteker. Gerald Dunger, Nigeria.

NATURAL HISTORY: Poorly known. Diurnal and terrestrial, often in holes. Said to live 'in wet lowlands'. Lays eggs. Diet includes amphibians and rodents. One was found in burnt grassland with its body held vertically, supposedly pretending to be a branch.

BEAKED SNAKES *RHAMPHIOPHIS*

An African genus of relatively large back-fanged, diurnal snakes with hooked or pointed snouts; they have a short skull and the rostral bone is braced by the nasal, an adaptation for shoving through soil or sand. They have strong muscular bodies. Big adults will sometimes bite vigorously but the venom does not appear to be particularly toxic to humans. Three species are known, inhabiting semi-desert, dry and moist savanna, all three occur in our area. They are under-represented in most museum collections, due to their secretive burrowing way of life. Before the building of the Kariba Dam on the Zambezi River, the Rufous Beaked Snake *Rhamphiophis rostratus* was represented in Zimbabwe museums by a handful of specimens but after the water rose more than 30 were collected, flooded out of their holes. Beaked snakes are fast moving, some species can spread a curious flattened hood.

KEY TO EAST AFRICAN MEMBERS OF THE GENUS RHAMPHIOPHIS

1a Midbody scale rows 19, head and neck red or rufous in adults, body grey......*Rhamphiophis rubropunctatus*, Red-spotted Beaked Snake [p. 449]
1b Midbody scales rows 17, medium-sized to small, head and neck not red in adults......2

2a Dark stripe through eye, in our area largely recorded in Kenya and Tanzania... ...*Rhamphiophis rostratus*, Rufous Beaked Snake [p. 448]
2b No dark stripe through eye, in our area, only in Uganda ......*Rhamphiophis oxyrhynchus*, Western Beaked Snake [p. 447]

WESTERN BEAKED SNAKE *Rhamphiophis oxyrhynchus*

IDENTIFICATION: A fairly large, muscular snake. The head is short, snout pointed, eye large, with a round pupil and a golden or red-brown iris, this is hard to see in normal light. The tongue is pink with a white tip. Body cylindrical, tail 27–32% of total length. The scales are smooth, in 17 rows at midbody, ventrals 170–196, subcaudals paired, 88–106. Maximum size about 1.5m, average 70cm–1.2m, hatchling 28–35cm. Colour quite variable: grey, pink, brown, yellow-brown or orange, in large specimens the dorsal scales are horizontally darkened in the centre, giving the body a finely striped appearance. The ventrals are immaculate white, cream or yellow, sometimes the throat is yellow and the belly white. Juveniles have a series of small rufous flank blotches.

Rhamphiophis oxyrhynchus

Taxonomic Notes: For a long time, this species and the Rufous Beaked Snake *Rhamphiophis rostratus*, were thought to be subspecies.

HABITAT AND DISTRIBUTION: Moist savanna, from about 500–1,300m altitude, elsewhere to sea level. In our area, known only from two Uganda localities: Ongijno and Bulisa, elsewhere west to Mali.

NATURAL HISTORY: Diurnal and terrestrial (although it will climb into bushes), spends much time in holes, looking for prey. Quick moving and alert, when moving through the

Rhamphiophis oxyrhynchus Western Beaked Snake. Left: Stephen Spawls, Ghana. Right: Gerald Dunger, Nigeria.

bush it often pauses, head up. When it sees prey it may jerk the head from side to side, targeting its prey. Copulation in captivity occurred in the open, taking between 20 minutes and 3 hours, in February and April. A Ugandan female was gravid in October, which could mean that hatchlings appear in the rainy season, March to April. Hatchlings in Ghana were captured in April and May, start of the rainy season. Multiple clutching has been recorded in captivity, up to four per year, whether this occurs in the wild is not known. Clutches of 6–18 eggs, roughly 2×4.5cm, recorded. These snakes eat a wide variety of prey; rodents, lizards, frogs and snakes are taken.

RUFOUS BEAKED SNAKE *Rhamphiophis rostratus*

IDENTIFICATION: A fairly large, muscular, rear-fanged snake. The head is short, snout pointed, eye large, with a broad dark line through it. The pupil is round, iris golden or red-brown, this is hard to see in normal light. Body cylindrical and muscular, tail 25–30% of total length. The scales are smooth, in 17 rows at mid-body, ventrals 148–192, subcaudals paired, 90–125, cloacal scale divided. Maximum size about 1.6m, average 80cm–1.2m, hatchling 25–32cm. Colour quite variable: the field impression is of a white snake with dark speckling; it may be grey, pink, brown, orange, white, in darker specimens the scales towards the tail have a light centre, the snake looks speckled. The ventrals are immaculate white, cream or yellow, the ventral scale sometimes finely dark-edged, Juveniles heavily speckled, irregular red-brown, more concentrated on the lower flanks, this fades at lengths around 60–70cm.

Rhamphiophis rostratus

HABITAT AND DISTRIBUTION: Semi-desert, dry and moist savanna, coastal thicket and woodland, from sea level up to about 1,500m. Widespread in northern and eastern Kenya, eastern and most of northern Tanzania, and between Lake Rukwa and Katavi National Park. Records lacking for most parts of south-western Tanzania probably due to undercol-

Rhamphiophis rostratus Rufous Beaked Snake. Left: Stephen Spawls, Kajiado. Right: Luke Verburgt, Northern Mozambique, juvenile.

lecting, it may be expected there. Just reaches eastern Uganda at Amudat. No Rwanda or Burundi records. Elsewhere, south to Botswana, north to Sudan and Ethiopia.

NATURAL HISTORY: Secretive, diurnal and terrestrial (although it will climb into bushes), spends much time in holes, looking for prey. When moving it often pauses, head up, and jerks the head from side to side like a chicken; and this behaviour is also shown in a more pronounced and continuous way by aroused males demonstrating to a possibly receptive female. Active all hours of the day, even at the hottest times, when inactive hides in holes, squirrel warrens, etc, often in abandoned termitaria; it seems to occupy a suitable refuge for a length of time, foraging around it. Fast moving and alert, it moves quickly away from danger, but if threatened and unable to escape it flattens the neck into a small hood and raised the forepart of the body. If picked up it will hiss, jerk convulsively and may bite. It can dig holes with its pointed snout, breaking the soil with the rostral and turning the head sideways to scoop it out. Clutches of 4–12 eggs laid, roughly 3.5×2cm, a Tanzanian specimen laid 15 eggs over a five-day period. Newly hatched juveniles collected in December in Tanzania. It eats a range of prey; rodents (including naked mole rats), lizards, frogs and snakes, reported eating beetles. Two captive specimens competing for a mouse showed agonistic behaviour, neck-wrestling with each other. Often seen dead on roads in southern Kenya and eastern Tanzania;

RED-SPOTTED BEAKED SNAKE *Rhamphiophis rubropunctatus*

IDENTIFICATION: A big, grey or grey-brown, orange-headed back-fanged snake. The common name is inappropriate, as only the juveniles are red-spotted. Head short, snout weakly pointed, tongue black, eye large and dark, pupil round, iris brown. Body elongate and cylindrical, the tail is about one-third of the total length. Scales smooth, in 19 rows at midbody, ventrals 207–245, subcaudals paired, 130–160, cloacal scale divided. A big snake, reaching 2.5m, average 1.4–2m, hatchling size unknown. Colour usually grey, with an orange head, sometimes the anterior part of the body is orange or yellow-orange. Ventrals cream, light brown or light orange. Juveniles are cream or white, heavily marked with red or red-brown sub-triangular spots, with a grey dorsal stripe; this colour fades at a length of 50–60cm to the adult phase. Interestingly, a colour phase of the Boomslang *Dispholidus typus* is identical, grey with a red head, and occurs in the same area.

Rhamphiophis rubropunctatus

Rhamphiophis rubropunctatus Red-spotted Beaked Snake. Left: Stephen Spawls, Watamu. Right: Tomas Mazuch, South Horr, juvenile.

HABITAT AND DISTRIBUTION: Coastal thicket and woodland, semi-desert and near-desert, moist and dry savanna, from sea level to about 1,200m. It is widespread in south-eastern Kenya and the Kenya coast, north to the northern bend of the Tana River, also in the rift valley at Olorgesaille; extending into a strip of extreme northeast Tanzania, from Lake Natron and Engaruka across to Mkomazi and Tanga. A population also occurs from Lake Baringo and the Kerio Valley north to Lake Turkana, and across the Dida-Galgalu Desert to the Ethiopian border. Probably more widespread in northern Kenya than the map indicates, but secretive and thus rarely collected. Elsewhere, to southern Sudan, southern Somali and through the Ogaden to northern Somalia, a species typical of the Somali-Maasai fauna. Conservation Status: IUCN Least Concern.

NATURAL HISTORY: Poorly known but probably similar to the Rufous Beaked Snake. Secretive, diurnal and mostly terrestrial, will climb into low trees and bushes. Often hunts prey in holes, when inactive shelters there; in squirrel warrens, abandoned termitaria, hollow logs etc. A specimen from the Dida-Galgalu Desert was active in a barren lava field in the heat of the day. They lay eggs. Diet includes rodents and squirrels, a captive juvenile took lacertid lizards. One of East Africa's biggest rear-fanged snakes, but little is known of its lifestyle, it is usually uncommon even in prime habitat, except in some places on Kenya's north coast. It is a big snake, so bites should be avoided.

SUBFAMILY **APARALLACTINAE**

A subfamily of small snakes, with around 46 species in nine genera. They are mostly African, a couple of species reach the Middle East. Twenty-one species, in five genera, occur in our area. Most are small, dark coloured burrowing snakes, with small, short heads, small eyes, no loreal scale and grooved rear-fangs (one centipede eater has no fangs) but are not dangerous to humanity.

KEY TO THE EAST AFRICAN GENERA OF *APARALLACTINAE*

1a Subcaudals in a single row......*Aparallactus*, centipede eaters [p. 451]
1b Subcaudals in a double row......2

2a No preocular present......3
2b Preocular present......4

3a Anterior temporal present......*Micrelaps*, two-coloured snake and desert black-headed snakes [p. 457]

3b Anterior temporal absent......*Amblyodipsas*, purple-glossed snakes [p. 466]

4a Two pairs of shields between rostral and frontal......*Polemon*, snake-eaters [p. 460]
4b A single shield between rostral and frontal......*Chilorhinophis*, black and yellow snakes [p. 464]

CENTIPEDE EATERS *APARALLACTUS*

A sub-Saharan genus of small slim snakes with a tiny head and eye, with a round pupil. Most are black, grey or brown, brown ones often have a black head and nape, It is speculated that these allow the snake to put its head into the sunshine and quickly warm the brain (black surfaces are efficient absorbers of heat) without exposing the body to predators. Unusually, they have single subcaudals. They have relatively large back fangs (except for the Western Forest Centipede Eater *Aparallactus modestus*). They live underground, in holes or soft soil, although they may emerge, especially on damp nights following rain and slide around, looking for prey or a mate. They eat centipedes. Most lay eggs but one species, Jackson's Centipede Eater *Aparallactus jacksoni*, gives live birth. Eleven species are known, with the majority in south-east Africa. Seven species occur in our area, two of these are endemic. Several species look very similar; in others there is considerable variety within the species, and some rather odd distribution patterns suggest misidentification; it is a group that could benefit from a thorough taxonomic revision.

KEY TO EAST AFRICAN MEMBERS OF THE GENUS *APARALLACTUS*

1a No grooved fangs, in western forest...... *Aparallactus modestus*, Western Forest Centipede Eater [p. 455]
1b Grooved fangs present......2

2a First pair of lower labials in contact behind the mental (rarely narrowly separate)......3
2b First pair of lower labials widely separated by the anterior sublinguals......6

3a Postoculars 1, separated from the anterior temporal......*Aparallactus lunulatus*, Plumbeou Centipede Eater [p. 454]
3b Postoculars 2, rarely 1, in contact with the anterior temporals......4

4a Supraoculars 5......*Aparallactus werneri*, Usambara Centipede Eater [p. 457]
4b Supraoculars 6, 7 or 8......5

5a Second and third upper labial enter orbit......*Aparallactus turneri*, Malindi Centipede Eater [p. 456]
5b Third and fourth upper labial enter orbit......*Aparallactus jacksoni*, Jackson's Centipede Eater [p. 453]

6a Nasal usually divided, grey or black dorsally......*Aparallactus guentheri*, Black Centipede Eater [p. 452]
6b Nasal usually entire, brown above...... *Aparallactus capensis*, Cape Centipede Eater [p. 451]

CAPE CENTIPEDE EATER *Aparallactus capensis*

IDENTIFICATION: A very small slim brown snake with a black head. Eye tiny, pupil round. Tail short, 16–24% of total length. Scales smooth, in 15 rows at midbody, ventrals 126–186, cloacal scale entire, subcaudals single, 29–63. Maximum size 41cm in South Africa but most East African specimens are 20–34cm, hatchlings 9–11cm. Dorsum brown, often with several fine dark longitudinal lines, top of head and nape black, light brown, pink or white below.

HABITAT AND DISTRIBUTION: Coastal thicket and moist savanna, from sea level to

1,600m altitude. In Kenya, known from the Shimba Hills, south into eastern Tanzania, inland along the Mozambique border, along the Rufiji/Ruaha river system and through the Usambara Mountains to the south side of Mount Kilimanjaro. In the west, occurs from Akagera National Park south though eastern Rwanda and Burundi to Ugalla and the Mahale area in Tanzania. Known also from the Ruzizi Plain north of Lake Tanganyika in the DR Congo. Elsewhere south to eastern South Africa. Conservation Status: IUCN Least Concern.

Aparallactus capensis

Aparallactus capensis Cape Centipede Eater. Stephen Spawls, Usambara Mountains.

NATURAL HISTORY: Lives underground, in holes and cracks, tree root clusters, soft soil, also shelters in rotting tree trunks and termitaria, it will sometimes emerge on wet nights and hunt or look for a mate on the surface. If picked up they writhe and twine around the fingers, they may bite but their mouths are so small they cannot do any damage. They lay 2–4 eggs, roughly 3×0.5cm, hatchlings collected in April in southern Tanzania. They eat centipedes, which are seized, chewed and swallowed head first, the venom of the centipede having no effect on the snake. A captive specimen ate a worm snake (*Leptotyphlops*). Note: a snake handler, who allowed a 26cm specimen of this snake to bite and chew him for a minute experienced mild burning pain and swelling.

BLACK CENTIPEDE EATER *Aparallactus guentheri*

IDENTIFICATION: A small, slim dark snake. Head slightly flattened, eye small and dark, pupil round. Body cylindrical, tail over 20% of total length, longest in females. Scales smooth in 15 rows at midbody, ventrals 150–173, cloacal scale entire, subcaudals single, 49–60. Maximum size about 50cm, average 30–45cm, hatchling size unknown. Adults are uniform grey or black, paler below, the throat may be white. Juveniles have two distinct white, cream or yellow crossbars on the nape and the neck. A curiously patterned individual from Mombasa had a pale head, broad pale neck collar and reticulated back.

Aparallactus guentheri

HABITAT AND DISTRIBUTION: Coastal bush, moist savanna and evergreen hill forest, from sea level to about 1,700m altitude. Occurs from the Tana Delta southwards along the East African coastline, inland in southern Tanzania to Tunduru and up the Rufiji and Ruaha Rivers, isolated inland records include Taita Hills in Kenya and North Pare Mountains In Tanzania.

Aparallactus guentheri Black Centipede Eater. Bill Branch, juvenile, Nguru Mountains. Inset: Stephen Spawls, northern Tanzania adult.

NATURAL HISTORY: Poorly known, probably similar to other centipede eaters, i.e. nocturnal, burrowing, sometimes on the surface, eats centipedes. A snail was found in the stomach of one specimen. Lays eggs.

JACKSON'S CENTIPEDE EATER Aparallactus jacksonii

IDENTIFICATION: A small, slim brown snake with a black head, eye small and dark, pupil round. Body cylindrical, tail 13–20% of total length, longest in females. Scales smooth in 15 rows at midbody, ventrals 134–166, cloacal scale entire, subcaudals single, 33–52. Maximum size about 28cm, average 20–26 cm, hatchling size 10–11cm. Dorsum pinkish, reddish or nut brown, head black above, usually a thin yellow crossbar on the nape, becoming a yellow blotch at the angle of the jaw, a broad black collar behind that. There are sometimes irregular black dots on the anterior third of the body, back sometimes with fine black longitudinal lines. Yellow, cream or pinkish below. Taxonomic Notes: Two subspecies occur. *Aparallactus jacksoni jacksoni* occurs in southern Kenya and eastern Tanzania, it has low ventral and subcaudal counts (ventrals 134–157, subcaudals 33–40), *Aparallactus jacksoni oweni* occurs in the north of our area, ventral counts 151–166, subcaudal counts 41–52.

Aparallactus jacksonii

HABITAT AND DISTRIBUTION: Coastal bush, moist and dry savanna and high grassland, from sea level to 2,200m. The northern race occurs at Malka Murri on Kenya's northern border and on the eastern side of Lake Turkana. Also reported from the southern Kerio Valley. The southern race occurs in the Kenya highlands east of the Rift Valley, south to Amboseli, Mt Kilimanjaro and its environs, thence to Tarangire, south-east along the border to the coast; north-west in Tanzania to as far as Olduvai Gorge and the central Serengeti, although records sporadic. Isolated records include the Udzungwa Mountains and Katavi in Tanzania, reported (probably in error) from the Rondo Plateau. Elsewhere, to South Sudan, southern Ethiopia and Somalia.

Aparallactus jacksonii Jackson's Centipede Eater. Anthony Childs, Meru National Park.

NATURAL HISTORY: Poorly known, probably similar to other centipede eaters, i.e. burrowing, sometimes on the surface, eats centipedes. Seven were collected in one evening at an Amboseli lodge following a rainstorm in November. They are the only species in the family to give live birth, a Naivasha female had two live young, 102 and 105mm, in early June, another female from Sagana had three young. Diet centipedes.

PLUMBEOUS CENTIPEDE EATER *Aparallactus lunulatus*

IDENTIFICATION: A small, slim dark snake. Usually grey but colour variable. Head short, eye small and dark, pupil round. Body cylindrical, tail 20-28% of total length, longest in females. Scales smooth in 15 rows at midbody, ventrals 140-176, cloacal scale entire, subcaudals single, 43-70. Maximum size about 52cm, average 30-40cm, hatchlings 9-10cm. Colour very variable; three main colour phases in East Africa are: (a) grey, each scale white-edged, head yellow, with a broad black collar, (b) uniform grey, black or plumbeous, lighter below, (c) brown or yellow, either uniform or with each dorsal scale black edged to give a reticulate effect, head brown or orange, a broad black neck collar. Juveniles of forms (a) and (c) often have a series of up to 30 dark crossbars on the neck, becoming shorter towards the tail, fading in adults. Taxonomic Note: Several subspecies have been described, most East African specimens are assigned to the subspecies

Aparallactus lunulatus

Aparallactus lunulatus Plumbeous Centipede Eater. Top left: Tomas Mazuch, Rumuruti. Top right: Anthony Childs, Meru National Park. Bottom: Lorenzo Vinciguerra, northern Tanzania.

Aparallactus lunulatus concolor. None are clear-cut, however, and the startling variation in colour suggest there are cryptic species concealed in this taxon.

HABITAT AND DISTRIBUTION: Coastal bush, dry and moist savanna and semi-desert, from sea level to 2,200m altitude, known from Rumuruti. It seems to be quite fond of stony country. Widespread in northeastern and eastern Uganda, parts of north-western Kenya, eastwards round Mt Kenya into most of south-eastern Kenya and extreme north-east Tanzania. Also occurs in central and south-eastern Tanzania. Isolated records from the Semliki River in Uganda, Wajir Bor and Malka Murri in northern Kenya. The main populations are probably interconnected. Elsewhere, west to Ghana, north to Eritrea, south to northern South Africa.

NATURAL HISTORY: Poorly known, probably similar to other centipede eaters, i.e. burrowing, sometimes on the surface on damp nights, eats centipedes. Some specimens are aggressive when handled, biting fiercely. Clutches of 2–4 eggs recorded, roughly 3×0.5cm; a female in southern Tanzania was gravid in August. A South African specimen ate a scorpion.

WESTERN FOREST CENTIPEDE EATER   *Aparallactus modestus*

IDENTIFICATION: A small, slim dark forest-dwelling snake. The eye is small and dark, pupil round. Body cylindrical, tail quite short and oddly stubby for a centipede eater, 13–16% of total length in females, 16–18% in males. Scales smooth in 15 rows at midbody, ventrals 134–169, cloacal scale entire, subcaudals single, 32–51. Maximum size about 65cm (big specimens usually females), average 30–50cm, hatchling size unknown. Adults are uniform light or dark grey or grey-brown, sometimes the scales are light-edged, giving a reticulate appearance, paler below, cream, yellow or grey, sometimes the tail underside is darker than the remaining ventrals. Juveniles have a distinctive white, cream

Aparallactus modestus

Aparallactus modestus Western Forest Centipede Eater. Konrad Mebert, juvenile, eastern Congo. Inset: Kate Jackson, Congo.

or yellow head patch or collar, which may extend from the eye to just behind the parietals or just encircle the neck, at lengths over 20 cm the patch darkens to brown and then disappears. Taxonomic Note: This is the only centipede eater with no fangs, which confused taxonomists into placing it in a separate genus, *Elapops*.

HABITAT AND DISTRIBUTION: Forest, forest-savanna mosaic and recently deforested areas at mid-altitude, from 600–1,300m in our area, to sea level elsewhere. It is sporadically recorded in Uganda from Kampala and the Mabira Forest, Budongo Forest, Bundibugyo and western Kabarole, and an odd record from Karita, north of Mount Elgon. Might be in western Kenya at Kakamega. No Rwanda or Burundi records but found just west of the border in the DR Congo. Elsewhere, west to Sierra Leone.

NATURAL HISTORY: Poorly known. Burrowing, living in soil holes and leaf litter. Most specimens found were dug up during farming operations in the forest. One was found crossing a forest path at night. A Mabira Forest female contained seven developing eggs, roughly 2.5×0.8cm. Diet centipedes.

MALINDI CENTIPEDE EATER *Aparallactus turneri*

IDENTIFICATION: Endemic to Kenya's coast. A very small, slim brown or yellow-brown snake with a black head, eye small and dark, pupil round. Body cylindrical, tail 16–18% of total length, longest in females. Scales smooth in 15 rows at midbody, ventrals 120–129 in males and 134–139 in females, cloacal scale entire, subcaudals single, 33–42 in males and 31–37 in females. Maximum size about 20cm, average 15–18cm, hatchling size unknown. Colour pinkish or yellowish brown above, pink or yellow below, the scales sometimes dark-edged, a fine dark vertebral line may be present. The head is black or dark grey above. In most specimens the entire snout is black, in front of the eyes there is usually a narrow white or yellow band, and a broad black collar behind this, extending almost down to the ventrals; the side of the head behind the eye, the chin and throat may be cream or vivid yellow.

Aparallactus turneri

Aparallactus turneri Malindi Centipede Eater. Bio-Ken archive, Watamu.

HABITAT AND DISTRIBUTION: Endemic to the coastal woodland and thicket of the Kenyan coast, at sea level or just above it. Recorded from Mkonumbi, near Lamu, south as far as Diani Beach (this is the only record south of Mombasa), known localities include Witu, the Tana Delta, Malindi, Arabuko-Sokoke Forest and Kilifi. A curious inland record from Voi, collected by C.J.P. Ionides, not shown on map. Conservation Status: IUCN Least Concern.

NATURAL HISTORY: Nothing known, probably similar to other Centipede Eaters, i.e. burrowing, sometimes on the surface, eats centipedes. The type series were collected under logs and rocks on sandy soil.

USAMBARA CENTIPEDE EATER Aparallactus werneri

IDENTIFICATION: A small, slim brown snake with a black or brown head, eye small and dark, pupil round. Body cylindrical, tail 15–18% of total length, longest in females. Scales smooth in 15 rows at midbody, ventrals 139–163, cloacal scale entire, subcaudals single, 32–45. Maximum size about 39cm but this was an unusually large Usambara animal, in a long series from southern Tanzania the largest was 27cm. Hatchling size 10–11cm. Colour quite variable, the top of the head is usually olive or brown, sometimes darker at the edges and extending down around the eye. There is usually a narrow yellow band or a couple of yellow patches on the nape, behind this a broad black collar. The dorsum is buff, brown or olive with or without fine dark longitudinal lines, the dorsal scales may be finely dark-edged; the belly is yellow or orange. In some individuals there is no collar and the top of the head is uniformly dark.

Aparallactus werneri

Aparallactus werneri Usambara Centipede Eater. Lorenzo Vinciguerra, Usambara Mountains.

HABITAT AND DISTRIBUTION: Endemic to Tanzania. Moist coastal and hill forests on eastern Tanzania, also moist savanna, from sea level to about 1,600m. Known localities include the North Pare Mountains, east and west Usambara Mountains, Magrotto Hill, Handeni, Uluguru Mountains, Kisarawe, Kiwengoma Forest Reserve and Liwale. Thus vulnerable in conservation terms as it occurs in restricted localities, many of which are being logged, but might be adaptable to agriculture, and distribution might be continuous between the known localities.

NATURAL HISTORY: Poorly known, probably similar to other centipede eaters, i.e. burrowing, living in leaf litter and soft soil, sometimes on the surface on damp nights. Specimens from the Uluguru and Usambara Mountains were found by hoeing up vegetation at the forest edge, also found under rocks, logs and bits of bark in and around the rain forest. They lay eggs, 17 out of a sample of 31 females from Amani in the Usambara Mountains contained eggs in November, clutch size varied from 2–4, the largest oviductal eggs were 3.9×0.6cm. Diet: centipedes. One specimen was found in the stomach of a purple-glossed snake (*Amblyodipsas*).

TWO-COLOURED AND BLACK-HEADED SNAKES
MICRELAPS

A genus of little burrowing snakes with blunt heads and round pupils, usually boldly marked. They are thin, with fairly short tails. They have large grooved back fangs but are small and not regarded as dangerous. The dorsal scales are smooth, in 15 rows at midbody. They live in holes or in soft sand or soil. Little is known of their diet and breeding details. Four species are known, two occur in our area, one is endemic; two species are found in the Middle East. They are provisionally included here in the Aparallactinae although a molecular analysis placed one species, *Micrelaps bicoloratus*, by itself outside even the Lamprophiidae.

KEY TO EAST AFRICAN MEMBERS OF THE GENUS *MICRELAPS*

1a Not grey, longitudinally striped, head not black, posterior chin shields usually not in contact......*Micrelaps bicoloratus*, Two-coloured Snake [p. 458]
1b Usually grey, no longitudinal stripes, head black, posterior chin shields usually in contact......*Micrelaps vaillanti* Guinea-fowl Snake/Desert Black-headed Snake [p. 459]

TWO-COLOURED SNAKE *Micrelaps bicoloratus*

IDENTIFICATION: A little striped burrowing snake with a short tail Head slightly broader than the neck, snout rounded. Body cylindrical, tail 10–15% of total length. Scales smooth, in 15 rows at midbody, ventrals 172–235, cloacal scale divided, subcaudals 16–30. Maximum size about 33cm, average 20–30cm, hatchling size about 11–12cm. There are three colour phases. One, the most common, has a broad brown, black or purplish dorsal stripe, 4–7 scales wide, the lower dorsal scale rows and ventrals may be white, cream or orange, top and side of head also brown, another phase has a dark brown or black dorsal stripe, a lateral stripe four scales deep of pale, black-edged scales, the third phase has a broad yellow or orange vertebral stripe, three or four scales wide, extending from snout to tail, below this a rich brown dorsolateral stripe three or four scales deep, which starts on the side of the snout, the lowest one or two dorsal scale rows and the ventrals white or cream; sometimes there is a rich brown spot on top of the head, on the parietals. This colour phase has so far only been recorded in northern Tanzania, and named as the subspecies *Micrelaps bicoloratus moyeri*. All phases may have a thin dark median line on the underside of the tail.

Micrelaps bicoloratus

HABITAT AND DISTRIBUTION: An East African endemic. Found in coastal bush, dry and moist savanna and high grassland, from sea level to about 2,000m. Known from the north end of Lake Manyara, thence east to Ol Donyo Sambu on the slopes of Mt Meru, round the southern slopes of Mount Kilimanjaro, widespread in south-eastern Kenya, up the Galana/Athi River to Nairobi Falls (Sukari Ranch) north-east of Nairobi and along the coast, through Lali Hills to the Lamu archipelago. A single record north of the equator from Ol Ari Nyiro ranch, north-west of Rumuruti. Conservation Status: IUCN Least Concern.

NATURAL HISTORY: Poorly known. Burrowing, active at night, hides in holes and under ground cover during the day. Specimens have been captured on roads at night in the rainy season, the Ol Donyo Sambu specimens were dug up in a field, one specimen was under a

Micrelaps bicoloratus Two-coloured snake. Left: Anthony Childs, Nairobi National Park. Top right: Anthony Childs, Watamu. Bottom right: Lorenzo Vinciguerra, northern Tanzania (subspecies *moyeri*).

collapsed fence post in a dry gully. Presumably lays eggs. Diet unknown but other *Micrelaps* eat snakes.

GUINEA-FOWL SNAKE/DESERT BLACK-HEADED SNAKE
Micrelaps vaillanti

IDENTIFICATION: A little speckled burrowing snake, usually with a black head. Head slightly broader than the neck, snout rounded. Body cylindrical, tail 5–10% of total length in females, 9–15% in males. Scales smooth, in 15 rows at midbody, ventrals 170–253, cloacal scale divided, subcaudals paired, 17–30. Maximum size about 50cm, but this is an unusually large Somali specimen, average 20–35cm, hatchling size 12–13cm. From a distance it looks speckled grey, often with a black head, close up each grey dorsal scale has a white spot at the free end. Northern Kenyan specimens have a uniform all-black head and collar, those from south-eastern Kenya and northern Tanzania have the top of the head dark but the lips and sides of the head are light, the lip scales black-edged. Underside white, pinkish or cream, southern specimens may have black edging on the ventrals. Occasional specimens are brown or purple brown, rather than grey, and so dark that the black head is not immediately obvious. Taxonomic Note: This species and *Micrelaps boettgeri* were recently shown to be the same species; the name *vaillanti* has priority.

Micrelaps vaillanti

Micrelaps vaillanti Guinea-fowl Snake. Michele Menegon, northern Tanzania.

HABITAT AND DISTRIBUTION: Semi-desert, dry and moist savanna, from 200m up to 1,700m altitude. Occurs from

Tarangire National Park north and east through the low country around Mt Kilimanjaro, Amboseli and Mkomazi to Tsavo, widespread in south-eastern Kenya and along the coastal strip. Isolated records north of the equator include Meru National Park, Kabartonjo in the Tugen Hills, Kalama Conservancy near the Samburu National Reserve, Wajir and Mandera, in Kenya, and Amudat in Uganda. Elsewhere, around the central Ethiopian massif to Sudan and northern Somalia.

NATURAL HISTORY: Poorly known. Burrowing, in soft soil and sand, secretive, active at night, hides in holes and under ground cover during the day. A specimen from Hunters Lodge was captured on a road at night, a Mandera specimen was hiding under a melon, Sudanese specimens were dug up in a field. A female from the Juba River in Somalia had six eggs in her oviducts in October, another from the same locality had eaten a blind-snake, *Typhlops*. A captive specimen ate a skink, *Panaspis*.

SNAKE-EATERS *POLEMON*

A genus of poorly-known, fairly small, usually dark-coloured snakes, some forms were originally placed in the genus *Miodon*. They have a very short, stubby tail, this is a good field character (bearing in mind that the venomous burrowing asps, *Atractaspis*, also have short tails). They have grooved rear-fanged and eat other snakes, but are not dangerous to humans. Snake-eaters have smooth scales, in 15 rows at midbody. They live mostly underground, in leaf litter and in holes. Twelve species are known, all in tropical Africa, most are forest dwellers. They are secretive snakes, there are few museum specimens, the significance of minor scale details in defining some species and subspecies is debatable. Five species occur in our area, all in the forests of the west. They can only be keyed out using colour and this may be unreliable, but a provisional key is given below. Only two species are at all widely distributed in our area (*Polemon christyi*, Christy's Snake-eater, and *Polemon graueri*, the Pale-collared Snake-eater). Some herpetologists take the view that all five forms belong to a single, rather variable species.

KEY TO ADULT EAST AFRICAN MEMBERS OF THE GENUS *POLEMON*

1a Dorsum and ventrum dark, ventral scale sometimes white-edged in adults......*Polemon christyi*, Christy's Snake-eater [p. 461]
1b Dorsum and ventrum not completely dark......2

2a Broad white crossbars or patches on the back of the head and nape...... *Polemon graueri*, Pale-collared Snake-eater [p. 462]
2b Either no neck crossbar or crossbar present but not white......3

3a Ventrals yellow......*Polemon gabonensis*, Gabon Snake Easter [p. 464]
3b Ventrals white...... 4

4a Ventrals less than 253......*Polemon collaris*, Fawn-headed Snake Eater [p. 461]
4b Ventrals more than 262......*Polemon fulvicollis*, Yellow-necked Snake Eater [p. 463]

CHRISTY'S SNAKE-EATER Polemon christyi

IDENTIFICATION: A small dark snake with a very short, stubby tail. Head slightly broader than the neck, eye very small, the pupil round. Body cylindrical, tail short and broad, tapering to a short spike, 3–5% of total length in females, 6–9% in males. Scales smooth, in 15 rows at midbody, ventrals 199–250, subcaudals paired, 15–24. Maximum length about 84cm, average 45–70cm, hatchling size unknown. Iridescent blue-black or lead black above (specimens from the eastern DR Congo were blue-grey), ventrals and subcaudals uniform black or black with white or silver edging. Juveniles grey-brown above, whitish below. Looks rather similar to a burrowing asp, so should not be handled.

Polemon christyi

Polemon christyi Christy's Snake Eater. Colin Tilbury, eastern Congo.

HABITAT AND DISTRIBUTION: Forest, thick woodland, well-wooded savanna and recently deforested areas, from about 600–1,760m altitude in our area. Two Kenya records, Kakamega and Netima, widespread in the forests of the northern Lake Victoria shore in Uganda, west of the Victoria Nile, thence south-west to Rwanda and the northern half of Burundi, around Lake George, also recorded from Murchison Falls and the Budongo Forest. Recorded from Ugalla and Tatanda in Tanzania, thence across to Mbala in Zambia. Elsewhere, widely distributed in the eastern DR Congo and northern and north-west Zambia.

NATURAL HISTORY: Poorly known. Lives mostly under the surface, in holes and leaf litter, emerges to prowl at night on the surface in the rainy season. Slow moving and inoffensive. Presumably lays eggs but no clutch details known. Diet other snakes (including, apparently, its own species), other recorded prey species include blind snakes, (*Typhlops*), worm snakes (*Leptotyphlops*), wolf snakes (*Lycophidion*) and White-lips (*Crotaphopeltis hotamboeia*).

FAWN-HEADED SNAKE EATER Polemon collaris

IDENTIFICATION: A small dark snake with a very short, stubby tail. Head not distinct from the neck, eye very small, the pupil round. Body cylindrical, tail short and broad, tapering to a short spike, 4–5% of total length in females, 5–6% in males. Scales smooth, in 15 rows at midbody, ventrals 181–252, subcaudals paired, 15–25. Maximum length about 86cm, average 40–70cm, hatchling size unknown. Iridescent grey or grey-brown, the top of the head is pale fawn, yellow or fulvous, orange in juveniles, with dark mottling on the snout; adults often have a broad yellow or cream collar. Uniform white below, the ventrals sometimes have a dark spot at the outer edge.

HABITAT AND DISTRIBUTION: Marginal in East Africa, in forest and forest remnants at altitudes below 1,200m. Known only from Bundibugyo and Kibale in extreme western Uganda, although presumably extends right around the base of the Rwenzori Mountains, also known from Rutshuru on the flood plain south of Lake Edward in the DR Congo. Elsewhere, south-west to the southern DR Congo, west to Cameroon.

NATURAL HISTORY: Poorly known. Lives mostly under the surface, in holes and leaf litter, emerges to prowl at night on the surface in the rainy season. Slow moving and inoffensive. Presumably lays eggs but no clutch details known. Diet other snakes, mostly blind snakes (*Typhlops*).

Polemon collaris

Polemon collaris Fawn-headed Snake Eater. Michele Menegon, Rwanda.

PALE-COLLARED SNAKE EATER *Polemon graueri*

IDENTIFICATION: A small black snake with a distinctive, broad white collar and very short, stubby tail. Head not distinct from the neck, eye very small, the pupil round. Body cylindrical, tail short and broad, tapering to a short spike, 4–5% of total length. Scales smooth, in 15 rows at midbody, ventrals 222–262, subcaudals paired, 13–21. Maximum length about 56cm, average 30–45cm, hatchling size unknown. Iridescent bluish black or grey above, the scales sometimes white-edged. There is a broad white collar extending from just behind the eyes to the fourth or fifth dorsal scale row. The ventrals are white, often with dark edging.

Polemon graueri

Polemon graueri Pale-collared Snake Eater. Harald Hinkel, Rwanda.

HABITAT AND DISTRIBUTION: Forest and recently deforested areas in Uganda and Rwanda, from 800–1,800m altitude, along the Albertine Rift, known localities include Kasese, Budongo Forest and Masindi, thence south from Bundibugyo along the western border, into western Rwanda, the Nyungwe Forest and extreme north-west Burundi. The holotype supposedly came from Entebbe, but subsequent work failed to discover it there (so not shown on map); there is also a curious record from Serere, just east of Lake Kyoga. Might be more widely distributed in Uganda. Elsewhere, known from the Virunga National Park, just across the border in the DR Congo.

NATURAL HISTORY: Poorly known. Lives mostly under the surface, in holes and leaf litter, emerges to prowl at night on the surface in the rainy season. Slow moving and inoffensive. Presumably lays eggs but no clutch details known. Diet mostly other snakes, particularly blind snakes (*Typhlops*), one specimen had eaten lizard eggs.

YELLOW-NECKED SNAKE EATER *Polemon fulvicollis*

IDENTIFICATION: A small black snake with a yellow collar (or just a patch) and a short, stubby tail. Head not distinct from the neck, eye very small, the pupil round. Body cylindrical, tail short and broad, tapering to a short spike, 4–6% of total length. The following notes apply to the eastern subspecies of this snake, *Polemon fulvicollis gracilis*. Scales smooth, in 15 rows at midbody, ventrals 262–285, subcaudals paired, 20–24. Maximum length about 50cm, average and hatchling size unknown. Iridescent black above, a yellow collar is present and broadens downwards, ventrals white.

Polemon fulvicollis

Polemon fulvicollis Yellow-necked Snake Eater. Eli Greenbaum, eastern Congo.

HABITAT AND DISTRIBUTION: Forest. There is only a single East African record, from the 'Bwamba Forest', which describes the patches of low altitude forest (an eastern extension of the Ituri Forest in the DR Congo) at 630m in north Bundibugyo, south of Lake Albert and east of the Semliki River, western Uganda. Elsewhere, in forests of the eastern DR Congo, the western subspecies occurs in Gabon.

NATURAL HISTORY: Poorly known. Lives mostly under the surface, in holes and leaf litter, emerges to prowl at night on the surface in the rainy season. Slow moving and inoffensive. Lays eggs, one female contained some very elongate eggs, 4×0.5cm. Diet: other snakes, mostly blind snakes (*Typhlops*).

GABON SNAKE EATER *Polemon gabonensis*

IDENTIFICATION: The following notes apply to the eastern subspecies of this snake, *Polemon gabonensis schmidti*. A small grey-brown snake with a short, stubby tail. Head not distinct from the neck, eye very small, the pupil round. Body cylindrical, tail short and broad, tapering to a short spike, 4–5% of total length.. Scales smooth, in 15 rows at midbody, ventrals 221–241 in males, 245–252 in females, subcaudals paired, 19–24 in males, 17–18 in females. Maximum length about 80cm, average 50–70cm, hatchling size unknown. Greyish-blue or iridescent browny-grey above, darkening towards the tail, head grey, the two lowest dorsal scale rows are light reddish-brown, the ventrals are yellow, this yellow extending up onto the side of the neck, it may extend up to form a pale yellow or fawn collar encircling the neck.

Polemon gabonensis

HABITAT AND DISTRIBUTION: Forest. There is only a single East African record, from the 'Bwamba Forest', which describes the patches of low altitude forest (an eastern extension of the Ituri Forest in the DR Congo) at 630m in north Bundibugyo, south of Lake Albert and east of the Semliki River, western Uganda. However, known also from Idjwi Island and just southwest of Lake Kivu, in the DR Congo, so might be in northern Rwanda. Elsewhere, widespread in forests of the eastern and northern DR Congo, the western subspecies occurs in Gabon, Equatorial Guinea and Cameroon, west to Togo.

Polemon gabonensis Gabon Snake Eater. Jean-Francois Trape, West Africa.

NATURAL HISTORY: Poorly known. Lives mostly under the surface, in holes and leaf litter, emerges to prowl at night on the surface in the rainy season. Slow moving and inoffensive. Lays eggs, one female contained some very elongate eggs, 4×0.5cm. Diet other snakes, one female had eaten a 35cm Angola Blind Snake *Afrotyphlops angolensis*.

BLACK AND YELLOW SNAKES *CHLORHINOPHIS*

An African genus of small, very elongate, cylindrical bodied snakes. They are striped black and yellow. They are rear-fanged, with relatively large fangs, but are so small that they are quite harmless to humans. They have little short heads with round or vertically elliptic pupils. Two species are known, both occur in our region.

KEY TO EAST AFRICAN MEMBERS OF THE GENUS *CHILORHINOPHIS*

1a Ventral scales more than 307 .....*Chilorhinophis gerardi*, Gerard's Black-and-yellow Burrowing-snake [p. 465]

1b Ventral scales less than 289.......*Chilorhinophis butleri*, Butler's Black-and-yellow Burrowing-snake [p. 465]

BUTLER'S BLACK-AND-YELLOW BURROWING-SNAKE
Chilorhinophis butleri

IDENTIFICATION: A very slender, black and yellow striped snake. The following notes apply to the Tanzanian subspecies *Chilorhinophis butleri carpenteri*. Head short and blunt, eye small, pupil round. Body cylindrical, tail very short and blunt, 5–10% of total length, males have longer tails than females. Scales smooth, in 15 rows at midbody, ventrals 216–288, males rarely more than 238, Tanzanian specimens rarely more than 256, subcaudals paired, 18–30, more than 25 in males. Maximum length about 36cm, average 20–30cm, hatchling size unknown. Head black above, body yellow with three longitudinal black stripes, ventrals yellow or orange, tail black, tail tip white. Taxonomic Note: Originally described as *Chilorhinophis carpenteri liwalensis*, discovered by C.J.P. Ionides.

Chilorhinophis butleri

HABITAT AND DISTRIBUTION: This subspecies occurs in low altitude moist savanna in south-east Tanzania, at around 400m altitude. Known from a number of localities including Liwale, Lindi, Ruponda (Luponda), Masasi and Songea; Tanzania, and also from Ancuabe in north-eastern Mozambique, this particular specimen has five stripes. The nominate subspecies is found at Mongalla on the White Nile in the Sudan.

NATURAL HISTORY: A secretive, burrowing snake, living in leaf litter, in holes and soft soil; may come to the surface at night after rain has fallen. It favours sandy soils. Presumably lays eggs, no details known. Newly-hatched juveniles were found in April in Newala. Feeds on worm lizards, might eat snakes, the other species in the genus does. Escapes by burrowing, but it has an interesting defence: if cornered it elevates the head and the tail and moves back and forward, presumably confusing a predator as to which end to attack, if further molested hides its head within its coils and waves its tail in the air; presumably in the hope that a predator will attack it.

GERARD'S BLACK-AND-YELLOW BURROWING-SNAKE
Chilorhinophis gerardi

IDENTIFICATION: A small, very slender, black and yellow striped snake. The following notes apply to the Tanzanian subspecies *Chilorhinophis gerardi tanganyikae*. Head short and blunt, eye small, pupil round. Body cylindrical, tail very short and blunt, 5–8% of total length. Scales smooth, in 15 rows at midbody, ventrals 308–310 in males, 375 in the only known female, subcaudals paired. Maximum length about 57cm, average and hatchling size unknown. Head black above, with two fine yellow crossbars or part bars, or blotches, body yellow with three longitudinal black stripes, chin yellow, ventrals bright yellow or orange, the tail tip is black above, resembling the head, blotched blue-grey below.

HABITAT AND DISTRIBUTION: This subspecies, with its high ventral counts, is known from only three localities; Ujiji in western Tanzania, Nyamkolo (= Mpulungu) in northern Zambia; both are on Lake Tanganyika at 770m altitude, and Lukonzolwa, on the west side of Lake Mweru in the south-east DR Congo, at 920m altitude. The other subspecies occurs in north and eastern Zambia and northern Zimbabwe.

Chilorhinophis gerardi

Chilorhinohis gerardi Gerard's Black and yellow Burrowing Snake. Colin Tilbury, eastern Congo.

NATURAL HISTORY: This subspecies is virtually unknown but the nominate subspecies is inoffensive and secretive, it lives in leaf litter, in holes and soft soil and comes to the surface at night after rain has fallen. It lays up to six elongate eggs, roughly 3×0.8cm. Feeds on worm lizards, small burrowing snakes, burrowing and ground-dwelling skinks. Escapes by burrowing, but if cornered hides its head within its coils and waves its tail in the air; it is believed that a predator seeing this will attack the tail, giving the snake a chance to escape.

PURPLE-GLOSSED SNAKES *AMBLYODIPSAS*

An African genus of small to medium, usually dark-coloured snakes, with a distinctive purple iridescence on the scales, this may be useful in the field. Several species have yellow flank stripes, including one in our area. Scales are smooth, in 15–21 rows at midbody. They are mostly burrowers, living underground, but will hunt on the surface on wet nights. They eat mostly other snakes, worm lizards and ground-dwelling lizards. Nine species are known, with the greatest concentration in the moist forests and woodland of south-east Africa, although two species live in near-desert. Five species occur in our area, two of which are endemic, there is also an endemic subspecies. They have long grooved rear fangs and although no symptoms have been recorded following bites, they sometimes bite quite savagely if restrained. They are very similar to burrowing asps, even showing similar defensive behaviour, so anyone attempting to handle a suspected purple-glossed snake should do so very cautiously, preferably with tongs.

KEY TO EAST AFRICAN MEMBERS OF THE GENUS *AMBLYODIPSAS*

1a Upper labials 4 or 5......2
1b Upper labials 6......3

2a Midbody scale rows 15......*Amblyodipsas katangensis* (Tanzanian subspecies), Ionides' Purple-glossed Snake [p. 468]
2b Midbody scale rows 17......*Amblyodipsas teitana*, Taita Hills Purple-glossed Snake [p. 470]

3a Flanks chrome yellow......*Amblyodipsas dimidiata*, Mpwapwa Purple-glossed Snake [p. 467]
3b Flanks not chrome yellow......4

4a Dorsal scale rows usually 15 on the neck, usually 17 at midbody, range eastern Tanzania and Kenya, not found west of the rift valley in Kenya......*Amblyodipsas polylepis*, (East African subspecies of) Common Purple-glossed Snake [p. 468]
4b Dorsal scale rows usually 17 on the neck, 17-19 at midbody, range western Kenya and Uganda, not east of the Rift Valley in Kenya......*Amblyodipsas unicolor*, Western Purple-glossed Snake [p. 470]

MPWAPWA PURPLE-GLOSSED SNAKE
Amblyodipsas dimidiata

IDENTIFICATION: A small shiny dark snake with a vivid yellow flank stripe and a long sharp snout. Head not distinct from the neck, eye very small, the pupil round but hard to see. Body cylindrical, tail short and blunt, about one twentieth of the total length. Scales smooth, in 17 rows at midbody, ventrals 196–205 in males, 215–219 in females, cloacal scale divided, subcaudals paired, 25–30 in males, 19–21 in females. Maximum length 51.5cm, average probably 30–40cm, hatchling size unknown. Iridescent blackish-brown or purple-brown above, the snout and upper lip are chrome yellow and this leads back to a chrome yellow flank stripe, an unmistakeable field character. The ventrals are pink, with scattered dark blotches.

Amblyodipsas dimidiata

HABITAT AND DISTRIBUTION: Endemic to Tanzania, recorded only from the Ugogo Region, from Dodoma and Mpwapwa, in dry savanna at 1,000m altitude, not seen for many years but rediscovered there in the 1990s. Conservation Status: IUCN Least Concern.

Amblyodipsas dimidiata Mpwapwa Purple-glossed Snake. Lorenzo Vinciguerra, Dodoma.

NATURAL HISTORY: Poorly known. Presumably similar to other purple-glossed snakes, i.e. lives mostly under the surface, in holes; its very sharp snout suggests that it shoves through soil and sand like a quill-snouted snake (*Xenocalamus*), probably emerges to prowl at night on the surface in the rainy season. One specimen was dug out of the roots of a rotting tree stump in sand on a dry river bank, others found in agricultural plots during the planting season, under ground debris and vegetation heaps around habitation; also in termite mounds. Probably lays eggs and eats worm lizards, three species of which occur in its habitat.

IONIDES' PURPLE-GLOSSED SNAKE *Amblyodipsas katangensis*

The notes and common name apply to the southern subspecies of this snake, *Amblyodipsas katangensis ionidesi*.

IDENTIFICATION: A small, slim sharp-nosed purple-glossed snake. The snout is distinctly pointed, with a very long rostral and underslung lower jaw. Head not distinct from the neck, eye very small, the pupil round but hard to see. Body cylindrical, tail short and blunt, feebly tapered, 6–10% of the total length, long in males. Scales smooth, in 15 rows at midbody, ventrals 164–184 in males, 179–206 in females, cloacal scale divided, subcaudals paired, 21–25 in males, 15–20 in females. Maximum length 40.5cm, average 25–35cm, hatchling size unknown. Black above, sometimes some white mottling on the snout, about two-thirds of specimens are uniform black below, one third chequered black and white, the white mostly on the neck and front half of the body.

Amblyodipsas katangensis

Amblyodipsas katengensis Ionides' Purple-glossed Snake. Luke Verburgt, Palma Northern Mozambique.

HABITAT AND DISTRIBUTION: In Tanzania, in moist savanna, woodland and coastal thicket, from sea level to about 400m altitude, in the south-east of the country; known localities include Kilwa, Lindi, Liwale, Nampugu and Tunduru. Also occurs in northern Mozambique across the Rovuma River.

NATURAL HISTORY: Poorly known. Lives mostly under the surface, in holes and leaf litter, particularly in sandy soil, emerges to prowl at night on the surface in the rainy season. Slow moving and inoffensive, but jerk convulsively when handled. Lays eggs, one Liwale female contained two very elongate eggs, 4.2×0.6cm in January, another held three eggs, 1.7×0.4cm, in April. The only known prey is a species of worm lizard, *Loveridgea ionidesi*.

COMMON PURPLE-GLOSSED SNAKE *Amblyodipsas polylepis*

Notes refer to the northern subspecies of this snake, *Amblyodipsas polylepis hildebrandtii*. The nominate subspecies, *Amblyodipsas polylepis polylepis* occurs further south.

IDENTIFICATION: A small dark stocky snake with a blunt tail. The head is short and rounded, not distinct from the neck, eye very small, the pupil round but hard to see. Body cylindrical, tail short and blunt, 5–10% of total length, long tails males. Scales smooth, in 17–19 rows at midbody, ventrals 161–169 in males, 176–203 in females, cloacal scale divided, subcaudals paired, 24–29 in males, 17–22 in females. Maximum length about 68cm. Females are larger than males, the nominate southern subspecies grows much larger, up to 1.1m. Average 40–60cm, hatchling size unknown. Purple-brown to pinkish brown, sometimes blackish-brown above and below, with a distinctive purple gloss on the scales, especially if freshly sloughed; individuals about to slough go very grey. The dorsal and ventral scale are sometimes pale edged, giving a netting effect.

Amblyodipsas polylepis

HABITAT AND DISTRIBUTION: Coastal woodland and thicket, moist savanna and evergreen hills forest, from sea level to about 1,500m. Occurs from Manda Island south along the East African coast to the Rovuma River, inland to the Usambara Mountains and across to Tunduru in south-east Tanzania; isolated inland records include the Nyambene Hills and Chuka in Kenya, the area between Arusha, Lake Manyara and the Serengeti in Tanzania, the Uluguru Mountains and the border country between the top of Lake Malawi (Lake Nyasa) and the bottom end of Lake Tanganyika. Probably more widespread in southern Tanzania. An enigmatic record from Nairobi is not shown on the map. Elsewhere, north to southern coastal Somalia, the nominate subspecies south to South Africa.

NATURAL HISTORY: Poorly known. Lives mostly under the surface, in holes and leaf litter or under ground cover, in hollow logs, etc., emerges to prowl at night on the surface in the rainy season. It is nervous and quite quick moving, if picked up it will turn and bite furiously and tenaciously. If molested, it will hide its head under its cols and raise the tail tip; a South African specimen elevated and flattened the tail, curling the tip down, possibly mimicking the hood of a cobra, to presumably dissuade a predator. It lays eggs, a specimen of the nominate subspecies laid seven eggs, roughly 3×1.5cm. Known prey items include snakes (centipede-eaters, wolf snakes, worm snakes and blind snakes), smooth-bodied skinks, worm lizards and caecilians. A specimen of the southern subspecies spent four hours killing a big Schlegel's Blind Snake *Rhinotyphlops schlegelii*.

Amblyodipsas polylepis Common Purple-glossed Snake. Left: Victor Wasonga, Shimba Hills. Right: Michele Menegon, Sloughing, Ruaha National Park.

TAITA HILLS PURPLE-GLOSSED SNAKE Amblyodipsas teitana

IDENTIFICATION: This snake is endemic to Kenya, and is known from a single female specimen, taken more than 60 years ago. It is a small dark snake with a very blunt tail. The head is short, snout quite pointed, eye very small, the pupil round but hard to see. Body cylindrical, tail short and blunt, 5.5% of total length. Scales smooth, in 17 rows at midbody, ventrals 202, cloacal scale divided, subcaudals paired, 16. Length 43.4cm. Black above and below except for the mental (scale at the front of the lower jaw), first and second lower labials, which are white; the preserved type has turned brown. Conservation Status: IUCN Data Deficient.

Amblyodipsas teitana

HABITAT AND DISTRIBUTION: Endemic to Kenya. The original catalogue entry gives the locality as 'Mt Mbololo, Teita Hills, 1,150m'. which presumably means the hills around Ndome Peak, on the north-eastern side of the Taita Hills, and source of the Mbololo River, which runs across the plain below to the Galana River. Ndome was forested at the time of collection, but currently the forest is mostly above this altitude. Conservation Status: IUCN Data Deficient. This snake, a rare endemic, inhabits a fragile ecosystem. The Taita Hills are host to a wide range of remarkable endemic species, there are probably other undiscovered species there, and they serve as a refuge for some interesting, more widespread forest species that occur nowhere near. It is important that some area of the hill forest that remains is given protection.

NATURAL HISTORY: Nothing known but presumably similar to other purple-glossed snakes, i.e. nocturnal, burrowing, active on the surface at night after rain, lays eggs, eats snakes. It would be interesting if some more specimens could be found. Might occur on Mt Sagalla near Voi, or on the North Pare Mountains in Tanzania; large hills in the same area.

WESTERN PURPLE-GLOSSED SNAKE Amblyodipsas unicolor

IDENTIFICATION: A medium-sized dark stocky snake with a blunt tail. The head is short, eye very small, the pupil round but hard to see, the snout rounded and protruding. Body cylindrical, tail short and tapering, 9–13% of total length in males, 7–9% females. Scales smooth, in 17 rows at midbody (sometimes 15), ventrals 165–179 in males, 192–207 in females, cloacal scale divided, subcaudals paired, 30–41 in males, 21–29 in females. Maximum length about 1.14m. Females are larger than males. Average 60–90cm, hatchling size unknown. Dark grey, purple-grey or black, above and below, often with a distinctive purple gloss on the scales, especially if freshly sloughed; West African specimens may be metallic blue-grey.

Amblyodipsas unicolor

HABITAT AND DISTRIBUTION: Dry and moist savanna, woodland and forest, in our area from about 1,000–1,500m, elsewhere to sea level. A western snake, with few records in East Africa; they include the Kerio Valley in Kenya (presumably on the wall rather than the valley floor), Uganda localities are Kampala, Jinja and Masindi. Elsewhere, west to Senegal through the Guinea savanna.

NATURAL HISTORY: Poorly known. Lives mostly under the surface, in holes and leaf litter or under ground cover, in hollow logs etc., emerges to prowl at night on the surface in the rainy season. It is nervous and quite quick moving, if picked up it will turn and bite furiously and tenaciously. If restrained above ground, it also shows a modified form of the head-pointing behaviour shown by the burrowing asp, arching the neck and pushing the nose into the ground. Lays eggs but no clutch details known. Diet similar to other purple-glossed snakes, includes small snakes, worm lizards and burrowing skinks.

Amblyodipsas unicolor Western Purple-glossed Snake. Stephen Spawls, Ghana.

SUBFAMILY ATRACTASPIDINAE

A subfamily of small dark dangerous snakes, with around 23 species in two genera. They are mostly African, a handful of species reach the Middle East. Five species occur in our area.

BURROWING ASPS *ATRACTASPIS*

A genus of 21 species of small (usually less than 70cm), dark, blunt-headed snakes. They have hinged front fangs, and were thus thought to be vipers but are now recognised as specialised members of the Lamprophiidae. They are widely distributed across Africa, two species occur in the Middle East; five species are known from East Africa. They are nocturnal and terrestrial, spending much time underground in holes; they feed largely on reptiles, amphibians, occasionally small mammals. They lay clutches of 3–11 eggs. Common names used for the genus, as well as burrowing asp, include burrowing vipers, mole vipers, stiletto snakes and side-stabbing snakes. They can be hard to identify; they closely resemble purple-glossed snakes (and to a lesser extent wolf snakes and dark centipede-eaters); even expert herpetologists have been bitten by mis-identified burrowing asps. They cannot be held safely by hand; their long fangs and short heads enable them to bite any restraining finger; the snake sliding a fang out of its mouth, sometimes without actually opening the mouth, the fang is then driven into the finger by a downwards jerk. The fangs are primarily used for stabbing and manipulating their prey, an adaptation that may be useful for feeding in holes, and they have been reported to enter a rodent nest and stab all the babies before commencing feeding. Apart from the fangs, burrowing asps are virtually toothless, and thus struggle to ingest prey, having to flex the neck and body to swallow (other snakes manipulate prey with their solid teeth; a process called cranial kinesis). They have very long venom glands; their venom is cytotoxic and of medical significance. In some areas they are a major cause of snakebite, primarily by people farming or gardening with short tools, treading on the snake after dark or people sleeping on the ground and rolling on the snake in the night. They tend not to strike forwards, even if approached closely, but if restrained will try to wrap around the restraining object and drive a fang down into it. No specific antivenom is obtainable so bites must be treated symptomatically. The venoms cause swelling, pain, blistering and necrosis. Nearly all victims recover without complications, but a few fatalities are recorded.

KEY TO EAST AFRICAN MEMBERS OF THE GENUS *ATRACTASPIS*

1a Cloacal scale divided, subcaudals scales paired......2
1b Cloacal scale scale entire, subcaudals single......3
2a Midbody scale rows 19, range north-eastern Kenya......*Atractaspis engdahli*, Engdahl's Burrowing Asp [p. 474]
2b Midbody scale rows 21-17, range the western side of East Africa......*Atractaspis irregularis*, Variable Burrowing Asp [p. 476]
3a Midbody scale rows 27 or more in East Africa, large adults stout......*Atractaspis fallax*, Eastern Small-scaled Burrowing Asp [p. 475]
3b Midbody scale rows fewer than 27 in East Africa, large adults slim or moderate......4
4a Midbody scale rows 21-25, ventrals 212-246 in males, 238-260 in females, widespread......*Atractaspis bibronii*, Bibron's Burrowing Asp [p. 473]
4b Midbody scale rows 19-23, ventrals 244-278 in males, 263-300 in females, rare......*Atractaspis aterrima*, Slender Burrowing Asp [p. 472]

SLENDER BURROWING ASP *Atractaspis aterrima*

IDENTIFICATION: A small, slender, fast-moving burrowing asp. Head blunt, eyes tiny. Tail short and blunt, 3–5% of total length, ending in a pointed cone. Scales in 19–21 (rarely 23) rows at midbody, ventrals 244–300, subcaudals single, 18–25. Maximum size about 70cm, average 30–50cm. Hatchling size unknown. Usually black or blackish-grey in colour (occasionally blackish-brown).

HABITAT AND DISTRIBUTION: An unusually wide choice of habitat, known from dry savanna, moist savanna, woodland and forest, from sea level to 2,000m altitude. However, rare and localised in East Africa; and known from only four localities: Udzungwa and Uluguru Mountains and Mwanihana Forest Reserve in Tanzania, and Wadelai on the White Nile in northern Uganda. Elsewhere, west to Senegal.

Atractaspis aterrima

NATURAL HISTORY: Poorly known. Lives and hunts in holes, but may emerge at night (especially during or after rain) and move around on the ground. If molested they have a distinctive response, arching the neck and pointing the head at the ground, looking like an inverted U. They may also release from the cloaca a distinctive-smelling chemical, which might serve to stop other burrowing asps eating them. If further threatened, they may wind the body into tight coils, or turn the head and neck upside down, lash from side to side or jerk violently. The defensive purpose of this behaviour may be significant in holes or underground, where most of the snake's activity occurs. Diet mostly smooth bodied reptiles in the west of its range, but Tanzanian specimens recorded eating caecilians, may also take mammals. No breeding details known. Swelling and pain and subsequent lymphadenopathy resulted from a bite.

Atractaspis aterrima Slender Burrowing Asp. Left: Michele Menegon, Sali Forest Reserve, Mahenge Mountain. Right: Michele Menegon, close-up of head.

BIBRON'S BURROWING ASP *Atractaspis bibronii*

IDENTIFICATION: A small, fairly slim dark snake, with a prominent snout. The tongue is white. Body cylindrical, tail short and blunt, ending in a pointed cone, 5–7% of total length. Slow moving. Scales in 19–25 rows at midbody (usually 21–25 in our area), ventrals 213–262 (usually 225–248 in East Africa), subcaudals single, 18–28. Maximum size about 70cm, average 30–50cm, hatchling size 20cm. May be black, brown, purplish-brown (especially Kenya coast animals) or grey in colour, often with a purplish sheen on the scales, the belly may be lead grey, brownish, white or pale with a series of dark blotches; animals with dark ventrals may have white patches on the chin and throat, or around the tail base. If the belly is pale, the pale colour may extend up to the lowest 2–3 scale rows on the sides and onto the upper labials.

Atractaspis bibronii

HABITAT AND DISTRIBUTION: In East Africa, mostly in coastal thicket, woodland and moist savanna, although extends into dry savanna and semi-desert in other areas. Found from sea level to about 1,800m altitude. Widely distributed in Tanzania (records lacking from the west-centre and north) and on Zanzibar, south-eastern Rwanda and eastern Burundi, in Kenya confined to the coastal strip as far north as Lamu, although their distribution in coastal Kenya is curious; no specimen has been collected in or around Malindi or Watamu, and yet several specimens are recorded at Kaloleni and the western Arabuko-Sokoke Forest. There is an inland record from Kitui and an unverified record (not shown on map) from Keekorok Lodge, Maasai Mara. Elsewhere, south-west to Angola, south to northern South Africa.

NATURAL HISTORY: Nocturnal and terrestrial, spending much time in holes, in rotting logs or under ground cover but active on the ground on wet nights. Clutches of three, four, six and seven eggs recorded, roughly 3.5×1.5cm. Known prey items for this species include rodents, shrews, burrowing skinks, skink eggs, snakes and worm lizards. Like the previous species, when molested they show the distinctive response, arching the neck, pointing the snout into the ground, jerking convulsively and releasing a distinctive chemical smell. In coastal South Africa and Tanzania this snake is responsible for many bites, but no fatalities are recorded.

Atractaspis bibronii Bibron's Burrowing Asp. Left: Michele Menegon, Kilombero Valley. Right: Thomas Hakannson, Kaloleni.

ENGDAHL'S BURROWING ASP *Atractaspis engdahli*

IDENTIFICATION: A small burrowing asp with a rounded snout. Body cylindrical, tail short, 5–6% of total length. Scales in 19 rows at midbody, ventrals 219–232, cloacal scale divided, subcaudals paired, 19–23. Maximum size about 45cm, average size 25–40cm, hatchling size unknown but probably about 15cm. Black, blue-black or red-brown in colour (juveniles can be quite light brown), paler below.

HABITAT AND DISTRIBUTION: Endemic to the horn of Africa. Coastal plain, thicket, grassland, dry savanna and semi-desert, from sea level to 250m altitude. In Kenya, known only from a single specimen from Wajir Bor, 50km east of Wajir in north-eastern Kenya; also known from the middle and lower Juba River and Kismayu in Somalia.

Atractaspis engdahli

NATURAL HISTORY: Poorly known, presumably similar to other *Atractaspis*;

Atractaspis engdahli Engdahl's Burrowing Asp. Preserved animal California Academy of Sciences; Wajir Bor.

living underground and eating largely smooth-bodied reptiles. Specimens have been captured in termitaria, in holes and prowling in semi-desert at night, the Wajir Bor specimen was in a dry well.

EASTERN SMALL-SCALED BURROWING ASP
Atractaspis fallax

IDENTIFICATION: A big burrowing asp, often stout, with a broad blunt head. Body cylindrical, tail short, 7–9% of total length. Scales smooth, in 25–37 rows at midbody (27–37 in East Africa), ventrals 210–258 (in East Africa usually 228–237 males, 239–252 females), subcaudals single (occasionally paired, or a mixture of both) 20–39. One of the biggest burrowing asps, reaching 1.1m in eastern Kenya, there are anecdotal reports of larger specimens, up to 1.3m, on the Kenyan coast. Average size 60–90cm, hatchling size unknown. Colour quite variable; black, grey, brown, often distinctly purple-brown, scales may be iridescent. Kenyan coastal specimens are often light purple-brown, darkening towards the head, with the head and the neck black. Juveniles often light brown. The belly is usually dark. Taxonomic Note: Originally called *Atractaspis microlepidota*, but that name is now reserved for West African snakes.

Atractaspis fallax

HABITAT AND DISTRIBUTION: Coastal bush and thicket, dry and moist savanna, grassland and semi-desert, from sea level to about 1,800m, but most common in dry savanna below 1,000m. Widespread in south-eastern Kenya, and enters north-east Tanzania along the border, west to the low country south of Mt Kilimanjaro; extends west to Emali in Kenya. Also known from the Mara and northern Serengeti, isolated records from Kenya include Garissa, west Laikipia, Lokori, Lodwar and along the Daua River from Malka Murri to Mandera in northern Kenya. It is also recorded just north of the Uganda border at Juba and Torit in South Sudan. Elsewhere, north to Ethiopia and Somalia.

Atractaspis fallax Eastern Small-scaled Burrowing Asp. Michele Menegon, northern Tanzania.

NATURAL HISTORY: Nocturnal and terrestrial; usually underground but active at night on the ground in the rainy season. Specimens have occasionally been found abroad by day. If molested they show the usual distinctive burrowing asp response, arching the neck and pointing the head at the ground, looking like an inverted U; then writhing and hiding the head. They may also release from the cloaca a distinctive-smelling chemical. Captive specimens have been photographed elevating the head and flaring the neck into a moderate hood, like a cobra. A Kenyan female laid eight eggs. Diet: snakes, smooth-bodied lizards and occasional rodents. Said to be relatively common in parts of its range, especially the north Kenya coast, from Kilifi to Malindi. In Somali areas it is known as *Jilbris*, 'the snake of seven steps', (meaning if you are bitten you take seven steps and die) or 'father of ten minutes' (for obvious reasons). A very difficult snake to catch, big adults are strong, active and bite furiously if restrained. This snake has caused a few fatalities, possibly due to unusual medical circumstances, but the large size of some adults means that bites may need effective symptomatic treatment (no antivenom is available). Occasional bites in Kenya have caused necrosis requiring excision or amputation.

VARIABLE BURROWING ASP *Atractaspis irregularis*

IDENTIFICATION: A fairly small burrowing asp. Body cylindrical, tail short, 5–8% of total length, ending in a pointed cone. Scales smooth, 23–27 rows at midbody (occasionally 21, usually 23 in the eastern side, 25–27 on the western side), ventrals 213–263, cloacal scale divided, subcaudals paired, 21–32, higher counts males. Maximum size about 66cm, average 30–50cm. Hatchling size unknown. Usually shiny black or blackish-grey in colour, the scales are iridescent. Ventrals dark grey, black or rufous, sometimes with white edging.

Atractaspis irregularis

HABITAT AND DISTRIBUTION: Moist savanna, woodland and forest, in East Africa from 600–1,800m altitude, elsewhere to sea level. In Kenya, occurs in the highlands, from Murang'a north-east and then west around the Mt Kenya massif, south to Naivasha and west across to Lake Victoria; Kenyan localities include Chuka, Nyambene Hills, Meru, Laikipia, Ol Ari Nyiro, Njoro, Kabatonjo, Kisumu and Kakamega. Widespread in Uganda, from the north-western shore of Lake Victoria north to Lake Kyoga and to the northern border at Nimule, south down the western side of Uganda to Rwanda and Burundi, to Kigoma and Kibondo in Tanzania. Elsewhere, west to Nigeria, and in Liberia west of the Dahomey gap. Conservation Status: IUCN Least Concern.

NATURAL HISTORY: Nocturnal and terrestrial, spending much time in holes or under ground cover but active on the ground on wet nights. Shows the nose-pointing response if molested. A female from the Mabira Forest in Uganda contained six eggs, mating observed there in September. Known to eat rodents as well as small reptiles. This is a species known to have caused human fatalities, albeit under unusual circumstances, one victim was bitten eight times after rolling on the snake whilst asleep.

Atractaspis irregularis Variable Burrowing Asp. Bill Branch, Gabon.

Philothamnus angolensis
Angolan Green Snake
Michele Menegon, Tanzania.

SUB-ORDER **SERPENTES**
FAMILY **COLUBRIDAE**
COLUBRID SNAKES

An intercontinental snake family, with around 1,900 species, found on every continent save Antarctica (although there are very few in Australasia), the colubrids might be loosely described as 'ordinary' or 'typical' snakes. None are very large or very small, most are between 50cm and 2m long. They do not have front fangs, most have obvious eyes, broad belly scales and their tails taper. However, at the time of writing there remains vigorous debate about what taxa should be placed in the family; even the Reptile Database is equivocal. We opt here for a recent analysis that suggests there are eight subfamilies within the Colubridae; three subfamilies occur in East Africa, the Colubrinae, Natricinae and Grayiinae. As before, no key is provided for the subfamilies, as the original key enabled identification of all the genera.

SUBFAMILY **COLUBRINAE**

A subfamily of around 100 genera and nearly 750 species; 15 genera and 51 species occur in our area. Most are harmless, but the group includes a number of rear-fanged snakes, including some that are dangerously venomous, namely the Boomslang *Dispholidus typus* and the three vine or twig snakes of the genus *Thelotornis* (and possibly the Dagger-tooth Vine Snake *Xyelodontophis uluguruensis*).

RACERS *PLATYCEPS*

A genus of diurnal fast-moving, slim, harmless egg-laying colubrid snakes, most are brown or grey, often banded or spotted. Originally placed in the genus *Coluber*. Twenty-five species are known from northern Africa, the Middle East and western Asia, ten occur in Africa, two in our area.

KEY TO EAST AFRICAN MEMBERS OF THE GENUS *PLATYCEPS*

1a 192 or more ventrals in males, 205 or more in females,......*Platyceps florulentus*, Flowered Racer [p. 478]

1b 189 or fewer ventrals in males, less than 200 in females......*Platyceps brevis*, Smith's Racer [p. 479]

FLOWERED RACER *Platyceps florulentus*

IDENTIFICATION: A slim, fast-moving, small diurnal dry-country snake. The eye is large, pupil round. Tail long, one-quarter of total length. Scales smooth and glossy, in 21–25 midbody rows, ventrals 192–228 (high counts in females), subcaudals 83–105. Maximum size about 1m, average 60–90cm, hatchling size unknown. Colour olive brown, grey-brown or rufous. Two subspecies occur in our area. *Platyceps florulentus florulentus* may be uniform or with one or two dark crossbars on the head and a series of cross-bars or darker spots on the body; these markings may fade in large adults, especially towards the tail. Occasional specimens are grey anteriorly and brown posteriorly. The belly may be white, cream or orangey-yellow. This form has scales in 21–23 rows at midbody. The subspecies from the Lake Baringo area (*Platyceps florulentus keniensis*) is usually warm brown or rufous, with a dark mark extending down from the crown to the side of the head, the body faintly or clearly spotted with rufous spots, fading towards the tail, the ventrals yellow, cream or white, it has 25 midbody scale rows.

Platyceps florulentus

HABITAT AND DISTRIBUTION: Low altitude dry savanna, between 370–1,200m altitude. The subspecies *Platyceps florulentus florulentus* is only known in our area from Nimule, on Uganda's northern border, in moist savanna, and on the eastern shores of Lake Turkana, elsewhere north to Egypt and Ethiopia. The subspecies *Platyceps florulentus keniensis* is known from Lake Baringo, where it appears to be most common on the islands. A specimen intermediate between the two has been taken at Lokori in the lower Kerio Valley. These latter two localities are in low altitude dry savanna. Conservation Status: IUCN Least Concern.

NATURAL HISTORY: Terrestrial, although they will climb into bushes and low trees.

Platyceps florulentus Flowered Racer. Left: Stephen Spawls, Lake Baringo. Right: Eduardo Razzetti, Lake Turkana.

They are fast moving and diurnal, although in parts of the Nile Valley they are crepuscular, moving at dawn and dusk. At night they shelter in holes and under ground cover. They lay eggs, but no clutch details are known. Captive Kenyan specimens fed readily on lizards, in Egypt recorded as taking frogs, rodents, lizards and birds. Captive specimens have been seen excavating holes in sand with their snouts. Racers will bite when first handled.

SMITH'S RACER *Platyceps brevis (smithi)*

IDENTIFICATION: A slim, harmless, little barred brown or rufous snake, usually in arid country. The eye is large and looks uniformly dark, but has a brown iris and a round pupil. The body is cylindrical, tail long, 25–30% of total length. Scales smooth, in 21 midbody rows, ventrals 176–199 (higher counts females), subcaudals 83–110. Maximum size about 70cm, average 40–60cm, hatchling size unknown but a Voi juvenile with an umbilical scar was 24cm long. Brown, rufous or yellow-brown, usually with three dark crossbars on the head and nape. The body is usually strongly barred in juveniles, in adults the bars fade to rufous spots or half-bars, in some the pattern completely disappears, although the head bars may remain. Sometimes a dark vertebral stripe may be present, or extensive fine white speckling. The belly is yellow, cream or white.

Platyceps brevis (smithi)

Platyceps brevis (smithi) Smith's Racer. Left: Stephen Spawls, Ithumba Hill. Right: Anthony Childs, Meru National Park.

HABITAT AND DISTRIBUTION: Dry savanna and semi-desert, at low altitudes in Kenya, from 100–1,300m. From Lokitaung and Lake Turkana south and east to Samburu, through the low country east of Mt Kenya and thence south through Ukambani and Tsavo, enters northern Tanzania just south of Mt Kilimanjaro. Isolated records from near Sololo and at

Malka Murri; may be more widespread in northern Kenya. Elsewhere, to southern and eastern Ethiopia and southern Somalia.

NATURAL HISTORY: Little known. Diurnal and terrestrial, although it will climb into low bushes. It secretive and fast-moving. Lays eggs, one Tsavo specimen laid three elongate eggs in March. Diet: probably diurnal lizards, possibly snakes and mammals, but poorly known. A captive specimen at Nairobi Snake Park fed readily on small lacertid lizards.

SMOOTH AND SEMI-ORNATE SNAKES *MEIZODON*

A sub-Saharan African genus of small, harmless colubrid snakes, mostly brown or grey in colour, with thin bodies, smooth scales and eyes with round pupils. They are diurnal and secretive; they lay eggs. Five species are known, four of which occur in East Africa, one is endemic to Kenya.

KEY TO EAST AFRICAN MEMBERS OF THE GENUS *MEIZODON*

1a	Scales in 19 rows at midbody......2	3a	Two anterior temporal scales, ventrals 159–204, usually barred, grey to brown, sometimes black, in colour......*Meizodon semiornatus*, Semi-ornate Snake [p. 482]
1b	Scales in 21 rows at midbody......3		
2a	Ventrals 166–176, only in the Tana Delta area......*Meizodon krameri*, Tana Delta Smooth Snake [p. 480]	3b	One anterior temporal, ventrals 201–235, no barring, red or pink in colour......*Meizodon plumbiceps*, Black-headed Smooth Snake [p. 481]
2b	Ventrals 177–201, only in western East Africa......*Meizodon regularis*, Eastern Crowned Snake [p. 481]		

TANA DELTA SMOOTH SNAKE *Meizodon krameri*

IDENTIFICATION: A small, slim colubrid snake. Pupil round, body cylindrical, tail 22–23% of total length. Scales smooth, with a single apical pit, in 19 rows at midbody, ventrals 166–176, subcaudals paired in 68–69 rows, cloacal scale divided. The two known specimens had lengths of 50 and 32cm, probably comparable in length to the Semi-ornate Snake *Meizodon semiornatus*. Uniform olive-brown, lip scales lighter, ventrals centrally light, dark at the edges. Taxonomic Note: This species was described in 1985 from two specimens originally assigned to the Eastern Crowned Snake *Meizodon regularis*, from which it differs in having 26 maxillary teeth as opposed to 23 or fewer, and a lower ventral count. It was collected in 1934, no further specimens have been found.

Meizodon krameri

HABITAT AND DISTRIBUTION: Endemic to Kenya, known only from the villages of Kau and Golbanti in the Tana River Delta. Presumably widespread in the delta area, might occur in the marshy regions south of Witu, or further upriver. Conservation Status: IUCN Data Deficient.

NATURAL HISTORY: Nothing known. Presumably similar to the other snakes of this genus, i.e. terrestrial and diurnal, lays eggs. Diet unknown. No living specimen has ever been photographed.

BLACK-HEADED SMOOTH SNAKE Meizodon plumbiceps

IDENTIFICATION: A slim, pink or red, black-spotted smooth snake. The eye is fairly large, the pupil round or slightly elliptic. The body is cylindrical, tail about a quarter of the total length. Scales smooth, in 21 midbody rows, ventrals 201–235 (higher counts usually females), subcaudals paired, 76–90 rows. Maximum size about 55cm, average 30–45 cm, hatchling size unknown. Red or pink, with four irregular rows of scattered black dots on the back. The top of the head is black, neck with a broad black collar. Chin and throat speckled and blotched with black, belly scales cream or pale brown. Taxonomic Note: Elevated to a full species in 1982, following a study of the species on the Juba River, originally regarded as a subspecies of the Semi-ornate Snake *Meizodon semiornatus*.

Meizodon plumbiceps

HABITAT AND DISTRIBUTION: Elsewhere, coastal thicket, riverine woodland, dry savanna and semi-desert; in Kenya known only from Malka Murri in dry savanna at 700m altitude. It probably occurs on the Kenya coastal plain north of Lamu Island and in north-eastern Kenya as it is widely distributed in southern Somalia and the Ogaden in Ethiopia. Conservation Status: IUCN Least Concern.

Meizodon plumbiceps Black-headed Smooth Snake. Jens Vindum/ California Academy of Sciences; Somalia, preserved specimen.

NATURAL HISTORY: Little known, but probably similar to other snakes in the genus, i.e., terrestrial, diurnal and secretive, laying eggs. The diet is unknown, but probably takes lizards. Some of the Juba River specimens were found in termitaria.

EASTERN CROWNED SNAKE Meizodon regularis

IDENTIFICATION: A slim, harmless smooth snake. Eye medium sized, pupil round. Body cylindrical, tail about 20% of total length. Scales smooth, in 19 rows at midbody, ventrals 177–201, cloacal scale divided, subcaudals 65–79 in East Africa. Maximum size about 65cm, average 35–50cm, hatchling size unknown. Juveniles are grey, the head and neck are black with four or five distinct, white or yellow crossbars. The eye is finely bordered with black, lips and chin white, throat white or yellow, the ventrals are grey. Adults are of two colour phases; uniform grey, or dark grey with regular light, broad dirty grey crossbars.

HABITAT AND DISTRIBUTION: Moist savanna, at medium altitude. A single Kenya record, Chemilil. Quite widespread in north-central Uganda, records include Lira, Gulu, Rhino Camp, Butiaba and Toro Game Reserve. Elsewhere, west to Guinea, north to Ethiopia and central South Sudan.

NATURAL HISTORY: Poorly known. Fast-moving, diurnal and terrestrial, often found in waterside vegetation. Hides under ground cover or in holes at night. Lays eggs, a Nigerian specimen had four eggs in the oviducts in June, roughly 3.5×0.6cm. Diet unknown, probably mostly lizards, possibly small amphibians; the closely related (possibly conspecific) West African form, *Meizodon coronatus*, eats only lizards.

Meizodon regularis

Meizodon regularis Eastern Crowned Snake. Left: Andre Zoersel, north-west Uganda, juvenile showing head stripes. Right: Jean-Francois Trape, West Africa.

SEMI-ORNATE SNAKE *Meizodon semiornatus*

IDENTIFICATION: A small, thin, harmless colubrid snake, juveniles are conspicuously barred on the front half of the body; hence the name. The head is slightly broader than the neck; eye fairly large, pupil round. The tail is about 25% of the total length. Scales smooth, 21 midbody rows, ventrals 159–204 (higher counts females), subcaudals paired, 74–88, cloacal scale divided. Maximum size about 80cm, average 40–70cm, hatchlings 18–20cm. Colour very variable, adults usually grey, brown, or olive, occasionally black or rufous-black; sometimes there is a yellow wash on the underside of the head and neck, the posterior flanks may be speckled grey, with a dark brown dorsal tail stripe. Juveniles are lighter, light grey, yellow-brown or blue-grey, and have a series of fine or broad dark crossbars or half-bars on the front two-thirds of the body, this fades totally or partially, from the top down as they grow. The ventrals are usually grey, grey-white or yellow, the chin and throat white. The top of the head is black in some juveniles.

Meizodon semiornatus

HABITAT AND DISTRIBUTION: Widespread in a range of habitats; coastal thicket

Meizodon semiornatus Semi-ornate Snake. Left: Anthony Childs, juvenile, Watamu. Right: Michele Menegon, Selous Game Reserve. Inset: Anthony Childs, Nairobi National Park.

and woodland, semi-desert, moist and dry savanna and high grassland, from sea level to 2,200m altitude, known from such diverse localities as Dar es Salaam, Lodwar, Kijabe, Liwale and Mitito Andei. Known from the south-west shore of Lake Turkana, and from Baringo and the upper Kerio Valley spreads east and south between Mt Kenya and the Aberdares, widespread across southeast Kenya and most of eastern and southern Tanzania. No Uganda, Rwanda or Burundi records. No records from north-east Kenya, but found at Dollo in Somalia, not far from Mandera. Elsewhere, north to Ethiopia and Somali, west to Chad, south to Botswana.

NATURAL HISTORY: Terrestrial, fast-moving, usually diurnal but occasionally active at dusk. When inactive, hides under ground cover, in holes etc., small aggregations of two or three individuals have been found under loose tree bark in Botswana. It is secretive and rare at high altitude (there are very few records from Nairobi) but more common at low altitude. Semi-ornate Snakes are inoffensive and rarely try to bite when handled. They lay two or three eggs, roughly 3.5×1cm. Male combat has been observed in the Selous Game Reserve in Tanzania, males neck-wrestling. Incubation times unknown but juveniles were collected in November and December in south-east Tanzania. Diet: lizards, especially skinks and geckoes; small frogs and rodents also recorded.

GREEN SNAKES *PHILOTHAMNUS*

An sub-Saharan African genus of 20 species of harmless, thin, diurnal, fast-moving agile tree snakes. Most are green, all-black (melanistic) forms of several species are known, and some other colour variants exist. Most live near water sources and eat amphibians. Few grow larger than 1m. They are often mistaken for green mambas or boomslangs and subsequently killed. They will bite if restrained and can cause deep lacerations; their saliva appears to be toxic to frogs, causing paralysis. Some have a curious, inexplicable display, holding the head and neck horizontally and sending a shimmy of transverse waves down the neck. If threatened, some inflate their necks, exposing startling hidden colours on the scales. Fourteen species are known in East Africa and Battersby's Green Snake *Philothamnus battersbyi* is probably highland Kenya's most commonly seen snake.

KEY TO EAST AFRICAN MEMBERS OF THE GENUS *PHILOTHAMNUS*

1a Cloacal scale entire......2
1b Cloacal scale divided......4

2a Dorsal sales in 13 rows at midbody......*Philothamnus carinatus*, Thirteen-scaled Green Snake [p. 487]
2b Dorsal scales in 15 rows at midbody......3

3a Ventral scales 141–166, usually green in life in East Africa......*Philothamnus heterodermus*, Forest Green Snake [p. 488]
3b Ventral scales 164–181, may be green, black, or barred yellow/black......*Philothamnus ruandae*, Rwandan Green Snake [p. 495]

4a Dorsal scales in 13 rows at midbody......5
4b Dorsal scales in 15 rows at midbody......6

5a Ventrals keeled, only in southern and eastern Tanzania and coastal Kenya ......*Philothamnus macrops*, Usambara Green Snake [p. 491]
5b Ventrals unkeeled, in our area only in Uganda ......*Philothamnus hughesi*, (part) Hughes' Green Snake [p. 491]

6a Subcaudals rounded or angular, not keeled......7
6b Subcaudals sharply angular and keeled like the ventrals......13

7a Body extremely slender, ventrals 168–194, no concealed white spots on dorsal scales......*Philothamnus heterolepidotus*, Slender Green Snake [p. 489]
7b Body moderate, ventrals 138–179, concealed white spots on dorsal scales......8

8a Two upper labials entering the orbit......9
8b Three upper labials entering the orbit......11

9a Green with black speckling and a bronze dorsal stripe......*Philothamnus hughesi*, (part) Hughes' Green Snake [p. 491]
9b Uniform green, with or without dark bars but not speckles......10

10a Subcaudals 60–104, in eastern and southern Tanzania, snout often yellow......*Philothamnus hoplogaster*, South-eastern Green Snake [p. 489]
10b Subcaudals 88–128, in northern Tanzania and central Kenya, snout not yellow......*Philothamnus battersbyi*, Battersby's Green Snake [p. 485]

11a A light-bordered, dark-centred dorsal stripe present......*Philothamnus ornatus*, Stripe-backed Green Snake [p. 493]
11b No light-bordered, dark-centred dorsal stripe......12

12a Ventrals keeled, maximum size 90cm......*Philothamnus bequaerti*, Uganda Green Snake [p. 487]
12b Ventrals not keeled, maximum size 1.2m......*Philothamnus angolensis*, Angola Green Snake [p. 484]

13a Temporals usually 1+2......*Philothamnus nitidus*, Loveridge's Green Snake [p. 492]
13b Temporals usually 2+2......14

14a Usually two upper labials entering the orbit, ventrals 157–188, concealed white spots on the dorsal scales lacking or of very low contrast......*Philothamnus punctatus*, Speckled Green Snake [p. 494]
14b Usually three upper labials entering the orbit, ventrals 170–209, concealed white spots present on the dorsal scales......*Philothamnus semivariegatus*, Spotted Bush Snake [p. 496]

ANGOLA GREEN SNAKE *Philothamnus angolensis*

IDENTIFICATION: A long thin green tree snake with a big eye and a golden iris, round pupil, the supraocular scale above the eye slightly raised. The body is cylindrical or sub-triangular in section. The tail is long, 27–35% of total length, males have the longer tails. Scales smooth, in 15 rows at midbody, ventral scales 143–167 in males, 148–184 in females, subcaudals 90–134 in males, 90–122 in females. Maximum size about 1.2m, average 70cm–1m, hatchling size 22–26cm. Usually emerald green, sometimes with a blue or yellowish tinge; occasionally black. Some specimens have black flecks or irregular blotches on the body. There is a white spot on the lower edge of each dorsal scale, visible when the snake inflates the body in defence; the skin between the scales is black. The ventrals are white, greenish-white or greenish-yellow.

HABITAT AND DISTRIBUTION: Moist savanna and woodland, from sea level to nearly 2,000m, usually in vegetation near water. Known from Lake Albert, south along the high country of the Albertine Rift through far western Uganda, Rwanda and Burundi, just reaching north-west Tanzania. A population also occurs in southern Tanzania, east of Lake Rukwa to the upper Rufiji and Ruaha Rivers, south to the border. Isolated records from the Mabira Forest in Uganda, and south of Smith Sound in Tanzania. Elsewhere, west to Cameroon, south to South Africa, north to South Sudan.

Philothamnus angolensis

NATURAL HISTORY: Arboreal and diurnal, spending the day hunting in waterside vegetation. It is a fast, secretive and graceful climber, going high, up to 8–10m in trees. Sleeps at night coiled up in high vegetation, often high up. Swims readily. When angry it inflates the body as described. Hisses and makes lunging strikes with open mouth, it has sharp teeth and draws blood when it bites. It lays from 4–16 eggs, roughly 2×3cm; communal nests containing up to 85 eggs of this species are recorded in Uganda, with embryos on the point of hatching in October and in February. Incubation time is about two months. Diet: mostly frogs, but known to take nestlings and birds caught in the nest, might eat lizards.

Philothamnus angolensis Angolan Green Snake. Left: Michele Menegon, Livingstone Mountains, Tanzania. Right: Michele Menegon, Livingstone Mountains, Tanzania, melanistic specimen.

BATTERSBY'S GREEN SNAKE *Philothamnus battersbyi*

IDENTIFICATION: A fairly long thin green tree snake with a big eye, golden brown iris and a round pupil. The body is cylindrical or sub-triangular in section. The tail is long, 26–32% of total length, males have the longer tails. Scales smooth, in 15 rows at midbody, ventral scales 152–176, subcaudals 88–129, higher counts usually males. Maximum size about 90cm, average 50–80cm, hatchling size unknown but probably 15–17cm. May be vivid emerald green or quite dull grey-green, there is a bluish or white spot on the front lower edge of each dorsal scale, visible when the snake inflates the body in defence; the skin between the scales is black. The ventrals are greenish-white, light green or greenish-yellow.

HABITAT AND DISTRIBUTION: Highland Kenya's most visible green snake and still common in most Nairobi suburbs. In moist savanna and woodland, often near rivers, lakes and dams. In East Africa most common at mid-altitude, between 1,300–1 800m, but may go as low as 500m (sea level elsewhere) and as high as 2,300m, for example at Limuru and on the Kinangop Plateau. It occurs from southeast Uganda, south and east through west and central Kenya, into northern Tanzanian west of the Rift Valley. Extends down the Galana River and environs into southeast Kenya, then west to Mt Kilimanjaro and the country south of the mountain, southwest to Mount Hanang. Isolated records from Mt Marsabit and South Horr in northern Kenya, some dubious records from central Rwanda and Burundi and Mwanihana Forest reserve, Udzungwa National Park in Tanzania. Elsewhere, to Somalia and highland Ethiopia.

Philothamnus battersbyi

NATURAL HISTORY: Arboreal and diurnal, spending the day in waterside trees and bushes, either waiting in a prominent position for prey to pass or actively hunting, investigating suitable spots for frogs. When motionless, it often shows the curious 'shimmy', as described in the generic introduction, sending sinuous waves down the body. It is a fast, secretive and graceful climber, but will descend to the ground to hunt or to move across country. Sleeps at night coiled up on the outer branches of waterside trees and big bushes, may also hide in hollow logs, or under vegetation heaps or other suitable ground cover. It emerges to bask in the early morning and once warm will hunt, it may also bask near sundown. It swims readily and it will also hunt fish and frogs under water. Although it usually disappears swiftly and silently if disturbed, if cornered and angry it inflates the body to expose the bright spots on the scales, it may hiss and makes lunging strikes with open mouth, it has sharp teeth and draws blood when it bites, jerking the head from side to side and causing lacerations. It lays from 3–11 eggs, roughly 2×3cm; communal nests containing over 100 eggs of this species are recorded from Lake Naivasha; laid in November and December; hatching in January and February and also July, a hatchling recorded at Athi River in September. During the egg-laying time, big concentrations of up to 40 snakes (gravid and non-gravid females, and males) have been seen in hedges and trees near the laying site on Lake Naivasha. Diet: mostly frogs and fish, hunted from trees and river banks but also in the water; other prey items include chameleons and skinks. Often called a 'green mamba' in the Kenya highlands and killed. Very tolerant or urbanisation, so long as some hedges, streams and frogs remain.

Philothamnus battersbyi Battersby's Green Snake. Left: Stephen Spawls, Kitengela. Right: Stephen Spawls, Naro Moru.

UGANDA GREEN SNAKE *Philothamnus bequaerti*

IDENTIFICATION: A slender green tree snake, eye fairly large, pupil round, body cylindrical, tail 28–30% of total length. Scales smooth, in 15 rows at midbody, ventrals keeled, 155–179 (higher counts usually females), subcaudals paired, 93–123, higher counts usually males. Cloacal scale divided, except in a type specimen from Niangara in the DR Congo, which has an entire cloacal scale. Maximum size about 90cm, average 40–70cm, hatchling size unknown. Uniform green above, lighter below, dorsal scales have a concealed white spot on the lower edge.

Philothamnus bequaerti

HABITAT AND DISTRIBUTION: Sporadically recorded from mid-altitude moist savanna in Uganda; localities include Entebbe, Kome Island, Bukataka, Lira, Gulu, Semliki National Park, Pakwach. Elsewhere, west across the north side of the forest to Cameroon, north to southern and south-east South Sudan and south-west Ethiopia, including the Omo River valley.

Philothamnus bequaerti Uganda Green Snake. Harald Hinkel, Kininya, Rwanda.

NATURAL HISTORY: Poorly known but presumably similar to other green snakes, i.e. diurnal, fast moving, largely arboreal, climbs and swims expertly, inhabiting waterside vegetation, lays eggs, eats amphibians.

THIRTEEN-SCALED GREEN SNAKE *Philothamnus carinatus*

IDENTIFICATION: A beautiful, barred green snake, often with a bronze or yellow sheen, in or near forest. Eye large, pupil round, yellow or orange. Body cylindrical, tail 22–26% of total length. Scales smooth, in 13 midbody rows in East Africa, ventrals 138–166 (higher counts females), subcaudal scales paired, 69–110 (higher counts males), cloacal scale entire. Maximum size about 85cm, average 50–70cm, hatchling size unknown. Colour green, blue-green, olive-green or blue, with a series of oblique darker cross-bars, the scales between the cross-bars edged with turquoise or blue. Occasionally black. Ventrals light or yellowish green.

Philothamnus carinatus

HABITAT AND DISTRIBUTION: Forest, forest islands and well-wooded savanna, from sea level to 2,300m altitude, most common at mid-altitude and not below 600m altitude in our area. Records include the Kakamega and western Mount Elgon forest and the environs, and south from the Budongo Forest along the Albertine Rift, right down the western border of Uganda, western Rwanda and Burundi and the northern half of the east shore of Lake Tanganyika; elsewhere into south-eastern DR Congo, west to Cameroon and Bioko Island.

Philothamnus carinatus Thirteen-scaled Green Snake. Michele Menegon, Rwanda.

NATURAL HISTORY: Not well known. Presumably diurnal. Mostly arboreal, although it will descend to the ground, in forest, not closely tied to water sources, often found in and around clearings. When angry, inflates the body, accentuating the bars and bright spots. Lays eggs, no clutch details known. One of only two genuine forest-dwelling *Philothamnus*. A specimen in the DR Congo was eaten by a Forest Vine-snake *Thelotornis kirtlandii*. Presumably feeds on amphibians.

FOREST GREEN SNAKE *Philothamnus heterodermus*

IDENTIFICATION: A slim green snake, from far western East Africa. The snout is slightly upturned, the pupil round. The tail is 24–27% of the total length in females, 26–29% in males. Scales smooth, in 15 midbody rows in East Africa, ventrals faintly keeled, 141–166, subcaudals 71–100, males with higher counts, cloacal scale entire. Maximum size about 76cm, average 45–70cm, hatchling size unknown but a Nigerian juvenile, caught in February, was 25cm. Colour either green, brown or metallic black (East African specimens green), body scales with a concealed white spot, green animals light green or yellow below, the belly of black specimens yellow brown or cream anteriorly, darkening to black posteriorly.

Philothamnus heterodermus

HABITAT AND DISTRIBUTION: Very sporadic in East Africa. Records include Kibale and Bwindi Impenetrable National Park in Uganda (so may occur in northern Rwanda). An enigmatic Tanzanian record from Dunda; this locality has been placed both northwest of Dar es Salaam and north of Lake Malawi (Lake Nyasa), the latter regarded as more likely. Recorded on the western shore of Lake Tanganyika in the DR Congo. Elsewhere, west to Guinea.

NATURAL HISTORY: Poorly known. Diurnal. A survey in Ghana found this species spent as much time in trees as on the ground; and during the dry season, green individuals were mostly in trees and brown ones on the ground. When inactive, coils up in the branches or hides in leaf litter. Presumably inflates the body when threatened to expose the white spots. Ghanaian and Nigerian specimens contained 1–4 eggs, roughly 1×3cm. Diet presumably frogs, toads (*Bufo* species) were ignored by captive specimens.

Philothamnus heterodermus Forest Green Snake. Robert Drewes, Bwindi Impentrable National Park.

SLENDER GREEN SNAKE   Philothamnus heterolepidotus

IDENTIFICATION: A large, very slender green snake, sometimes with a bronzy sheen. Eye quite large, pupil round, iris yellow, golden-brown or red-brown. Body cylindrical, tail long, 33–38% of total length in males, 28–33% in females. Scales smooth, in 15 midbody rows, ventrals always unkeeled, 164–194, subcaudals paired, 101–144, male counts highest. Cloacal scale divided. Maximum size just over 1m, average 60–80cm, hatchling size unknown. Colour quite variable; immaculate vivid green in Rwanda and Burundi, greenish white or yellowish white below, the outer edges of the ventrals sometimes finely black-edged, Uganda specimens often have a bronzy sheen. The interstitial skin is black, chin and throat white or bluish-white.

Philothamnus heterolepidotus

Philothamnus heterolepidotus Slender Green Snake. Left: Max Dehling, Rwanda. Right: Max Dehling, Rwanda.

HABITAT AND DISTRIBUTION: Woodland and moist savanna, at altitudes of 600–2,000m, elsewhere down to sea level. Occurs from Mumias in Kenya west along the northern Lake Victoria shore, into south-west Uganda and most of Rwanda and Burundi, extending to the northern shore of Lake Tanganyika in Tanzania, also found north of the Nile between Lakes Kyoga and Albert in Uganda. Three enigmatic records are Tunduma, south of Lake Rukwa in Tanzania, Dar es Salaam and the Lamu archipelago; these might prove to be misidentified animals, being so far from the centre of distribution. Elsewhere, north to Khartoum, west to Sierra Leone, south-west to Angola.

NATURAL HISTORY: Probably similar to other green snakes, i.e. partially arboreal, diurnal, climbs and swims well, lives near water sources. In Bugesera, Rwanda, several were collected asleep at night in tall grass. The absence of the white warning spots on the dorsal scales may mean it does not inflate the body in defence. Lays eggs, three ova recorded in a Uganda female in March. Diet: frogs, fond of tree frogs (*Hyperolius*).

SOUTH-EASTERN GREEN SNAKE   Philothamnus hoplogaster

IDENTIFICATION: A fairly large green snake with a short head. Pupil round, iris golden. Body cylindrical, tail 27–33% of total length, males have longer tails. Scales smooth, in 15 midbody scale rows (occasionally 13 or 11, but nearly always 15 on the neck), ventrals 138–167, subcaudals paired, 60–118, higher counts males, cloacal scale divided. Maximum size about 96cm, average 40–70cm, hatchlings 15–20cm. Colour vivid green, Kenyan coastal specimens sometimes bright blue on the lower flanks; the snout may be orange, yellow or greenish-yellow. Occasionally a series of black or rufous crossbars or blotches on the neck and front third of the body, especially in southern Tanzanian specimens. Interstitial skin black. A black specimen is known from Nandi Hills in Kenya.

Philothamnus hoplogaster

HABITAT AND DISTRIBUTION: From sea level to about 1,800m altitude, in coastal forest, thicket and moist savanna, often near watercourses. Occurs from Lamu south along the Kenya coast, up the Tana River to Tana River Primate Reserve and inland in south-east Kenya to the Taita Hills; it occurs in eastern Tanzania, and inland through the Usambara and Pare Mountains to the southern side of Mt Kilimanjaro. Widespread in most of south-eastern and southern Tanzania, and westwards through Lake Rukwa, up the eastern shore of Lake Tanganyika to the Burundi shore. Isolated records from Nandi Hills and the Kakamega area, Kampala and Rumanyika Game Reserve in north-western Tanzania. Elsewhere, in the Imatong Mountains in the South Sudan, west to Cameroon, south to South Africa.

Philothamnus hoplogaster South-eastern Green Snake.
Top: Stephen Spawls, Watamu. Centre: Michele Menegon, southern Tanzania. Bottom: Stephen Spawls, Watamu.

NATURAL HISTORY: Diurnal, active on the ground and in trees, sleeps on the outer edges of vegetation at night. Climbs and swims well. Lays clutches of 3–8 eggs, roughly 1×3cm. Male combat has been observed in Zambia. Diet: mostly frogs but will take fish, it will hunt in the water, known to take lizards and juveniles recorded eating insects. Doesn't appear to inflate its neck, unlike most *Philothamnus*, but if restrained will hold the neck and head out rigidly, away from the hand, and then strike.

HUGHES' GREEN SNAKE Philothamnus hughesi

IDENTIFICATION: A slim, fairly large green snake. The eye is fairly large, tail about 30% of total length. The scales are smooth, in 15 rows (occasionally 13) at midbody, ventral scales 152–163, unkeeled, subcaudals 93–105, unkeeled. Maximum size 93.3cm, average size 60–80cm, hatchling size unknown, but probably around 15cm. The head and neck are vivid green, the neck scales heavily blotched with blue. A broad brownish-bronze stripe extends the length of the back, edged by a lighter bronze stripe; the front half looks heavily speckled on account of the white and blue scale margins and black interstitial skin. The underside is light yellow or bluish green.

Philothamnus hughesi Hughes' Green Snake. Laurent Chirio, Central African Republic.

Philothamnus hughesi

HABITAT AND DISTRIBUTION: In our area, only known from Sango Bay at 1,130m on Lake Victoria's western, Uganda shore. Elsewhere, known from Congo-Brazzaville, Cameroon, the Central Africa Republic, Chad and Gabon.

NATURAL HISTORY: Unknown but presumably similar to other *Philothamnus*, i.e. diurnal, partially arboreal, climbs and swims well, lays eggs, eats frogs.

USAMBARA GREEN SNAKE Philothamnus macrops

IDENTIFICATION: A slim, fairly large green snake with a remarkable range of colours. Eye fairly large, pupil round, eye orange or yellow. Body cylindrical or subtriangular, tail long, 27–33 % of total length, longer in males. Scales smooth, in 13 rows at midbody, ventrals 135–146, keeled, subcaudals 73–96, higher counts usually males. Cloacal scale usually divided but entire in a Zanzibar specimen. Maximum size 92cm, average 50–80cm, hatchlings 18–20cm. Colour very variable, known colour phases include: (a) green with irregular black crossbars, blotches or spots; (b) green or bluish-green, uniform or with fine black spotting; (c) dull red, maroon, orange-brown or rufous, with many fine turquoise or green half-bars or crossbars consisting of light-edged scales, these bars may be black-edged; (d) olive-green with crossbars consisting of yellow spots; (e) uniform brown; (f) black, with extensive yellow, blue or green blotching, especially on the head, neck and ventrals. The ventral colour is also variable, sometimes yellow with each scale dark-edged, sometimes yellowish, light green or bluish-white. However, usually identified as the only *Philothamnus* within its eastern range with 13 midbody scale rows.

Philothamnus macrops

Philothamnus macrops Usambara Green Snake. Top left: Bill Branch, Usambara Mountains. Top right: Michele Menegon, Udzungwa Scarp Nature Reserve. Centre right: Lorenzo Vinciguerra, Usambara Mountains. Bottom left: Lorenzo Vinciguerra, juvenile, Usambara Mountains. Bottom right: Lorenzo Vinciguerra, Usambara Mountains.

HABITAT AND DISTRIBUTION: Coastal thicket, woodland, moist savanna and hill forest, from sea level to over 2,000m in the Usambara Mountains. Probably widespread but distribution at present appears patchy; found from the Shimba Hills in Kenya to the northern Tanzanian coast and inland on the Usambara, Nguu, Nguru and Udzungwa Mountains and at Tukuyu. It is on Zanzibar, and occurs from the Rufiji Delta southwards to Lindi and the Rondo Plateau. Elsewhere to northern Mozambique.

NATURAL HISTORY: Terrestrial and partially arboreal, diurnal, often near water sources. Inflates the body when threatened. Lays 3–14 eggs, roughly 3–3.5×1cm; eggs laid in October in the Usambara Mountains, 19cm hatchlings collected on the Rondo Plateau in February. Diet presumably amphibians but might be a lizard feeder; a Tanzanian snake ate a Bearded Pygmy Chameleon *Rieppeleon brevicaudatus*.

LOVERIDGE'S GREEN SNAKE *Philothamnus nitidus*

IDENTIFICATION: A long-tailed, emerald or blue-green snake. Two subspecies are recognised, a western form *Philothamnus nitidus nitidus* and an eastern form *Philothamnus nitidus loveridgei*; the following notes apply only to the eastern form. Snout notably pointed. Eye fairly large, pupil round, iris orange or yellow. Supraocular scales slightly raised to give the impression of an eyebrow. Body cylindrical, tail very long, 36–39% of total length in males, 33–36% in females. Scales smooth, in 15 rows at midbody, 154–179 keeled ventrals, subcaudals 112–154. Maximum size about 93cm, average 50–80cm, hatchling size unknown. Colour emerald to blue-green, lighter green below, throat white.

Philothamnus nitidus

Philothamnus nitidus Loveridge's Green Snake. Bill Branch, Gabon.

HABITAT AND DISTRIBUTION: A forest snake, but also known from medium-altitude moist savanna in our area. Sporadic records from western East Africa include Kakamega area, Kome Island in Lake Victoria, Bundibugyo in western Uganda, Kigali in Rwanda and Rutanda in Burundi; widely distributed in western Kivu in the DR Congo. An enigmatic record from Lamu in Kenya. Elsewhere, west to the lower Congo River.

NATURAL HISTORY: Not well known. Diurnal and largely arboreal, lays eggs, eats amphibians. The western subspecies in Ghana eats tree frogs, *Hyperolius*.

STRIPE-BACKED GREEN SNAKE *Philothamnus ornatus*

IDENTIFICATION: A small green snake, identified by its red or brown, yellow-edged vertebral stripe. Eye large, pupil round, iris yellow or orange. Body cylindrical, tail fairly long, about one-third of the total length. Scales smooth, in 15 rows at midbody, ventrals usually keeled, 147–174 (high counts usually females), subcaudals paired, 85–106. Maximum size about 80cm, average 50–70 cm, hatchling size unknown. Emerald or olive-green above, the back half of the body usually bronzy, with a broad red brown or rich brown, yellow-edged vertebral stripe, extending from the neck to the tail. Ventrals white, greenish-white or yellow, often with a bronzy tint.

Philothamnus ornatus Stripe-backed Green Snake. Bill Branch, northern Zambia.

Philothamnus ornatus

HABITAT AND DISTRIBUTION: In our area, known only from moist savanna at Tatanda, south-western Tanzania, near the Zambian border. Elsewhere, south to Angola, northern Botswana and eastern Zimbabwe, two isolated populations in north-eastern DR Congo and central Cameroon.

NATURAL HISTORY: Diurnal, partially arboreal, often in reedbeds and waterside vegetation, swamps and flooded valleys, but has been found some distance from permanent water. Inflates the body and strikes when threatened. Lays eggs, no clutch details known. Diet mostly amphibians.

SPECKLED GREEN SNAKE *Philothamnus punctatus*

IDENTIFICATION: A big, slim green (sometimes turquoise or blue) snake with a raised 'eyebrow' (supraocular scale) and prominent eyes, usually with black speckles but not crossbars. The snout is upturned, the pupil round, iris golden or pale yellow. Body cylindrical, tail long, 33–37% of total length in males, 28–33% in females. Scales smooth, in 15 rows at midbody, ventrals strongly keeled, 157–188, subcaudals paired, 126–170 rows, keeled. Maximum size about 1.2m, average 75cm–1.1m, hatchling size 21–24cm. Ground colour bright green or yellow-green, usually finely speckled black. Occasional blue or turquoise specimens occur, particularly around Watamu, some specimens have blue heads and green bodies, animals from the Daua River in northeast Kenya were uniform light green, without markings. Three colour varieties occur on

Philothamnus punctatus

Philothamnus punctatus Speckled Green Snake. Top left: Dennis Rodder, Pemba. Top right: Stephen Spawls, Watamu, turquoise phase. Bottom: Stephen Spawls, Watamu, green phase.

Pemba; bronzy with head scales black-edged, dull green, and black; they will probably turn out to be a different species. Ventrals cream, white, bluish-green or pale green.

HABITAT AND DISTRIBUTION: Coastal forest, woodland and thicket, moist and dry savanna and semi-desert, from sea level to about 1,200m altitude. Unlike most green snakes, it is not tied to water sources (although often found in waterside vegetation) and ranges widely into dry and open country, provided there are trees or large bushes. Known from the Daua River and Laisamis in northern Kenya, where it is probably more widespread but overlooked. Occurs on the lower Tana River and south along the coast from Lamu to the Ruvuma River, inland to Tsavo Station, and recorded from Kitobo Forest, south and west across south-eastern Tanzania to the shores of Lake Malawi (Lake Nyasa). Elsewhere, in northern and southern Somalia, and south to Mozambique.

NATURAL HISTORY: A diurnal, fast-moving, agile snake, largely arboreal, climbs high in trees and coastal thickets, using its keeled belly scales to ascend vertical tree trunks. Sleeps at night high up on the outer branches of vegetation. When threatened and unable to escape it inflates its neck and will make lunging strikes. It lays eggs, details in the past confused with the spotted bush snake but Kenyan females had clutches of 3–6 eggs, two Somali females had five eggs each in the oviducts in June and October, of size 1×3cm. Diet: mostly lizards, unusually for a *Philothamnus*, major prey items include dwarf geckoes (*Lygodactylus*), other geckoes and chameleons, will also take frogs and reputed to eat nestling birds. Often mistaken on the East African coast for a green mamba and needlessly killed. The blue form has become increasingly common around Watamu in recent years, it is speculated that a curious selection process is occurring, whereby the blue ones are recognised as not being mambas and are thus spared, the green ones not.

RWANDAN GREEN SNAKE *Philothamnus ruandae*

IDENTIFICATION: An unusual green snake, native to the Albertine Rift Valley, very variable in colour. The eye is large, the pupil is round, the iris is yellow but darkly mottled, thus hard to see. The body is sub-triangular, with a suggestion of a vertebral ridge. Tail short for a green snake, 24–27% of total length in females, 27–30% in males. Scales smooth, in 15 rows at midbody, ventrals 164–181 (high counts usually females), subcaudals paired, 84–102, males more than 94, females less, cloacal scale entire. Maximum size about 96cm, average 60–90cm, hatchling size unknown. Colour very variable; may be bright or dull green; metallic or grey green, with or without dark cross bars, uniform black or brown, black with fine or broad yellow cross-bars. The chin, throat upper labials and snout often yellow, to a greater or lesser extent. The flanks are sometimes spotted yellow. Ventrals light green, dark or metallic green, or grey green, sometimes with yellow spots on the outer edges.

Philothamnus ruandae

HABITAT AND DISTRIBUTION: Medium to high altitude forest, woodland and moist savanna of the southern Albertine Rift, from 700–2,300m altitude. Occurs from north Bushenyi down through south-western Uganda to the western side of Rwanda and north-west Burundi, elsewhere on the valley slopes in the eastern DR Congo.

NATURAL HISTORY: Diurnal, largely arboreal, active in trees and bushes, often near water sources, especially swamps and bogs, favours thick vegetation. Sleeps at night in the branches or in a vegetation clump. Inflates the neck when threatened. Lays eggs, no clutch details known. Eats amphibians.

Philothamnus ruandae Rwandan Green Snake. Top left: Michele Menegon, Rwanda. Top right: Max Dehling, Rwanda. Bottom: Harald Hinkel, Rwanda.

SPOTTED BUSH SNAKE/SPOTTED WOOD SNAKE
Philothamnus semivariegatus

IDENTIFICATION: A big, slim green snake, usually with black cross-bars. The eye is large, pupil round, iris golden or orange. The tongue is bright blue with a black tip. The supraocular is raised, giving a 'raised eyebrow' look. The body is cylindrical, tail 27–31% of total length in females and 32–36% in males. Scales smooth, in 15 rows at midbody, ventrals 170–209, keeled, subcaudals 98–166, paired and keeled, high counts usually males, cloacal scale divided. Maximum size about 1.3m; the biggest form in the genus, average 70cm–1.1m, hatchlings 22–25cm. Colour various shades of green, yellow-green or turquoise, usually with distinct, broad or narrow black crossbars along the back, the flanks are spotted black, sometimes black spots on the head. Specimens from eastern Tanzania often have no black markings at all. Ventrals yellowish or light green. Some animals have a turquoise or blue head, and in parts of its range the tail has no spots and is bronzy-green. Black specimens recorded from the Chyulu Hills, and a curious deep green-black specimen with blue ventrals recorded from the Mathews Range in Kenya.

Philothamnus semivariegatus

Philothamnus semivariegatus Spotted Bush Snake. Top left: Stephen Spawls, Kerio Valley. Right: Stephen Spawls, Bagamoyo, uniform animal. Inset: Stephen Spawls, melanistic animal, Chyulu Hills.

HABITAT AND DISTRIBUTION: Widely distributed, although often uncommon, in coastal bush, woodland, moist and dry savanna, and semi-desert, will move into forest clearings and fringes, from sea level to 1,500m altitude, possibly higher. Occurs virtually throughout Tanzania and south-eastern Kenya (but not right on the Kenya coast, where the Speckled Green Snake *Philothamnus punctatus* occurs). Widespread in Rwanda and Burundi but absent from the centre and higher areas. In Uganda along the Albertine Rift and the Lake Victoria shore, across into Kenya. Isolated Kenya records from Mount Kulal, Lokitaung and on the north-east border, from Malka Murri across to Mandera. Elsewhere, north to South Sudan and Ethiopia, south to eastern South Africa, west to Senegal.

NATURAL HISTORY: A diurnal, fast-moving, agile snake, largely arboreal, but will descend to cross open ground, climbs high in trees, using its keeled belly scales to ascend vertical tree trunks, it is a superb climber, able to use the smallest projections. Sleeps at night high up on the outer branches of vegetation. Not tied to water sources (although it will hunt there and it swims well), ranges widely in dry and open country. Alert and secretive, its camouflage is superb, it is hard to spot even in the smallest tree, and if approached moves swiftly and silently away from danger. When threatened and unable to escape it inflates its neck, exposing the concealed blue-white lower edges of its scales and will make lunging strikes. When resting in a tree it will exhibit the curious 'shimmy' shown by other green snakes; it will do this on the ground as well, even while moving. Captive specimens, unlike other green snakes, seem to remain irascible for a long time. Clutches of 3–12 eggs, roughly 3–4×1cm are laid; a Kigoma (Tanzania) specimen laid eight eggs in April, a Liwale specimen laid six eggs in December, a Kisumu specimen laid ten eggs in July. Diet: mostly arboreal lizards, unusually for a *Philothamnus*, major prey items include dwarf geckoes (*Lygodactylus*), other geckoes and chameleons, will also take frogs.

KEELED GREEN FOREST SNAKES *HAPSIDOPHRYS*

A genus of slim green forest tree snakes, three species are known, one from the island of Príncipe, the other two from the central African forests, both mainland species occur in western East Africa. Both are harmless. They are similar to some *Philothamnus*, but differ in having strongly keeled dorsal scales. The Emerald Snake was originally placed in the genus *Gastropyxis*.

KEY TO THE MEMBERS OF THE GENUS *HAPSIDOPHRYS*

1a Subcaudals 93–115, tail less than 35% of total length......*Hapsidophrys lineatus*, Black-lined Green Snake [p. 498]

1b Subcaudals 119–172, tail more than 35% of total length......*Hapsidophrys smaragdinus*, Emerald Snake [p. 499]

BLACK-LINED GREEN SNAKE   *Hapsidophrys lineatus*

IDENTIFICATION: A slim, very large-eyed tree snake, green with fine black stripes. The head is quite rounded, tongue light blue with a black tip, the pupil of the eye round, iris golden or orange. The body is cylindrical, tail 27–33% of total length. Dorsal scales strongly keeled, in 15 midbody rows, ventrals keeled, 150–175, subcaudals paired, 93–115. Cloacal scale entire. Maximum size about 1.1m, average 70–90cm, hatchling size unknown. Emerald or dark green, each dorsal scale edged black on the upper and lower margin, thus the snake appears finely striped in green and black. The ventral scales are pale or yellow-green, sometimes a fine black or blue-black median line under the tail. The head scales may be dark edged. It bears some resemblance to the black and green form of the Boomslang *Dispholidus typus*, so should be treated with caution.

Hapsidophrys lineatus

HABITAT AND DISTRIBUTION: Forest and thick woodland, often near water sources, at altitudes from 700–1,800m in our area, elsewhere down to sea level. In East Africa, recorded from the Kakamega area (there is an anecdotal report from Kisumu). In Uganda, occurs in lakeshore woodland and forest from Jinja and the Mabira Forest westwards, quite widespread in south-western Uganda, and also known from the Budongo Forest. Not recorded from Rwanda but may be expected there, found on the western slope of the Albertine Rift on the DR Congo side. Elsewhere, south-west to northern Angola, west to Guinea, in forest.

NATURAL HISTORY: Poorly known. Presumably diurnal, although there are anecdotal reports of night activity. Arboreal, a fast-moving agile climber, its keeled ventrals are a climbing aid. Reputed to be timid, but captive specimens from Kakamega settled down

Hapsidophrys lineata Black-lined Green Snake. Konrad Mebert, eastern Congo.

quite quickly in captivity. Lays eggs, a Kakamega female had four eggs in her oviducts in mid-February, another was gravid in October, a Nigerian female had four eggs roughly 2×0.8cm in oviducts in June. Diet poorly known in East Africa, in Nigeria mostly amphibians, especially tree frogs, and lizards, including chameleons, skinks, geckoes and agamas, one specimen ate a shrew.

EMERALD SNAKE *Hapsidophrys smaragdinus*

IDENTIFICATION: A slim, green tree snake with a bright yellow belly and a distinct dark stripe through the eye. The head is long, the snout quite pointed, eyes large and prominent, pupil round, iris yellow or golden-brown. Body cylindrical, tail very long, 33–42% of total length. Dorsal scales heavily keeled, giving the body a distinctly ridged appearance, 15 midbody scale rows, ventrals 150–174, heavily keeled; the belly is square and box-like (originally placed in the genus *Gastropyxis*, meaning 'box-belly'), subcaudals paired, 129–172, cloacal scale scale divided. Maximum size about 1.1m average 70cm–1m, hatchling size unknown. Light to dark green or turquoise, the outer edge of the ventrals is also green, centrally yellow, chin, lips and throat yellow. There is a dark line through the eye, tapering away from the orbit.

Hapsidophrys smaragdinus

HABITAT AND DISTRIBUTION: In woodland, forest and deforested areas, even suburbia, (and swamp forest elsewhere), at altitudes between 600–1,300m, from Jinja west along the northern and western Lake Victoria shore, and on the Sese Islands, south to Rubondo Island and Biharamulo Game Reserve, Tanzania. West along the Uganda/Tanzania border and thence north to the Uganda/Congo border country between Lakes Edward and Albert. Not recorded from Rwanda or Burundi but may be expected as it occurs on the watershed west of Lake Kivu. Elsewhere, west to Guinea-Bissau, south-west to northern Angola.

NATURAL HISTORY: Diurnal and largely arboreal, a fast moving agile climber, able to ascend vertical tree trunks using its keeled ventral scales. Basks in forest clearings. Sleeps at night out on a branch, often near water. Not known to inflate its throat when threatened. Lays 2–6 eggs, a specimen from the DR Congo contained three elongate eggs rough-

Hapsidophrys smaragdinus Emerald Snake. Konrad Mebert, eastern Congo.

ly 5.6×1.2cm, a Mabira Forest animal contained three oviductal eggs, 3×0.7cm, in April; Nigerian females were gravid between October and January. Diet largely arboreal lizards (although terrestrial skinks recorded), known prey items include geckoes, skinks, agamas, and tree frogs. If seized by the tail it struggles violently and the tail may break off, it is not regenerated.

BLACK TREE SNAKES *THRASOPS*

A tropical African genus of large, forest-dwelling, harmless black tree snakes with keeled or smooth scales. They inflate the neck as a threat display. Four species are known, two in East Africa, one of these is endemic. They look like black Boomslangs, and recent molecular analysis reveals they are actually very close to them.

KEY TO EAST AFRICAN MEMBERS OF THE GENUS *THRASOPS*

1a	Dorsal scales usually 19 rows at mid-body (sometimes 17–21), ventrals 187–214, occurs west of the Rift Valley...... *Thrasops jacksonii*, Jackson's Tree Snake [p. 500]	1b	Dorsal scales usually 17 rows at mid-body (rarely 19), ventrals 172–184, occurs east of the Rift Valley in Kenya...... *Thrasops schmidti*, Meru Tree Snake [p. 501]

JACKSON'S TREE SNAKE *Thrasops jacksoni*

IDENTIFICATION: A big black tree snake with a large dark eye and short head. It smells strongly of liquorice, especially if freshly sloughed. The pupil is round but the iris is black, so the whole eye just looks black. The body is laterally compressed, the tail is long, a third of the total length. The body scales are usually unkeeled in East African specimens, in 19 rows at midbody (sometimes 17–21), ventrals 187–214, subcaudals 129–155, cloacal scale divided. Maximum size about 2.3m, average 1.4–1.8m, hatchlings 32–35cm. Adults are uniform glossy black above and below, the throat may be grey or white (a dull green specimen reported from Zambia). Juveniles are checked black and orange or yellow, above and below, the tail spotted yellow, the head and neck olive green or brown; the juvenile colour changes when the snake is between 40–60cm long, some young adults may retain traces of the juvenile pattern. The adult is virtually indistinguishable from the black colour form of the Boomslang, so treat with caution.

Thrasops jacksoni

HABITAT AND DISTRIBUTION: Forest, forest islands, woodland and riverine forest, from 600m to over 2,400m altitude. Widespread in much of the southern half of Uganda, although records lacking from the centre; also in western and southern Rwanda (Nyungwe and Cyamudongo Forests) and north-western Burundi. In Kenya from Mt Elgon (also on the Cherang'any Hills and Kapenguria) south through the western forests to the north shore of the Winam Gulf. Also found south along the western flank of the Mau escarpment to the Soit Ololo escarpment, and into the Maasai Mara National Reserve along the riverine forest of the Mara River, to at least as far as its junction with the Telek River. Might occur on the eastern Mau Escarpment and in the Loita Hills, although unrecorded there. In Tanzania confined to the extreme north-west (Kabare, Rubondo Island). Elsewhere, extends to South Sudan, Congo and Central African Republic, south to north-west Zambia.

Thrasops jacksoni Jackson's Tree Snake. Harald Hinkel, Rwanda.

NATURAL HISTORY: Diurnal and arboreal, a superb climber, ascending high into trees, to 30m or more. If approached or threatened in a tree and unable to slide away, it will launch itself off into space and its long light body enables it to move sideways while falling; when it hits the ground it quickly slides away. Mara River specimens were seen to jump from trees into the water. If cornered, it will inflate its neck and anterior body like a boomslang, move sideways and make huge, lunging strikes, looking very large and threatening. It lays between 7–12 eggs, roughly 1.5×3cm. It is a generalist feeder, known prey items include arboreal lizards (especially chameleons), mammals including bats; it raids nests for eggs, nestlings and adult birds, it has been seen to drop out of a tree to catch a frog.

MERU TREE SNAKE *Thrasops schmidti*

IDENTIFICATION: A big black tree snake with a large dark eye and short head. The pupil is round, the iris looks black but is brown in good light. The body is laterally compressed, the tail is long, a third of the total length. The body scales are usually weakly keeled but may be a mixture of keeled and unkeeled, in 17 rows at midbody (sometimes 19), ventrals 172–184, subcaudals 121–147, cloacal scale divided. A big snake, maximum size about 2.3m, average 1.4–1.8m, hatchlings 32–35cm. Adults are uniform glossy black above and below, although in good light they appear brown. Juveniles are reportedly greyish white or black, thus differing from the boldly marked juvenile Jackson's Tree Snakes. The adult is virtually indistinguishable from the black colour form of the Boomslang, so treat with caution.

Thrasops schmidti

HABITAT AND DISTRIBUTION: A Kenyan endemic, largely found in mid-altitude forest east of the Gregory Rift, between 1,400–2,400m altitude. Known localities include Nairobi area (forest in Nairobi National Park, Ngong Forest, Karen, Karura Forest, known from Muthaiga although not seen there for many years), Ngong Hills forest, Castle Forest Station and the surrounding Mt Kenya Forest, Meru and the Nyambene Hills. Might occur in the forests and woodland around the Aberdares. Not confined to closed forest, or even large forest areas, known from little hilltop forests in the Nyambene Hills, and also in the acacia woodland fringing the rivers running west from the Nyambene Hills into the savanna, at altitudes of 1,400m or lower. Conservation Status: IUCN Endangered B1ab(iii). It has a very restricted range, and much of the mid-altitude forest it inhabits is being rapidly felled, so may need monitoring.

Thrasops schmidti Meru Tree Snake. Stephen Spawls, Langata.

NATURAL HISTORY: Similar to Jackson's Tree Snake, diurnal and arboreal, a superb climber, ascending high into trees, to 30m or more. If threatened it will jump into space. Sometimes hunts on the ground. If cornered, it will inflate its neck and anterior body like a Boomslang, move sideways and make huge, lunging strikes, looking very large and threatening. No breeding details known but probably lays eggs. Captive specimens readily took chameleons.

LARGE-EYED TREE-SNAKES *RHAMNOPHIS*

A tropical African genus, comprising two fairly big, large-eyed harmless tree snakes. One species occurs in western East Africa.

LARGE-EYED GREEN TREE SNAKE/SPLENDID TREE SNAKE *Rhamnophis aethiopissa*

IDENTIFICATION: A large slim snake, yellow or green and black, with big, prominent eyes. The following notes apply to the eastern subspecies, *Rhamnophis aethiopissa elgonensis*. The head is short, the supraocular raised, pupil round, iris golden or greeny-yellow. The mouth curves markedly upwards. Body cylindrical, tail about one-third of the total length. Dorsal scales smooth, in 15–17 rows at midbody, the vertebral row of scales distinctly enlarged. Ventrals 159–167, subcaudals 127–133, paired, cloacal scale divided. Maximum size about 1.4m, average 90cm–1.1m, hatchlings 23–24cm. Green or yellow-green, all dorsal scales are black-edged, giving a chequered effect, the lip scales are yellow edged with black, the belly scales are uniform green or black-edged, sometimes with a small transverse bar on the outer edge of each ventral, a dark median line beneath the tail. The tongue is blue with a black tip. Very similar in appearance to the green/black colour form of the Boomslang, so treat with caution.

Rhamnophis aethiopissa

HABITAT AND DISTRIBUTION: Primary forest or forest patches at medium altitude; in West Africa more common in secondary forest and bushed farmland, rather than primary forest. Records include the Kakamega and Mt Elgon forests in Kenya, Uganda records

Rhamnophis aethiopissa Large-eyed Green Tree Snake. Left: Fabio Pupin, Rwanda. Right: Steve Russell, Kasese.

include the Mabira Forest, forests around Kampala, Budongo Forest, Fort Portal area and Kigezi, in Nyungwe and Cyamudongo forests in Rwanda, Tanzanian records are Minziro Forest and the Mahale peninsula. This subspecies seems to be endemic to East Africa, another (*R. a. ituriensis*) occurs in the Ituri Forest west of the Albertine Rift watershed.

NATURAL HISTORY: Diurnal and arboreal, a fast graceful climber. Secretive and well camouflaged, it spends the day in thick vegetation. If threatened and unable to slide away, it inflates its neck and anterior body in the manner of a boomslang. It lays eggs, no clutch details known in East Africa but the western subspecies lays from 4–17 eggs, roughly 3.5×1.5cm; females were gravid in May and June (early wet season), and hatchlings collected in September and October, at the end of the wet season. Diet in East Africa reported as frogs, but adult Nigerian specimens of the western subspecies ate mostly birds; small mammals, chameleons and agamas were also taken, juveniles had eaten arboreal geckoes.

BOOMSLANG *DISPHOLIDUS*

A large, widespread, rear-fanged tree snake. A single species in the genus, found virtually throughout sub-Saharan Africa; it will probably turn out to represent a species complex. Unlike most rear-fanged snakes, it has a deadly venom.

BOOMSLANG *Dispholidus typus*

IDENTIFICATION: A big, highly venomous, back-fanged tree snake, usually green, grey or brown, but may be several other colours, with a huge eye, short egg-shaped head and keeled dorsal scales. The pupil is almost round but elongate at the front (sometimes dumbbell shaped). Iris yellow, black-veined in adults, light green in juveniles. The body is laterally compressed, tail long, 25–30% of total length in males, 22–28% in females. The dorsal scales are strongly keeled, in 17–21 (usually 19) rows at midbody, ventrals 161–201, with double keels, cloacal scale divided, subcaudals 91–142. Maximum size about 2.1m (anecdotal reports of larger ones exist), average 1.2–1.5m, hatchlings about 30cm. Colour very variable: males are usually shades of green, females usually

Dispholidus typus

brown or grey. Other colour phases in East Africa include green with all scales black edged (characteristic of woodland), uniform black, black and yellow, brown with a rufous head, striped black and grey-white, light yellow-brown with white spots and blue-grey. Juveniles have a different colour and pattern to adults; they have a grey or brown, blue-spotted vertebral stripe, the flanks are light grey or pinkish, speckled black, the top of the head brown, olive or grey, chin and throat light, often with a big yellow blotch on the side of the neck; the eye apple green. This pattern gradually changes to adult coloration between lengths of 60–90cm; occasional adults retain the vertebral stripe. Large adults sometimes darken; captive green Boomslangs have gradually become black.

HABITAT AND DISTRIBUTION: Throughout East Africa, in coastal thicket, woodland, moist and dry savanna and semi-desert, around forest edges and clearings but not usually within dense forest or in high grassland. Absent from areas above 2,200m, few records from dry northern Kenya and parts of western Tanzania, but this is probably due to under-collecting. Also absent from high central Kenya, not found in Nairobi (although known at Kajiado, Olorgesaille and the eastern slopes of the Kikuyu Escarpment Forest near Kijabe). Elsewhere, north to South Sudan, west to Senegal, south to South Africa.

NATURAL HISTORY: A diurnal snake, almost totally arboreal although it will descend to cross open areas, seize prey and lay eggs. Fast moving, alert and graceful, climbing effortlessly. When prey is spotted it pauses, whilst curious lateral waves run down its body, it then darts forward to seize the animal. It appears to have binocular vision and can spot prey animals before they move, unusually for a snake. It is non-aggressive and if approached will slip quietly away. However, if threatened, it has a remarkable threat display, inflating the neck and forepart of the body to reveal the skin between the scales, it also flicks the tongue up and down in a deliberate manner, if further molested or restrained it will strike.

Dispholidus typus Boomslang. Top left: Stephen Spawls, Watamu. Top right: Stephen Spawls, threat display, Botswana. Bottom left: Michele Menegon, Tanzania. Centre right: Stephen Spawls, 'kivuensis' phase, Maasai Mara. Bottom right: Caspian Johnson, juvenile, Ugalla.

Boomslangs lay 8–25 eggs, roughly 2×4cm, in deep moist holes, damp tree hollows or in vegetation heaps. Mating observed in March, April, May and July in southern Tanzania, and in July in Kisumu, Kenya. A southern Tanzanian female laid ten eggs in September; a female from Voi laid 11 eggs in August. Hatchlings captured in Kakamega in September and in January in southern Tanzania. Incubation time 3–4 months in East Africa. Prey items include arboreal lizards such as chameleons, agamas and big geckoes, birds (often raids nests) and their eggs, frogs, and rodents; a captive female ate a dead bird which had been disgorged by another freshly-caught Boomslang. They are often mobbed by birds. Boomslangs are said to sometimes congregate in huge numbers, the purpose of which is unknown. This snake has a deadly, slow-acting venom, causing non-clotting blood and eventually renal failure, although being totally unaggressive, bites from these snakes are very rare. A specific serum is available from the South African Vaccine Producers.

VINE OR TWIG SNAKES *THELOTORNIS*

An African genus of medium-sized, diurnal tree snakes, with a curious dumbbell or keyhole-shaped eye pupil. They are slim, with a long tail. They are rear-fanged and although gentle and non-aggressive, they have a deadly venom that leads to non-clotting blood and renal failure. Four species are known, three in our area. The slim body and pointed head have lead to the legend in some parts of Africa that they can shoot through humans like an arrow. In Swahili, vine snakes are sometime called *'mfunga kuni'* ('tie up the firewood'), the suggestion being that when women are gathering firewood, they mistake the snake for a vine that they use to tie the wood bundles.

KEY TO THE EAST AFRICAN MEMBERS OF THE GENUS *THELOTORNIS*

1a	Top of head and temporals brown or green, but both the same colour......2		on the sixth labial, inhabits forest west of 33°E......*Thelotornis kirtlandii*, Forest Vine Snake [p. 505]
1b	Top of head usually green, temporals brown with black speckling......*Thelotornis mossambicanus*, Eastern Vine Snake [p. 507]	2b	Top of head green or brown, ventrals 156–169, nearly always a black triangle or diagonal or black spotting on the sixth labial, inhabits forest east of 36°E......*Thelotornis usambaricus*, Usambara Vine Snake [p. 508]
2a	Top of head always green, ventrals 162–189, never a black triangle or spotting		

FOREST VINE SNAKE/FOREST TWIG SNAKE
Thelotornis kirtlandii

IDENTIFICATION: A long, very thin tree snake, with a long arrow-shaped head, and a large pale eye with a horizontal, keyhole shaped pupil, this pupil is diagnostic of vine snakes, no other African snake has a pupil of this shape. Iris green and yellow. A narrow groove runs from each eye to the tip of the snout, giving the snake binocular vision. The tongue is bright red with a black tip. The body is cylindrical, the tail is very long and thin, 33–42% of total length. The scales are feebly keeled, in 19 rows at midbody; ventral

162–189, subcaudals 139–161. Maximum size about 1.6m, average size 90cm–1.4m, hatchling size unknown but probably around 25cm. Grey-brown in colour, heavily speckled with darker grey or black, the underside also distinctly speckled. From a distance this snake appears silvery-grey. The head is green above, the lips and chin are white, sometimes with black spotting. Specimens from Rwanda have deep grey-green heads, and the lips are extensively mottled with green or grey; the females have orange throats.

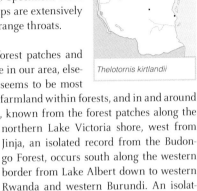
Thelotornis kirtlandii

HABITAT AND DISTRIBUTION: In forest, forest patches and woodland, from 600m to about 2,200m altitude in our area, elsewhere to sea level. Within the forest itself, it seems to be most common around natural glades. Often found in farmland within forests, and in and around parks and gardens in forest towns. In Uganda, known from the forest patches along the northern Lake Victoria shore, west from Jinja, an isolated record from the Budongo Forest, occurs south along the western border from Lake Albert down to western Rwanda and western Burundi. An isolated Tanzanian population on the Mahale peninsula; having crossed from the Congo when the lake was low. Elsewhere, north to Torit in South Sudan, west to Sierra Leone, south-west to northern Angola.

Thelotornis kirtlandii Forest Vine Snake. Michele Menegon, Rwanda.

NATURAL HISTORY: Diurnal and arboreal, in trees, bushes, thickets and reed beds. Climbs quickly and elegantly, but rarely seems to go very high, preferring lower branches. It will descend to ground to pursue prey or cross to other trees. Can move quickly on the ground and in trees. They rely on camouflage for defence, and spend much time sitting totally motionless in trees, sometimes with the front part of the body sticking out and swaying back and forth, resembling a grey branch with a green leaf at the end, moving in the wind, and may remain motionless even if closely approached or poked. However, if disturbed they will often flick the tongue up and down in a deliberate and highly visible manner, and if sufficiently molested and unable to escape, they will inflate the neck to a considerable size, as described for the Eastern Vine Snake *Thelotornis mossambicanus*, and make a determined strike. Their fangs are set well back, but they have a wide gape, and should they be determined to bite, they may seize and chew the victim. They lay eggs, clutches of 4–12, size 35×15mm. Eats mostly lizards, especially arboreal ones such as chameleons, agamas and geckoes, but will also ambush ground-dwelling species from a perch, or drop down and pursue them. Also known to take other snakes and amphibians, reputed to take eggs. Another name for the snake (although little used and inaccurate) is 'Bird Snake', and there is a curious legend that it hypnotises birds with its flickering red and black tongue. This snake has binocular vision and is one of the few species that can detect prey by sight before it moves. In some areas of its forest range it is regarded as totally harmless, which it is unless severely molested. In other areas, it is greatly feared, and believed to be able to fly through humans like an arrow. May be abundant in areas without being seen, due to its excellent camouflage, in a collection from the Ivory Coast it was the second most common species. Along with other *Thelotornis*, it has a deadly slow-acting venom for which no antivenom is produced.

EASTERN VINE SNAKE *Thelotornis mossambicanus*

IDENTIFICATION: A very thin, medium-sized grey tree snake. The head is long, snout pointed. The eye is large, pupil keyhole-shaped, iris yellow. The tongue is red with a black tip. Body thin, tail long, 35–40% of total length. Body scales keeled, in 19 rows at midbody (occasionally 17 or 21), ventrals 144–169, subcaudals paired, 123–158, cloacal scale divided. Maximum size about 1.4m, average 90cm–1.3m, hatchlings 23–30cm. From a distance, this snake appears silvery-grey or grey-brown, close-up it is pale grey or whitish with intense dark speckling. The temporal region behind the eye is brown, heavily speckled with black. Below, slightly paler but also speckled with black.

Thelotornis mossambicanus

HABITAT AND DISTRIBUTION: Savanna, woodland and coastal thicket, between sea level and 1,700m altitude. It occurs in the country around Mt Kilimanjaro, including Kitobo Forest, the North Pare Mountains and Arusha National Park, and between Lakes Eyasi and Manyara, and throughout most of southeastern Tanzania, extending up the east side of Lake Tanganyika almost to the Burundi border. Elsewhere to Mozambique and eastern Zimbabwe (and reported from Somalia, although the status of Somali specimens is debated).

NATURAL HISTORY: Diurnal, arboreal and secretive. Waits motionless in the branches for prey, often with the forepart of the body sticking out like a twig. It will descend to the ground to cross open areas and to seize prey, it may also strike down at passing terrestrial prey animals. Gentle and non-aggressive, moving away if threatened, but if cornered or seized it inflates the neck and front half of the body, exposing vivid black and pale grey and blue crossbars, and flicks its bright tongue up and down slowly. If further molested, it may strike, sometimes it bites, sometimes it is deliberately off-target, or may bang the aggressor with its snout, mouth closed. In the breeding season, male combat has been observed, with rivals trying to force their opponent's heads down. Mating observed in August in southern Tanzania. Vine snakes lay 4–16 eggs, roughly 1.5×3.5cm. Hatchlings collected in February in southern Tanzania. Diet: mostly lizards, including chameleons (taken in trees) lacertids

Thelotornis mossambicanus Eastern Vine Snake. Michele Menegon, Pare Mountains.

and agamas (often ambushed from above); when prey is seized from a tree the snake hangs down and swallows the animal upwards if feasible. Odd prey items include other snakes, frogs and fledglings. Often mobbed by birds; one attacked in southern Tanzania by a Black-crowned Tchagra *Tchagra senegala* fell from the tree and died shortly afterwards. Vine snakes have a deadly, but very slow-acting venom, causing incoagulable blood and internal bleeding, for which no antivenom is produced, but they are so inoffensive that the only known bites were to snake handlers.

USAMBARA VINE SNAKE *Thelotornis usambaricus*

IDENTIFICATION: A very thin, medium-sized grey tree snake. The head is long, snout pointed. The eye is large, pupil key-hole-shaped, iris yellow centrally, edged with green and sometimes white. The tongue is red with a black tip. Body thin, tail long, roughly 37–40% of total length. Body scales keeled, in 19 rows at midbody, ventrals 156–169, subcaudals paired, 143–175, cloacal scale divided. Maximum size about 1.4m (reports of larger specimens exist), average 90cm–1.3m, hatchlings size unknown but probably around 25cm From a distance, this snake appears silvery-grey or grey-brown, close-up it is pale grey or whitish with intense dark speckling. The top of the head can be any shade of green, brown or rufous, and some captive snakes have undergone a change of head colour. The lips are white, and nearly always the sixth labial is speckled with black, often in the form of a triangle or diagonal. Below, slightly paler but also speckled with black.

Thelotornis usambaricus

HABITAT AND DISTRIBUTION: Savanna, woodland and coastal thicket, between sea level and 2,000m altitude. Occurs along the length of the Kenyan coast, and into north-eastern Tanzania (but slightly away from the coastal strip), including the Usambara, Nguu and Nguru Mountains, south to Dar es Salaam. Inland in Kenya to the Taita Hills and Mt

Thelotornis usambaricus Usambara Vine Snake. Left: Stephen Spawls, Watamu. Top right: Stephen Spawls, Watamu. Bottom right: Bio-Ken Archive, Watamu.

Kasigau, and across the border to the southern side of Kilimanjaro, although some of the southern Kenya animals seem intermediate between this species and *Thelotornis mossambicanus*. Conservation Status: IUCN Vulnerable B2ab(iii)

NATURAL HISTORY: Similar to the previous species: diurnal, arboreal and secretive. Waits motionless in the branches for prey, often with the forepart of the body sticking out like a twig. It will descend to the ground to cross open areas and to seize prey, it may also strike down at passing terrestrial prey animals. Gentle and non-aggressive, moving away if threatened, but if cornered or seized it inflates the neck and front half of the body. Diet as for other members of the genus. Along with other *Thelotornis*, it has a deadly slow-acting venom for which no antivenom is produced.

DAGGER-TOOTH VINE SNAKE *XYELODONTOPHIS*

A recently described, large rear-fanged tree snake. There is a single species in the genus, known only from Tanzania at present.

DAGGER-TOOTH VINE SNAKE *Xyelodontophis uluguruensis*

IDENTIFICATION: A medium to large, rather thin arboreal snake, with elongate head and large eyes with horizontal, pear shaped pupil. In general it resembles the species in the genus *Thelotornis*, but it lacks the distinctive keyhole-shaped pupil typical of that genus and it differs in the development of strongly curved, rear maxillary teeth, which have sharp flanges similar to those of *Rhamnophis*, but larger and narrower at the base. Dorsal scales are elongated, moderately to feebly keeled, in 19 rows at midbody, with a single, large apical pit. Ventrals 166–169, subcaudals 132–159. Maximum recorded length is over 1.9m, other specimens were over a metre, hatchlings 37cm. The head is clearly bicoloured, with the top completely bronze, brown or green and the lower part (labials, throat and chin) white to yellow and immaculate. Body brown to blue-grey or black, speckled with light brown, green and yellow. When the snake inflates the neck for defensive purposes, the neck appears bright yellow with three large black patches on both sides of the neck; the body appears barred yellow and brown. Taxonomic Note: Recent molecular investigation shows that the records of *Thelotornis kirtlandii* from the Udzungwa mountains of Tanzania are probably assignable to this species and that the genus *Xyelodontophis* is nested within *Thelotornis* and will probably be put in synonymy with the latter.

Xyelodontophis uluguruensis

HABITAT AND DISTRIBUTION: A montane and sub-montane forest species. Originally known from the Uluguru mountains only, recent investigations revealed its presence in the forests of the Nguru, Udzungwa and Mbarika (Mahenge) mountains, from 1,000 to about 2,000m altitude. There is a curious report of a dead specimen from Manda Island, on the north Kenya coast, not shown on the map.

Xyelodontophis uluguruensis Dagger-tooth Vine Snake. Lorenzo Vinciguerra, Udzungwa Mountains.

NATURAL HISTORY: Diurnal, individuals were found both on trees and moving on the ground. When disturbed, it inflates its neck in the manner of a Boomslang, enhancing the vivid barring on its throat. It bites readily. A female, length 1.47m, laid ten eggs. Little is known about its diet but probably similar to *Thelotornis*. One of the type specimens had eaten a chameleon, *Rhampholeon uluguruensis*. Nothing is known about the toxicity or composition of the venom, but it is presumably similar (and thus equally toxic) to the species in the genus *Thelotornis*, so bites must be avoided.

BROAD-HEAD TREE SNAKES
TOXICODRYAS (FORMERLY BOIGA)

An African genus containing two species of broad-headed, big-eyed, rear-fanged tree snakes, both found in East Africa. Originally in the genus *Boiga* (a diverse Asian genus, with more than 30 species), and some authorities still classify the African species as *Boiga*. Fascinatingly, molecular analysis indicates that the egg eaters, *Dasypeltis*, are their closest relatives. Laboratory studies have shown that their venom is toxic and produced in relatively large quantities, so they might represent a danger to humans; however bites to humans by tree snakes are very rare.

KEY TO EAST AFRICAN MEMBERS OF THE GENUS *TOXICODRYAS*

1a Midbody scale rows 19......*Toxicodryas pulverulenta*, Powdered Tree Snake [p. 512]

1b Midbody scale rows 21–25......*Toxicodryas blandingii*, Blanding's Tree Snake [p. 511]

BLANDING'S TREE SNAKE *Toxicodryas blandingii*

IDENTIFICATION: A very big, stocky tree snake, with a thin neck, a short, broad, flattened head and prominent eyes set well forward, with vertical pupils. Nostrils large. Body triangular in section and rubbery in texture, the head is also rubbery and flexible, this is particularly noticeable when the snake is held by the head. Tail long and thin, 20–26% of total length. Scales large and velvety, in 21–25 (21–23 in East Africa) rows at midbody, ventrals 240–289, subcaudals 120–141. Maximum size about 2.8m, average size 1.4–2m, hatchling size unknown. There are two main colour phases, one is glossy black above, yellow underneath, the yellow may extend up the sides or be confined to a narrow stripe in the middle of the belly, the lips are yellow, bordered with black. The eye is dark. Black specimens are usually male. The second colour phase is brown, grey or yellow-brown, yellow-brown below, with faint or clear darker cross bars or diamonds on the flanks. The skin between the scales is bluish-grey, and this is obvious when the snake inflates the body. The head is brown, the lips may be yellow-brown, the eye is yellowish or brown with a vertical pupil. Brown specimens are usually female. Juveniles are all brown (both sexes) with clear irregular black bars that are roughly diamond-shaped, sometimes with pale edging and/or lighter centres. Black adults are startlingly similar in colour to Gold's Tree Cobra *Pseudohaje goldii*, but have a more velvety skin.

Toxicodryas blandingii

Toxivodryas blandingii Blanding's Tree Snake. Top left: Stephen Spawls, Kakamega, female. Top right: Konrad Mebert, threat display, male. Bottom: Olivier Pauwels, juvenile, Gabon.

HABITAT AND DISTRIBUTION: Forest, woodland, forest-savanna mosaic and riverine woodland, in East Africa from 700m to about 2,200m altitude, elsewhere down to sea level. In Kenya, known from the Kakamega Forest and Serem, occurs in the forests of the northern Lake Victoria shore and the Sesse Islands; in Uganda, known from the Budongo forest, and the forest along the Albertine Rift on the western Ugandan border, from the south end of Lake Albert to the Rwanda border (although not in Bwindi Impenetrable National Park), thence west in the Congo. Two Tanzanian records are known; Rubondo Island and the Mahale peninsula. Elsewhere, west to Sierra Leone, south-west to northern Angola, also in the Imatong Mountains in South Sudan.

NATURAL HISTORY: Arboreal, it climbs quickly but ponderously, but its great length enables it to go very high, up to 30m or more in big forest trees. It will descend to ground to cross open spaces and roads. Active at night (one of Africa's few nocturnal tree snakes). When inactive (mostly during the day) rests in leaf clumps, hollows in trees, etc. When threatened, this snake responds with a distinctive display, inflating the body to a huge extent, flattening the head and lifting the forepart of the body off the ground into wide C-shaped coils. If further molested, will open the mouth enormously wide, exposing the pink lining, and make a huge, lunging strike. Active specimens prowl about trees, crawling along branches, investigating nests, recesses and hollows in the tree. They seem able to smell sleeping birds, especially in nests, and when one is detected, will make a very slow, careful approach and try to catch the bird in the nest. Quite often found around human habitation, and in parks and gardens in forest cities, where it has been found in hedges and even quite small ornamental trees. Known to enter buildings, often in search of bats, and also known to frequent trees outside caves where bats roost, where it can catch bats as they emerge at dusk. Lays 3–14 eggs, approximately 20×40mm. Eats a range of prey; birds, birds eggs, arboreal lizards such as chameleons and agamas, frogs, arboreal rodents and bats; juveniles eat largely lizards, adults eat mostly birds and mammals. Its venom is a powerful neurotoxin, and although there are no documented human fatalities it should be treated with care and bites avoided.

POWDERED TREE SNAKE   *Toxicodryas pulverulenta*

IDENTIFICATION: A fairly big, broad-headed, rear-fanged tree snake, usually pinky or red-brown. The head is broad, eye large and prominent, pupil vertical. The body is laterally compressed, with a vertebral ridge. The tail is 20–25% of total length. Scales smooth, 19 rows at midbody, ventrals 236–276, subcaudals 96–132. Maximum size about 1.25m, average 80cm–1.2m, hatchling size unknown. Colour pinkish to brown, red-brown or pinky-grey, with darker crossbars alternatively narrow and uniform or broadening to subtriangular, enclosing a pale spot. The back is usually very finely dusted with brown or black specks (hence the common name), as is the belly, which is pale pink with dashed dark lines on the side of each ventral scale. The tongue is pink with a white tip.

Toxicodryas pulverulenta

HABITAT AND DISTRIBUTION: Forest and woodland of western East Africa at medium to high altitude. Kenyan localities include Mumias, Kakamega and Serem; Ugandan local-

Toxicodryas pulverulenta Powdered Tree Snake. Left: Konrad Mebert, eastern Congo, brown. Right: Konrad Mebert, eastern Congo, pink.

ities include the Budongo Forest, Mabira and other forests of the northern Lake Victoria shore, including around Entebbe, Kibale and Itwara Forest, and the forest patches of western Kibale, Kabarole and Bundibugyo, thence west into the DR Congo, elsewhere west to Guinea, south-west to northern Angola.

NATURAL HISTORY: Arboreal, it is an elegant, careful climber, and nocturnal, which is unusual for a big tree snake. During the day shelters in holes and cracks in trees, creeper and epiphyte tangles, disused plant pods and bird nests and sleeps curled up, will also sleep in a fork or on a branch. No documented threat display. Lays eggs, a Ghanaian specimen contained three oviductal eggs, roughly 3×1cm, in September (end of rainy season), a specimen from the DR Congo contained eggs in June, a Nigerian specimen contained two eggs in January. Diet apparently rodents and arboreal lizards. Its venom is unstudied but this snake should be treated with care.

EGG-EATING SNAKES *DASYPELTIS*

An African genus of small to medium-sized harmless snakes, with blunt bullet-shaped heads and small eyes, with vertical pupils. They have rudimentary teeth but are adapted for eating birds eggs; some vertebral processes protrude into the gullet, acting like saw blades to cut through the eggshell, which is then regurgitated and the egg contents swallowed. As well as the 'gular saw' they have a very flexible, extensible skin, a pleated buccal membrane which consists of accordion-like folds of gum tissue and a very loose attachment between the scales and bones of the lower jaw, all are adaptations enabling egg eaters to swallow eggs several times wider than their heads.

They show a wide range of colours even within each species, which has created taxonomic problems. At present there are 12 species in the genus, several of which were described recently from West and Central Africa. The situation in East Africa is not totally clear, and specimens will need to be re-examined in light of the West and Central African taxonomic work. At present there appear to be five species in East Africa; and a provisional key is given below. As a result of variation, some species appear more than once.

KEY TO EAST AFRICAN MEMBERS OF THE GENUS *DASYPELTIS*

1a A distinct, not faint, rhombic pattern extends the length of the body......2
1b No distinct rhombic pattern extending the length of the body...4

2a Vertical dark bars on the flanks are parallel with the dark blotches along the spine......*Dasypeltis confusa*, Confusing Egg Eater [p. 517]
2b Vertical dark bars on the flanks are parallel with the light blotches along the spine......3

3a Less than 80 pattern cycles between the nape and base of tail, usually at altitudes below 2000m......*Dasypeltis scabra*, Common Egg Eater (part) [p. 516]
3b More than 85 pattern cycles between nape and base of tail, usually at altitudes above 2000m......*Dasypeltis atra*, Montane Egg Eater (part) [p. 514]

4a 3-7 V-shapes on the neck.......*Dasypeltis medici*, Rufous Egg Eater (part) [p. 519]
4b No V-shapes on neck....5

5a Bright orange or red, with or without markings......6
5b Not bright orange or red......7

6a Usually in high altitude areas, over 1,500m altitude......*Dasypeltis atra*, Montane Egg Eater (red form only) (part) [p. 514]

6b In lowland, below 1,500m......*Dasypeltis medici lamuensis*, Lamu subspecies of Rufous Egg Eater (part) [p. 519]

7a Black in colour......*Dasypeltis atra*, Montane Egg Eater (part) [p. 514]
7b Not black in colour......8

8a Grey, with or without markings...... 9
8b Brown, with or without markings......10

9a Pattern consists of white dots along the spine, but not on flanks, occurs at low altitude......*Dasypeltis medici lamuensis*, Lamu subspecies of Rufous Egg Eater [p. 519]
9b Pattern non-existent, or consisting of white, solid or spotted cross bars that extend down the flanks, not at low altitude......*Dasypeltis scabra*, Common Egg Eater [p. 516]

10a Body with dark crossbars present on the skin, not the scales, only in Uganda......*Dasypeltis fasciata*, Western Forest Egg Eater [p. 518]
10b Body unicolored, if markings present (often white spots) then on the scales......11

11a One or two postoculars, usually at altitudes over 1,500m......*Dasypeltis atra*, Montane Egg Eater (part) [p. 514]
11b Two postoculars, at altitudes below 1,500m......*Dasypeltis scabra*, Common Egg Eater (part) [p. 516]

MONTANE EGG EATER *Dasypeltis atra*

IDENTIFICATION: A small snake with a bullet-shaped head, eyes small but prominent, pupil vertical, iris golden, flecked with black, tongue black. Body cylindrical, tail 12–18% of total length (longest in males). Scales strongly keeled and serrated, in 22–27 rows at midbody, ventrals 202–237, subcaudals paired, 49–72. Maximum size about 1.1m, average 50–80 cm, hatchlings 22–25 cm. There are several colour forms: (a) uniform black, above and below (sometimes the throat is grey); (b) uniform red-brown, red or pink, paler below; (c) uniform brown; (d) brown with a row of many darker blotches along the centre of the back, often white between the blotches, faint dark lateral bars may be present, pale brown or yellow-brown below. Taxonomic Note: A recent molecular analysis indicates *Dasypeltis atra* is polyphyletic; it might be just a melanistic or dark morph of *Dasypeltis scabra*, rather than a discrete species.

HABITAT AND DISTRIBUTION: High altitude savanna, grassland, woodland and forest, from 1,000–2,800m, but usually above 1,500m. It occurs from Musoma anticlockwise around Lake Victoria, north through Kakamega and Mt Elgon, in a broad swathe across central Uganda, south to Bukoba, into western Rwanda and Burundi. In Kenya known from the high country of the Eastern Rift Valley, from Nairobi north to Nakuru, on the wall and floor of the rift, also from Chuka and the northeast flank of Mt Kenya. Sporadic records from the Taita Hills in Kenya, southern and eastern Tanzania (Mbizi Forest Reserve near Sumbawanga, Oldeani at Ngorongoro, Moshi (black specimens); Kleins Camp in northern Serengeti, and Ibaya Camp, Mkomazi (brown specimens)). Known also from the Imatong Mountains in South Sudan and the high areas to the west of the Albertine Rift in the DR Congo.

Dasypeltis atra

Dasypeltis atra Montane Egg Eater. Top left: Michele Menegon, Rwanda. Top right: Daniel Hollands, Mt Elgon. Inset: Harald Hinkel, red phase, Rwanda.

NATURAL HISTORY: Terrestrial and semi-arboreal, hunting on the ground but also climbing into trees and bushes to find birds' nests. Nocturnal, sheltering by day in abandoned birds' nests, tree holes and cracks, under ground cover or in holes. Egg eaters detect occupied bird nests by smell, they crawl up to the nest, when the bird flees they swallow the eggs; this is a remarkable process. They carefully smell the egg, and move around it, touching it with their nose, judging its size. They then proceed to swallow it, working their jaws over the egg; a small egg eater with a big egg in its mouth is a remarkable sight. The egg is swallowed until it is well into the throat, the snake's head is just clear of the egg. They then move the head and neck up and down, using the vertebral processes to saw through the shell. Once punctured, the fluid begins to escape and the snake uses the neck muscles to compress the shell, when it is effectively emptied the snake lifts its head, writhes about and regurgitates the shell, crushed into a boat-shaped mass. Egg eaters eat only eggs; as well as freshly-laid eggs they will swallow those with partially-developed embryos, if the embryo is small it is also swallowed, if not the snake squeezes what fluid remains out of the shell and then regurgitates the remaining embryo with the shell. If threatened, egg eaters have a remarkable threat display (although not all individuals show it), forming the body into

C-shaped coils; these coils are rotated against each other, the friction between the keeled scales producing a remarkable hissing or crackling noise, like water falling on a red-hot plate. At the same time the snake opens its mouth very wide to expose the black interior, lowering and broadening the lower jaw, if further molested they will hiss loudly and simultaneously strike; a tremendous threat display but it is all bluff. Their teeth are so tiny that they can do no damage, consequently they usually strike to miss. Montane Egg Eaters lay from 7–14 eggs, roughly 3×1.5cm; in their montane habitat there seems to be no particular breeding season; laying recorded in January, April, October and September in Uganda.

COMMON OR RHOMBIC EGG EATER *Dasypeltis scabra*

IDENTIFICATION: A small to medium-sized snake with a bullet-shaped head, eyes small but prominent, pupil vertical, iris golden, flecked with black, tongue black. Body cylindrical, tail 12–18% of total length (longest in males). Scales strongly keeled and serrated, in 20–27 rows at midbody, ventrals 180–263, subcaudals paired, 38–94. Maximum size is about 1.1m, average 50–80cm, hatchlings 20–26cm. Colour quite variable but the most common form is the 'rhombic' form, which is brown, rufous or grey, with a series of darker, oval or rhombic shapes along the centre of the back, often interspersed with lighter blotches, there are often one or two V-shapes on the neck, the flanks are usually marked with fine irregular, oblique darker bars. The ventrals may be white, cream, either uniform or blotched to a greater or lesser extent. Specimens with faint patterns, or unicolored specimens, which may be reddish, any shade of brown or grey, also occur. Note: in parts of its range it closely resembles both the North-east African Carpet Viper *Echis pyramidum* and the Rhombic Night Adder *Causus rhombeatus*, so great care should be taken with identification; the C-shaped coil threat display shown by this species is also shown by *Echis pyramidum*.

Dasypeltis scabra

HABITAT AND DISTRIBUTION: Occupies a very wide range of habitats, from coastal forest and thicket to moist and dry savanna, semi-desert, near-desert, high grassland and woodland, absent only from areas above 2,600m and closed forest where it is replaced by the Montane Egg Eater. Occurs almost throughout East Africa, absent only from dry eastern and northern Kenya (except Wajir and Buna and Mount Kulal). In some high altitude areas it is replaced by the Montane Egg Eater, but in others both species occur. Swims well and has colonised many Lake Victoria islands. Elsewhere, north to the South Sudan and Ethiopia, south to South Africa, populations further west are probably a different species, as presently defined. Conservation Status: IUCN Least Concern.

NATURAL HISTORY: As for other egg eaters, see our notes on natural history for the Montane Egg Eater *Dasypeltis atra*. This species is nearly always willing to show the C-coil threat display with its accompanying sizzling noise; do not confuse it with a Carpet Viper. Common Egg Eaters lay from 6–25 eggs, roughly 3.5×1.5cm. A specimen lived 31 years in captivity. They can tolerate long periods of starvation, a captive specimen went 12 months without food before it commenced feeding; this is presumably an adaptation since in some areas of Africa birds have a short breeding season, in between times the snake will not be able to find food.

Dasypeltis scabra Common Egg Eater. Top left: Stephen Spawls, Dodoma. Top right: Stephen Spawls, Tarangire National Park. Centre left: Stephen Spawls, Uganda. Centre right: Harald Hinkel, threat display, Rwanda. Bottom left: Anthony Childs, Nairobi National Park.

CONFUSING EGG EATER *Dasypeltis confusa*

IDENTIFICATION: A small to medium snake with a bullet-shaped head, eye quite large and prominent, pupil vertical, iris golden, or silvery yellow. Body cylindrical, tail 12–17% of total length (longest in males). Scales strongly keeled and serrated, in 23–26 rows at midbody, ventrals 213–242, subcaudals paired, 53–73. Maximum size about 97cm, average 50–80cm, hatchlings around 23cm. Usually brown or grey, along the spine is a chain of around 40–60 sub-rectangular dark markings, in some specimens these may be linked by lighter patches, and on the flanks are irregular dark vertical or oblique bars, but the top of these bars is always parallel with the dark spinal markings, never the pale ones or the gaps between the dark markings. There are often narrow dark cross-bars

Dasypeltis confusa

Dasypeltis confusa Confusing Egg Eater. Jean-Francois Trape, West Africa.

on the head, and a narrow V-shape on the nape, followed by a broad V-shape. The underside is beige, unmarked or with black edging.

HABITAT AND DISTRIBUTION: Parts of north western Uganda, and western Kenya; in mid-altitude savanna.

NATURAL HISTORY: Similar to other egg eaters, i.e. nocturnal, largely terrestrial, eats birds eggs, lay eggs; see the notes on the Montane Egg Eater *Dasypeltis atra*.

WESTERN FOREST EGG EATER *Dasypeltis fasciata*

IDENTIFICATION: A small to medium snake with a bullet-shaped head, eye quite large and prominent, pupil vertical, iris golden, or silvery yellow sometimes suffused with black. Body cylindrical, tail 14–20% of total length (longest in males). Scales strongly keeled and serrated, in 20–25 rows at midbody, ventrals 227–254 (usually 239–248 in Uganda) subcaudals paired, 64–88 (usually 64–68 in Uganda) Maximum size in Uganda 86cm, but larger elsewhere, average 50–80cm, hatchlings in Ghana were 23–26cm. Usually brown or yellow-brown, sometimes greenish brown, with a series of slightly darker saddles along the back, these are more obvious in juveniles, and slightly darker irregular flank bars; the pattern being on the skin, not the scales, this is a valuable clue to the identity of this egg eater. The head is strikingly reticulated with deep and light brown, often in a vague broad Y-shape. The belly is yellow or light brown. The upper labials may be dark edged.

Dasypeltis fasciata

HABITAT AND DISTRIBUTION: Forest, moist savanna and dry savanna, from sea level to over 2,000m, but in our area known only from forest patches in the Semliki National Park and Forest Reserve, Bundibugyo, western Uganda, at 700–800m altitude. Elsewhere, west to Sierra Leone.

Dasypeltis fasciata Western Forest Egg Eater. Jean-Francois Trape, Cameroon.

NATURAL HISTORY: Similar to other egg eaters, see the notes on the Montane Egg Eater *Dasypeltis atra*. In Ghana this species lays from 5–9 eggs, roughly 3.5×1.2 cm, hatchlings collected in April at the start of the rainy season.

RUFOUS EGG EATER/EAST AFRICAN EGG EATER
Dasypeltis medici

IDENTIFICATION: A small snake with a bullet-shaped head, eyes quite large and prominent, pupil vertical, iris yellow, orange or grey, depending on the colour pattern, tongue black. Body cylindrical, tail 15–21% of total length (longest in males). Scales strongly keeled and serrated, in 23–27 rows at midbody, in the southern subspecies (*Dasypeltis medici medici*) ventrals 235–259, in the northern subspecies (Lamu Red Egg Eater *Dasypeltis medici lamuensis*) ventrals 226–237. The subcaudals are paired, 71–109. Maximum size about 1m, average 50–80cm, hatchlings 23–27cm. Colour is quite variable; may be delicate pink, orange, red, grey, fawn or brown. In the southern subspecies there is usually a darker vertebral line, about five scales wide, interrupted by white patches or spots, usually there are 3–7 distinctive V-shapes on the neck, and narrow oblique bars along the flanks. The ventrals are a paler version of the main dorsal coloration. The northern subspecies, the Lamu form, is commonly red, vivid pink or orange, other colours include grey and blue-grey, sometimes there are fine white dot clusters along the spine but no other markings, although close examination will show each scale is finely stippled, sometimes with white. In some areas (Shimba Hills, Usambara Mountains) intermediate

Dasypeltis medici

Dasypeltis medici Rufous Egg Eater. Left: Luke Verburgt N Mozambique. Right: Michele Menegon, Udzungwa Mountains. Bottom: Stephen Spawls, Watamu (subspecies *lamuensis*).

specimens occur with the V-shapes but no other markings.

HABITAT AND DISTRIBUTION: The northern form occurs from Lamu south along the coast to Diani beach area, inland to Tsavo East National Park, the Taita Hills, south-eastern Kenya and Mount Kilimanjaro, usually in red soil areas. The typical form occurs from the Usambara Mountains south along the length of the Tanzanian coast, also on Zanzibar and Mafia but not Pemba, inland to Morogoro, the Uluguru Mountains and up the Rufiji River. An interesting recent record of the typical form from Rhino River Camp, Meru National Park, and an old record from Murang'a. Elsewhere, south to the north coast of South Africa, inland to Malawi and eastern Zimbabwe.

NATURAL HISTORY: Similar to other egg eaters, see our notes on the Montane Egg Eater *Dasypeltis atra*. Lays 6–28 eggs, roughly 3×1.5cm; a juvenile was taken in May at Vipingo on the Kenyan coast.

TIGER, LARGE-EYED AND CAT SNAKES
TELESCOPUS

A genus of small to medium-sized, large-eyed, broad-headed, rear-fanged snakes. Although venomous, they are not dangerous to humans. All are nocturnal, some are partially arboreal, they feed on a range of prey, favouring lizards. They lay eggs. At least fourteen species are known, in Africa, the Middle East, Asia and southern Europe. Nine occur in Africa, two species are found in East Africa. They are bad-tempered snakes, some species if handled will inflict slow deliberate bites on the handler.

KEY TO EAST AFRICAN MEMBERS OF THE GENUS *TELESCOPUS*

1a Orange or red, with faint or dark saddles or crossbars along the back...... *Telescopus semiannulatus*, Tiger Snake [p. 522]

1b Pink, grey, black or orange, without saddles on the back, sometimes fine lighter crossbars......*Telescopus dhara*, Large-eyed Snake [p. 520]

LARGE-EYED SNAKE *Telescopus dhara*

IDENTIFICATION: A medium-sized, broad-headed snake. The snout is square, eye prominent, pupil vertical, iris yellow. The tongue is pink. Neck very thin, body triangular or sub-triangular in section, tail 15–20% of total length. Scales smooth, in 19–25 rows at midbody, ventrals 200–244, subcaudals paired, 56–97 rows. Maximum size of the East African subspecies about 1m, average about 50–80cm, hatchling size 17–21cm. Colour very variable: may be black, slate grey, various shades of brown, orange, pink or golden yellow, mono-coloured; occasional specimens have narrow, lighter crossbars. Often several phases will occur in one area; at Wajir orange, black and pink specimens are found.

Telescopus dhara

Freshly-sloughed specimens are often quite iridescent. Lighter below, usually cream, pearly or pink. Taxonomic Notes: Kenyan specimens belong to the subspecies *Telescopus dhara somalicus*, with low ventral counts, 200–219. Some authorities treat *Telescopus dhara obtusus* as a subspecies, others as a full species, which probably occurs in parts of north-east Kenya, as it is known from Lugh, in Somalia, it has higher ventral counts, 230–278.

HABITAT AND DISTRIBUTION: Semi-desert, dry savanna and coastal thicket, from sea level to about 1,400m. Widespread in dry northern and eastern Kenya and on the coast, from the Tana Delta south to Malindi, thence inland and west across to the dry country of northern Tanzania, to Mkomazi Game Reserve and south around Mount Kilimanjaro to Tarangire National Park, also across southern Kenya to Olorgesaille, Magadi and Lake Natron. Just reaches eastern Uganda at Amudat, but might be in northern Uganda as known from Torit and Fula Rapids in South Sudan. Elsewhere, this subspecies reaches Somalia, other subspecies reach Algeria and the Arabian Peninsula.

NATURAL HISTORY: Nocturnal and semi-arboreal, hunting at night, crawling across the ground, climbing bushes and trees and investigating holes, By day, shelters under ground cover and in holes, but may sometimes rest coiled up in trees or bushes, or in bird nests. Slow moving and ponderous. It is tolerant of dry conditions and low night-time temperatures. If threatened, it will raise the head and forepart of the body in a loop, hiss loudly and strike, if restrained it turns and bites slowly and deliberately. Large-eyed snakes lay clutches of 5–20 eggs, average size 1×2.5cm; a Voi animal was gravid in November. They eat a range of prey, including lizards, birds, bats and rodents. They take chameleons, a Wajir specimen had eaten a lark, a Voi animal had eaten an *Agama lionotus*. Greatly feared in Somalia, where it is called *Mas Gadut* ('red snake'), but possibly confused with the Red Spitting Cobra, *Naja pallida*.

Telescopus dhara Large-eyed Snake. Top: Stephen Spawls, Malindi. Centre: Malcolm Largen, Ethiopia. Bottom: Anthony Childs, Meru National Park.

TIGER SNAKE *Telescopus semiannulatus*

IDENTIFICATION: A medium-sized, broad-headed, orange and black snake. The eye is prominent, pupil vertical, iris orange or yellow. The tongue is pink. Neck very thin, body cylindrical or sub-triangular in section, tail 15–20% of total length. Scales smooth, in 17–21 rows at midbody, ventrals 190–244, subcaudals paired, 51–85 rows. Maximum size about 1.05m, average 50–80cm, hatchlings 17–22cm. Colour orange or orangey pink, with black, brown or dull rufous saddles all along the back, these may be dark (inland) or quite light (coastal); oval or narrow. Lighter orange or pink below. Sometimes black speckling on the flanks. The tongue is pink. In southern Africa, where there are several similar-looking members of the genus, this snake is called the Eastern Tiger Snake.

Telescopus semiannulatus

HABITAT AND DISTRIBUTION: A south-east African species, in dry and moist savanna, coastal thicket and woodland, from sea level to about 1,700m. Widespread throughout most of Tanzania, save in the north-west and parts of the north, extends into Kenya from the Serengeti into the Maasai Mara and along the Kenya coast to north of Malindi. Kenyan inland records include the edge of the Tharaka Plain below Chuka, Kindaruma, Kima and Elmau Hill, just west of the northern Chyulu Hills, but might be more widespread in the south-east. Reported from Akagera National Park in Rwanda, without a specimen (not shown on map). Elsewhere, south to north-eastern South Africa.

NATURAL HISTORY: Slow-moving, nocturnal and semi-arboreal, hunting at night, crawling across the ground, climbing bushes and trees and investigating holes and bird nests. By day, shelters under ground cover and in holes, but may sometimes rest coiled up in trees or bushes, in tree holes or in bird nests, on the coast often hides in makuti thatch. Slow moving and ponderous. If threatened, it will raise the head and forepart of the body in a loop, hiss loudly and strike, if restrained it turns and bites slowly and deliberately. Lays 6–20 eggs, roughly 3×1.5cm. East African details lacking but in South Africa eggs are laid in the summer and hatch in 71–85 days. Diet quite varied, it eats lizards, especially geckoes and chameleons, rodents, other snakes, birds and bats; large prey items are killed by a mixture of constriction and envenomation.

Telescopus semiannulatus Tiger Snake. Left: Stephen Spawls, Serengeti, dark phase. Right: Stephen Spawls, Watamu, light phase.

HERALD AND WHITE-LIPPED SNAKES
CROTAPHOPELTIS

A tropical African genus of small (less than 1m), dark, rear-fanged snakes with large eyes and vertical pupils, scales at midbody in 17–21 rows. They have blade-like rear-fangs, but their venom is not dangerous to humans. They are terrestrial and nocturnal, most eat amphibians. Six species are known, four occur in East Africa, none are endemic. Although effectively harmless, they are fierce little snakes, with a distinctive threat display; they flatten the head into a triangle, flare the lips, hiss and strike readily if threatened.

KEY TO EAST AFRICAN MEMBERS OF THE GENUS *CROTAPHOPELTIS*

1a Lower dorsal scale rows and belly orange or yellow......*Crotaphopeltis degeni*, Yellow-flanked Snake [p. 524]
1b Lower dorsal scale rows and belly not yellow or orange......2
2a Dorsal scales in 17 (rarely 19) rows at midbody, has vivid orange or red eyes......*Crotaphopeltis tornieri*, Tornier's Cat Snake [p. 526]
2b Dorsal scales in 19 (occasionally 21) rows at midbody, eyes not vivid orange or red......3
3a Usually white speckling on the body, males with short hemipenes, not extending beyond the 13th subcaudal......*Crotaphopeltis hotamboeia*, White-lipped Snake [p. 525]
3b No white speckling on body, males with long hemipenes, extending beyond the 16th subcaudal......*Crotaphopeltis braestrupi*, Tana Herald Snake [p. 523]

TANA HERALD SNAKE *Crotaphopeltis braestrupi*

IDENTIFICATION: A small dark snake, head sub-triangular, broader than the neck, pupils vertical, iris dark. Tail slender, 20% of total length. The dorsal scales are mostly smooth, faintly keeled towards the tail, in 19 rows at midbody, ventrals 156–174, subcaudals paired, 47–67 (high counts males), cloacal scale entire. Maximum size 67cm, average 30–55cm, hatchling size unknown but probably around 15–18cm. Colour quite variable; usually black, grey or grey-brown, sometimes with a dark horseshoe shape on the nape; the snout is sometimes yellow. Juveniles often red-brown. Ventrals cream or yellow. It closely resembles the White-lipped Snake *Crotaphopeltis hotamboeia*, but it is never white-speckled (as the White-lipped Snake usually is); see also the key.

Crotaphopeltis braestrupi Tana Herald Snake. Bio-Ken archive, Watamu.

Crotaphopeltis braestrupi

HABITAT AND DISTRIBUTION: Coastal thicket, floodplains, riverine forest and moist and dry savanna, from sea level to about 600m altitude. In Kenya, known from the north coastal plain, Malindi north to the Tana Delta, in Somalia known from the lower Juba and Webi Shebelle Rivers, the only inland record from Oddur.

NATURAL HISTORY: Poorly known. Nocturnal and terrestrial. Juba River specimens were found along irrigation canals, in swamps and gardens. It probably has the same threat display as the White-lipped Snake. The holotype was caught in a swamp, hunting tree frogs (*Hyperolius pusillus*), the diet is largely amphibians. The species was only recently described in 1985, although first collected in the Tana Delta in the 1930s.

YELLOW-FLANKED SNAKE/DEGEN'S WATER SNAKE
Crotaphopeltis degeni

Crotaphopeltis degeni

IDENTIFICATION: A small, dark, broad-headed rear-fanged snake. Eye large, pupil vertical. Body cylindrical, tail short, 10% of total length. Scales smooth, in 19 rows at midbody, ventrals 168–183, subcaudals paired, 28–40, cloacal scale entire. Maximum size about 60cm, average 30–50cm, hatchling size unknown. Black or blue-black above, ventrals vivid orange or yellow, this colour extending up onto the dorsal scales to a greater or lesser extent (Kenyan specimens seem to be more orange than Ugandan animals), and above the region of uniform yellow the scales may be yellow-tipped. Chin and lips yellow, orange, white or dark. The tongue is dark with a white tip.

HABITAT AND DISTRIBUTION: Moist savanna, in swamps and on floodplains, usually near water, at medium altitude, from 600–2,700m in East Africa. In Kenya occurs from Moiben, Iten and Sergoit on the Uasin Gishu Plateau south and west through Eldoret to the Lake Victoria shore, probably occurs right along the northern Lake Victoria shore but in Uganda recorded only from the north-western shore, from Entebbe round to Masaka, known also from the eastern shore of Lake Kyoga and Uganda shore of Lake Albert; probably more widespread but overlooked. There is a rather strange record from extreme southern Tanzania, over 1,000km further south, from the Rukwa Valley (not shown on map). Elsewhere, in southern South Sudan and western Ethiopia.

NATURAL HISTORY: Terrestrial and nocturnal (although Kisumu specimens observed moving by day), active in damp areas; hides by day under ground cover, in waterside vegetation or holes. It swims well and will hunt in water. Fairly slow-moving on land. It has

Crotaphopeltis degeni Yellow-flanked Snake. Left: Konrad Mebert, Entebbe. Right: Konrad Mebert, threat display.

a similar, but sometimes more dramatic threat display to the white lip; inflating the body, hissing, flattening the head to a triangle, flaring the lips, raising the forepart of the body in a sinusoidal curve and striking with a loud hiss. A clutch of six eggs recorded. Diet: amphibians, it might eat small fish.

WHITE-LIPPED SNAKE/WHITE-LIP *Crotaphopeltis hotamboeia*

IDENTIFICATION: One of East Africa's most common snakes, found virtually throughout East African savannas, a small, dark, broad-headed rear-fanged snake. Eyes prominent, pupil vertical, iris dark. The body is cylindrical, tail short, 11–15% of total length. The dorsal scales are mostly smooth (keeled posteriorly), in 19 (occasionally 21) rows at midbody, ventrals 139–181, subcaudals paired, 29–57, cloacal scale entire. Maximum size about 80cm, possibly larger, big specimens usually females, average 40–70cm, hatchlings 14–18cm. Black, grey or olive-green, the lips usually white, although sometimes with dark blotches or the head colour encroaching on the lip scales; the head scales iridescent. Nearly always has a series of crossbars, consisting of fine white dots, particularly visible when the snake inflates itself. Olive specimens often have a black patch on each temple. Ventrals cream or white, the snout sometimes yellowish. The tongue has a purple base, then darkens, and the tips are white. This snake is called the Herald Snake in southern Africa (its discovery was reported in a newspaper, the *Eastern Province Herald*) or the Red-Lip (some southern African specimens have red lips).

Crotaphopeltis hotamboeia

HABITAT AND DISTRIBUTION: Abundant almost throughout the moist savannas and woodland of East Africa, from sea level to 2,500m altitude, one of East Africa's most common nocturnal snakes, and highly adaptable, readily takes to suburbia and farmland so long as frogs and ground cover persist. Absent from much of Kenya's dry north and east, with isolated records from Mt Marsabit, Mt Kulal, Wamba and Lokitaung. No records from west-central Tanzania but this is probably due to undercollecting, it almost certainly occurs there. Elsewhere, south to southern South Africa, north to Eritrea, west to Senegal.

NATURAL HISTORY: Nocturnal and terrestrial, prowling at night, hunting for frogs. Hides by day under ground cover, in holes etc. Slow moving, but it has a fierce and distinctive threat display; it flattens the head into a triangular shape and flares the lips, lifts the

Crotaphopeltis hotamboeia White Lip. Left: Stephen Spawls, Lake Nakuru. Right: Stephen Spawls, Lolldaiga Hills.

forepart of the body off the ground, hisses and puffs, if the aggressor gets close it will strike and hiss loudly at the same time, if it connects it will sometimes bite viciously but it may strike to miss; the threat display is so effective that in Botswana its local name is the same as that for the cobra, *kake*. It looks dangerous, and is thus often needlessly killed. It has big, blade-like fangs, and can cause frightful lacerations to the frogs it is chewing on, but the venom has no effect on humans. White-lips lay 6–19 eggs, roughly 3×1.5cm, which take 50–80 days to hatch, depending on the temperature. Hatchlings collected in March, April and September in Nairobi; a Nairobi female laid seven eggs in December. Diet largely amphibians, but will take lizards; a Rwandan specimen took a gecko, a South African snake ate a fish; Nigerian specimens had eaten agamas, skinks and lizard eggs.

TORNIER'S CAT SNAKE *Crotaphopeltis tornieri*

IDENTIFICATION: A small to medium sized snake of the Eastern Arc forests, looks like a White-lip with red eyes. Head broad, body cylindrical, tail short, 10% of total length. 17 or 19 midbody scale rows; dorsal scales keeled posteriorly, and with increased size also anteriorly, Ventrals smooth, 146–177 in males, 144–186 in females, cloacal scale entire; subcaudals paired; 39–56 subcaudals in males, 35–54 in females. Largest male, 63cm; largest female 58cm; average 30–50cm, hatchlings about 13cm. Dorsum pale grey, plumbeous, to almost black, freshly sloughed anmals show iridescence; ventrum white or cream in juveniles, becoming a paler shade of the dorsal colour in adults, the pigment extending progressively further forward as size increases; underside of tail more or less densely pigmented. Lip colour white, cream or even reddish. Eyes red or orange-red with a vertical black pupil.

Crotaphopeltis tornieri

HABITAT AND DISTRIBUTION: This snake's preferred habitat is moist forest and its edges, as well as tracks and clearings, at low to medium altitude. It is known from the Eastern Arc forests of Tanzania, Ufipa and the Southern Highlands, including Usambara Mountains, Uluguru Mountains, Rubeho Mountains, Udzungwa Mountains and Nguru Mountains. Elsewhere, known from the Misuku Mountains, northern Malawi. Conservation Status: IUCN Least Concern. A near-endemic, forest-dependent species, so under threat from habitat loss.

Crotaphopeltis tornieri Tornier's Cat Snake. Michele Menegon, Madehani, southern Highlands.

NATURAL HISTORY: Terrestrial and nocturnal. Usually found coiled under logs, fallen trunks of trees or in leaf litter. Also found in insect tunnels and holes in forest at the edge of tea plantations. Nine small eggs taken in a 51cm-long female; females of lengths of 50, 51 and 53cm, had 10, 11, and 12 eggs, respectively. Seems to form breeding aggregations, or at least lays eggs communally. A juvenile, 78 large eggs and three adult females were found under a stone patio in December; an additional clutch of 68 eggs with full term embryos was found in Mufindi in February, indicating an extended breeding period.

MARBLED AND GREEN TREE-SNAKES
DIPSADOBOA

An African genus of slim-bodied, broad headed, large-eyed elongate tree snakes. They are rear-fanged, but are not known to be dangerous to humans. There are ten species in the genus, in west, central and southern Africa, mostly in forest but also moist savanna; seven species occur in our area. Maximum size about 1.4m but most are less than 1m. They are secretive and nocturnal, and little is known of their habits.

KEY TO EAST AFRICAN MEMBERS OF THE GENUS *DIPSADOBOA*

1a	Subcaudals single......2	6a	Dorsum not marbled in red-brown and yellow or white, cloacal scale scale divided, loreal does not enter eye...... *Dipsadoboa shrevei*, Shreve's Tree-snake (part, northern subspecies) [p. 529]
1b	Subcaudals paired......5		
2a	Midbody scale rows 19, underside of tail blackish......*Dipsadoboa weileri*, Black-tailed Tree-snake (part) [p. 531]		
2b	Midbody scale rows 17......3	6b	Dorsum marbled in red-brown and yellow or white, cloacal scale scale undivided, loreal enters eye......7
3a	Body slender, tail moderate to long, subcaudals 76–112 in males and 71–106 in females......*Dipsadoboa viridis*, Laurent's Green Tree-snake [p. 530]	7a	Dorsum with 38–57 dark-edged, pale crossbars, fading posteriorly in adults, tongue white......*Dipsadoboa aulica*, Marbled Tree-snake [p. 527]
3b	Body stout, tail short, subcaudals 54–78 in males, 52–77 in females.....4	7b	Dorsum with 58–82 transverse rows of confluent pale, yellow or white spots which persist through life, tongue white at the tip, with a black band before the fork......*Dipsadoboa flavida*, Cross-barred Tree-snake [p. 528]
4a	Upper labials and belly usually unpigmented, colour of the latter strongly contrasting with grey or black colour of tail, dorsum usually black......*Dipsadoboa weileri*, Black-tailed Tree-snake (part) [p. 531]		
4b	Upper labials and belly usually pigmented, blue or green, not contrasting with colour of tail, dorsum brown or green......*Dipsadoboa unicolor*, Gunther's Green Tree-snake [p. 530]	8a	Ventrals 221–229 in males, 212–221 in females, subcaudals 102–111 in males and 98–106 in females......*Dipsadoboa werneri*, Werner's Tree-snake [p. 532]
		8b	Ventrals 203–219 in males, 199–215 in females, subcaudals 74–91 in males and 74–96 in females......*Dipsadoboa shrevei*, Shreve's Tree-snake (part, southern subspecies) [p. 529]
5a	Midbody scale rows 17......6		
5b	Midbody scale rows 19.....8		

MARBLED TREE-SNAKE *Dipsadoboa aulica*

IDENTIFICATION: A small slender tree-snake from east and south-east African forests, just reaching southern Tanzania. The head is broad, eyes prominent. Dorsal scales are smooth and with single (rarely double) apical pits. Midbody dorsal scale rows 17; ventrals are 172–190 in males, 167–189 in females. The cloacal scale scale is usually entire. Subcaudal scales are paired, 75–97 in males, 74–86 in females. Largest male 72cm, largest female 62cm. The tongue is white. Juveniles and subadults are red-brown above with a pattern of 38–57 white to cream dark-edged cross-bands which begin just behind the head. The white pattern on the head gives a 'marbled' appearance from which the common name is derived. The light cross bands tend to disappear from behind with age, as the size of the individual increases. In larger, older specimens the dorsal colour is an almost uniform brown. The ventrum is white to cream with occasional red-brown speckles.

HABITAT AND DISTRIBUTION: Recorded from Liwale, southern Tanzania. Elsewhere, in low to medium altitude moist savanna, south through Mozambique to Zululand, west to southern Malawi, south-eastern Zimbabwe, eastern Transvaal and Swaziland. Conservation Status: Because it is widely distributed in woodland, it is not considered threatened.

Dipsadoboa aulica

Dipsadoboa aulica Marbled Tree-snake. Stephen Spawls, South Africa.

NATURAL HISTORY: This nocturnal, arboreal species hides during the day under loose bark or in hollow trees; it emerges from its hiding place at night to feed on geckos and skinks. When threatened, it flattens its head, puts the body into a coil and strikes with its mouth wide open. It lays 7–9 eggs, 2×1cm in December. Although rear-fanged, it is not dangerous to humans.

CROSS-BARRED TREE-SNAKE *Dipsadoboa flavida*

IDENTIFICATION: A small, slender tree-snake with a broad head and prominent yellow eyes. Two subspecies are known, the notes here apply to the subspecies *Dipsadoboa flavida broadleyi*. Dorsal scales smooth, in rows of 17 at midbody. Cloacal scale entire, ventrals in males 177–197, in females 170–194. Subcaudals paired, 90–100 in males, 79–100 in females. The tip of the tail is usually rounded. Largest male about 63cm, largest female 61cm, hatchlings about 20cm. Rich brown or red-brown, marbled with yellow, particularly on the front half of the body, becoming brown with narrow yellow cross bars towards the tail. The head is mottled yellow-brown, often with a brown eye stripe. Those from the northern Kenya coast are often very yellow. The underside is pinkish, white or cream. The tongue is white towards tip with a black base, or a black band near the fork.

Dipsadoboa flavida

HABITAT AND DISTRIBUTION: Coastal thicket, woodland and moist savanna, from sea level to around 1,200m. Occurs the length of the East African coastline, from the Somali border to the Ruvuma River. Inland records in Kenya up the Tana to the Primate Reserve, along the Galana to the Lali Hills, isolated records from Kibwezi and Umani Springs. In Tanzania, inland to Chanzuru and across the south-east corner, also on Zanzibar. Elsewhere, north to Somalia, south to northern South Africa.

NATURAL HISTORY: This species is nocturnal and partially arboreal; it is usually found

Dipsadoboa flavida Cross-barred Tree-snake. Michele Menegon, southern Tanzania.

under bark and in crevices. Feeds on geckos and frogs, particularly tree frogs. Specimens collected in the Arabuko-Sokoke Forest at night were mostly on bushes around a swamp, chasing tree frogs. When angry, they raise the forepart of the body off the ground, flatten the head and strike. They lay eggs.

SHREVE'S TREE-SNAKE *Dipsadoboa shrevei*

IDENTIFICATION: A medium to large, broad-headed, slim-bodied tree snake of moist savanna and forest. Two subspecies are recognised: a southern Tanzanian form, *D. s. shrevei*, Shreve's Tree-snake, and a northern Tanzanian form, from near Mt Kilimanjaro, *D. s. kageleri*, Kageler's Tree-snake. In the southern subspecies, the body is elongate, tail very long. The dorsal scales are smooth in 19 rows at midbody; ventrals 199–219, cloacal scale usually entire (rarely divided); subcaudals 71–99. Maximum size 1.2m, average 70–90cm, hatchling size 27–28cm. The northern form has scales in 17 rows at midbody, ventrals 191–195, and no specimens larger than 70cm have been found. Shiny black, dark grey or blue-grey above, grey below; the eyes of the northern subspecies are reticulated grey and rufous.

Dipsadoboa shrevei

Dipsadoboa shrevei Shreve's Tree-snake. Tomas Mazuch, Arusha.

HABITAT AND DISTRIBUTION: The southern subspecies *Dipsadoboa shrevei shrevei* is recorded from Mtene, Rondo Plateau, Tanzania, elsewhere west to the Congo, south-west to Angola through Zambia. The northern subspecies *Dipsadoboa shrevei kageleri* is apparently endemic to the xeric (dry) savanna at Sanya Juu in the vicinity of Mt Kilimanjaro, 1km from the border of its rainforest.

NATURAL HISTORY: The species appears to be arboreal; it prefers gallery forest or savannas with open woodlands. Presumably nocturnal. A bad-tempered species, when molested coils up and strikes repeatedly. Chameleons are an important food item, including *Trioceros melleri*, also takes birds. Lays eggs but no clutch details known. Hatchlings collected in January on the Rondo Plateau.

GUNTHER'S GREEN TREE-SNAKE *Dipsadoboa unicolor*

IDENTIFICATION: A green or blue-green tree snake of western East Africa. Head broad, eyes large, body slim. Midbody scales smooth, in 17 rows. Ventrals 180–211, cloacal scale entire; subcaudals single (first sometimes divided), 52–78. Maximum size about 1.3m, average 70cm–1m, hatchlings about 17cm. The colour changes with age. Juveniles and subadults brown or grey, at around 40cm length they change, the adults may be uniform green or green with a mottled blue-turquoise tail; the tail can have black spotting; the head is iridescent. Oddly, when very large, they may become shiny olive-black. The tongue is pink with white tip.

Dipsadoboa unicolor

Dipsadoboa unicolor Gunther's Green Tree-snake. Max Dehling, Rwanda.

HABITAT AND DISTRIBUTION: Medium to high altitude forest, from 1,500–3,000m. The species occurs along the Albertine Rift; down the western border of Uganda, into western Rwanda and Burundi. Isolated Uganda records include the Budongo and Mabira Forest, also Mahale Peninsula in Tanzania

BIOLOGY: Nocturnal and arboreal, and cold adapted, recorded active at temperatures capable of activity at night at a temperature of 7°C. It feeds on frogs; prey species recorded include *Phrynobatrachus*, *Arthroleptis*, *Hyperolius* and *Ptychadena*, toads and tadpoles. Presumably lays eggs.

LAURENT'S GREEN TREE-SNAKE *Dipsadoboa viridis*

IDENTIFICATION: A very slim and elongate tree snake, usually green. In our area, the subspecies *Dipsadoboa viridis gracilis* occurs, the following notes largely apply to this form only. The head is broad, the eyes large and prominent, the neck and body tremendously slim. Scales smooth, in 17 rows at midbody, ventrals 208–238, subcaudals single, 87–112. Colour in life various shades of green (the juveniles may be shiny grey or grey-green), with a curious coppery iridescence, specimens from further west can be quite blue, or have the dorsal scales blue-edged. The chin and throat are usually white (the lip scales may have fine blue edging); the underside is yellow. The irises are grey with fine black reticu-

Dipsadoboa viridis

Dipsadoboa viridis Laurent's Green tree-snake. Left: Kate Jackson, eastern Congo.

lations; the tongue pinky-grey with a pale tip.

HABITAT AND DISTRIBUTION: Medium-altitude forest in our area, elsewhere in low altitude forest, rarely over 1,200m. In East Africa, known only from Rwanda (Nyansa) and Burundi (without a definite locality). Elsewhere, extends west across the Congo; the nominate subspecies occurs from Gabon westwards to Liberia.

NATURAL HISTORY: Poorly known. Arboreal, the large eye and vertical pupil suggest that like other members of its genus, it is nocturnal. Occurs up to 1,400m but is more usually found in forest below 1,000m. It feeds on toads, frogs, and geckos. Copulating animals have been observed to leave trees and were observed in small depressions on the forest floor, partly covered by leaves. A 70cm female contained four eggs.

BLACK-TAILED TREE-SNAKE *Dipsadoboa weileri*

IDENTIFICATION: A moderate-sized broad headed, elongate tree snake of the West and Central African forests. Midbody dorsal scale rows 17 (rarely 19), ventrals 181–205, subcaudals single, 56–73. Largest male 96cm, largest female 88cm, average 60–80cm, hatchling size unknown. Adults grey or black dorsally yellow underneath, except the tail which is grey below.

HABITAT AND DISTRIBUTION: Medium-altitude forest, although up to 2,100m on Sabinyo volcano. In East Africa, known only from Nyansa in Rwanda and the Mabira Forest in Uganda, but occurs just across the Uganda border in the Semliki area of the Congo. Elsewhere, west to Nigeria, and across the Dahomey gap to Ghana, north to the Imatong Mountains, South Sudan, southwest to Angola. Status: Widely distributed and unlikely to be threatened by human activity.

Dipsadoboa weileri

NATURAL HISTORY: Virtually unknown. Probably arboreal. It has moderate sized eyes with vertical pupils, indicating a nocturnal way of life. Largely limited in distribution to equatorial forest and grassy fields, where it feeds mostly on forest frogs and toads. Pregnant females vary from 59–82cm and may contain 4–8 eggs.

Dipsadoboa weileri Black-tailed Green Tree-snake. Left: Eli Greenbaum, eastern Congo. Right: Eli Greenbaum, showing underside.

WERNER'S TREE-SNAKE   *Dipsadoboa werneri*

IDENTIFICATION: A medium to large broad-headed rufous, grey or yellow-green tree snake. Eyes large and prominent, pupil vertical, iris yellow or yellow-green. Body elongate. Midbody scale rows 19, ventrals 212–229, cloacal scale entire, subcaudals 98–111. Maximum size about 1.2m, average size probably around 60–90cm. Colour quite variable, adults may be green or rufous-brown, uniform or barred. The green colour form has each scale black edged so the snake looks finely striped black on greeny-yellow, the underside is vivid yellow, the posterior ventrals black edged. Rufous-brown individuals are either uniform yellow or white with yellow edging below, and the tail darkens; some specimens are pinky-brown with darker half-bars. Juveniles are vivid orangey-brown, white below, sometimes with a dotted black line on either side of the ventrals; although some juveniles are described as having much darker grey-brown spots which are arranged into bands.

Dipsadoboa werneri

HABITAT AND DISTRIBUTION: A Tanzanian endemic, known only from the medium altitude forest of the Usambara, Nguru and Udzungwa Mountains, at altitudes between 1,000–1,600m. Conservation Status: Not assessed by IUCN but probably vulnerable to forest alteration, destruction and fragmentation.

NATURAL HISTORY: Little is known of this apparently rare species. The long, thin body and the angulated ventrals indicate an arboreal way of life; the relatively large eye and vertical pupil indicate that it is nocturnal. Individuals from Nguru and Udzungwa mountains were active at night in trees and bushes, up to 4m above ground level. If threatened, it flattens the head and may elevate the body into a pile of sinusoidal curves. It is known to feed on the blade-horned chameleons of the Eastern Arc Mountains. Presumably lays eggs.

Dipsadoboa werneri Werner's Tree-snake. Top left: Michele Menegon, Usambara Mountains. Top right: Michele Menegon, Nguru Mountains. Bottom left: Lorenzo Vinciguerra, Usambara Mountains. Bottom right: Michele Menegon, Nguru Mountains.

HOOK-NOSED SNAKES *SCAPHIOPHIS*

An African genus of fairly big, robust, harmless terrestrial snakes, living mostly in burrows or soft soil and sand. They are well adapted to a subterranean life, having slit-shaped recessed nostrils, a huge tough projecting rostral scale and a mouth opening under the head, when the mouth closes it fits with valve-like precision into the flanged upper lip. The body is stout and muscular. They have small eyes, with round pupils. The mouth is black inside, the teeth are strangely tiny. They lay eggs. Two species are known, both occur in East Africa

KEY TO MEMBERS OF THE GENUS *SCAPHIOPHIS*

1a Midbody scales 19–25, ventrals 170–201 in males, 189–221 in females, dorsal scales broad, 1.5 times as long as wide.......*Scaphiophis albopunctatus*, Hook-nosed Snake [p. 534]

1b Midbody scales 27–31, ventrals 204–216 in males, 225–243 in females, dorsal scales narrow, more than twice as long as wide.......*Scaphiophis raffreyi*, Ethiopian Hook-nosed Snake [p. 535]

HOOK-NOSED SNAKE *Scaphiophis albopunctatus*

IDENTIFICATION: A fairly big, thick-bodied harmless snake. Head short and blunt, with a large projecting rostral scale. The last upper labial is huge. The eye is small, pupil round, iris golden or golden-brown. The tongue is black. The tail is fairly short, 14–18% of total length. The body scales are smooth, details above in key. Subcaudals 49–76. Maximum size about 1.6m, average 90cm–1.3m, hatchling size unknown but a Malindi juvenile was 34cm. Colour quite variable; may be shades of grey or pinky grey, uniform black, brown or orange with a broad grey vertebral stripe; often with a sprinkling of black or dark brown spots on the head and back; this may be extensive, ventrals orange, pink, cream or ivory-white. The juveniles are grey or blue-grey with white spots, in some specimens the head is greenish, this colour changes, initially to pinky grey and then to the adult phase around 60–80cm.

Scaphiophis albopunctatus

HABITAT AND DISTRIBUTION: Coastal thicket and forest, dry and moist savanna and woodland, from sea level to about 1,500m. The East African populations seem to be disjunct. Known from south and west of Lake Rukwa, north-central Tanzania from Dodoma north to Smith sound and central Serengeti, south-eastern Kenya from Meru National Park south through Ukambani into Tsavo and the coast, just reaches Mkomazi in northern Tanzania, it is also on the Victoria Nile (Kamali and Jinja), and north from Nimule on Uganda's northern border. Known from Uvira, at the northern tip of Lake Tanganyika, so might be

Scaphiophis albopunctatus Hook-nosed Snake. Top left: Stephen Spawls, Watamu. Top right: Tim Spawls, threat display, Watamu. Bottom left: Stephen Spawls, hatchling, Watamu. Bottom right: Anthony Childs, black specimen, Meru National Park.

in Burundi. Elsewhere, west to Ghana, south around the lower edge of the central African forest to Gabon.

NATURAL HISTORY: Poorly known. Spends much time underground, in holes, sometimes emerges and moves about in cover. Probably diurnal, it has been seen crossing roads in the day. Two specimens were found in a burrow at Gede. Occasional large concentrations of this species have been found in suitable refuges, a number were found in the foundations of a church in Jinja. Juveniles at Ol Donyo Sabuk were under rocks during the day. If threatened, these snakes (especially the juveniles) have a spectacular threat display; the lower jaw is dropped and spread, exposing the black mouth, the snake raises the forepart of the body into an elevated loop and if further molested makes a huge, lunging strike where most of the body comes off the ground, but it does not actually try to bite its target, it is pure bluff. If its bluff is called it will try to jam its head under its coils in the manner of other burrowing species. Forty-eight eggs were found in one female; eggs laid in October in Uganda, no further details known. It eats mammals, which it catches in holes and kills by slamming them against the walls of the tunnel. Captive specimens in Nairobi Snake Park refused to feed until provided with plastic pipes in which to hide; they then killed and ate mice there. The curious pleated black buccal membrane has lead to suggestions that they might also eat eggs.

ETHIOPIAN HOOK-NOSED SNAKE *Scaphiophis raffreyi*

IDENTIFICATION: A fairly big, thick-bodied harmless snake. Head short and blunt, with a large projecting rostral scale. The last upper labial is huge. The eye is small, pupil round The tongue is black. The tail is fairly short, 15–20% of total length. The body scales are smooth and elongate, details above in key. Subcaudals 55–68 in females, 72–79 in males. Maximum size about 1.5m, average 90cm–1.3m, hatchling size unknown. Colour dull grey, brown or red-brown, often with a sprinkling of black spots on the head and back; this may be extensive, ventrals orange, cream or white. The juveniles are grey or blue-grey with white spots, this colour changes to the adult phase around 60–80cm. Similar species: See our notes on the Hook-nosed Snake.

Scaphiophis raffreyi

HABITAT AND DISTRIBUTION: Moist and dry savanna and semi-desert, at low to medi-

Scaphiophis raffreyi Ethiopian Hook-nosed Snake. Left: Dietmar Emmrich, Ethiopia. Right: Dietmar Emmrich, juvenile, Ethiopia.

um altitude. In East Africa, known only from Moroto in Uganda, specimens from Baringo and the Kerio Valley are probably of this form. The Ethiopian type locality is at 2,500m, but this might represent the locality from which it was despatched rather than where it was caught. Also known from Eritrea, north-east and south-east South Sudan.

NATURAL HISTORY: Poorly known but probably similar to the hook-nosed snake. An Ethiopian juvenile was caught during ploughing. An adult attempted to cross a road around mid-morning, so it is probably diurnal. Presumably has a similar threat display to the hook-nosed snake. A Baringo specimen was dug out of a termite hill. It lays eggs, no clutch details known. Diet presumably rodents.

SUBFAMILY **GRAYIINAE**

A tiny subfamily containing four species of central African water snake, genus Grayia (although at least one undescribed species occurs in Cameroon). Two species occur in East Africa.

KEY TO THE EAST AFRICAN MEMBERS OF THE GENUS *GRAYIA*

1a Midbody scale count 15, tail long, over 40% of total length......*Grayia tholloni*, Thollon's Water Snake [p. 537]

1b Midbody scale count 17, tail shorter, less than 36% of total length......*Grayia smythii*, Smyth's Water Snake [p. 536]

SMYTH'S WATER SNAKE *Grayia smythii*

IDENTIFICATION: A big stocky water snake, conspicuously banded when young. The head scales are dark-edged. The eye is quite small, with a little round pupil (good field character), the iris is yellow or brown. Tail fairly long, over 30% of total length, but is often truncated, presumably by predators; males have longer tails. Dorsal scales smooth, in 17 rows at midbody; ventrals 149–167, subcaudals 90–102. Maximum size about 1.7m, although larger animals, up to 2.5m, reported from West Africa. Average size 1–1.5m, hatchling size unknown. The colour is quite variable, the ground colour varies from yellowish-brown to olive or black. The chain of broad dark blotches along the spine makes juveniles appear banded, as they become adult the bands usually fade, leaving a mottled brown or olive snake, with vague dark cross bars and blotches, often darkening towards the tail. The head is usually light brown or pinkish brown, chin and lips yellow, head scales black-edged. The ventrals are orange, yellow or cream, usually spotted, the spots forming thin longitudinal lines. Taxonomic Note: there is much debate about the spelling of the specific name, *smythii*, since it was named after Norwegian botanist Chris-

Grayia smythii

Subfamily: Grayiinae

ten Smith, it is suggested that *smithii* or *smithi* is more appropriate. The debate continues.

HABITAT AND DISTRIBUTION: In waterside vegetation or in the water, usually in savanna country. In our area, at altitudes between 600–1,200m. Found all round the shores of Lake Victoria, the Victoria Nile, Lakes Kyoga, Albert, Edward and the Albert Nile. Might be in some of the smaller Uganda rivers, but not recorded. Elsewhere, west to Senegal, north into southern South Sudan, south-west to northern Angola.

NATURAL HISTORY: Diurnal and aquatic, spends time in the water but shelters in thick waterside vegetation, will also live in stone jetties and other man-made waterside constructions, if there are places to hide. Moves slowly and ponderously on land, but swims well. Often found in fish traps, it eats adult frogs (particularly *Xenopus*, clawed frogs), fish and tadpoles. It is placid in temperament but will hiss if threatened and may threaten with an open mouth. Females become fertile at around 75cm length. They lay a clutch of 8–20 eggs, roughly 5×2.5cm, and they are laid in batches, 3–4 eggs at a site. Enemies include herons, Nile Monitors *Varanus niloticus* and other snakes.

Grayia smythii Smyth's Water Snake. Top: Kate Jackson, eastern Congo. Bottom: Kate Jackson, eastern Congo, hatchling.

THOLLON'S WATER SNAKE *Grayia tholloni*

IDENTIFICATION: A fairly large, brown, banded water snake. The eye is quite large, pupil round, iris yellow centrally. Original tail very long, over 40% of total length. Scales smooth, in 15 rows at midbody, ventrals 133–147, subcaudals 110–129. Maximum size about 1.1m, average 70–90 cm, hatchling size unknown. Colour shades of brown or grey-brown, the colour lightening down the flanks, the body with indistinct bands of light and dark scales, prominent in juveniles, they may fade in adults but do not disappear. Like many aquatic or semi-aquatic snakes, the lip scales are cream, yellow or orange with distinct black sutures. Belly yellow or cream, with black spots on the ends of the ventral scales.

Grayia tholloni

HABITAT AND DISTRIBUTION: Lakes, rivers and streams in moist savanna and woodland, at altitudes of 500–1,500m in our area. It is not so tied to big water bodies as Smyth's Water Snake, possibly because it is smaller, and may be found in quite small streams. In East Africa occurs right around the northern and most of the western shores of Lake

Grayia tholloni Thollon's Water Snake. Left: Konrad Mebert, Entebbe. Right: Harald Hinkel, Rwanda.

Victoria, and north along the Niles in Uganda up to Lake Albert; north on the Nile to the South Sudan border. Also occurs from the southern end of the Rwenzoris south into central Rwanda and most of Burundi, reaching Kibondo in north-west Tanzania. Elsewhere, south-west to Angola, north to Ethiopia, west to Nigeria.

NATURAL HISTORY: Not well known, but presumably similar to Smyth's Water Snake, i.e. diurnal, aquatic and terrestrial, sheltering in waterside vegetation. Presumably lays eggs, no details known, it might give live birth. It eats fish, frogs and tadpoles.

SUBFAMILY NATRICINAE

There is some debate about whether this group represents a full family or subfamily. There are around 40 genera and 230 species in the group, occurring in the Americas, Europe, and Asia; four genera occur in sub-Saharan Africa, one genus occurs in our area.

KEY TO EAST AFRICAN MEMBERS OF THE GENUS *NATRICITERES*

1a Dorsal scale rows 19 anteriorly, 17 posteriorly, usually five lower labials in contact with the anterior sublinguals......*Natriciteres olivacea*, Olive Marsh Snake [p. 539]

1b Dorsal scale rows 13-17 anteriorly, 13-15 posteriorly, usually four lower labials in contact with the anterior sublinguals......2

2a Ventrals 120-126, subcaudals 50-62, Pemba Island only......*Natriciteres pembana*, Pemba Marsh Snake [p. 541]

2b Ventrals 125-143, subcaudals 60-84, in southern Tanzania......*Natriciteres sylvatica*, Forest Marsh Snake [p. 540]

Subfamily: Natricinae

MARSH SNAKES *NATRICITERES*

An African genus of small harmless snakes, with small heads with round pupils, inhabitants of sub-Saharan Africa, the greatest diversity is in the moist savannahs of south-east Africa. They have fairly short tails which they can shed usually easily, but they are not regenerated, hence many specimens have damaged or truncated tails. Their taxonomy is debated; at present there are six species, three of which occur in our area.

OLIVE MARSH SNAKE *Natriciteres olivacea*

Natriciteres olivacea

IDENTIFICATION: A medium sized slender marsh-snake, with distinctive black-edged lip scales, of variable coloration, but usually with a broad dorsal stripe, usually dull green and sometimes hard to see but may be another bright colour. Body cylindrical, tail short, about 25% of total length, dorsal scales smooth, in 19 rows at midbody (rarely 17). Ventrals 120–153. Cloacal scale divided. Subcaudals 51–87. Maximum size about 55cm, average 25–35cm, hatchling size unknown. The dorsum may be various shades of brown, olive green, grey or blue black; all-black (melanistic) specimens recorded from the Congo. There is usually a vertebral band of scales five scales wide. This may be darker than the dorsal colour, or a rich maroon shade, Mount Meru (Tanzanian) specimens were bright green with such a maroon stripe. The dorsal stripe may also be bordered by a series of tiny white dots. The ventral scales are usually yellow with the ends bordered with olive, grey, or red; in other areas mauve or pale blue. The upper labials are usually yellow with distinctive black edging. The chin and throat are white. The ventrals are yellow with the ends widely bordered with olive, grey, red, mauve or pale blue.

HABITAT AND DISTRIBUTION: Coastal bush, moist savanna and in dry savanna around watercourses (e.g. from Lake Baringo); from sea level to over 2,200m. Occurs all around the Lake Victoria basin, and south through Rwanda and Burundi to the north-eastern shores of Lake Tanganyika. Also in southern and eastern Tanzania and on Zanzibar. Isolated populations in the savanna and forest around Mts Kilimanjaro and Meru, in Kidepo National Park in Uganda and on the north Kenya coast from the Arabuko-Sokoke Forest north to Witu. Elsewhere, west to Guinea, south-west to Angola, south to Zimbabwe. Conservation Status: IUCN Least Concern.

Natriciteres olivacea Olive Marsh Snake. Top left: Max Dehling, Rwanda. Right: Stephen Spawls, Ngurdoto Crater.

NATURAL HISTORY: A diurnal species which lives at the edge of streams and other water sources. Slow-moving and inoffensive. Lays a clutch of 4–11 eggs, roughly 2×1.5cm. Diet: amphibians and fish. Members of this genus often seem to be predated upon by birds such as Hammerkops *Scopus umbretta*. Their tips of their tails are easily 'dropped' (autotomised) and many animals are found in nature with the tip missing.

FOREST MARSH SNAKE *Natriciteres sylvatica*

IDENTIFICATION: A small dark olive or grey-black snake, often with a vertebral stripe, short head, pupil round. Body cylindrical, tail short, often truncated, like the previous species, if undamaged then 25–30% of total length. Dorsal scales in 13–17 rows anteriorly, 13–15 posteriorly. Ventrals 125–143, subcaudals 60–84. Maximum size about 41cm, average 25–35cm, hatchling size unknown. There is often a faint indication of a pale yellow or cream collar on the upper surface of the neck area. Colour very variable, may be olive, chestnut brown, shiny blue-black, light or dark grey, some specimens have a maroon, dull green or black dorsal stripe. Upper labials are nearly always yellow with distinctive black sutures (borders between the scales). The ventrum is yellow, with the outer ends of the ventral and subcaudal scales a dark grey. Taxonomic Note: formerly regarded as subspecies of *Natriciteres variegata*.

Natriciteres sylvatica

Natriciteres sylvatica Forest Marsh Snake. Left and Right: Michele Menegon, southern highlands, Tanzania.

HABITAT AND DISTRIBUTION: The species is found at the edge of montane and lowland evergreen forest, and in formerly forested areas where moist conditions persist, from about 600–2,000m altitude. It is a forest-dependent species. Occurs in forest areas of southern Tanzania; from Katavi and Tatanda south along the border, and north to Kilombero and the Udzungwa Mountains. Elsewhere, south through Mozambique, Malawi, and eastern Zimbabwe to northern Zululand. Conservation Status: Widely distributed, but dependent on continued existence of forests.

NATURAL HISTORY: Poorly known. Diurnal and terrestrial, lays eggs. Probably feeds on frogs, especially tree frogs, and small ranids.

PEMBA MARSH SNAKE Natriciteres pembana

IDENTIFICATION: A small dark olive snake, on Pemba Island, with a short head, pupil round. Body cylindrical, tail short often truncated, if undamaged then 25–29% of total length. Dorsal scales in 17 rows anteriorly, 15 posteriorly. Ventrals 120–126, cloacal scale scale divided, subcaudals 50–62. Maximum size about 28.5cm, average 20–25cm, hatchling size unknown. Olive or red-brown above, with a brown or dull green, sometimes dark-edged vertebral stripe, the back extensively spotted with white. Lip scales yellow or cream with black edging. Underside yellow.

Natriciteres pembana

HABITAT AND DISTRIBUTION: Endemic to Pemba island, where it lives in marshy areas. Conservation Status: IUCN Least Concern. A very limited range but probably able to adapt well to agriculture, especially rice farms, provided they are not polluted, as it feeds on amphibians, which are sensitive to pollution.

NATURAL HISTORY: Poorly known. Diurnal and terrestrial, lays eggs. Diet amphibians, might take small fish.

Natriciteres pembana Pemba Marsh Snake. Frank Glaw, Pemba.

Dendroaspis polylepis
Black Mamba
Johan Marais

SUB-ORDER **SERPENTES**
FAMILY **ELAPIDAE**
COBRAS, MAMBAS AND RELATIVES

A family of dangerous snakes, many of international medical significance, with short poison fangs at the front of the upper jaw. Over 360 species and over 50 genera are known world-wide in tropical regions. There are none in Europe but there has been a remarkable radiation of elapids in Australia, where they are the largest group. Members of the family include the mambas, cobras, king cobra, kraits, sea snakes, coral snakes, African garter snakes and the taipan.

KEY TO THE EAST AFRICAN ELAPID SUBFAMILIES

1a	Tail shaped like an oar, living in the sea or stranded on the foreshore...... subfamily Hydrophiinae, sea snakes [p. 565]	1b	Tail not shaped like an oar, not living in the sea......subfamily Elapinae, land elapids [p. 542]

SUBFAMILY **ELAPINAE**

At present there are around 300 species of elapid, in around 40 genera. There are around 40 species of land elapid in Africa, in seven genera. Fifteen species in four genera occur in East Africa, one is endemic. They range from large

Family Elapidae: Cobras, Mambas and relatives

(Black Mamba *Dendroaspis polylepis*) to small (garter snakes). All East African species have round pupils, smooth scales, lack a loreal scale and lay eggs. Most are large snakes, reaching more than 1.5m and are highly dangerous, the exception being the small garter snakes, which are not known to have caused deaths. The key below will aid technical identification of an elapid, but they are usually identifiable by eye.

KEY TO EAST AFRICAN ELAPID GENERA

1a Midbody scales in 13 rows, adults small, less than 80cm, juveniles always with broad bands......*Elapsoidea*, garter snakes [p. 543]
1b Midbody scales in more than 13 rows, big, adults more than 1m, juveniles not banded or with narrow bands......2

2a Three preoculars, head long and thin......*Dendroaspis*, mambas [p. 561]
2b One to two preoculars, head more square......3

3a Dorsal scales in 15 (rarely 17) rows, glossy black above, black and yellow below, hood small, eye huge......*Pseudohaje*, one species only, *Pseudohaje goldii*, Gold's Tree Cobra [p. 559]
3b Dorsal scales in 17–27 rows, not black and bright yellow, hood large, eye not huge......*Naja*, true cobras [p. 548]

GARTER SNAKES *ELAPSOIDEA*

A genus of small, ground dwelling and burrowing venomous elapids (not to be confused with the harmless North American garter snakes, genus *Thamnophis*). Ten species occur in sub-Saharan Africa, four in our area. They have small blunt heads and short tails. Most have vivid dorsal bands when juvenile but some lose the bands on growth; the banding may serve to stop other garter snakes eating them. They are venomous, they have front fangs, but are reluctant to bite, no fatalities have been recorded from African garter snake bites and no generic antivenom is available; a bite should be treated symptomatically. Due to the colour change as they grow, we provide two keys below; one for juveniles, the other for adults. They are equivocal in parts, but these snakes have mutually exclusive ranges in East Africa; so the locality (if known) may be used to distinguish the species.

KEY TO JUVENILE EAST AFRICAN MEMBERS OF THE GENUS *ELAPSOIDEA*

1a Black crossbars narrower than grey bands, head orange, (adult bands fade to fine white lines and then disappear), in hill forest of north-east Tanzania and the Shimba Hills......*Elapsoidea nigra*, Usambara Garter Snake [p. 547]
1b Black crossbars wider than pale bands, head not orange, not in hill forests of north-east Tanzania or Shimba Hills......2

2a Eight to 21 light bands, (adults with that number of fine paired white lines)......3
2b Juveniles (and adults) with 17–34 light bands, bands not faded in adults......*Elapsoidea loveridgei*, East African Gar-

ter Snake [p. 546]

3a Range north-western Uganda......*Elapsoidea laticincta*, Central African Garter Snake [p. 545]

3b Range south-eastern and north-western Tanzania......*Elapsoidea boulengeri*, Boulenger's Garter Snake [p. 544]

KEY TO ADULT EAST AFRICAN MEMBERS OF THE GENUS *ELAPSOIDEA*

1a Tail 7% or less of total length,...... *Elapsoidea nigra*, Usambara Garter Snake [p. 547]
1b Tail usually more than 7% of total length......2
2a Specimens of more than 20cm length distinctly banded, no fading of the band centre...... *Elapsoidea loveridgei*, East African Garter Snake [p. 546]
2b Specimens of more than 20cm length not distinctly banded, band centre faded......3
3a Adults small, 25–40cm (maximum 56cm), belly yellowish or dark orange...... *Elapsoidea laticincta*, Central African Garter Snake [p. 545]
3b Adults large, 40–70cm (maximum 76cm), belly dark grey...... *Elapsoidea boulengeri*, Boulenger's Garter Snake [p. 544]

BOULENGER'S GARTER SNAKE *Elapsoidea boulengeri*

IDENTIFICATION: A small, glossy, snake, the head short, slightly broader than the neck, eyes set well forward, with round pupils. Body cylindrical, tail fairly short, 6–9% of total length. Scales smooth, in 13 rows at midbody, ventrals 140–163, subcaudals 18–27 in males, 14–22 in females. Maximum size 76.6cm, average size 40–70cm, hatchling size 12–14cm. Juveniles have 8–17 white, cream or yellow bands on a black or chocolate-brown body, the head is usually white with a black Y-shaped central mark. As the snake grows the bands fade to grey, with white edging, then to a pair of white irregular rings, finally they fade completely leaving a uniformly dark grey or black adult, sometimes a few scattered white scales remain. Chin and throat white, the rest of the belly dark grey.

Elapsoidea boulengeri

Elapsoidea boulengeri Boulenger's Garter Snake. Top: Bill Branch, adult, southern Africa. Bottom: Bill Branch, juvenile.

HABITAT AND DISTRIBUTION: In moist savanna, from sea level to 1,500m. In our area there are two (at present) disjunct populations in Tanzania; in the west from Kibondo south to the western side of

Lake Rukwa, and also in the south-east corner, from Kilwa south and west to the Ruvuma River at Tunduru. Elsewhere, in eastern DR Congo, south to northern South Africa and Botswana.

NATURAL HISTORY: A burrowing snake, living in holes, it emerges on warm wet nights and moves around, looking for prey or a mate. Slow-moving and inoffensive, doesn't usually attempt to bite if picked up, but may hiss, inflate the body and jerk convulsively. Clutches of 4–8 large eggs recorded, in January and February in southern Tanzania; a hatchling was collected in June. Diet: other snakes (including its own species), lizards, small rodents and frogs; Zimbabwean snakes had eaten a rubber frog (*Phrynomerus*) and a rain frog (*Breviceps*), both of which have poisonous skin secretions; Tanzanian specimens recorded eating snout burrowers (*Hemisus*) and squeakers (*Arthroleptis*). It has a neurotoxic venom which is not life-threatening. One recorded bite caused swelling, pain and transient nasal congestion.

CENTRAL AFRICAN GARTER SNAKE *Elapsoidea laticincta*

IDENTIFICATION: A small, glossy, snake with a short head; eyes fairy large, pupils round. Body cylindrical, tail fairly short, 6–9% of total length. Scales smooth, in 13 rows at midbody, ventrals 140–150, subcaudals 17–25 (more than 21 in males). Maximum size about 56cm, average size 25–40cm, hatchling size unknown, probably about 12cm. Young specimens have 8–17 pale brown or reddish-brown bands on a black body, the bands about half as wide as the black spaces between. The belly is brownish yellow or dull orange. As the snake grows the bands darken in the centre, but the scales at the outer edges of the bands remain pale, so that adults appear to be black, with a series of fine white double rings along the body; occasional specimens lose all their markings.

Elapsoidea laticincta

HABITAT AND DISTRIBUTION: In moist and dry savanna and woodland (and in forest/savanna mosaic at Banangai in South Sudan), at 500–700m altitude. In our area, known

Elapsoidea laticincta Central African Garter Snake. Left: Stephen Spawls, Central African Republic. Right: Stephen Spawls, preserved juvenile, South Sudan.

only from Nimule on the South Sudan-Uganda border, also known from Mahagi Port, just inside the DR Congo on Lake Albert. Elsewhere, west across the top of the forest to western Central African Republic.

NATURAL HISTORY: Poorly known. Terrestrial. Fairly slow moving. Nocturnal, emerging at dusk, hides in holes or under ground cover during the day. Will prowl on the surface at night, especially on damp nights, after rain, sometimes active in the dry season. Totally inoffensive, and many specimens will let themselves be handled freely, making no attempt to bite, but if teased or molested may flatten and inflate the body, showing the bands prominently, and may lift the front half of the body off the ground and jerk from side to side. Only likely to bite if restrained, however. Lays eggs, but no information on size and number. Diet includes snakes, smooth-bodied lizards and frogs.

EAST AFRICAN GARTER SNAKE *Elapsoidea loveridgei*

IDENTIFICATION: A small, glossy garter snake, from highland East Africa. The head is short, eyes set well forward, with round pupils. Body cylindrical, tail fairly short, 6–9% of total length. Scales smooth, in 13 rows at midbody. There are three subspecies. The typical, eastern form (*Elapsoidea loveridgei loveridgei*) occurs in central Kenya (east of the Rift Valley) and northern Tanzania and has 16–23 bands on the body, 152–163 ventrals and 19–30 subcaudals. The many-banded central form (*Elapsoidea loveridgei multicincta*) occurs in western Kenya and most of Uganda (not the south-west); it has 23–34 bands or pairs of white transverse lines, 151–171 ventrals and 17–26 subcaudals. Collet's East African Garter Snake (*Elapsoidea loveridgei colleti*) occurs in south-west Uganda and Rwanda (and probably northern Burundi); it has 18–25 white bands, 161–170

Elapsoidea loveridgei

Elapsoidea loveridgei East African Garter Snake. Top left: Michele Menegon, Arusha. Top right: Stephen Spawls, white-banded phase. Bottom left: Lorenzo Vinciguerra, northern Tanzania, yellow-banded phase. Bottom right: Tim Spawls, hatchling, Kitengela

ventrals, and 18–26 subcaudals, and seems in parts of its range to be associated with the mountains of the western rift valley. Maximum size is about 65cm, average size 30–55cm, hatchling size unknown, probably about 12–14cm. Juveniles vividly marked with 19–36 narrow white, red, pinkish or white edged grey-brown bands on the grey or black body, occasional individuals have yellow bands. Unlike many other garter snakes, the bands usually persist in the adults and the band colour may vary a lot, some adults have vivid pink or red bands (particularly those from east of the Rift Valley in Kenya and northern Tanzania), others may have white edged black bands, in others the centre of the band will darken, leaving two fine white bands. Very rarely specimens may become totally dark grey or black.

HABITAT AND DISTRIBUTION: Mid-altitude woodland, moist savanna and grassland, from 600–2,200m. Occurs from Kijabe and Nairobi south into the Crater Highlands and vicinity of Kilimanjaro, west into the Serengeti. Widespread in high western Kenya, the northern lakeshore in Uganda and west from Lake Victoria to the border, south into Rwanda. Reported from Burundi, but no locality given, not shown on map. Apparently isolated populations at Moroto in Uganda; on Marsabit, north and west of Mt Kenya and the Taita Hills in Kenya, and Sekenke in central Tanzania. Elsewhere, this species occurs in southern Somalia and Ethiopia and eastern and north-eastern DR Congo.

NATURAL HISTORY: Terrestrial. Fairly slow moving. Nocturnal, emerging at dusk, hides in holes or under ground cover during the day. Believed to spend much of its activity time in holes, but will prowl on the surface at night, especially on damp nights, after rain, which acts as an activity trigger, a number were found dead on the road just south of Nairobi in December following a heavy, unseasonable storm the previous night. May emerge in the day, a Nairobi specimen was crawling through the grass in the afternoon, following a rainstorm. Inoffensive, many specimens will not attempt to bite when handled gently, but if molested may flatten and inflate the body, showing the bands prominently, and may lift the front half of the body off the ground and jerk from side to side; if restrained may try to bite. Lays eggs, but no information on size and number; hatchlings collected in Nairobi in April, May and July. Diet includes small snakes, smooth-bodied lizards and frogs, reptile eggs and rodents. This snake is venomous; a bite caused local pain, swelling and pain in the lymph nodes.

USAMBARA GARTER SNAKE *Elapsoidea nigra*

IDENTIFICATION: A small, glossy, snake, the head blunt and short (3–4% of total length, less than most other garter snakes), eyes set well forward, with a yellow iris and round pupils. Body cylindrical, tail short, 5–7% of total length. Scales smooth, in 13 rows at midbody, 151–168 ventrals, 13–24 paired subcaudals (males usually more than 18, females less than 18. Maximum size about 60cm, average size 30–50cm, hatchling size unknown. Juveniles are beautiful, with an orange head, the first three or four body bars are orange with black in between, the light bars then become grey with white edging. There are a total of 18–23 light bands on the body and two or three bands on the tail; the bands are the same width as the spaces between. The pale bands darken

Elapsoidea nigra

Elapsoidea nigra Usambara Garter Snake. Left: Lorenzo Vinciguerra, Usambara Mountains. Right: Stephen Spawls, juvenile, Usambara Mountains.

as the snake grows, so that adults simply show a series of thin white bands, where the white-edged scales that marked the outer edges of the bands are all that remain of the bands, in some the bands seem to disappear. However, if a uniformly dark adult is molested it inflates the body, exposing the concealed white scale tips.

HABITAT AND DISTRIBUTION: An East African endemic. Mostly in high evergreen forest, but possibly in moist savanna as well, at altitudes of 200–1,900m. Confined to north-east Tanzania and extreme south-east Kenya, known from the Shimba Hills in Kenya, and Tanga, the North Pare Mountains, Usambara, Nguru, and Uluguru mountains in Tanzania. Conservation Status: IUCN Endangered B1ab(iii). At risk as it inhabits a restricted area of forest, but might cope with deforestation.

NATURAL HISTORY: Poorly known. Terrestrial, slow-moving, burrowing, living and hunting in holes, soft soil and leaf litter, taking shelter in vegetation, rotten logs, under rocks, etc. Sometimes active in the day following storms, but mostly hunts at night, emerging at dusk, especially on damp nights after rain, sometimes active in the dry season. Inoffensive and gentle, but if molested it hisses, flattens the body and lashes about, sometimes raising coils off the ground. Only likely to bite if restrained forcibly. Lays 2–5 eggs, roughly 1×4cm, females with eggs in their oviducts captured mostly between October and December. Eats mostly (or perhaps exclusively) caecilians (legless, worm-like amphibians). Nothing is known of its venom.

TYPICAL COBRAS NAJA

Famous for their ability to spread a hood, worshipped and feared in the ancient kingdoms of Egypt, the cobras, *Naja*, are a genus of just under 30 species of highly venomous and medically significant snakes, some very large, found in Africa and Asia. Seventeen species occur in Africa, seven in our area. Four of these have an unusual defence; they can spit (spray might be a better word) venom to a distance of 2–3m. Their taxonomy has changed significantly in recent years; new species have been identified, and the DNA of the water cobras, originally in their own genus, *Boulengerina*, shows they cluster within *Naja*, true cobras; thus the Banded Water Cobra *Boulengerina annulata*, has become *Naja annulata*.

KEY TO THE EAST AFRICAN COBRAS OF THE GENUS *NAJA*

1a Hood narrow, dorsal scales not oblique, obvious black bands across back of neck.....*Naja annulata*, Banded Water Cobra [p. 549]
1b Hood broad, dorsal scales oblique, no obvious black bands (or at most only one) across back of neck......2

2a Upper labials separated from the eye by subocular scales......*Naja haje*, Egyptian Cobra [p. 551]
2b One or more upper labials enter the eye......3

3a Sixth upper labial largest and in contact with the postocular scales, a single preocular......*Naja melanoleuca*, Forest Cobra [p. 552]
3b Sixth upper labial not the largest, not in contact with the postocular scales, usually two preoculars......4

4a Dorsal surface of entire body usually rufous, red or orange (very rarely grey or yellow), no dorsal pattern save a single black band that encircles the neck (this may be faint or invisible in large specimens), usually 25–27 scale rows at midbody......*Naja pallida*, Red Spitting Cobra [p. 558]

4b Dorsal surface of the entire body not rufous, red or orange, dorsal colour may vary, various bars present on throat but these do not encircle the neck, often under 25 scale rows at midbody......5

5a Usually brown, olive or grey, never red or deep pink on the throat, underside usually light at midbody......6
5b Usually black, sometimes grey, often with red or pink or yellow bars interspersed with black on the throat, underside usually black at midbody...... *Naja nigricollis*, Black-necked Spitting Cobra [p. 555]

6a Usually a single broad throat bar which is never black in adults, lip scales never black-edged, reaches 2.4m length, in dry northern Tanzanian, dry northern, eastern and coastal Kenya......*Naja ashei*, Ashe's Spitting Cobra [p. 557]
6b Usually several throat bars which are usually black, lip scales usually black-edged, reaches 1.5m in length, in savanna of southeastern Tanzania and on the big islands......*Naja mossambica*, Mozambique Spitting Cobra [p. 554]

BANDED WATER COBRA/STORM'S WATER COBRA
Naja annulata

IDENTIFICATION: A big water-dwelling cobra, in Lake Tanganyika. It is a large, heavy-bodied snake (weighing up to 4kg), with a broad flat head, a medium-sized dark eye with a round pupil. The body is cylindrical, the tail is long, about a quarter of total length. The scales are smooth and glossy, in 21–23 rows at midbody. Ventrals 192–218, paired subcaudals 67–78, cloacal scale entire. Maximum size about 2.7m, average size 1.4–2.2m. Size of hatchlings unknown, but a juvenile from Lake Tanganyika measured 43cm. Colour quite variable. Specimens from south and central Lake Tanganyika are grey-brown, brown or yellow-brown, darkening towards the tail, the tail is glossy black. There are two or three narrow but distinct black bars on the back of the neck

Naja annulata

(sometimes rings), the belly is pale cream or yellow. The head scales may be black-edged, sometimes there is a black blotch on the side of the neck. As one moves north and west through the range, the ground colour changes and the number of bands increase; those from the northern part of the lake may have seven or more bands or blotches on the neck and are brown in colour. Those from eastern DR Congo are warm brown or orange-brown, bright orange below, and the entire body is banded with broad jet-black bands that go all the way around the body, this colour form might occur in Rwanda. Taxonomic Note: The

Lake Tanganyika form has been described as a subspecies, Storm's Water Cobra, originally *Boulengerina annulata stormsi* now *Naja annulata stormsi*.

HABITAT AND DISTRIBUTION: Associated with lakes and rivers, in forest and well-wooded savanna, usually where there is enough waterside cover to conceal a big snake, but will venture out onto open beaches and sand bars. In our area Storm's Water Cobra occurs in and on the shores of Lake Tanganyika; there is a single record from Lake Mweru on the Zambian/Congo border. Might also be in associated rivers. Elsewhere, the nominate subspecies occurs west to southern Cameroon, south-west to Angola. Conservation Status: Widely distributed, not likely to be threatened by human activities, unless severe pollution were to affect the lake ecosystem and have negative effects on its fish and other prey.

Naja annulata Banded Water Cobra. Left: Michele Menegon, Mahale, Lake Tanganyika. Right: Wolfgang Wuster, captive.

NATURAL HISTORY: An aquatic snake, spends most of its time in the water, hunting fish, by day and by night. May bask in the morning and later afternoon. Moves somewhat ponderously on land, but a superb, fast swimmer, spends much time under water, recorded as staying down more than 10 minutes and diving to 25m depth. Groups of these cobras can be seen in the lake at a distance with their nostrils just out of water, seemingly hanging suspended vertically in the water. Hunting snakes also investigate recesses like rock cracks, mollusc shells, underwater holes etc, looking for concealed fish. Small fish captured this way are rapidly swallowed, without waiting for the venom to take effect. The Congo subspecies is often caught in fish traps in small rivers, coming downstream, leading fishermen to suggest it may follow fish movements from large rivers to small rivers at night, and return at dawn. When not active, tends to hide in shoreline rock formations, in holes in banks or overhanging root clusters, or in holes of waterside trees, in such refuges it may live for long periods if not disturbed. Often makes use of man-made cover such as jetties, stone bridges and pontoons to hide among, will also use buildings and fish-drying huts. Juveniles have been found under boats or riverside debris, logs etc. When approached in water, simply swims away, and on land will attempt to escape into water, but if cornered will rear, spread a prominent hood and demonstrate with open mouth. They lay eggs, clutches of 22–24 eggs laid in August and September. A 48cm juvenile was captured in July. Diet: only fish recorded in wild snakes but captives took amphibians. Note: this is a venomous snake. Although bites are unlikely (it usually moves rapidly away from humans), venom analysis indicates it is a highly potent neurotoxin, with a low lethal dose, and thus should be treated with caution. South African polyvalent antivenom was effective in laboratory tests, although not prepared specially for this snake.

EGYPTIAN COBRA *Naja haje*

IDENTIFICATION: A big thick-bodied cobra, with a broad head, distinct from the neck, eye fairly large, pupil round. This is the only species of cobra with a subocular scale, separating the eye from the upper labials. The body is cylindrical, the tail fairly long, 15–18% of total length. Scales smooth, in 21 (very occasionally 19) rows at midbody. Maximum size 2.5m (possibly larger), average size 1.3–1.8m, hatchlings 25–30cm. Colour very variable. In East Africa the dorsal colour is usually brown, red-brown, greyish-brown, sometimes black or yellow (especially Ugandan animals). Yellow or cream underneath, often with a broad grey or brown throat bar, 6–20 scales deep, visible when the hood is spread. There are often irregular brown or yellow blotches and speckles on the neck, back and underside. Lips yellow, often a dark patch under the eye.

Naja haje

Naja haje Egyptian Cobra. Top left: Stephen Spawls, Amboseli. Right: Anthony Childs, juvenile, Nairobi National Park. Centre left: Stephen Spawls, sub-adult, Meseraní. Bottom left: Vincenzo Ferri, Uganda phase, Mbale.

Some brown specimens are patterned with yellow scales on the back, and the cream ventral colour may extend onto the flanks in irregular blotches (especially specimens from southern Kenya and northern Tanzania. Juveniles may be yellow, grey, orange or reddish, with a distinctive dark ring on the neck, some juveniles have fine dark back bands; others have yellow bands, with a fine dark ring in the centre of the band.

HABITAT AND DISTRIBUTION: Mostly in moist and dry savanna, woodland (never forest) or grassland, at medium altitude (800m to about 1,600m) in East Africa. Two main populations in East Africa. Widely distributed in north-west, north-central and south-eastern Uganda, east from Lake Albert and the Albert Nile to the Kenya border, south to the environs of Mt Elgon and Bungoma in Kenya. A population also occurs from Naivasha south round the Ngong Hills, east to Nairobi and Thika, south through Kajiado to Amboseli, around Kilimanjaro and into Tsavo West, thence west in Tanzania, through Arusha district, north into the central Serengeti and west past the lakes to Shinyanga. There is also an isolated record from Kilosa, west of Morogoro, and an isolated population reported from around Samburu National Reserve, the western Uaso Nyiro and eastern Laikipia. Elsewhere, sporadically distributed over huge areas of Africa; north to the Mediterranean coast west to Senegal. Its fragmented range is curious, possibly it has been displaced from some areas by more successful snakes, but might be present and simply overlooked.

NATURAL HISTORY: Mostly terrestrial. Clumsy but quick-moving. Active by day and night, often basks in the day, but will hunt at any time. Often lives in termite holes, rock fissures, hollow trees or holes; in Kenya known to share its refuge with Black Mambas. Quick to rear up (usually only to 30cm or so, but sometimes rears quite high) and spread a broad hood when threatened; if very angry will hiss loudly, rush and strike at aggressor. Sometimes shams death. Lays up to 23 eggs, averaging 50–60×30–35mm; a Kitengela snake laid 23 eggs in early March, and these hatched at the end of May. Eats a wide range of prey, fond of toads but will take eggs (a notorious chicken run raider) small mammals and other snakes. Note: this is a highly dangerous snake, and kills many people every year in Africa. The neurotoxic venom causes progressive descending paralysis, starting with ptosis (drooping eyelids) paralysis of the neck and head muscles, proceeding to paralysis of the lungs. A bite is a medical emergency and will probably need treatment with a respirator and antivenom in a well-equipped hospital.

FOREST COBRA *Naja melanoleuca*

IDENTIFICATION: A big, fairly thick-bodied cobra, with a big head and a large dark eye with a round pupil. Body cylindrical, tail fairly long and thin, 15–20% of total length. Scales smooth and glossy, in 19–21 rows at midbody (sometimes 17 in eastern Tanzania and Zanzibar), ventrals 197–226, subcaudals 55–74. Maximum size about 2.7m, average size 1.4–2.2m, hatchlings 25–40cm. Several colour forms exist. Those from the western forest or forest fringe are glossy black, sometimes with white spotting, the chin, throat and anterior part of the belly are white, cream or pale yellow, with broad black cross bars and blotches, although in some specimens most of the underside is black. The sides of the head are strikingly marked with black and white, giving the

Naja melanoleuca

impression of vertical black and white bars on the lips. The second colour form, from the coastal plain of east Africa, inland to Zambia and the south-eastern Albertine Rift is brownish or blackish-brown above, sometimes spotted darker brown, paler below, the belly is yellow or cream, heavily speckled with brown or black, specimens from the southern part of the range have black tails. Mt Kenya snakes are yellow-brown anteriorly, becoming dark or black towards the tail, and the head lacks the traditional black lip edging. Another colour form, banded yellow and black, is confined to West Africa. Taxonomic Note: The different colour forms (forest black, coastal brown) might receive recognition as full species in the future, research continues.

HABITAT AND DISTRIBUTION: Forest, woodland, coastal thicket, moist savanna and grassland, from sea level to over 2,500m altitude. Widespread in Uganda, except the dry north-east. Widespread in Rwanda wherever there is water, especially in the vicinity of Lake Kivu and around the many wetlands of Bugesera, south of Kigali, but also in the high wet forests such as Nyungwe and Cyamudongo, even above the fog level. Equally widespread in Burundi, save the south-west strip. Occurs in western Kenya, from the southern border with Uganda eastwards to the western edge of the rift, north beyond Eldoret but not recorded on the Mau. At Nakuru it has descended the escarpment and is found around Rongai in grassland, on the slopes of Mt Menengai and the escarpment at the western edge of the Laikipia Plateau, at Ol Ari Nyiro Ranch. The brown form occurs in the forests of south-east Mt Kenya and the Nyambene Hills, and on the coast; known from the forest of the lower Tana River, and from the Arabuko-Sokoke Forest south through Kilifi and the Mombasa hinterland to Diani Beach. In Tanzania, some isolated records from around Mt Kilimanjaro (Ngorongoro, Arusha, and Kitobo Forest on the Kenyan eastern side), occurs from the Usambaras south along the entire coastal plain to the Ruvuma River, and inland in the extreme south to Liwale; also recorded in the Udzungwa Mountains. It is found on Zanzibar and Mafia but not Pemba. Records from the west and south-west include the northern end of Lake Malawi (Lake Nyasa), southern end of Lake Tanganyika, Mahale peninsula, Rubondo and Ukerewe islands and Mwanza, might be more widespread around the

Naja melanoleuca Forest Cobra. Top Left: Stephen Spawls, juvenile, Arabuko-Sokoke Forest. Centre: Stephen Spawls, Mwanza. Right: Stephen Spawls, Usambara Mountains. Bottom Left: Stephen Spawls, Chuka.

southern shore of Lake Victoria. Elsewhere, south to eastern South Africa, west to Senegal.

NATURAL HISTORY: Terrestrial, but it is a fast graceful climber, known to ascend trees to a height of 10m or more. Quick moving and alert. It swims well and readily takes to the water, in some areas it eats mostly fish and could be regarded as semi-aquatic. Active both by day and by night, mostly by day in uninhabited areas and by night in urban areas. When not active, takes cover in holes, brush piles, hollow logs, among root clusters or in rock crevices, or in termite hills at forest fringe or clearings, in some areas fond of hiding along river banks, in overhanging root systems or bird holes, in urban areas will hide in junk piles or unused buildings. If molested it rears up to a considerable height and spreads a long narrow hood; if further stressed and unable to retreat it may rush forward and bite. Male–male combat has been observed; one fracas in Nigeria involved five males. Lays from 15–26 eggs, roughly 30x60mm, incubation times of around two months recorded. Eats a wide variety of prey; fond of mammals (squirrels, mice, shrews, elephant shrews) and amphibians, will also take fish (including mudskippers), other snakes, monitor lizards and other lizards, birds and their eggs. A Ghanaian specimen had eaten a Gifford's giant shrew, a noxious smelling insectivore. Captive specimens can seem almost cunning in their timing of escape bids, and have been reported as trying to attack rather than escape; when restrained it will jab the sharp tail tip into the handler. A zoo specimen lived 28 years in captivity. Note: this is a dangerous snake, and like the Egyptian Cobra the venom causes a progressive paralysis, which can be rapid, fatalities are recorded, a bite must be treated with antivenom and respiratory support in hospital.

MOZAMBIQUE SPITTING COBRA *Naja mossambica*

IDENTIFICATION: A relatively small cobra of eastern Tanzania. The head is blunt, the eye medium sized, with a round pupil. Body cylindrical, tail quite long, 15–20% of total length. Scales smooth, in 23–25 rows at midbody, ventrals 177–205, subcaudals 52–69. Maximum size about 1.5m, average size 80cm–1.3m, hatchlings 23–25cm. Back colour usually some shade of brown, occasionally pinkish, juveniles may appear olive-green, large adults may be grey. Underside pale brown, pinkish or grey, on the neck, throat and anterior third of the belly there is a mixture of black bars, half-bars, blotches and spots, some specimens only have a few small markings, others have the throat heavily mottled with black. The skin between the scales is blackish and visible, giving a 'net' appearance in some specimens, and the scales on the side of the head (especially the lips) may be black edged. Pemba specimens are olive or pinky-grey, the dorsal scales are increasingly black-edged downwards on the flanks, the snout can be pinkish and the amount of black edging on the lip scales varies a lot.

Naja mossambica

HABITAT AND DISTRIBUTION: In our area, in coastal forest, thicket and moist savanna at low altitude, below 1,000m. Confined to eastern and south-eastern Tanzania, known from Pemba (where it is common) and Zanzibar (rare), and from Morogoro south along the coast, inland to Liwale and Tunduru. Elsewhere, south to northern South Africa and north-east Namibia, in semi-desert in some areas and up to 1,800m altitude.

Naja mossambica Mozambique Spitting Cobra. Left: Luke Verburgt, Northern Mozambique. Right: Michele Menegon, southern Tanzania.

NATURAL HISTORY: Mostly terrestrial, but able to climb well, adults readily ascend low trees and even sleep in them. Adults active mostly by night, but sometimes by day, juveniles often active during the day, presumably to avoid competition with adults and/or to avoid being eaten by them. When not active, shelter in termite hills, holes, rock fissures or under ground cover such as logs and brush piles. Very quick and alert, if molested may rear up (sometimes quite high) spread a hood and spit readily, sometimes without spreading a hood. Sometimes pretends to be dead. Lays between 10–22 eggs, size roughly 35×20mm, in December/January (summer) in southern Africa. Diet quite varied, fond of amphibians but also takes lizards, rodents, other snakes (including its own kind) and even insects have been recorded. These snakes spit in self-defence, the venom has no effect on unbroken skin but if it enters the eye, agonising pain results, and if untreated, corneal lesions and complications can occur. The affected eye must be gently irrigated with large quantities of water. A venomous bite from one of these snakes causes pain and local necrosis, without neurotoxic effects, but often with long term complications; a bite must be treated in hospital.

BLACK-NECKED SPITTING COBRA *Naja nigricollis*

IDENTIFICATION: A big spitting cobra, widespread in the savannas of East Africa. It has a broad head, a cylindrical body and smooth scales, the tail is 15–20% of total length. Midbody scale count 17–23, ventrals 175–212, subcaudals 50–70. Maximum size seems to vary; east of the Rift Valley rarely larger than 1.8m, averaging 1.2–1.5m, hatchlings 33–36cm. Snakes from immediately west of the Rift Valley seem to get much larger, reaching 2.3m or even more. Colour very variable, in much of Tanzania adults are black or steely grey, the head and neck are very black; the lower throat is barred back and red, in some specimens the red is replaced by orange or yellow. Juveniles are grey, with a black

Naja nigricollis

Naja nigricollis Black-necked Spitting Cobra. Left: Caspian Johnson, Ugalla. Right: Anthony Childs, Maasai Mara.

head and neck, the grey darkening at lengths of 80cm–1m; the throat bars may be pale grey or pink. In highland Kenya, adults may be copper coloured, pinky-grey, rufous or rufous-brown, but most have red and black throat bars. An all black adult (above and below) colour form occurs west of the central Rift Valley in Kenya and in north-west Tanzania. Some individuals of this form have some white blotching on the underside, and it seems to reach a much larger size, up to 2.4m (possibly larger).

HABITAT AND DISTRIBUTION: Moist and dry savanna and woodland (but not forest or semi-desert), from sea level to above 1,700m altitude. Widespread in Uganda, except the north-east. Extends from Ukambani west into south-west Kenya. Widespread in most places below 1,700m in Rwanda and Burundi, even in the capital cities, and most of western and central Tanzania, and reaches the coast around Dar es Salaam. Isolated Kenya records from Kiasa Hill in Tsavo East National Park, the Chyulu and Taita Hills. Absent from most of southern and northeast Tanzania, and northern and eastern Kenya, where other spitting cobras occur. Elsewhere, west to Senegal, south and west to Angola.

NATURAL HISTORY: Terrestrial, but they climb readily, and will even attempt to escape up trees or up into rocks. Quick moving and alert. Large adults mostly active by night, but will also hunt or bask during the day, especially in uninhabited areas. Juveniles often active during the day. When not active, takes cover in termitaria, holes, hollow trees, old logs, brush piles, etc. When threatened, rear and spread a hood and will spit if further molested, may also spit without actually spreading a hood. Big adults can spit 3m or more. Tend to stand their ground when threatened, not rush forward as the Forest Cobra and Egyptian Cobra may do. Male–male combat has been observed, snakes intertwined and neck-wrestling. They lay 8–20 eggs, size roughly 25×40mm, often in termitaria. Hatchlings collected in Nairobi in April and May. They eat a wide variety of food, including amphibians, various mammals, lizards and snakes; they will raid chicken runs and will take eggs and chicks.

Note: this is a dangerous snake, and like the Mozambique Spitting Cora, its venom is damaging to the eyes and a bite can cause major local necrosis, with long-term damage; hospital treatment is required.

ASHE'S SPITTING COBRA/LARGE BROWN SPITTING COBRA Naja ashei

IDENTIFICATION: A recently described big cobra of the coast and dry country. It has a large eye, with a round pupil and orange-brown iris. The body is cylindrical, the tail fairly long, 15–18% of total length. Scales smooth, in 20–23 rows at midbody, ventrals 192–207, subcaudals 55–65. Maximum size about 2.4m (maybe larger), average 1.2–1.8m, hatchling size 35–38cm. Dorsal colour various shades of brown or olive, sometimes grey, some specimens can be quite yellow-brown; some specimens uniform, others have a sprinkling of yellow scales, and these may form narrow yellow dorsal bars. Underside dull light yellow, each ventral scale edged with light brown; there is usually a deep, dark brown bar on the underside of the neck, but this may be fragmented into several bars with yellow scales between. Inland juveniles are light grey-brown, the throat bar is dark brown or black, the head deep brown; juveniles from Kenya's north coast have a shiny black head and neck, this fades in their second year. Named for James Ashe, charismatic Kenyan herpetologist.

Naja ashei

HABITAT AND DISTRIBUTION: Dry savanna, semi-desert and coastal thicket, from sea level to about 1,200m altitude. Occurs right along the coastal strip from the Somali border to the Shimba Hills, widespread in Tsavo, south-east Kenya, westwards to Amboseli, and north through Ukambani and the Tharaka Plain, through Samburu and eastern Laikipia across to Baringo and the northern Kerio Valley, just reaching Uganda at Amudat in the east. Isolated records from northern Kenya from the east side of Lake Turkana, Mt Kulal, Buna and Mandera. In Tanzania, it is known only from Ndukusi, a village near Lolkisale Mountain, south-south-east of Arusha, so possibly in Tarangire National Park. It should also be on the extreme north Tanzanian coast as it is recorded from Lunga-Lunga. Elsewhere, occurs in eastern and southern Ethiopia (including the Omo Delta), Somalia and South Sudan.

NATURAL HISTORY: Terrestrial, but may climb into bushes and low trees; a specimen on the road in Meru National Park climbed an acacia when pursued. Quick moving and alert. Adults in undisturbed habitat seem to be both nocturnal and diurnal, one was found mid-morning pursuing frogs on the edge of a dam near Mt Kasigau, and when approached also climbed an acacia tree; several have been photographed active by day on Laikipia. Juveniles in the Shimba Hills and at Lunga-Lunga were active by day. When inactive, shelters in holes (particularly in termite hills), old logs or underground. When disturbed rears up ,spreads a broad hood and may spit twin jets of venom. Lays eggs. Eats a range of prey, takes mammals, birds, amphibians and reptiles; known prey items include a domestic cat, chickens and whole nests of hen's eggs, Red Spitting Cobras *Naja pallida*, White-throated Savanna Monitors *Varanus albigularis* and Puff Adders *Bitis arietans*; a Watamu specimen ate a Puff Adder so large it could subsequently only crawl in a straight line. They will also

Naja ashei Ashe's Spitting Cobra. Left: Stephen Spawls, Watamu. Top right: Stephen Spawls, Watamu. Bottom right: Bio-Ken archive, hatchling.

take carrion, including chicken heads and road killed snakes; one ate a dead egg-eating snake that was maggot ridden and decomposing. Note: this snake spits readily, the venom is produced in huge quantities and is damaging to the eyes. Bites can cause local necrosis; hospital treatment will be needed.

RED SPITTING COBRA *Naja pallida*

IDENTIFICATION: A relatively small cobra with a smallish head and a large eye, with a round pupil. The body is cylindrical, the tail fairly long, 15–19% of total length. Scales smooth, in 21–27 rows at midbody, ventrals 192–228, subcaudals 56–81. Maximum size about 1.5m (possibly larger), most adults 70cm–1.2m, hatchlings about 16–20 cm. Colour variable, specimens from northern Tanzania, east and southern Kenya (especially those from red soil areas) are orange or red, with a broad black throat band, several scales deep, and a black 'suture' on the scales below the eye, giving the impression of a tear-drop. Reddish below, sometimes with a white chin and throat. Specimens from other areas may be pale red, pinkish, pinky-grey, red-brown, yellow or steel grey. Most, however, have the dark throat band. In large adults, this band may fade or disappear. Red specimens become dull red-brown with increasing size.

Naja pallida

HABITAT AND DISTRIBUTION: Dry savanna and semi-desert, from sea level to about 1,200m altitude. Occurs almost throughout northern and eastern Kenya, although does not reach the coast. From Tsavo West extends south into Mkomazi, in north-east Tanzania, and from Amboseli it occurs west in Kenya's southern Rift Valley, through Magadi and Olorgesaile to Mt Suswa, south into Tanzania past Lake Natron to Olduvai Gorge. Elsewhere, occurs in eastern and southern Ethiopia, Somalia and South Sudan.

Naja pallida Red Spitting Cobra. Top left: Stephen Spawls, Voi. Right: Stephen Spawls, juvenile, Olorgesaile. Centre: Tim Spawls, juvenile with nape spot. Bottom left: Johan Kloppers, Kalama.

NATURAL HISTORY: Terrestrial, but may climb into bushes and trees. Quick moving and alert. Adults are mostly nocturnal, hiding in holes (particularly in termite hills), brush piles, old logs or under ground cover during day, but juveniles often active during the day. When disturbed rears up relatively high and spreads a long narrow hood, and may spit twin jets of venom. Lays 6–15 eggs, size about 50×25mm. A juvenile was taken on the eastern Tharaka plain in October. Amphibians are its favourite food, when the storms arrive in the dry country where this snake lives, the frogs all come out and the snake gorges itself. A specimen in Tsavo was found at night up in a thorn tree, eating foam-nest tree-frogs, *Chiromantis*. It also takes rodents and birds, known to raid chicken runs, and probably eats other snakes. It will come around houses in the dry country looking for water, and frogs that live in or near water tanks. Note: this snake spits readily, the venom is damaging to the eyes, and bites can cause local necrosis; hospital treatment will be needed. A surprising number of bites from this cobra result from the snake entering buildings at night.

TREE COBRAS *PSEUDOHAJE*

Tree cobras occur in forested tropical Africa. Two species are known, one of which reaches western East Africa. They are large, agile snakes, they look like a cross between a mamba and a cobra.

GOLD'S TREE COBRA *Pseudohaje goldii*

IDENTIFICATION: A big, shiny, thin-bodied tree cobra, with a short head and a huge dark eye, with a round pupil. Body cylindrical, tail long and thin, ending in a spike, about

20% of total length; the spike probably a climbing aid; if held by the head this snake will drive the spike into the restraining arm. Scales smooth and very glossy, in 15 rows (occasionally 13 or 17) on the front half of the body, reducing to 13 rows slightly past midbody, ventrals 185–205, subcaudals 76–96. The skin is very fragile, and tears or abrades easily. Glossy black above, scales on the side of the head, chin and throat yellow edged with black. Juveniles have several yellow cross bars or band, narrowing towards the tail, becoming scattered yellow scales on the back half. Maximum size 2.7m (perhaps slightly larger), most adults average 1.5–2m, hatchlings 40–50cm.

Pseudohaje goldii

HABITAT AND DISTRIBUTION: In our area, in forest or forest islands, at medium to high altitude, from about 600–1,700m, elsewhere to sea level; in West Africa it lives in Guinea savanna, primary and altered forest, swamp and mangrove forest and, surprisingly, in suburbs. In Kenya, known only from the west, in the Kakamega area, might occur on Mt Elgon; a skin supposedly from this species reported from the North Nandi Forest. A handful of Uganda records include Mabira Forest, a lakeshore island near Entebbe, Bundibugyo and Kibale areas and Bushenyi, south of the Kazinga channel and Lake George. Also known from the Congo side of Virunga National Park and Kalehe, north-west of Lake Kivu in the DR Congo, so might be in Rwanda, although not recorded. Elsewhere, south-west to northern Angola, west to Nigeria and jumps the Dahomey gap to appear in Ghana.

Pseudohaje goldii Gold's Tree Cobra. Left: Stephen Spawls, Kakamega. Right: Stephen Spawls, spreading narrow hood.

NATURAL HISTORY: Very poorly known. Arboreal, but will descend to the ground. Quick and alert, climbs fast and well. Moves rapidly on the ground, often with the head raised. Active during the day, might also be nocturnal. In Kenya, specimens have been found in squirrel traps, in trees along rivers, specimens in the eastern Congo are reportedly common in oil palm plantations, in West Africa has been found in fish traps, and a survey in Nigeria found most specimens fairly close to water bodies. Gold's Tree Cobra is not an aggressive species, but if molested or restrained, can flatten the neck into a very slight hood, nothing like as broad as a cobras, and may move forward with head raised and try to bite, specimens have been seen to make rapid direct strikes. Male–male combat has been observed in Nigeria in January. If disturbed will pause with the head up, moving it from side to side like a metronome. Lays 10–20 eggs, approximately 50×25mm. A hatchling was captured in September in the Ituri Forest in the Congo. Nigerian specimens had eaten frogs, mudskippers and other fish, a small hinged tortoise and rats; specimens were taken in squirrel traps in Kakamega, suggesting it eats arboreal mammals. It has been reported as descending to the ground to feed on terrestrial amphibians, but ascending to then digest its prey. Prized as a human food in eastern Congo, but in the same area it has a fearsome reputation, it is said to readily retaliate if molested. The venom is a potent neurotoxin, a snake collector in eastern Congo was bitten during a demonstration and died within 12 hours. South African polyvalent antivenom, although not prepared for this snake, is reportedly effective against the venom.

MAMBAS *DENDROASPIS*

Mambas occur in sub-Saharan Africa. Four species are known, three occur in East Africa. Mambas are large, agile, slender diurnal elapid snakes. They are greatly feared, but their reputation is largely unjustified, and based mostly on legend; they are not aggressive. However, a mamba bite is a medical emergency and needs prompt and thorough treatment.

KEY TO THE EAST AFRICAN MEMBERS OF THE GENUS *DENDROASPIS*

1a Olive, grey or brown, mouth lining dark, midbody scales in 21–25 rows......*Dendroaspis polylepis*, Black Mamba [p. 564]
1a Green, mouth lining pink, midbody scales in 15–19 rows......2
2a Uniform light green, lip scales immaculate; midbody scales 17–19, in eastern Kenya and Tanzania......*Dendroaspis angusticeps*, (Eastern) Green Mamba [p. 561]
2b Darker green, tail black, lip scales black-edged, midbody scales 15–17, in western east Africa......*Dendroaspis jamesoni*, Jameson's Mamba [p. 562]

GREEN MAMBA *Dendroaspis angusticeps*

IDENTIFICATION: A big, slender, tree snake with a long head and a small eye, with a round pupil, the iris is yellow. Tail long and thin, 20–23% of total length. Scales smooth, in 17–19 (sometimes 21) rows at midbody, ventrals 201–232, subcaudals 99–126. Maximum size 2.3m (possibly more), average size 1.5–2m, hatchlings 30–40cm. Bright uniform green above, sometimes a sprinkling of yellow scales, pale green below. Juveniles less than 60cm long are bluish-green.

Dendroaspis angusticeps

HABITAT AND DISTRIBUTION: Coastal bush and forest, moist savanna and evergreen hill forest, mostly at low altitude (less than 500m), but up to 1,700m in parts of its range. Tolerant of coastal agriculture in East Africa, often in cashew nut and coconut plantations and mango trees. It occurs from the Somali border and the Boni Forest south along the entire East African coastal plain to the Ruvuma River, inland to the Usambaras, along the Rufiji River system and across south-eastern Tanzania to Tunduru. Isolated populations in the Nyambene

Dendroaspis angusticeps Green Mamba. Left: Elvira Wolfer, Arusha National Park. Right: Stephen Spawls, Arusha.

Hills and Meru National Park and around Kibwezi in Kenya, and the forests around Mt Meru and southern Mt Kilimanjaro in Tanzania, it can be found in and around Arusha and Moshi. The connections between these isolated populations are unknown, but might be due to periods of higher rainfall and thus more extensive woodland cover in the past. Might be in the Chyulu Hills, the Taita Hills and North and South Pare Mountains. Elsewhere, sporadically through Malawi, Mozambique and eastern Zimbabwe to eastern South Africa.

NATURAL HISTORY: A fast moving, diurnal, secretive tree snake, climbs expertly, but will descend to ground, sometimes seen crossing roads on the East African coast. Not aggressive, if threatened tries to escape, rarely ever tries to bite. May climb very high in trees. Sleeps at night in the branches, not seeking a hole to sleep in but coiled up in a thick patch. May emerge to sunbathe in the early morning. Will shelter in and under *makuti* (coconut thatch) roofs; anecdotal evidence indicates they are most active in the morning. An even-natured snake, not known to spread a modified hood or threaten with open mouth like the black mamba. Male combat has been observed in South Africa in June. Mating observed in April, May and July and hatchlings collected December to February in southern Tanzania. Lays up to 17 eggs, roughly 6×3cm, these take 2–3 months to hatch. Hatchlings collected in Watamu in June and July. Feeds mostly on birds, their nestlings, rodents and bats; occasionally lizards. Tanzanian snakes took a Lilac-breasted Roller *Coracias caudatus* and a Lesser Blue-eared Glossy Starling *Lamprotornis chloropterus*. Ground-dwelling rodents have been found in their stomachs, indicating they will descend to feed or hunt at low level. Captive specimens will take chameleons. A recent study around Watamu indicates that green mambas spend much time in ambush, rather than actually foraging. Has been found in huge concentrations within parts of its range (especially coastal Kenya and south-east Tanzania), concentrations of 2–3 snakes per hectare (i.e. 200–300 per square kilometre) have been noted. Tolerant of coastal urbanisation and agriculture, living in coconut and cashew nut farms, in thick hedges and ornamental trees. Note: this is a dangerous snake. Its venom contains unusual neurotoxins called dendrotoxins, causing pins and needles, muscular contractions and a progressive paralysis. Being a non-aggressive tree snake, bites are rare but need rapid hospital treatment with antivenom and respiratory support.

JAMESON'S MAMBA *Dendroaspis jamesoni*

IDENTIFICATION: A big, slender, tree snake with a long head and a small eye, with a round pupil. Tail long and thin, 20–25% of total length. Scales smooth, in 15–17 rows at midbody, ventrals 202–227, subcaudals 94–113. Maximum size 2.64m (possibly more), big specimens usually males. Average size 1.5–2.2m, hatchling size unknown but probably around 30cm. Dull green above, mottled with black and yellow-green, pale green below, neck and throat yellow. In some specimens (particularly juveniles) the black and yellow dorsal scales form narrow bands. The scales on the head and body are narrowly edged with black. Specimens from Uganda and Kenya (subspecies *kaimosae*) have a black tail, those from the centre and west have a yellow tail and the tail scales are black-edged, giving a netting effect, specimens of this colour form might occur in the far west of our area.

Dendroaspis jamesoni

HABITAT AND DISTRIBUTION: Forest, woodland, forest-savanna mosaic, thicket and deforested areas, from 600m to about 2,200m (i.e. up into montane woodland), elsewhere to sea level; in swamp-forest in Nigeria. Will persist in areas after forest has been felled, providing there are still thickets and trees to hide in, and will move across open country to reach isolated tree clumps and thickets. Quite often found around buildings, and in and around towns within the forest, and in parks and farms. Being a green, secretive tree snake, it often clings on in well-inhabited areas without being seen. In Kenya, known from the Kakamega area (there is a sight record from Lolgorien, above the Mara escarpment). Widespread in southern and central Uganda in suitable habitat, widespread but sporadically recorded in Rwanda (records include Akagera National Park, Nasho, Huye district, Cyamudongo Forest) and the lakeshore woodland in south-west Burundi. It is absent from the forest in Bwindi Impenetrable National Park, suggesting it is not cold tolerant, but is found at altitudes well over 2,000m in Rwanda. In Tanzania, known from the Minziro Forest near Bukoba and the Mahale Peninsula. Elsewhere, west to Ghana, south-west to Angola, also in South Sudan.

NATURAL HISTORY: A fast moving, diurnal tree snake, climbs expertly, but will descend to ground. Not aggressive, if threatened tries to escape but occasional specimens may flatten the neck to a narrow hood. It will reluctantly strike if provoked. It may shelter in hollow trees, cracks, tangles of vegetation. Male to male combat and mating was observed in Nigeria in the dry season, December to January; mating observed in Kakamega Forest in September. Lays eggs, clutches of 7–16 in Nigeria, laid mostly in April, May and June; females collected in June in Uganda showed no yolk deposition but females collected

Dendroaspis jamesoni Jameson's Mamba. Left: Bio-Ken Archive, Kakamega. Right: Mike Perry, hatchling.

in November contained oviductal eggs. Diet in Kenya known to include rodents, birds and arboreal lizards; studies in Nigeria found that the adults largely ate birds and some rodents, juveniles also cold-blooded prey; known prey items included toads, agamas, several bird species including woodpeckers, cisticolas and doves, squirrels, shrews and mice. Captive specimens readily take rodents but are nervous feeders. A mark-recapture study in Nigeria found this snake was sedentary, only moving a short distance between captures. Note: as with the Green Mamba, its venom is a potent neurotoxin, and although bites are unlikely they need hospital treatment; a man collecting firewood in Nasho, Rwanda, was bitten by a 'green snake with a black tail' and died in three hours.

BLACK MAMBA *Dendroaspis polylepis*

IDENTIFICATION: A long, slender, fast moving snake, with a long, narrow, 'coffin-shaped' head, with a fairly pronounced brow ridge and medium sized eye with a round pupil. The inside of the mouth is bluish-black. Body cylindrical, tail long and thin, 17–25% of total length. Scales in 23–25 (rarely 21) rows at midbody. Ventrals 239–281, subcaudals 105–132. Maximum size probably about 3.2m (unsubstantiated reports of bigger specimens, up to 4.3m, exist), adults average 2.2–2.7m, hatchlings 45–60cm. The juveniles grow rapidly and may reach 2m in a year. Olive, brownish, yellow-brown or grey in colour, sometimes khaki or olive-green, but never black as the name suggests. Juveniles are greeny-grey. The scales are smooth and have a distinct purplish bloom in some adult specimens. The belly is cream, ivory or pale green. The back half of the snake is often distinctly speckled with black on the flanks, especially in animals from dryer areas. Some snakes have rows of lighter and darker scales towards the tail, giving the impression of oblique lateral bars of grey and yellow.

Dendroaspis polylepis

HABITAT AND DISTRIBUTION: In coastal bush, moist and dry savanna and woodland, from sea level to about 1,600m, very rarely above this altitude, most common in well-wooded savanna or riverine forest, especially where there are rocky hills and big trees. Found along the length of the East African coast, but where it is sympatric with the Green Mamba, it tends to be in more open country rather than in forest or woodland. Occurs virtually throughout Tanzania (few records from the south-west around Lake Malawi (Lake Nyasa)). Widespread in south-east, central and western Kenya, in suitable habitat, but absent from the high centre between Lake Victoria and Mt Kenya, although it occurs in the Mara and also at Kisumu. The northern population comes as far south as Mogotio but not Nakuru. 'Near Nairobi' records include Karen, Lukenya Hill and Olorgesaille. Few records from northern and eastern Kenya, save the hills and rivers along the Ethiopian border east of Sololo. Occurs in a broad band across central Uganda, also known from Torit in South Sudan. Few Burundi records, but occurs in low eastern Rwanda, in Akagera National Park and Nasho. Elsewhere, south to Namibia and South Africa, sporadic records from central and West Africa. Conservation Status: IUCN Least Concern.

NATURAL HISTORY: A big, fast-moving, alert, agile diurnal snake, equally at home on the ground or climbing rocks or in trees, often moving with the head and neck raised high. Usually moves away if approached, but if threatened it may rear up, flatten the neck into a narrow hood, hiss loudly and open the mouth to display the black interior, while shaking the head from side to side. If further antagonized, may respond with a long upwards strike. Black Mambas shelter, often for several months, in holes, termitaria, rock fissures and tree cracks, hammerkop nests and even in tree beehives and in roof spaces. In South Africa they frequently basked between 7–10am and from 2–4pm. They often share their refuges with other snakes, including cobras and pythons; at Kimana in Kenya two Black Mambas shared their hole with two Egyptian Cobras. Black mambas lay from 6–17 eggs (although 12–18 more usual) measuring approximately 65×30mm, which usually hatch after 2–3 months incubation, at the beginning of the rainy season. They take a variety of prey, all sizes of mammals from mice to squirrels, bats, bushbabies, mongooses and hyrax, they may eat birds (especially nestlings, one ate a peregrine falcon chick) and snakes, one specimen was observed eating winged termites as they left the nest. Males indulge in combat, neck wrestling with bodies intertwined; this has been mis-identified as courtship. They are often mobbed by birds and also by squirrels and vervet monkeys. Note: the venom is a potent neurotoxin, causing a rapid paralysis, and the venom yield can be many times the lethal dose. Any Black Mamba bite is a medical emergency; rapid treatment in an advanced hospital with antivenom and respiratory support will be needed.

Dendroaspis polylepis Black Mamba. Top: Daniel Liepack, Diani beach. Centre: Stephen Spawls, showing fangs. Bottom: Stephen Spawls, spreading hood, Malindi.

SUBFAMILY **HYDROPHIINAE**
VIVIPAROUS SEA SNAKES

A family of front-fanged, venomous aquatic snakes, living in the sea. Their taxonomy is still under investigation, but about 62 species, in seven genera, are recognised at present; they are most closely related to Australian land elapids. A single species, the Yellow-bellied Sea Snake *Hydrophis platurus* occurs in the western Indian Ocean and occasionally comes ashore on the East African coast. No other sea snakes are known from our area; reports of other sea snakes on the East African coast are invariably either snake eels (Ophichthidae) or Moray eels (Muraenidae); these may be told from sea snakes by their pointed tails, the sea snake has an oar-like tail.

YELLOW-BELLIED SEA SNAKE *Hydrophis platurus*

IDENTIFICATION: A medium-sized, black and yellow sea snake. The head is long and tapering, the eye is small, with a round pupil, nostrils on top of the head. The body is laterally compressed, the tail is short, vertically flattened and oar-like, about 10% of total length. Scales smooth, 49–76 midbody rows, ventrals 264–406, maximum size about 90cm, average 60–85cm, hatchling size 24–25cm. It has bright warning colours, the broad dorsal stripe is usually black, greeny-black or very dark brown, the belly is yellow, the tail spotted black and cream or yellow. The dorsal stripe may be straight-edged (most East African specimens are this colour), wavy, break up into black saddles or totally disappear.

Hydrophis platurus

HABITAT AND DISTRIBUTION: Widespread across the Indian and Pacific Oceans; in all tropical seas save the Atlantic. In our area, recorded from Malindi, Watamu, Kilifi, from 20km north of Dar es Salaam and Masoro near Kilwa. Not known from the Red Sea, but known from Djibouti, Madagascar and the Seychelles, elsewhere eastwards across the Indian and Pacific Oceans.

NATURAL HISTORY: Lives mostly in open water, not around reefs, but usually within 300km of land. Usually only washed ashore when dying or accidentally in storms; it can only wriggle in an ungainly fashion on land. It is a superb swimmer, moving with side-to-side undulations, like a snake on land, but it can swim backwards and forwards, and is capable of very rapid bursts of speed. Often found in the vicinity of sea slicks (long narrow lines of floating vegetation and accumulated debris on the sea surface), where it waits quietly to ambush fish; it may attack individual fish or rush into a small shoal, indiscriminately biting everything it can grab. It is afraid of humans and moves off if approached by swimmers. Three to eight young are born alive, gravid females have been collected between March and October off the South African coast. Concentrations of up to ten snakes have been seen in East African waters, a group of six were found in Turtle Bay, Watamu. The sea snake sloughs its skin as do land snakes, but since it doesn't have convenient objects to rub against, it forms tight loops, rubbing coil against coil, this way it also frees itself from sea animals like barnacles. Note: this is a dangerously venomous snake, although bites are extremely rare (and often not noticed). The venom is a powerful neurotoxin, leading to paralysis, muscle destruction and renal failure. No antivenom is produced.

Hydrophis platurus Yellow-bellied Sea Snake. Royjan Taylor, Watamu.

Bitis nasicornis
Rhinoceros Viper
Michele Menegon

SUB-ORDER **SERPENTES: SNAKES**
FAMILY **VIPERIDAE**
VIPERS

A family of dangerous snakes, with long, perfectly tubular, folding poison fangs; and of international medical significance. The vipers or adders (the names are interchangeable, although it is suggested that the name adders be restricted to those that give live birth) are split into three subfamilies: the Viperinae, Old World vipers; the Crotalinae, pit vipers of America and Asia and the Azemiopinae, two species in one genus, curious slim Asian vipers.

SUBFAMILY **VIPERINAE**
OLD WORLD VIPERS

A subfamily of 13 genera and around 100 species, found in Asia, Europe and Africa, five genera are confined to Africa and five are in both African and Asia. Just under 60 species occur in Africa, 23 species in six genera occur in our area. A key to the Viperinae is provided below, but the various genera, and most of the species can be readily identified in the field using a combination of appearance, locality and behaviour.

KEY TO THE EAST AFRICAN VIPER GENERA AND SOME SPECIES

1a Small horn above the eye, on the ground in open country of central Kenya......*Bitis worthingtoni*, Kenya Horned Viper [p. 581]
1b Does not have the above combination of characters......2
2a Pupil round, nine large scales on top of the head......*Causus*, night adders [p. 568]
2b Pupil vertical, many small scales on top of the head......3
3a Subcaudal scales paired......4
3b Subcaudal scales single......6
4a On the moorlands of Aberdares or Mt Kenya, slim, brown or grey with black markings......*Montatheris hindii*, Kenya Montane Viper [p. 594]
4b Not on the moorlands of Aberdares or Mt Kenya, not slim, brown or grey with black markings......5
5a A big supraocular shield present, black saddles on the back, southern Tanzania only......*Proatheris superciliaris*, Floodplain Viper [p. 596]
5b No big supraocular shield, back markings V-shaped, rectangular or sub-rectangular, widespread......*Bitis*, African vipers (except Kenya Horned Viper) [p. 575]
6a Head pear-shaped, in dry country and semi-desert of northern Kenya...... *Echis pyramidum*, North-east African Carpet Viper [p. 582]
6b Head not pear-shaped, in woodland...... *Atheris*, bush vipers [p. 584]

NIGHT ADDERS CAUSUS

A genus of small stout frog-eating vipers, found in sub-Saharan Africa in savanna and forest. Seven species are known, six occur in East Africa. Unusually for vipers, they have a round pupil, nine large scales on top of the head and lay eggs. Despite their common name, they are active by day and night. They have a somewhat primitive fang rotation movement and their fangs are relatively short. Several species have exceptionally long venom glands that extend well down the neck, and in some areas they are responsible for a high proportion of snakebite cases, but these largely cause local pain and swelling; no fatalities are recorded and no antivenom is available.

KEY TO THE EAST AFRICAN MEMBERS OF THE GENUS *CAUSUS*

1a Vivid green when adult......2
1b Brown, grey or dull olive green when adult......3
2a Subcaudals single, eyes large, in forest, rostral not upturned......*Causus lichtensteinii*, Forest Night Adder [p. 571]
2b Subcaudals paired, eyes moderate, not in forest, rostral slightly upturned...... *Causus resimus*, Velvety-green Night Adder [p. 573]
3a Snout clearly upturned, subcaudals less than 19......*Causus defilippii*, Snouted Night Adder [p. 570]
3b Snout not upturned, subcaudals more than 14......4
4a Head narrow, usually a pair of pair of narrow pale dorsolateral stripes present, only in northern Rwanda and south-west Tanzania......*Causus bilineatus*, Two-striped Night Adder [p. 561]
4b Head broad, no dorsolateral stripes......5
5a Upper labials not dark-edged, ventrals 118-150, no dark bar between the eyes......*Causus maculatus*, West African Night Adder [p. 572]
5b Upper labials dark-edged, ventrals 134-166, a dark bar between the eyes......*Causus rhombeatus*, Rhombic Night Adder [p. 574]

TWO-STRIPED NIGHT ADDER *Causus bilineatus*

IDENTIFICATION: A small adder, head narrow and long for a night adder, tapering to a narrow but rounded, not upturned, snout, the head is slightly distinct from the neck, a medium-sized eye with a round pupil, the top of the head covered with nine large scales (unlike most vipers). The body is cylindrical or slightly depressed, the tail is short, 9–12% of total length. The scales are soft, velvety and feebly keeled, in 17 (occasionally 15, 18 or 19) rows at midbody, ventrals 119–149, subcaudals 18–35, higher counts usually males. Maximum size about 65cm, average size 30–50cm, hatchling size about 13–14cm. The back is brown (it may be pinkish or greyish brown), there are a number of irregular or vaguely rectangular black patches all along the back, interspersed with rufous or brown blotches. There are usually two distinct, narrow pale dorso-lateral stripes that run the length of the body. There is a sprinkling of black scales and oblique black bars on the flanks. On the head there is a characteristic V-shaped mark, which may be black, or black-edged with a brown infilling. The belly is dark to dark cream.

Causus bilineatus

HABITAT AND DISTRIBUTION: Within our area, known only from Gabiro and nearby Akagera National Park in eastern Rwanda, and at Tatanda, in south-western Tanzania, in moist savanna at 1,000–1,800m altitude and also recorded just across the border there in northern Zambia. Elsewhere, known from Zambia, across the huge plateau of south-central Africa, to the southern DR Congo and Angola.

NATURAL HISTORY: Little known, probably rather similar to other night adders, i.e. terrestrial, slow-moving, active by day and night, inflates body when angry and hisses and puffs, will strike if further molested. May be locally abundant. Presumably, like other night adders, eats frogs and toads; it is known to feed on clawed frogs (*Xenopus*), which might mean it is more aquatic than other night adders. It lays eggs. Regarded at one time as simply a colour variant of the Rhombic Night Adder *Causus rhombeatus*, but the range of the two species overlap throughout the Two-striped Night Adder's range, and this species also has slightly lower ventral counts (the number of scales along the belly) than the Rhombic Night Adder. Note: a venomous snake but no fatalities recorded, and no antivenom available.

Causus bilineatus Two-striped Night Adder. Left: Andre Zoersel, northern Zambia. Right: Philipp Wagner, northern Zambia.

SNOUTED NIGHT ADDER Causus defilippii

IDENTIFICATION: A small, stout viper with a fairly short head slightly distinct from the neck, a medium-sized eye with a round pupil, the top of the head covered with nine large scales (unlike most vipers), the tip of the snout quite distinctly upturned. The body is cylindrical or slightly depressed, the tail is short, 7–9% of total length in males, 5–7% in females. The scales are soft, velvety and feebly keeled, in 13–17 (usually 17) rows at midbody, ventrals 109–130 (higher counts usually females), subcaudals 10–19. Maximum size about 42cm, average size 20–35cm, hatchlings about 10cm. The back is brown (it may be pinkish or greyish brown), there are 20–30 clear, dark, pale-edged rhombic blotches all along the back, and a sprinkling of black scales and oblique black bars on the flanks. On the head there is a characteristic V-shaped mark, usually solid black, a black bar is often present behind the eye and the upper lip scales are usually black-edged. The belly is cream, pearly white or pinkish-grey, it may be glossy black or grey in juveniles.

Causus defilippii

HABITAT AND DISTRIBUTION: Moist and dry savanna and coastal thicket, from sea level to about 1,800m. Occurs from Malindi (northernmost record) south along the coast to Tanga, inland to the Usambara Mountains, also on Zanzibar, then south-west across the southern half of Tanzania, to the south end of Lake Tanganyika and northern Lake Malawi (Lake Nyasa). Elsewhere, south to northern South Africa, an unequivocal inhabitant of the eastern and southern Africa coastal mosaic.

NATURAL HISTORY: Terrestrial, but may climb into low bushes in the pursuit of frogs. Fairly slow-moving, but strikes quickly. Despite its name, seems to be active during day, as well as in the twilight and at night. Known to bask. When inactive, hides in holes, bush piles, under ground cover, etc. If angered, it hisses loudly, inflates the body, elevates the head and may make a determined (but often clumsy) strike. It may also lift the front part of the body off the ground, flatten the neck and move forward with tongue extended, looking like a little cobra. Eats frogs and toads. Lays from 3–9 eggs, roughly 23×15mm. Hatchlings collected in the Usambara Mountains in May and June. In the breeding season, males are

Causus defilippi Snouted Night Adder. Left: Stephen Spawls, Usambara Mountains. Right: Andre Zoersel, northern Zambia.

known to engage in combat, rearing up and wrestling until the weaker male is forced to the ground. Note: a venomous snake, bite cases characterised by swelling, local pain and sometimes lymphadenopathy.

FOREST NIGHT ADDER Causus lichtensteinii

IDENTIFICATION: A small, relatively slim night adder of the western forest. The head is slightly more rounded than the usual night adder head, distinct from the neck. It has a prominent eye with a round pupil, the top of the head covered with nine large scales (unlike most vipers). The body is cylindrical or slightly depressed, and (particularly in juveniles) slimmer than other night adders. The tail is short and blunt, 6–8% of total length in females, 8–11% in males. Unusually the subcaudal scales are single, they number from 14–23, higher counts in males. The scales are soft, velvety and feebly keeled, in 15 rows at midbody, ventrals 134–156, higher counts in females. Maximum size about 70cm but this is exceptional, average size 30–55cm, hatchlings about 15cm. The back is usually green, of various shades, from vivid pale green to dark or olive green, sometimes brown, some individuals have two or three orange bars across the tail. There is sometimes a series of vague, pale-centred rhombic back markings, that may be obscured and appear as chevrons, facing forwards or backwards; sometimes there is a white V-shape on the neck. The belly scales are yellowish, cream or pearly. Juveniles very variable; deep green, brown or turquoise; uniform or with dark crossbars; the lips are white. Hatchlings have a distinct white V-shape on the neck, this shrinks in large adults to a very fine V-shape, sometimes with black edging. The lips are yellow, the throat is yellowish or white, usually with two or three distinct black cross-bars, the tongue is blue at the base, then black with an orange band.

Causus lichtensteinii

HABITAT AND DISTRIBUTION: In forest and woodland, swampy areas associated with forest and recently deforested areas, from 500m to about 2,100m altitude, elsewhere to sea level. In our area, occurs in western Kenya in the north Nandi Forest, Kakamega area and the Yala River valley, sporadic Uganda records (Serere, Budongo Forest, north shore of Lake

Causus lichtensteinii Forest Night Adder. Left: Konrad Mebert, Eastern Congo. Right: Harald Hinkel, Eastern Congo, juvenile.

Victoria), and along the Albertine Rift from south of Lake Albert to Kasese and northern Bushenyi. No records from Rwanda or Burundi, but known not far south-west of Lake Kivu in the DR Congo. Elsewhere, west to Sierra Leone, south-west to northern Angola, isolated records from the Imatong Mountains in South Sudan.

NATURAL HISTORY: Terrestrial and secretive. Fairly slow-moving, but strikes quite quickly. Despite its name, seems to be most active during day among the shade and leaf litter of the forest floor. Swims well and has colonised islands in Lake Victoria. When inactive, hides in holes, bush piles, tree root clusters, under ground cover, etc. If angered it responds by inflating the body with air, which makes the markings stand out, and hissing and puffing, it may then raise the forepart of the body off the ground, into a coil with the head back, and strike. Eats frogs and toads. Lays from 4–8 eggs. No details of venom known

WEST AFRICAN NIGHT ADDER *Causus maculatus*

IDENTIFICATION: A small, stout viper with a fairly short head slightly distinct from the neck, a rounded snout and a medium-sized eye with a round pupil, the top of the head covered with nine large scales (unlike most vipers). The body is cylindrical or slightly depressed, the tail is short, 7–9% of total length in females, 9–11% in males. The scales are soft and feebly keeled, in 17–22 rows at midbody, ventrals 118–154, subcaudals 15–26, higher counts usually males. Maximum size about 70cm (possibly slightly larger), average size 30–60cm, hatchlings 13–16cm. Usually some shade of brown, pinkish-brown or grey above, with dark cross-bars or rhombs down the centre of the back and dark bars or spotting on the flanks (occasional individuals have virtually no markings). There is usually a V-shape on the neck, extending onto the head, uniformly dark or with dark edges and a light centre. The underside may be white, cream or pinkish-grey, uniform or with each scale black-edged so the belly looks finely barred.

Causus maculatus

HABITAT AND DISTRIBUTION: A handful of records from Uganda that include Kidepo National Park, Moyo in the north-west, Serere and all along the eastern shore of Lake Albert, south as far as Toro game reserve. Elsewhere, west to Mauritania, north to south-east Ethiopia, from sea-level to 1,800m altitude, in forest, moist and dry savanna and even semi-desert.

Causus maculatus West African Night Adder. Left: Steve Russell, Kidepo National Park. Right: Stephen Spawls, Ghana.

NATURAL HISTORY: Terrestrial, but will climb into bushes in pursuit of frogs. Active by day, by night and at twilight. In West Africa most active in the rainy season, and tends to disappear in the dry season, although Uganda habits little known. If molested it inflates the body, hiss loudly and makes swiping strikes; like other night adders it may flatten the body, looking like a little cobra, and demonstrate with an open mouth. Eats frogs and toads. Lays from 6–20 eggs, roughly 26×16mm, which in West Africa are laid in February–April, hatchlings appear in May–July. Quite a common snake in parts of its range. As with other night adders, the venom causes mild symptoms such as pain, moderate swelling, local lymphadenitis and slight fever.

VELVETY-GREEN NIGHT ADDER *Causus resimus*

IDENTIFICATION: A small, stout viper with a fairly short head slightly distinct from the neck, a medium-sized eye with a round yellow pupil, the top of the head covered with nine large scales (unlike most vipers), the tip of the snout slightly upturned. The body is cylindrical or slightly depressed, the tail is short, 7–9% of the total length in females, 8–10% in males. The scales are soft, velvety and feebly keeled, in 19–21 rows at midbody, ventrals 131–152, subcaudals 15–25, males usually more than 20, females usually less than 20. Maximum size about 75cm, average size 30–60cm, hatchlings 12–15cm. A beautiful snake, the back is usually vivid green (sometimes brown elsewhere) of various shades. The chin and throat are yellow, the belly scales yellowish, cream or pearly. Often on the head black scales form a V-shaped outline, (especially in juveniles) and scattered black scales may form indistinct rhomboid or V-shapes on the back and oblique dark bars on the flanks. The tongue is pale blue and black. The hidden margin of the scales is often a vivid blue, and this appears when the snake inflates its body in anger.

Causus resimus

HABITAT AND DISTRIBUTION: Coastal savanna and thicket, dry and moist savanna and woodland, from sea level to 1,800m altitude. It has a patchy distribution in East Africa. Occurs on the Kenyan coastal plain, from the Somali border south to Diani and the Shimba Hills, and a little way up the Tana River. There is an anecdotal report of an individual at Makindu Station (but no supporting specimen), so might extend up the Galana River. It occurs throughout the Lake Victoria basin, from the Kerio Valley and western Kenya through the southern half of Uganda, Rwanda and eastern Burundi, and south through north-western Tanzania to Katavi National Park. Also known from Kidepo National Park, and from Lokitaung on the northern shores of Lake Turkana through to the lower Omo River in Ethiopia. Elsewhere, to coastal southern Somalia, the eastern DR Congo, South Sudan, also known

Causus resimus Velvety-green Night Adder. Vincenzo Ferri, Maasai Mara.

from north-west Angola and on the Cameroon–Chad border, although these might prove to be another species..

NATURAL HISTORY: Terrestrial, but will climb low vegetation in pursuit of frogs. Active by day and night, and will bask. Fairly slow-moving, but strikes quite quickly. Swims well. When inactive, hides in holes, bush piles, under ground cover, etc. If angered it responds like other night adders, hissing and puffing and striking. Diet: amphibians. Lays from 4–12 eggs, captive specimens have been recorded producing clutches at two month intervals, without any noticeable breeding season, but Somali females laid from 4–11 eggs, in July. Little is known of its venom, presumably similar to other night adders.

RHOMBIC NIGHT ADDER *Causus rhombeatus*

Causus rhombeatus

IDENTIFICATION: A small, stout viper with a fairly short head slightly distinct from the neck, a rounded snout and a medium-sized eye with a round pupil, the top of the head covered with nine large scales (unlike most vipers). The body is cylindrical or slightly depressed, the tail is short, 9–12% of total length. The scales are soft and feebly keeled, in 15–23 rows at midbody, ventrals 134–166, subcaudals 21–35, higher counts in males. Maximum size about 95cm (possibly slightly larger), average size 30–60cm, hatchlings 13–16cm. Usually some shade of brown on the back (may be pinkish or greyish brown), occasionally olive green, and some snakes show a limited ability to change colour. There are usually 20–30 dark, pale-edged rhombic blotches all along the back, and a sprinkling of black scales and oblique black bars on the flanks; the rhombic patches may lack white edging; some snakes are quite patternless. On the head there is a characteristic V-shaped mark, this may be solid black, or simply a black outline, with brown inside. The belly is cream to pinkish-grey, the ventral scales may be uniform, but sometimes each scales grades from light to dark, the belly thus looks finely barred.

HABITAT AND DISTRIBUTION: In moist savanna, grassland and woodland, from about 600–2,200m altitude, nearly always in the vicinity of water sources. Widespread in central Kenya, from Nairobi north round the west and north sides of Mt Kenya to the wetter parts

Causus rhombeatus Rhombic Night Adder. Left: Stephen Spawls, Nairobi. Right: Gary Brown, patternless individual.

of the Nyambene Hills, west through the Rift Valley to Kakamega. All around the northern Lake Victoria shore in Uganda and along the Niles to Lake Albert, thence south down the Albertine Rift into eastern Rwanda and Burundi, along the Lake Tanganyika shore past Lake Rukwa and eastwards to the Udzungwa Mountains. Isolated records include the Arabuko-Sokoke Forest, the Chyulu Hills in Tsavo and around Kibwezi in Kenya; also known from the lower southern slopes of Mt Meru and Mt Kilimanjaro, probably occurs in the crater highlands but not recorded there. Elsewhere, north to Ethiopia, south to eastern South Africa, west to Nigeria.

NATURAL HISTORY: Like other night adders: terrestrial, but will climb low vegetation in pursuit of frogs. Active by day and night, and will bask. Fairly slow-moving, but strikes quite quickly. Swims well. When inactive, hides in holes, bush piles, under ground cover etc. If angered it responds like other night adders, hissing and puffing and striking. Diet: amphibians (one specimen, oddly, had eaten a bird). It may also lift the front part of the body off the ground, flatten the neck and move forward with tongue extended, looking like a little cobra. Eats frogs and toads, often swallowing them alive and suffering little obvious initial effect from the venom. Lays from 7–26 eggs, roughly 26–37×16–20 mm, in southern Africa these take about two and a half months to hatch. A venomous snake, the venom causes pain and swelling. In the early days in East Africa, this snake had a fearsome reputation (one of its early names was 'death adder'), but no human fatalities have been reliably reported.

AFRICAN VIPERS *BITIS*

A genus of stout-bodied, broad-headed nocturnal African vipers, all beautifully marked and fairly easily identified in the field. Eighteen species are known; six large, the rest small. All give live birth. They all move slowly but strike quickly, from ambush, the stout body lending stability to the strike. These vipers have a curious small pocket behind the nostril, the supranasal sac. This sac is similar to the pit organ of the rattlesnake and other pit vipers. It may be able to detect radiant heat, and thus serve to help the snake target warm prey animals in the dark, recent experiments have shown that blinded puff adders can still detect and strike at warm-blooded prey.

The small *Bitis* vipers, although venomous, are not medically significant, and have not caused any deaths. But the big ones are dangerous, delivering a large dose of a highly toxic venom in a bite; a bite from one is a medical emergency. The Puff Adder *Bitis arietans* is one of Africa's most dangerous snakes; it is widespread, large, common and irascible, it bites a lot of rural people and their stock.

KEY TO EAST AFRICAN MEMBERS OF THE GENUS *BITIS*

1a	A small horn above each eye......*Bitis worthingtoni*, Kenya Horned Viper [p. 581]	1b	No horn above the eye......2
		2a	A pale line between the eyes, V-shapes

2b	on the back, usually in savanna or semi-desert......*Bitis arietans*, Puff Adder [p. 576] No pale line between the eyes, rectangular or sub-rectangular back marking, usually in forest or woodland......3	**3b**	big black arrowhead in the centre, long horns on the nose of the adult......*Bitis nasicornis*, Rhinoceros Viper [p. 579] Head pale, white or cream, no arrowhead marking, horns absent from nose or relatively short......*Bitis gabonica*, Gaboon Viper [p. 577]
3a	Head green, turquoise or blue, with a		

PUFF ADDER *Bitis arietans*

IDENTIFICATION: Africa's biggest viper, a stout snake, with a broad, flat, triangular head, covered in small overlapping strongly-keeled scales. The small eye, with a vertical pupil, is set far forward, nostrils upturned. Neck thin, body fat and depressed, tail fairly short, 10–15% of total length in males and 6–9% in females. Dorsal scales keeled, in 27–41 rows at midbody, ventrals 123–147, subcaudals 14–39 (usually more than 25 in males). Maximum size depends on locality, in most of East Africa rarely larger than 1.2m, average 70cm–1.1m. Much larger examples occur in northern Uganda, northern and eastern Kenya; a Lolldaiga Hills female was 1.85m. Hatchlings 15–23cm. Colour variable; the ground colour may be brown, grey, orange or yellow, with a series of yellow or cream, dark-edged V-shapes or crossbars along the back. There is a dark oblique bar under each eye, a pale line between the eyes and a broadening pale line from the eye to the angle of the jaw. Below yellow or white, with short irregular black dashes on the outer edge of the ventrals. Males are usually brighter than females, specimens from highland areas often vivid yellow-orange. Occasional odd colour morphs appear (one is illustrated); individuals without, or with subdued, markings, with vertebral stripes, bars instead of chevrons etc.). Taxonomic Note: the doubtfully distinct subspecies *B. a. somalica*, with keeled subcaudal scales, occurs in northern Kenya.

Bitis arietans

HABITAT AND DISTRIBUTION: Throughout East Africa, in all types of country from semi-desert and near-desert to woodland, from sea level to 2,200m altitude, sometimes higher. Absent only from high altitude over 2,400m (although an enigmatic specimen was supposedly reported from 2,740m in the Aberdares) and closed forest (e.g. in south-west Uganda), although specimens were found in Cyamudongo Forest in Rwanda, having entered from nearby farmland. Elsewhere, south to the Cape, north to Eritrea, west to Senegal, isolated populations in southern Morocco and the south-west Arabian peninsula.

NATURAL HISTORY: Terrestrial (occasionally climbs into in low bushes or trees). Nocturnal, sometimes active by day in the rainy season. In high altitude areas it may bask. Hides in thick grass, under ground cover, down holes, in leaf drifts, under bushes, etc. Often poorly concealed under little bushes or in grass tufts but well camouflaged, and thus liable to be trodden on. It hunts by ambush, waiting for prey, which is ambushed with a rapid strike. Crawls in a straight line if unhurried ('caterpillar crawl'), leaving a single broad track; if stressed it can move in the usual serpentine, side-to-side movement, and semi-desert specimens may sidewind. When threatened, inflates the body, hisses loudly (hence the common name) and raises and draw back the front third of the body. If a target is within

Bitis arietans Puff Adder. Top left: Stephen Spawls, Arusha. Top right: Stephen Spawls, Kajiado. Bottom left: Michele Menegon, Ruaha National Park juvenile. Bottom right: Stephen Spawls, Ogaden, somali subspecies. Inset: Caspian Johnson, Ugalla, specimen without head markings.

range, a rapid strike follows. The withdrawal is equally rapid, the snake may overbalance as it does so (hence the dangerous belief that Puff Adders only strike backwards and it is safe to stand in front of one). Captive individuals often remain bad-tempered and noisy. Males indulge in combat, neck-wrestling. Females produce a pheromone which attracts males; a female in season at Malindi was being followed by seven males. They give live birth. The clutch size depends on female size and altitude, usually 10–35 are born to females from medium to high altitude, much larger numbers at low altitude, a Laikipia female had 147 young and a captive Kenyan female had 156 (this is a record number of living offspring for any vertebrate). Diet varied; includes a range of mammals (from rats up to dik-dik) and birds, other known prey includes frogs, toads, lizards, other snakes and even small tortoises. Large prey animals are struck, released and followed; small prey may be simply seized and swallowed. Enemies include monitor lizards, large savanna cobras (although it appears to have some resistance to their venom), various birds, mongooses and caracals have been seen to kill them. Note: this is a very dangerous snake; and its relative abundance, camouflage and disinclination to move if approached means rural people are often bitten. Bite symptoms include pain, progressive and massive swelling, blood blister formation and local necrosis; often resulting in permanent damage or death. A Puff Adder bite is a medical emergency and the victim should be transported immediately to hospital.

GABOON VIPER *Bitis gabonica*

IDENTIFICATION: A fat viper, with a broad, white, flat, triangular head and a body with a remarkable geometric pattern. Between the raised nostrils is a pair of tiny horns. The little eyes, with a vertical pupil and a cream, yellow-white or orange pupil are set far forward. The tongue is black with a red tip. The body is very stout and depressed, tail short, 9–12% of total length in males, 5–8% in females. The scales are heavily keeled, 35–46 rows at midbody, ventrals 124–140, subcaudals 17–33 (higher counts in males). Maximum size about 1.75m, possibly larger. Average 90cm–1.5m (the mean size of a big Nigerian collection was 1.3m), females usually larger than males, hatchlings 25–37cm. The head is white or cream with a fine dark central line and black spots on the rear corners, and a dark blue-black triangle behind and below each eye. Along the centre of the back are a series of pale, sub-rectangular blotches, interspaced with dark, yellow-edged, hourglass markings, on the flanks a series of fawn or brown rhomboidal shapes, with light vertical central bars. Belly pale, with irregular black or brown blotches.

Bitis gabonica

HABITAT AND DISTRIBUTION: Coastal forest and thicket, woodland, forest-savanna mosaic, well-wooded savanna and forest, from sea level to 2,100m. Distribution somewhat sporadic in the west of our area; in Uganda known from the east side of Lake Albert and from Kasese south to the Rwanda border; a single Rwandan record, from the foot of Mt Muhavuru. There are several anecdotal reports from Burundi, including on the Bugarama Plain, but without supporting museum specimens. Also in the forest and woodland of the north Lake Victoria shore. In Kenya known from Kakamega Forest and Nandi Hills, a sight record from Lolgorien, west of the Mara, it might be in the north and south Nandi Forest. An anecdotal report from the Shimba Hills (not shown on map). In Tanzania, south from the Usambara Mountains along the coast, inland to Liwale; isolated populations are known to occur in the Nguu, Nguru and Udzungwa Mountains, and around Tatanda in the south-west. Also recorded just across the south-west border at Mbala in Zambia. Elsewhere, west to Nigeria, south to eastern South Africa.

NATURAL HISTORY: A slow-moving, usually placid, nocturnal viper, spending much of its time motionless, hidden in leaf litter, thick vegetation, under bushes or in thickets, often near a game trail, sometimes climbs into the understory. Hunts by ambush, waiting, often for a long time (a South African specimen remained motionless for 90 days), until suitable prey passes. Males were more active than females, especially after rain, even on cool nights. With big animals, the snake strikes (the fangs may be up to 4cm long), recoils, waits for the prey to die and follows the scent trail, but smaller animals may be simply seized and held until they stop struggling. A study in South Africa in low altitude forest and thicket indicat-

Bitis gabonica Gaboon Viper. Left: Stephen Spawls, Kakamega. Right: Michele Menegon, Usambara Mountains.

ed that Gaboon Vipers avoided dense forest, where there were few small mammals and instead favoured clearing, forest fringes and sunlit thickets. They sometimes basked during the day, and went into damp grasslands at night to hunt. Most hunting occurs in the first six hours of the night. In Kumasi, in Ghana, juvenile Gaboon Vipers were regularly killed in grassland more than 500m from the forest, indicating they were foraging in the open; in Nigeria juveniles were more likely to have newborn mice in their stomachs, indicating they actively foraged. In Tanzania, they have been reported as more frequently encountered than usual in disturbed forest at the edge of oil-palm plantations which have high densities of rats. They give live birth (and the females do not eat while they are heavily gravid), in southern Tanzania broods varied from 11–34, but clutches of up to 60 are recorded in West Africa. A 35cm juvenile was taken in Kakamega Forest in September, hatchlings in mid-December to February in southern Tanzania. The diet is mostly small mammals; shrews and rodents, including rope squirrels, large specimens have taken genets, banded mongooses, brush-tailed porcupines, royal antelope and juvenile duikers; birds are sometimes taken (several Tanzanian specimens had eaten chickens), amphibians and lizards very occasionally recorded. A captive specimen ate several Puff Adders. Males fight in the mating season, neck-wrestling, one male trying to force the other's head down, while hissing continuously and striking with closed mouths. Sometimes hybridise with the Puff Adder in areas where the habitat overlaps. Note: this is a very dangerous snake, with a deadly venom, although bites are rare, due to its placid nature. Symptoms include spontaneous systemic bleeding, swelling, blistering, necrosis and cardiovascular abnormalities. Bites must be treated in hospital with antivenom and clinical support.

RHINOCEROS VIPER/RHINO-HORNED VIPER   Bitis nasicornis

IDENTIFICATION: A large stout viper, with a narrow, flat, triangular head, covered in small strongly-keeled scales, on the end of the nose is a cluster of two or three pairs of horn-like scales, the front pair may be quite long. The small eye is set well forward, the pupil is vertical, the iris green or gold, with black flecks. Neck thin, body triangular in section, tail short, 13–18% of total length in males, 7–10% in females. Body scales rough and heavily keeled, the keels so hard and prominent that they have been known to inflict cuts on snake handlers when the snake struggles. Scales in 30–43 rows at midbody, ventrals 117–140, cloacal scale entire, subcaudals paired 12–32 (higher counts in males). Maximum size about 1.4m, average size 80cm–1.1m, hatchlings 18–25cm. The

Bitis nasicornis

colour pattern is complex. Down the back runs a series of 15–18 large oblong blue or blue-green markings, each with a yellow line down the centre. Irregular black rhombic blotches enclose these markings. On the flanks is a series of dark crimson triangles, sometimes narrowly bordered with blue or green. Many lateral scales are white-tipped. The top of the head is blue or green, with a vivid black arrow mark. The belly is dirty white to dull green, extensively marbled and blotched in black and grey. Specimens from the centre and west of its range tend to be more blue, those from the east more greenish. The tongue is black with a purple base.

HABITAT AND DISTRIBUTION: Forest, woodland and forest-savanna mosaic, from 600–2,400m altitude (elsewhere down to sea level). More of a forest snake than the Ga-

Bitis nasicornis Rhinoceros Viper. Top: Stephen Spawls, Kakamega. Bottom: Stephen Spawls, specimen with some Gaboon viper DNA.

boon Viper, more sensitive to habitat destruction and usually doesn't cling on in deforested areas. Found in south-western and western Rwanda (Ntendezi, Nyungwe Forest, and also Idjwi island, in the DR Congo), all along the western Uganda border to the Budongo Forest, occurs on the Uganda Lake Victoria shore from Bukoba and Minziro Forest in north-west Tanzania, north to Sango Bay, east to Jinja, and also on several of the associated islands, especially those with swamp forest. Kenya records include Kakamega Forest, Nandi Hills, Serem and the Yala River valley, so probably in the north and south Nandi Forest, might be on the lower western Mau Escarpment. A curious Tanzanian record (two specimens in the National Museum, Nairobi) from 'Ulugula Mountains, Usambara' (sic); these are different places and no subsequent specimens have been found. Elsewhere, west to Liberia, isolated records from the Imatong Mountains in South Sudan.

NATURAL HISTORY: Terrestrial, but it regularly climbs into thickets and trees; in a series of 66 from Kakamega, over a third were taken in trees or bushes, sometimes up to 2.5m above ground level. Slow moving, but strikes quickly. It swims, and has been found in shallow pools. Nocturnal, tends to hide during the day among leaf litter, around fallen trees, in holes or among root tangles of big forest trees, will also climb into thicket, clumps of leaves or cracks in trees. It hunts by ambush, and probably spends most of its life sitting motionless, waiting for a suitable prey animal to pass within striking range. More bad tempered than the Gaboon Viper but less than the Puff Adder. If threatened, it can produce an astonishingly loud hiss, almost a shriek. The diet consists mostly of mammals (rodents and shrews) but amphibians are also eaten regularly. They give live birth in March–April (start of the rainy season) in West Africa, few details in East Africa but a Kakamega female had 26 young in March. From 6–38 young have been recorded, the juveniles measuring 18–25cm. Note: although bites are rare, this is a dangerous snake, its bite causing massive local swelling and necrosis; bite victims need to be rapidly transported to a major hospital and treated.

KENYA HORNED VIPER Bitis worthingtoni

IDENTIFICATION: A small, stout horned viper, with a broad, flat, triangular head, covered in small overlapping strongly-keeled scales. The small eye, with a vertical pupil, is set far forward, the iris is silver, flecked with black. There is a single horn on a raised eyebrow. Neck thin, body stout, tail thin and short, 19–33 undivided subcaudals. Scales rough and heavily keeled, in 27–29 (occasionally 31) rows at midbody. Maximum size about 50cm (possibly larger), average size 20–35cm, hatchlings 10–12cm. The ground colour is grey, along each flank is a dirty white or cream dorsolateral stripe and above and below this, on each side, a series of semicircular, triangular or square black markings. There is a dark arrow on top of the head. The belly is dirty-white, heavily stippled with grey.

Bitis worthingtoni

Bitis worthingtoni Kenya Horned Viper. Left: Stephen Spawls, Naivasha. Right and inset: Daniel Hollands, Naivasha.

HABITAT AND DISTRIBUTION: A Kenyan endemic, found in the high grassland and scrub of the Gregory Rift Valley in Kenya. It favours broken rocky country and scrub-covered hill slopes along the edge of the escarpment, right up to the forest's edge, but has also been found on the valley floor, and at the edges of acacia woodland. It is restricted to high altitudes (usually over 1,500m, up to 2,400m) along the high central rift valley. The southernmost record is from the north-west Kedong valley, from where it extends north along the floor and eastern wall of the rift valley through Naivasha and Elmenteita to Njoro. It then extends up the western wall and out of the rift to Kipkabus and Eldoret (most northerly record). It occurs on the Kinangop, around Kijabe, on the hills west of Lake Naivasha, and probably occurs on the eastern Mau escarpment. Might be more widespread, suitable habitat exists on the slopes at the southern end of the Mau, around Narok and south-west of Mt Suswa. Conservation Status: Not assessed by IUCN but it is at major risk from habitat loss; its entire range lies within prime agricultural country; aerial observation shows that much of its range is now used for small-scale farms where this snake is not tolerated; in addition it is collected for the pet trade, particularly in the Naivasha area.

NATURAL HISTORY: Poorly known. Terrestrial. Slow moving, but can strike quickly. Mostly nocturnal, prowling after dusk, but will strike from ambush at any time. Often found sheltering in leaf litter among the stems of the Leleshwa shrub (*Tarchonanthus camphoratus*) which grows mainly along the lower rocky slopes of the escarpment edge, but may also hide under rocks, logs, etc. Bad-tempered when disturbed, constantly hissing and puffing, and struggling wildly when restrained. Captive specimens feed readily on rodents and lizards, struck from ambush or stalked. Gives live birth, 7–12 young are born in March or April, at the start of the rainy season. Note: it is venomous but small, bites can be treated symptomatically.

THE CARPET OR SAW-SCALED VIPERS *ECHIS*

A genus of 11 species of small dangerous vipers, widespread across northern Africa and the Middle East, extending to India; they are often relatively abundant and of major medical significance. Their taxonomy is not yet fully resolved. Seven species occur in Africa, one reaches our area.

NORTH-EAST AFRICAN CARPET VIPER/SAW-SCALED VIPER *Echis pyramidum*

IDENTIFICATION: A small, fairly stout snake, with a pear-shaped head and a thin neck. Top of head covered with small scales, pale yellowish prominent eyes with vertical pupils set near the front of the head, tongue reddish. Body cylindrical or sub-triangular in section, tail short, 10–11% of total length. Scales rough and heavily keeled. Scale counts for Kenyan animals: 25–31 rows at midbody, ventrals 155–182, subcaudals single, 27–41, higher counts usually males. Maximum size about 65cm (possibly slightly larger), average size 30–50cm, hatchlings 10–12cm. The ground colour may be yellowish, brown, grey or rufous. There is usually a series of oblique pale cross bars along the back, with dark spaces between, and along each side there is usually a row of triangular, sub-triangular or circular dark markings, with a pale or white edging. Specimens with very faded or almost invisible markings are known. The belly is pale, finely spotted brown

Echis pyramidum

or red. Taxonomic Notes: Several subspecies of this snake have been described; two from Kenya, those from Wajir and its vicinity as *Echis carinatus aliaborri*, a rufous animal, those from further west as *Echis carinatus leakeyi*. The status of subspecies is uncertain.

HABITAT AND DISTRIBUTION: Near-desert, semi-desert and dry savanna, most Kenyan records from areas with less than 500 mm annual rainfall, between altitudes of 250–1,250m. Probably distributed throughout most of northern Kenya, but at present its distribution is disjunct. Occurs from Lake Baringo north to Lake Turkana, north-west through Kakuma and Lokitaung, north-east through North Horr. Also known from Samburu National reserve north through Laisamis and Marsabit (but not actually on the mountain) to the southern edge of the Dida-Galgalu Desert, east to Kula Mawe, Habaswein and Wajir. There is also a population in the Garissa-Sankuri area on the Tana River. Elsewhere, north to Egypt and Libya, thence east into the Arabian Peninsula. Conservation Status: IUCN Least Concern.

Echis pyramidum North-east African Carpet Viper.
Top: Eduardo Razzetti, Lake Turkana. Centre: Eduardo Razzetti, Lake Turkana. Bottom: Stephen Spawls, Wajir.

NATURAL HISTORY: Terrestrial, although it occasionally climbs into low bushes to avoid hot or wet surfaces. Moves relatively quickly. Nocturnal, active from twilight onwards. During the day hides in holes, under or in logs, under rocks or brush piles, may partially bury itself in sand or coil up in or around grass tufts. A spirited snake, when angry it forms a series of C-shaped coils, these coils are shifted against each other in opposite directions, and this friction between the scales produces a sound like water falling on a very hot plate, at the same time the snake may be moving backwards or forwards. The purpose of this stridulation is to enable the snake to make a warning noise without hissing; which causes water loss in a dry area. If further agitated will strike continuously and vigorously, to the point of overbalancing. And it can strike a relatively large distance. It may, most unusually, even move towards a perceived threat. It can also sidewind at considerable speed. Lays from 4–20 eggs; hatchlings found in July in Wajir, August near Ololokwe. Eats a huge variety of prey, especially invertebrates; a study of 90+ specimens from Wajir in northern Kenya found mostly remains of solfugids, scorpions, centipedes and grasshoppers in the stomachs, only three specimens contained non-invertebrate remains; two had writhing-skink (*Lygosoma*) tails and one contained mammalian hair; but of 15 specimens examined at Ololokwe; fourteen had eaten lizards (mostly *Latastia*), and one contained a scorpion. Other known prey items include birds, snakes (even its own kind) and amphibians. This may explain its relative abundance in certain areas; in the Moille hill area of northern Kenya, in an area of 6,500km^2, nearly 7,000 of these snakes were collected in just under four months. However, in other parts of its range it is uncommon; around Cairo two specimens were found in 25 years. Note: this is a dangerous snake and many people are bitten every year in northern Kenya. Symptoms include local pain, swelling, blistering and a generalised bleeding tendency; antivenom therapy may be necessary.

BUSH VIPERS *ATHERIS*

A tropical African genus of distinctive small, broad-headed vipers, most are arboreal. Bush vipers are venomous, one species has caused fatalities, no antivenom is produced against their bite. Sixteen species are known, nine occur in East Africa. Several species of bush-viper can be instantly identified to species by their locality. The key below is largely based on a mixture of colour and locality.

KEY TO EAST AFRICAN MEMBERS OF THE GENUS *ATHERIS*

1a   Tail very short, subcaudals 14-22, only in southern Tanzania, brown in life......*Atheris barbouri*, Udzungwa Viper [p. 585]
1b   Tail long, subcaudals 38-65, widespread in suitable mid-altitude woodland and forest, usually some mixture of black, yellow and green in life......2

2a   In hill forests of central Kenya......*Atheris desaixi*, Mt Kenya Bush Viper [p. 589]
2b   Not in central Kenya......3

3a   Usually horns above the eye.....4
3b   No horn above eye......5

4a   Midbody scale rows 21-23......*Atheris ceratophora*, Usambara Bush Viper [p. 586]
4b   Midbody scale rows 26-27......*Atheris matildae*, Matilda's Bush Viper [p. 588]

5a   In south-western and western Tanzania, colour yellow and green, sometimes with black markings......*Atheris rungweensis*, Mt Rungwe Bush Viper [p. 592]

5b   Not in south-western and western Tanzania, colour not necessarily yellow and green......6

6a   Midbody scale rows 14, western Uganda......*Atheris acuminata*, Acuminate Bush Viper [p. 586]
6b   Midbody scale rows more than 14, not confined to western Uganda......7

7a   Scales strongly lanceolate (prickly) on the front half of the body, body very slim......*Atheris hispida*, Rough-scaled Bush Viper [p. 590]
7b   Scales not lanceolate on the front half of the body, body not very slim......8

8a   Stout, green with heavy black markings, along the Albertine Rift, midbody scales 23-34......*Atheris nitschei*, Great Lakes Bush Viper [p. 591]
8b   Not noticeably stout, green, usually no black markings, not necessarily along Albertine Rift. Midbody scales 15-25......*Atheris squamigera*, Green Bush Viper [p. 593]

BARBOUR'S SHORT-HEADED VIPER/UDZUNGWA VIPER
Atheris barbouri

IDENTIFICATION: A small forest viper endemic to the Udzungwa and Ukinga mountains, Tanzania. Unusually for a bush viper, it is terrestrial, and thus adapted by not having a prehensile tail and by laying eggs. It has a triangular head with a distinctly short and rounded snout; the tail is short and non-prehensile. The scales of head and body are strongly keeled and in 19–23 rows at midbody. Ventrals 115–127, the cloacal scale is entire; subcaudals single, 15–23. Total length of largest male 35.2cm, largest female 36.9cm. The body is brown to dark olive in colour above, with a pair of pale yellow, zigzag dorsolateral stripes which extend from the back of the head to the end of the tail. An irregular chain of darker rhombic blotches may be present on the back. Faint black chequering may be present on the tail. The ventrum is greenish-white to olive. Females may be more speckled than males.

Atheris barbouri

HABITAT AND DISTRIBUTION: An enigmatic Tanzanian endemic, from the hills of southern Tanzania, found in mid-altitude woodland of the Udzungwa and Ukinga Mountains, at around 1,700–1,900m altitude, in thick bush and bamboo undergrowth, but also found in gardens of tea farms. Known localities include: Udzungwa Mountains: Bomalang'ombe, Dabaga Kifulilo, Lugoda, Massisiwe and Mufindi; in the Ukinga Mountains localities include Madehani and Tandala. Conservation status: IUCN Vulnerable B1ab. Very little is known about the ecology and habitat requirements of this species; it would appear to be forest dependent and while it may be found in forest edge situations, such as tea plantations, it is unlikely to survive in the absence of forest nearby, nor is it likely to survive extensive forest destruction.

NATURAL HISTORY: Very poorly known. This species has a large eye with a vertical pupil suggesting a nocturnal life style. Some specimens were collected during daylight hours, usually after rain when the sun was shining. Specimens have been collected in January–March, June, September and October. Preferred habitat seems to be moist forest at 1,800–1,900m. All recent specimens have been collected on the forest floor, but earlier specimens

Atheris barbouri Udzungwa Viper. Left: Michele Menegon, Ukinga Mountains. Right: Michele Menegon, Ukinga Mountains.

were recorded among agricultural plots and gardens at the edge of forest or formerly forested land. This species possesses no known specialisations for burrowing or for climbing, and it would appear that it is mainly an animal of the forest floor leaf litter. The little data available suggest that earthworms constitute the main diet of juveniles and adults; small frogs may also be taken. Three females collected in February 1930 each contained 10 eggs, the largest of which measured 10×6mm; a fourth was non-gravid. Three females collected at Mufindi in June 1983 held fairly large follicles, as did a female in October. The species appears to lay eggs. Nothing is known of the venom but unlikely to be life-threatening.

ACUMINATE BUSH VIPER Atheris acuminata

IDENTIFICATION: A small, slim bush viper, described from a single specimen from Uganda, it is virtually identical to the Rough-scaled Bush Viper and might just be an aberrant example of that species. Head sub-triangular, eye fairly large, pupil vertical, iris mottled green with gold bordering the pupil. Many small scales on top of the head. Body sub-triangular in section, tail 18% of total length. Scales keeled, 14 midbody rows, the dorsal scales are very elongate and taper to a point (the name acuminate translates as tapering to a point), ventrals 160, subcaudals 54. The specimen was 44cm long, tail 8.1cm; yellow-green above, with a vague H-shaped black marking on the crown and a short black blotch behind the eye. The tail is black-blotched, the belly is pale greenish-yellow, black-blotched posteriorly.

Atheris acuminata

HABITAT AND DISTRIBUTION: A Ugandan endemic, known from a specimen from Kyambura game reserve, just south of Lake George, altitude 950m, in riverine forest. Little data in conservation terms, but it was found in a protected area.

Atheris acuminata Acuminate Bush Viper Don Broadley, preserved type.

NATURAL HISTORY: Nothing known. It was found on a path. Its similarity to the Rough-scaled Bush Viper might mean it has much the same habits, i.e. nocturnal, arboreal, eats small rodents and frogs, lays eggs.

HORNED BUSH VIPER/USAMBARA BUSH VIPER
Atheris ceratophora

IDENTIFICATION: A small 'horned' bush viper of the Eastern Arc Mountains in Tanzania. It has a noticeably triangular head covered with small, keeled scales, with a little cluster of 'horns', two or three elongate scales above the eye. The neck is narrow; the dorsal scales are strongly keeled; 19–27 midbody scale rows, ventrals 134–152, subcaudals 46–58. Maximum size 55cm, average 40–50cm, hatchlings 12–16cm. The tail is prehensile. Colour very variable; this snake may be uniform olive green, black (with or without light cross bars), orangey-yellow with black mottling near the tail, golden yellow with increasing dull green and black mottling towards the tail, barred black and yellow, dull brownish-yellow or bright yellow with black mottling. Hatchlings are either shiny or dull black, with a vivid yellow tail tip (which may be sued as a caudal lure, waved to attract an insect-eating prey). Juveniles gain their adult colours after the third moult of the skin and at this stage their horns become noticeable.

HABITAT AND DISTRIBUTION: Lives in forest and woodland at low to medium altitude, a forest dependent species; at below 700m in the Udzungwa Mountains, but found at over 2,000m in the Usambara National Park. Endemic to Tanzania: known only from the Usambara, Nguru, Uluguru and Udzungwa Mountains. Conservation Status: Conservation status: IUCN Vulnerable B1ab. A forest-dependent species but much of its habitat is in protected areas.

Atheris ceratophora

NATURAL HISTORY: Theoretically nocturnal, certainly arboreal, most specimens have been collected during the day time when animals are either basking or are detected by their movement as they crawling on the forest floor. In captivity, some individuals remain motionless for weeks on the same branch. Usually found on the ground, or coiled in a clump of vegetation up to 2m above ground. Most specimens have been found between January and April, when it is warm; during the colder months, when minimum temperatures may drop to 6°C, the animals may go into a period of inactivity, if not true hibernation. Few details are available on its feeding, but amphibians probably make up a major portion of its diet. Limited evidence suggests that breeding takes place in September and October; young have been found in April, and captured females gave birth in March. The specific name refers to the horn-like scales (*cerato* = horn). A venomous snake but unlikely to be dangerous, the venom causing local pain.

Atheris ceratophora Usambara Bush Viper/Horned Bush Viper. Top left: Michele Menegon, Udzungwa Mountains. Top right: Michele Mwenegon, Udzungwa Mountains. Bottom left: Michele Menegon, juvenile. Bottom right: Lorenzo Vinciguerra, black individual.

MATILDA'S BUSH VIPER *Atheris matildae*

IDENTIFICATION: A recently-described, relatively large, heavy-bodied tree viper from south-western Tanzania; adults often larger than 70cm. The head is broad and triangular, covered in small scales. The small eye, with a vertical pupil, is set far forward. The tail is long and prehensile. Dorsal scales strongly keeled, in 26 or 27 rows at midbody, ventrals 142–150, subcaudals 44–50. Maximum size is probably around 80cm. Its coloration, though quite variable, is generally black with bright yellow dorsolateral zigzag lines running its length. A black patch around nasal, rostral, mental and first infralabials is present in most of the observed males but also in few females and immature individuals. Males in general tend to be darker with belly suffused with black.

Atheris matildae

Adult females tend to be more yellow, in some cases with immaculate throat and belly; horn-like scales are yellow with black outer edges. The side of the head can be completely yellow or with black patches at the tip of the scales. It is distinguished from most other *Atheris* vipers by the presence, in both males and females, of enlarged, supraciliary scales that create the viper's namesake horns. Closely resembles the only other horned bush viper *A. ceratophora*, but it is larger and heavier and has differences in scalation.

HABITAT AND DISTRIBUTION: Known from just two forest fragments in the southern highlands of Tanzania, north of the northern tip of Lake Malawi (Lake Nyasa).

Atheris matildae Matilda's Bush Viper. Top left: Michele Menegon, southern highlands, Tanzania. Top right: Michele Menegon, southern highlands, Tanzania. Bottom: Michele Menegon, southern highlands, Tanzania.

NATURAL HISTORY: Poorly known. Individuals were found both climbing on bushes and trees and moving on the ground; it appears that larger individuals become more terrestrial and hide in rodent burrows. Presumably feds largely on small vertebrates and gives live birth. Nothing is known of the toxicity or composition of the venom, but as with other larger bush viper species, a bite from this species might be serious.

MT KENYA BUSH VIPER *Atheris desaixi*

IDENTIFICATION: A large, thick-bodied bush viper, found only in the forests of high central Kenya, discovered in 1967. The head is broad and triangular, covered in small strongly-keeled scales. The small eye, with a vertical pupil, is set far forward. The tail is long and prehensile, 13–15% of total length in females, 16–18% in males. Scales keeled, in 24–31 rows at midbody, ventrals 160–174, subcaudals 41–54 (higher counts in males). Maximum size about 70cm, average size 40–60cm, hatchlings size 17–22cm. The body is greeny-black to charcoal black in colour, each scale edged with yellow or yellowy-green, creating either a speckled effect or a series of yellow loops. On the hind-body and tail the speckles may fuse into yellow zigzags. The belly is yellow on the front half, becoming progressively suffused with purplish black to the rear and under the tail, the tail tip is blotchy yellow. Occasional yellow-brown or dull green individuals occur, they have faint darker festoons. Hatchlings predominantly yellow, with a white tail tip, but gradually darken to adult colour at around 30cm length.

Atheris desaixi

HABITAT AND DISTRIBUTION: Endemic to the central highlands of Kenya, in mid-altitude evergreen forest between 1,000–1,800m. There are three populations; one in the Ngaia (Ngaya) forest, at 1,000m on the north-east flank of the Nyambene Range, one around Igembe, at 1,300m in the northern Nyambene range and one in forest just west of Chuka, south-eastern Mt Kenya, between 1,500–1,800m. Might be more widespread; snakes that match the description of this viper have been reported from mid-altitude forest in the lower salient in the Aberdares, from near Meru and in the Kikuyu Escarpment forest northeast of Kijabe. Conservation Status: Not assessed by IUCN but it is of conservation

Atheris desaixi Mt Kenya Bush Viper. Left: Stephen Spawls, Chuka, dark phase. Right: Stephen Spawls, Chuka, light phase.

concern; two of its known localities are unprotected, although Ngaia Forest is nominally protected. This snake might prove to be more widespread, but its known habitats are being rapidly felled; the Igembe Forest has virtually disappeared, the Chuka population was recently surveyed and is being decimated by collection for the pet trade. Its distribution needs investigation and, if feasible, an area of its habitat needs rigorous protection.

NATURAL HISTORY: Arboreal, slow-moving snakes. Activity patterns not known, might be diurnal or nocturnal, or both. Usually found draped in low vegetation, around 2–3m from the ground, around the edges of small clearings in forest, but has been seen 15m up a tree. The Ngaya specimen was in a lantana bush at the forest edge. Some Nyambene specimens were in yam plantations. They are perfectly camouflaged and difficult to observe. They are very willing to strike when first caught, they will form C-shaped coils, like carpet vipers, the coils are shifted against each other in opposite directions, producing a hissing sound, like water falling on a hot plate. They struggle fiercely in the hand, but soon tame in captivity. Little is known of their biology. They appear to feed on small mammals. A female from the Nyambene range gave birth to 13 young in August, the smallest was 17.5cm and the largest 21.1cm. The venom is unstudied, but a bite victim suffered swelling and pain.

ROUGH-SCALED BUSH VIPER *Atheris hispida*

IDENTIFICATION: A long slender bush viper with bizarrely long scales. Head sub-triangular, eye fairly large, pupil vertical, iris brown, heavily speckled with black (looks uniformly dark under most conditions). Many small scales on top of the head. Body cylindrical in section, tail long, 17–21% of total length. The scales are heavily keeled, prickly and leaf-shaped (this is a good field character), in 15–19 midbody rows, ventrals 149–166, subcaudals 49–64 (high counts usually males). Maximum size 73.5cm, but this is unusual, average 40–60cm, hatchlings 15–17cm. Males are usually olive-green in colour, with a black mark on the nape (an H, V, or W shape, or just a blotch), sometimes a dark line behind the eye. The ventrals are greenish, darkening towards the tail, females are usually yellowy or olive brown, with a similar dark nape mark, yellow-brown below.

Atheris hispida

Atheris hispida Rough-scaled Bush Viper. Left: Bio-Ken archive, Kakamega. Right: Philipp Wagner, Kakamega.

HABITAT AND DISTRIBUTION: Associated with forest, woodland and thicket, sometimes in waterside vegetation at altitudes from 900–2,400m. It has a bizarre, disjunct distribution. One Tanzanian (Minziro Forest, north-west) and one Kenyan (Kakamega Forest) record, in south-west Uganda in the south-west Rwenzoris, in Kigezi game reserve and Bwindi Impenetrable Forest. No Rwandan or Burundi records, but taken at Rutshuru, in the DR Congo, so probably in the Bufumbira Range in north-west Rwanda, also known from just west of Lake Kivu in the DR Congo. Elsewhere; sporadic records from north-east of the DR Congo.

NATURAL HISTORY: Poorly known. Arboreal, an expert climber, moving relatively quickly through the branches. Lives in tall grasses, papyrus, bushes, creepers and small trees. In Kakamega forest, these snakes were found slightly higher up than the sympatric Green Bush Vipers, in dryer bushes, but were also found in reeds. They will bask on top of small bushes or on flowers. Although probably nocturnal, they will opportunistically strike from ambush if prey passes by. They are irascible snakes, struggling furiously if held, and forming the body into C-shaped coils like the Mt Kenya Bush Viper. Their mouths are often full of small back mites. They give live birth, clutches of 2–12 young recorded; two Kakamega females had clutches of 2 and 9 babies in mid-April, the time of highest rainfall there. Diet not well known, the holotype had a snail in its stomach. Captive specimens at Nairobi Snake Park took small rodents and tree frogs (*Hyperolius*) at night; when the snake saw a movement on the ground it would descend and either strike down from a branch or go onto the cage floor to stalk. Population numbers of this species in Kakamega seem to peak and crash, for reasons as yet unexplained, in some years they are abundant, in others cannot be found. The venom is unstudied and no bite cases documented.

GREAT LAKES BUSH VIPER *Atheris nitschei*

IDENTIFICATION: A stout, black and green, large bush viper. Head triangular, neck thin, eye fairly large, set well forward with a vertical pupil but this is hard to see, the whole eye just looks black, although the iris is brown. Many small scales on top of the head. Body cylindrical and stout, tail fairly long and strongly prehensile, 15–17% of total length in males, 13–16% in females. Scales keeled, 23–34 midbody rows, ventrals 140–162, subcaudals 35–59 (higher counts in males). Maximum size about 75cm, average 45–65cm, hatchlings 15–18cm. Colour various shades of green, blue-green, yellowy-green or olive, heavily speckled, blotched or barred black; there is usually a conspicuous black bar behind the eye, a dark blotch or V-shape on top of the head. The belly is yellow or greeny-yellow. There is an ontogenic colour change; hatchlings are deep green (almost black), brown or grey-brown, with a white or yellow tail-tip, after three or four months they become uniform green, black blotches then appear gradually.

Atheris nitschei

HABITAT AND DISTRIBUTION: A characteristic species of the Albertine Rift. It lives in medium- to high-altitude moist savanna and woodland, montane forest and bamboo, sometimes associated with lakes and swamps, in waterside vegetation, at altitudes between 1,000–2,800m. In Uganda, occurs along the eastern flanks of the Rwenzoris and in Rukungiri, Kabale and Kisoro, in the volcano country, south-west Uganda (fairly common

Atheris nitschei Great Lakes Bush Viper. Top left: Michele Menegon, striking. Centre: Max Dehling, Rwanda. Right: Michele Menegon, Rwanda.

in Bwindi Impenetrable National Park) thence down the western side of Rwanda (where largely in primary forest or tea plantations), and Burundi, to the Lake Tanganyika shore (but doesn't enter Tanzania). South of there it is replaced by the Mt Rungwe Bush Viper. Elsewhere, in the high country of the eastern DR Congo, to about 5°S on the west shore of Lake Tanganyika.

NATURAL HISTORY: Arboreal, living in papyrus and reedy swamp vegetation, elephant grass, bushes, small trees and bamboo. It will descend to the ground to hunt. Probably nocturnal, but known to bask, at heights of 3m or more in creepers, elephant grass and papyrus. If disturbed it slides quickly downwards or simply drops off its perch. In Bwindi, it was found waiting in ambush for diurnal lizards such as Jackson's forest lizard, concealed in grass at the base of roadcut walls. Hatchlings used caudal luring, waving their bright tail tips to attract insect-eating prey. Some specimens are placid, but often it is bad-tempered, hissing, striking and forming C-shaped coils when angry. It gives live birth, clutches of 4–13 young, a female from Mt Karissimbi gave birth to 12 young in February, in south-west Uganda gravid females were collected in October and November, and in January in the Rwenzoris. Diet includes small mammals, amphibian, lizards (including chameleons); captive specimens descended to the ground to stalk rodents. A bite by this species in Rwanda caused extreme pain, swelling, oedema and necrosis resulting in the loss of the top joints of a finger; another caused clotting abnormalities. It would seem it is a typical viper venom; no antivenom is produced.

MT RUNGWE BUSH VIPER *Atheris rungweensis*

IDENTIFICATION: A large bush viper of south-western Tanzania. The large head is triangular in shape and rather flat in profile, with a distinct neck. Dorsals keeled and pointed, but with keels ending before the tip, 22–33 rows at midbody; ventrals 150–165; subcaudals 46–58. Maximum size about 65cm, average 35–55cm, hatchlings 15–17cm. Colour predominantly green, light or dark, usually with oblique yellow crossbars, festoons or zigzags, in some specimens the upwards curve of the festoon has a black infilling. The underside is greenish-yellow. New-born animals are a dark brown or grey, with a bright yellow tip to the tail. They may use this brightly coloured tail as a 'lure' to attract prey. After several moults, they develop a uniform green and then more patterned adult colouration.

Atheris rungweensis

HABITAT AND DISTRIBUTION: Most records from moist savanna, woodland and hill

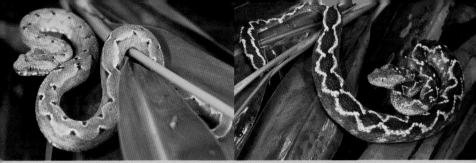

Atheris rungweensis Mt Rungwe Bush Viper. Left: Michele Menegon, Mbizi Forest. Right: Michele Mwenegon, Mt Rungwe.

forest of south-western Tanzania, at altitudes between 800–2,000m. Occurs southwards from the Gombe Stream National Park and Kigoma down the eastern shores of Lake Tanganyika to the Mahale Peninsula, and inland to the Ugalla River, it also occurs in forest patches in the high country around Sumbawanga, between the north end of Lake Rukwa and the south end of Lake Tanganyika, and on Mt Rungwe, just north of the north end of Lake Tanganyika, thence south into northern Malawi and Zambia.

NATURAL HISTORY: Poorly known. Arboreal, and the vertical pupil suggests it is nocturnal. An expert climber, found in bushes 1–3m above the ground, or on the ground at edge of forest from 800–2,000m altitude. Might hunt on the ground or ambush from trees. Gives live birth. Diet includes small frogs. Nothing is known about its venom.

GREEN BUSH VIPER *Atheris squamigera*

IDENTIFICATION: A large green bush viper, with fine yellow cross-bars, sporadically recorded in western East Africa. The head is broad and triangular, covered in small strongly-keeled scales. The small eye, with a vertical pupil, is set far forward. The tail is long and prehensile, 15–17% of total length in females, 17–20% in males. Scales keeled, in 15–25 rows at midbody, ventrals 133–175, subcaudals 45–67 (higher counts in males). Maximum size about 80cm, average size 40–65cm, hatchlings size 16–22cm. Colour in East Africa nearly always green or yellow-green (very occasionally brown), often becoming turquoise towards the tail, sometimes yellow-tipped scales form a series of fine yellow cross bars. The belly is greenish-blue, with yellow blotches. Juveniles are olive,

Atheris squamigera

green or yellow-green, with olive, dark-edged V-shapes on the back. The hatchling are more strongly banded than adults, but have a bright yellow or pink tail tip, which may be used as a lure. A wider range of colours known elsewhere includes orange, yellow and grey specimens.

HABITAT AND DISTRIBUTION: All types of forest, also in well-wooded savanna, from 700–1,700m altitude in East Africa, to sea level elsewhere; in Nigeria it is found in a wide range of habitats including suburbia, agriculture and mangroves. It seems to be separated by altitude from the Great Lakes Bush Viper in Uganda, the Green Bush Viper usually below 1,600m, the Great Lakes Bush Viper above that. In Kenya, known from Kakamega Forest, an odd record from Chemilil and a possible sight record from the woodland above the Soit Ololol escarpment above the Mara. Uganda records include the forest of western Mt Elgon, the lakshore woodland east from Entebbe to Jinja, Budongo Forest and Masindi,

Atheris squamiger Green Bush Viper. Left: Stephen Spawls, Kakmega. Right: Michele Menegon, captive.

Semliki (Bwamba Forest) and the Kibale area. Not at high altitude, found just outside the Bwindi Impenetrable National Park, at lower middle elevations. The only Tanzanian record is from the Rumanyika National Park, in the north-west. No Rwanda or Burundi records. Elsewhere, west to Ivory Coast, south-west to northern Angola.

NATURAL HISTORY: Largely arboreal, slow-moving snakes, but they will descend to the ground at night. Largely nocturnal, most active between sundown and 2am. They may bask on top of vegetation in clearings during the day; Kakamega animals sometimes found on the ground. Usually found in bushes and small trees, but may climb up to 6m or more above ground level. They are willing to strike when first caught and form C-shaped coils, like carpet vipers, the coils are shifted against each other in opposite directions, producing a sizzling sound, like water falling on a hot plate. At night, they descend to low level, and wait in ambush with the head hanging down. They can also drink in this position, sipping water from condensing mist or rain running down the body. They feed largely on small mammals, also known to take birds, lizards, amphibians and small snakes; adults eat mostly warm-blooded prey; juveniles ate lizards and shrews. Mating observed in October in Uganda, seven to nine young born in March–April. In areas where the Rough-scaled Bush Viper also occurs, this species is ecologically separated, living in flowering, low, thicker bushes, the rough-scaled bush-viper lives in taller thinner shrubs. Note: this bush viper has killed people; a fatal bite case showed massive swelling and incoagulable blood; a Nigerian victim lapsed into a coma (but recovered). Antivenom is not available, so bites may need symptomatic treatment.

MONTANE VIPER *MONTATHERIS*

A genus containing a single species that lives at high altitude in central Kenya.

KENYA MONTANE VIPER *Montatheris hindii*

IDENTIFICATION: A small, slender viper from the moorlands of the Aberdare Mountains and Mt Kenya. It has an elongate head, covered in small strongly-keeled scales. The small eye, with a brown iris and vertical pupil, is set far forward. The body is cylindrical, the tail short, 10–13% of total length. Scales keeled, in 24–28 rows at midbody, ventrals 127–144, subcaudals paired, 25–36. Maximum size about 35cm, average size 20–30cm, hatchlings 10–14cm. The dull grey or brown body has a paired series of pale-edged, black, triangular blotches along the back; on the Aberdares males are more brown, females grey, but on Mt

Kenya both sexes seem to be predominantly warm grey. The underside is yellow, cream or grey-white, but heavily mottled black, to the extent of appearing black in places. There is an irregular, dark brown, arrow or V-shaped mark on the crown of the head. A wide dark stripe passes through the eye to the temporal region. The upper and lower labials are white.

Montatheris hindii

HABITAT AND DISTRIBUTION: This beautiful, tiny, unique Kenyan endemic is found only at high altitude (2,700–3,800m) in treeless montane moorland. There is an isolated population on the Aberdare Mountains and another on Mt Kenya. Probably the smallest range of any dangerous African snake. Conservation Status: IUCN Not Assessed; it has a tiny range, but fortunately it all lies within two national parks. Might be at risk from global warming; it is cold adapted and there is a limit to how much further up it can go.

NATURAL HISTORY: Poorly known. Terrestrial. Somewhat sluggish unless warmed up. It is active by day, as the night-time temperatures in its mountain habitat are usually below freezing. Usually active between 10am and 4pm. It shelters in thick grass tufts that provide cover and insulation from the extreme cold, but may also hide under ground cover (rocks, vegetable debris) or in holes. Due to the cold winds and rarefied air, much time is spent

Montatheris hindii Kenya Montane Viper. Top, Left and Right: Stephen Spawls, Aberdares.

inactive in shelter. On days of warm sunshine it emerges to bask on patches of warm soil or a grass tussock, and it will then hunt. It is irascible and willing to bite if threatened. A female from the Aberdares gave birth to two live young, of 13.1 and 13.5cm length, in late January; a captive female produced three young (length 10.2–10.9cm) in May; small juveniles (15–17cm) were collected in February. Lizards, including chameleons and skinks, and small frogs are eaten, it may take small rodents as well; a Mt Kenya specimen ate a shrew whilst in a pitfall trap. It is quite common in suitable habitat, and females seem to be found more often than males, probably because they need to bask more frequently when gravid. The enemies of this little viper will include predatory birds, Augur Buzzards *Buteo augur* have been seen to take them. The relationships of this little viper, and its evolutionary history are mysteries. Nothing is known of the venom.

LOWLAND/FLOOD PLAIN VIPERS   *PROATHERIS*

A genus containing a single, medium-sized terrestrial viper, found on a river floodplain in south-eastern Africa.

FLOODPLAIN VIPER   *Proatheris superciliaris*

IDENTIFICATION: A small, moderately robust terrestrial viper, pale with black saddles or spots. It has a distinct, rather elongate triangular head which is covered with small, keeled, overlapping scales. In East Africa, known only from the floodplains of the northern end of Lake Malawi (Lake Nyasa), southern Tanzania. The tail is short and distinct in females, but longer and less distinct in males. Dorsal scales in 27–29 (rarely 26 or 30) rows at midbody, strongly keeled and overlapping, the outermost row enlarged and feebly keeled to smooth. Ventrals smooth, 131–156; cloacal scale entire; subcaudals smooth, in 32–45 pairs. Maximum size about 60cm in females, 55cm in males, average 25–45cm, hatchlings 13.5–15.5cm. The dorsum is grey-brown with three rows of dark brown spots separated laterally by a series of elongate, yellowish bars which form an interrupted line on either side of the body. Three dark, chevron-shaped marks cover the front of the head. The belly is off white, with numerous black blotches in irregular rows. The

Proatheris superciliaris

Proatheris superciliaris Floodplain Viper. Stephen Spawls, Malawi.

lower surface of the tail is straw yellow to bright orange. Its colour and pattern make it extremely difficult to detect unless it is in motion.

HABITAT AND DISTRIBUTION: In East Africa, known only from grasslands bordering floodplains, and floodplains in southern Tanzania at the northern end of Lake Malawi (Lake Nyasa). The species ranges south through Malawi to central Mozambique and Beira. It may be more widely distributed in northern Mozambique, since it is known from Cape Delgado in the northeast of that country, might be in south-east Tanzania. Conservation Status: Not known to be threatened.

NATURAL HISTORY: Terrestrial. Just before the breeding season, Floodplain Vipers can be seen basking at the mouths of rodent burrows. As is suggested by its vertical pupil, this snake is most active in the early evening, when its prey, mostly amphibians, but also small rodents, are also active. During the cold season (April–July) when night temperatures drop to 6°C, animals may bask in front of their retreat during daylight hours. When threatened, Floodplain Vipers throw their bodies into C-shaped coils, and can strike rapidly. Mating occurs in July and from 3–16 young are born November–December; new-born measure 13.5–15.5cm in length. Diet includes small amphibians and rodents. The venom causes pain, swelling and blistering, but is unlikely to be life-threatening.

GLOSSARY

Allopatric: occurring in separate regions, not overlapping.

Autotomy: the loss of a part of the body (for example the tail in lizards) as a defence mechanism by an animal under threat.

Bicuspid: having two points or peaks.

Brille: the transparent scale that covers a reptile's eye.

Buccal: relating to the mouth.

Caniculate: having longitudinal channels.

Canthal: any scale that lies along the canthus rostralis.

Canthi rostrales: (singular, canthus rostralis): in chameleons, the ridges or crests that connect the ridge above the eye to the snout.

Casque: the raised helmet-like structure on the back of the head of some chameleons.

Colubrid: a 'typical' snake of the family Colubridae (a family now much reduced); in its older sense a snake without front fangs and with broad ventral scales that span the width of the belly.

Costals: plates or shields on the carapace of a chelonian, between the margins and vertebrals (see Figure 3).

Crepuscular: active at twilight.

Diapsid: animals that developed two holes in each side of their skulls, although some have now lost them; the group includes all reptiles.

Frontonasal: scale at the front of a lizard's snout (see Figure 7).

Frontoparietals: paired scales in the upper centre of a lizard's head (see Figure 7).

Gular: relating to the throat.

Holotype: a single type specimen that the description of a new species is based upon.

Immaculate: in terms of appearance, uniformly coloured, without any spots or other markings.

Inguinal: relating to the area just in front of the point where the hind limbs connect to the body, the groin.

Intergular: a small scale between the gular shields at the front of a turtle plastron.

Internasal: the scales between the nostrils of a snake (see Figure 26).

Interparietal: a scale at the back of a reptile head (see Figure 7).

Interstitial: between the scales (usually refers to the thin skin between a reptile's scales).

Lamellae: any thin plate-like scale or scale structure, usually refers to the small scales across the underside of lizards' digits.

Loreal: a scale between the eye and the nostril (but not touching either) of a snake; its presence or absence is a useful diagnostic tool.

Mental: a scale right at the front of the lower jaw of a reptile (see Figure 8).

Monophyletic: of a group of organisms that are descended from a single common ancestor, not shared with any other group.

Nuchal: relating to the nape of the neck; in snakes the nuchals scales are small scales at the back end of the enlarged head scales, in a tortoise the term refers to the small scale at the front centre of a carapace.

Occipital: relating to the back of the head or the skull.

Occiput: the back part of the skull or the head, where it joins the vertebral column.

Ocelli: eye-like spots or markings.

Orbit: the bony socket of the eye, or the skin bordering the eye of a reptile.

Osteoderms: very small bones within the scales of some reptiles, toughening them.

Parapatric: of two species, occurring immediately adjacent to one another but their ranges do not overlap.

Paravertebral: usually refers to scales on either sided of the vertebral column.

Parietals: relates to the crown of the head; in squamate reptiles it usually refers to the large scales just behind the eye on top of the head (see Figures 7 and 26).

Pentadactyl: Having limbs with five digits.

Plantar: to do with the sole or underside of the (hind) feet.

Plastron: the lower surface of a chelonian's shell.

Precloacal: in front of the cloaca.

Quadricarinate: having four ridges or keels.

Rostral: related to the nose, also reptile scale at the front of the head (see Figures 7 and 26).

Rugose: wrinkled, ridges or corrugated.

Rupicolous: living on or around rocks.

Scansors: specialised pad on the toe tips (usually of a gecko), consisting of minute hairs, setae, that by their gripping properties enable the animal to climb vertical surfaces.

Scutes: scales.

Solifugids: order of relatively large, hard biting but non-venomous arachnids.

Spectacle: see brille.

Spinose: bearing spines.

Stylet: small slender pointed degenerate limb.

Subcaudals: scales under the tail, usually refers to such relatively broad scales in snakes.

Subimbricate: weakly overlapping.

Sulcus: a furrow or faintly depressed linear groove.

Supra-orbital: above the eye.

Supraciliaries: relating to the region above the eye, scales above the eye (see Figure 7); usually small and present in certain lizards.

Supranasals: scales above the nose.

Supraocular: scales above the eye (see Figures 7 and 26).

Sympatric: of two or more species, occurring in the same geographical areas or regions.

Syntopy: of two species, occurring in the same place at the same time.

Temporals: scales behind the eye but behind the postocular scales (see Figures 6 and 25).

Temporo-lateral: the rear side of the head.

Tricarinate: of scales, having three keels or ridges.

Tridactyle: having three digits.

Tubercles: small rounded protuberances or projections, usually refers to enlarged button-like scales in geckoes.

Tympanum: the eardrum.

Type: the specimen(s) on which the original description of a species is based.

Unicuspid: having a single cusp.

Venter: refers to the underside.

Vertical: a group of structures arranged in a circle, usually refers to a cluster of small scales.

LINE DRAWINGS

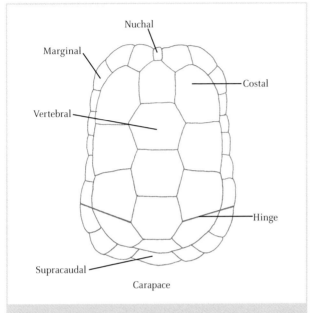

1. *The upper shield (carapace) of a tortoise shell (after Branch)*

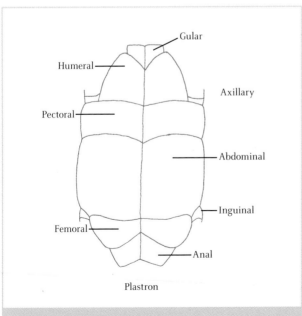

2. *The under shield (plastron) of a tortoise shell*

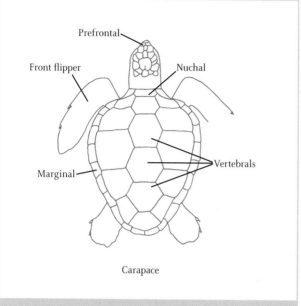

3. *The upper shield (carapace) of a turtle shell*

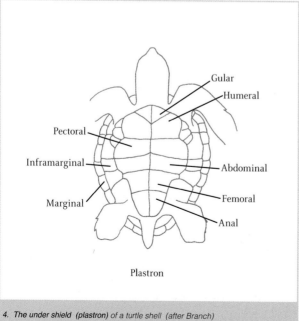

4. *The under shield (plastron) of a turtle shell (after Branch)*

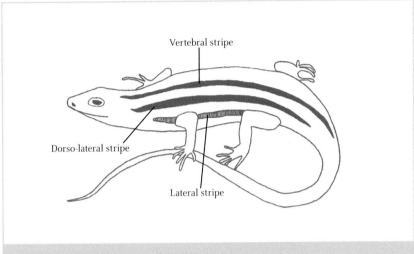

5. *Lizard Stripes*

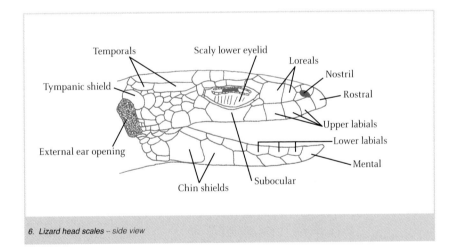

6. *Lizard head scales – side view*

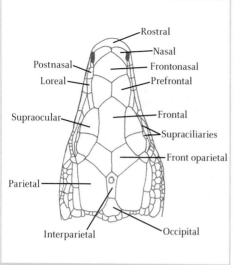

7. **Lizard head scales** - *from above*

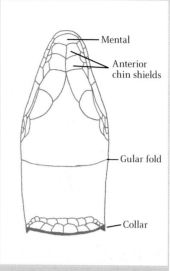

8. **Lizard head scales** – *from below*

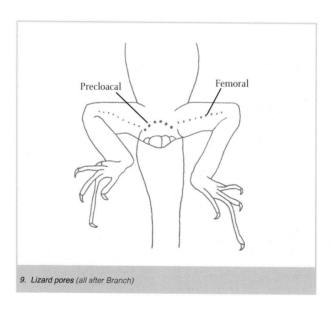

9. **Lizard pores** *(all after Branch)*

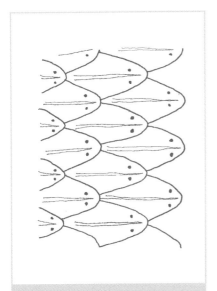

10. **Keeled imbricate lizards scales** with pores

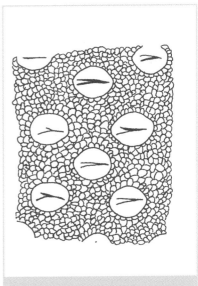

11. **Granular scales** with enlarged tubercular scales

12. **Mucronate scales**

13. **Cycloid scales**, tricarinate and smooth (all after Branch)

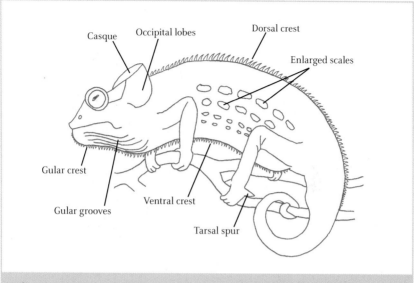

14. Chameleon anatomy (after de Witte)

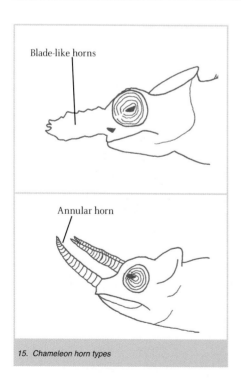

15. Chameleon horn types

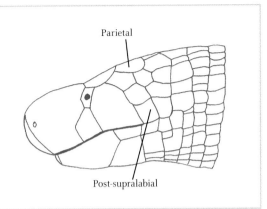

16. Head scales of Chirindia, side view

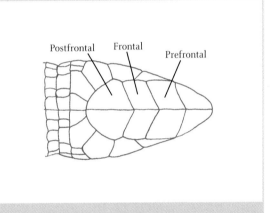

17. Head scales of Loveridgea, from above

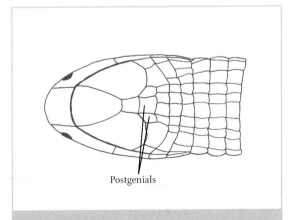

18. Head scales of Chirindia, from below

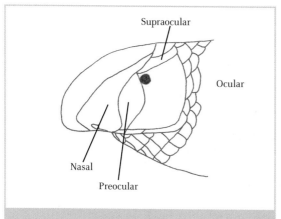

19. *Head scales of a blind snake,* side view

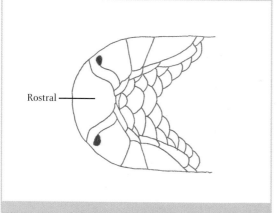

20. *Head scales of a blind snake,* from below

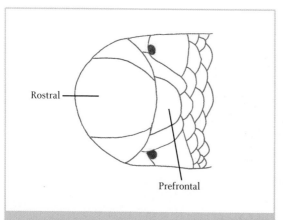

21. *Head scales of a blind snake,* from above

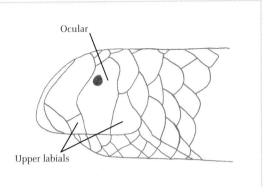

22. Head scales of a worm snake, side view

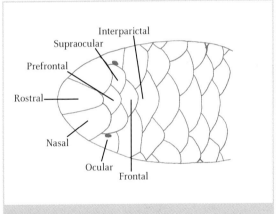

23. Head scales of a worm snake, from above

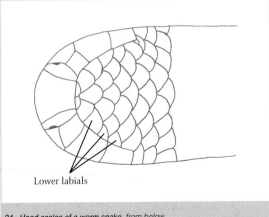

24. Head scales of a worm snake, from below

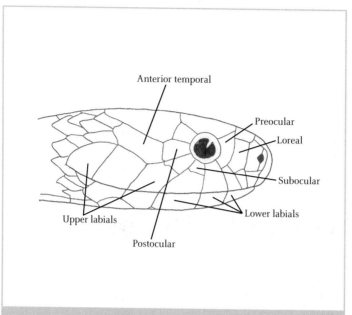

25. Head scales of a colubrid snake, side view

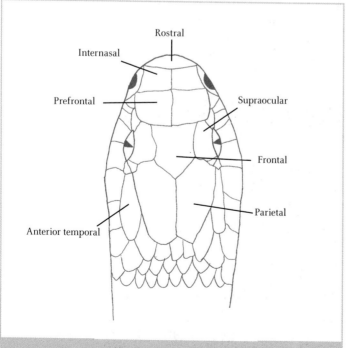

26. Head scales of a colubrid snake, from above

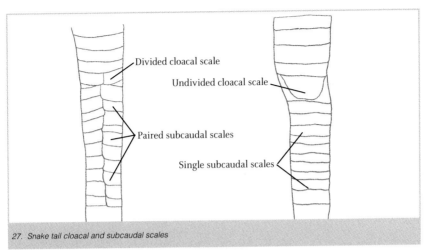

27. Snake tail cloacal and subcaudal scales

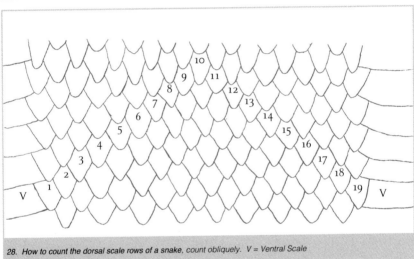

28. How to count the dorsal scale rows of a snake, count obliquely. V = Ventral Scale

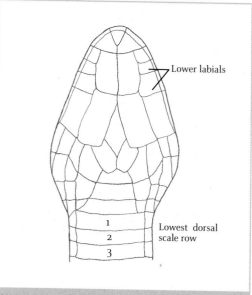

29. *How to start the ventral count for a snake*; start with the fist ventral that touches the lowest dorsal scale row.

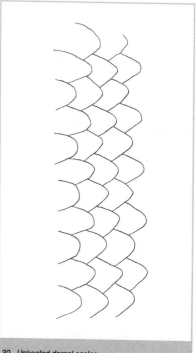

30. *Unkeeled dorsal scales*

31. *Keeled dorsal scales*

INDEX

A

Acanthocercus annectans 221–2
  atricollis 222–4

Acontias percivali 129

Adder, Forest Night 571–2
  Puff 576–7
  Rhombic Night 574–5
  Snouted Night 570–1
  Two-striped Night 569
  Velvety-Green Night 573–4
  West African Night 572–3

Adolfus africanus 181
  alleni 182
  jacksoni 182–4
  masavaensis 184

Afrotyphlops angolensis 343–4
  brevis 344–5
  calabresii 345
  congestus 345–6
  gierrai 346–7
  kaimosae 347
  lineolatus 348
  mucruso 348–9
  nanus 349–50
  nigrocandidus 350–1
  platyrhynchus 351
  punctatus 351–2
  rondoensis 352
  tanganicanus 352–3
  usambaricus 353–4

Agama, Blue-headed Tree 222–4
  Brian Finch's 229–30
  Dodoma Rock 227–8
  Elmenteita Rock 233–4
  Eritrean Rock 221–2
  Hanang 230–1
  Kakamega 231–2
  Kirk's Rock 238
  Malaba Rock 229–30
  Montane Rock 239–40
  Mozambique 240
  Mwanza Flat-headed 232–3
  Ngong 228
  Red-headed Rock 225–7
  Ruppell's 235–7
  Somali Painted 234–5
  Tropical Spiny 237–8

Agama amata 237–8
  caudospinosa 233–4
  dodomae 227–8
  finchi 229–30
  hulbertorum 228
  kaimosae 231–2
  kirkii 238
  lionotus 225–7
  montana 239–40
  mossambica 240
  mwanzae 232–3
  persimilis 234–5
  ruppelli 235–7
  turuensis 230–1

Aldabrachelys gigantea 36–7

Amblyodipsas dimidiata 467–8
  katangensis 468
  polylepis 468–9
  teitana 470
  unicolor 470–1

Ancyclocranium barkeri 323–4
  ionidesi 324

Aparallactus capensis 451–2
  guentheri 452–3
  jacksonii 453–4
  lunulatus 454–5
  modestus 455–6
  turneri 456–7
  werneri 458

Asp, Bibron's Burrowing 473–4
  Eastern Small-scaled Burrowing 475–6
  Engdahl's Burrowing 474–5
  Slender Burrowing 472–3
  Variable Burrowing 476

Atheris acuminata 586
  barbouri 585–6
  ceratophora 586–7
  desaixi 589–90
  hispida 590–1
  matildae 588–9
  nitschei 591–2
  rungweensis 592–3
  squamigera 593–4

Atractaspis aterrima 472–3
  bibronii 473–4
  engdahli 474–5
  fallax 475–6
  irregularis 476

B

Bitis arietans 576–7
  gabonica 577–9
  nasicornis 579–81
  worthingtoni 581–2

Boa, Kenya Sand 376–7

Boaedon fuliginosus 391–3
  olivaceus 393–4
  radfordi 394

Boomslang 503–5

Bothropthalmus lineatus 388–9

Broadleysaurus major 214–15

Buhoma depressiceps 414–15
  procterae 415–16
  vauerocegae 416–17

Burrowing-snake, Butler's Black-and-yellow 465
  Gerard's Black-and-yellow 465–6

C

Calotes versicolor 241

Caretta caretta 41–2

Causus bilineatus 569
  defilippii 570–1
  lichtensteinii 571–2
  maculatus 572–3
  resimus 573–4
  rhombeatus 574–5

Centipede Eater, Black 452–3
  Cape 451–2
  Jackson's 453–4
  Malindi 456–7
  Plumbeous 454–5
  Usambara 458
  Western Forest 455–6

Chalcides bottegi 166–7

Chamaeleo anchietae 258
  dilepis 258–60
  gracilis 260–1
  laevigatus 261–2

Chamaelycus fasciatus 404

Chamaesaura miopropus 206–7
  tenuior 207–8

Chameleon, Aberdare Mountains Dwarf 295
  Angola 258
  Bearded Pygmy 246–7
  Beardless Pygmy 245–6
  Boulenger's Pygmy 250–1
  Carpenter's 267
  Crater Highlands Side-striped 304–5
  Eastern Arc Sharp-nosed 272–3
  Flap-necked 258–60
  Giant East Usambara Blade-horned 270–1
  Giant One-horned 298–9
  Goetze's Whistling 287–8
  Graceful 260–1
  Hanang Hornless 276–7
  High-casqued 289–90
  Ituri 264
  Ituri Forest 291–2
  Jackson's 292–3
  Kenya Pygmy 247–8
  Kikuyu Three-horned 292–3
  Magombera Forest 269–70
  Mahenge Pygmy 250
  Montane Side-striped 286
  Msuya's Forest 271
  Mt Hanang Dwarf 288–9
  Mt Kenya Hornless 267–8
  Mt Kenya Side-striped 303–4
  Mt Kilimanjaro Two-horned 273–4
  Mt Kulal Stump-nosed 299–300
  Mt Marsabit 297
  Mt Mtelo Stump-nosed 301
  Mt Nyiro Bearded 265
  Mt Nyiro Montane 300
  Nchisis Pygmy 253
  Nguru Blade-horned 268–9
  Nguru Spiny Pygmy 249–50
  Pare Pygmy 256–7
  Pitless Pygmy 253
  Pokot 301
  Poroto Single-horned 277–8
  Poroto Three-horned 287
  Rosette-nosed 253–4
  Rwenzori Plate-nosed 279–80
  Rwenzori Side-striped 301–2
  Rwenzori Three-horned 293–4
  Schouteden's Montane Dwarf 302–3

Side-striped 282-3
Slender 260-1
Smooth 261-2
South Sudanese Unicorn 283-4
Spiny-flanked 296-7
Strange-horned 279-80
Taita Hills Blade-horned 265-6
Udzungwa Double-bearded 305-6
Udzungwa Pygmy 252
Ukinga Hornless 290-1
Uluguru One-horned 272-3
Uluguru Pygmy 255-6
Uluguru Two-horned 276
Usambara Flap-nosed 274-5
Usambara Pitted Pygmy 255
Usambara Soft-horned 274-5
Usambara Spiny Pygmy 253-4
Usambara Three-horned 285
Von Höhnel's 289-90
Vosseler's Blade-horned 278
Werner's Three-horned 306-7
West Usambara Blade-horned 272

Chelonia mydas 43-4

Chilorhinophis butleri 465
    gerardi 465-6
Chirindia ewerbecki 320-1
    mwapwaensis 321
    rondoensis 322
    swynnertoni 322-3

Chondrodactylus turneri 117-18

Cnemaspis africana 77-8
    barbouri 79
    dickersonae 79-80
    elgonensis 80-1
    quattuorseriata 81-2
    uzungwae 82
Cobra, Ashe's Spitting 557-8
    Banded Water 549-51
    Black-necked Spitting 555-7
    Egyptian 551-2
    Forest 552-4
    Gold's Tree 559-61
    Large Brown Spitting 557-8
    Mozambique Spitting 554-5
    Red Spitting 558-9
    Storm's Water 549-51

Congolacerta vauereselli 185

Cordylus beraduccii 210-11
    tropidosternum 211-12

ukingensis 212

Crocodile, Dwarf/Broad-fronted 333-4
    Nile 329-32
    Slender-snouted 327-9

Crocodylus niloticus 329-32

Crotaphopeltis braestrupi 523-4
    degeni 524-5
    hotamboeia 525-6
    tornieri 526

Cryptoblepharis africanus 163-4

Cycloderma frenatum 48-9

D

Dasypeltis atra 514-16
    confusa 517-18
    fasciata 518-19
    medici 519-20
    scabra 516

Dendroaspis angusticeps 561-2
    jamesoni 562-4
    polylepis 564-5

Dermochelys coriacea 45-6

Dipsadoboa aulica 527-8
    flavida 528-9
    shrevei 529-30
    unicolor 530
    viridis 530-1
    weileri 531-2
    werneri 532

Dispholidus typus 503-5

Duberria lutrix 418-19

E

Ebenavia inunguis 75

Echis pyramidum 582-4
Egg Eater, Common 516
    Confusing 517-18
    East African 519-20

Montane 514–16
Rhombic 516
Rufous 519–20
Western Forest 518–19

Elapsoidea boulengeri 544–5
  laticincta 545–6
  loveridgei 546–7
  nigra 547–8

Elasmodactylus tetensis 119
  tuberculosus 119–20

Epacrophis boulengeri 365
  drewesi 365–6

Eretmochelys imbricata 42–3

Eryx colubrinus 376–7

Eumecia anchietae 130–1

F

Feylinia currori 169–70

G

Gastropholis prasina 186
  vittata 187–8

Gecko, Angolan Dwarf 100
  Angulate Dwarf 100–1
  Archer's Post 88–9
  Banded Velvet 121
  Barbieri's Turkana 86–7
  Barbour's 85–6
  Boulenger's 91–2
  Broadley's Dwarf 101
  Bunty's Dwarf 103–4
  Cape Dwarf 102
  Chevron-throated Dwarf 105–6
  Conradt's Dwarf 103
  Copal Dwarf 106–7
  Cross-marked Sand/Sempahore 71–2
  Dar Es Salaam Dwarf 107
  Dickerson's Forest 79–80
  Dull Green Day 123–4
  Dutumi 97–8
  East African House 84–5
  Electric Blue Dwarf 115–16
  Elegant 76
  Elgon Forest 80–1
  Four-lined Forest 81–2
  Grote's Dwarf 105
  Howell's Dwarf 106–7
  Jens Rasmussen's Tail-pad 126
  Kaya 92–3
  Kenya Dwarf 107–8
  Kim Howell's Dwarf 108–9
  Lanza's 89–90
  Madagascan Clawless 75
  Mann's Dwarf 110
  Mt Kenya Dwarf 116–17
  Nyika 96–7
  Ogaden 98
  Pacific 88
  Pemba Day 122–3
  Prince Ruspoli's 96
  Richardson's Forest 94–5
  Scheffler's Dwarf 113
  Scortecci's Dwarf 113–14
  Side-spotted Dwarf 109–10
  Somali Banded 87
  Somali Dwarf 114–15
  Somali-Maasai Clawed 70
  Somali Plain 95
  Tana River 92
  Tete Thick-toed 119
  Tree 93–4
  Tropical House 90–1
  Tsavo Dwarf 115
  Tuberculate Thick-toed 119–20
  Turner's Thick-toed 117–18
  Turquoise Dwarf 115–16
  Udzungwa Forest 82
  Udzungwa Tail-pad 126
  Uluguru Forest 79
  Uluguru Tail-pad 124–5
  Uniform-scaled 89–90
  Usambara Dwarf 104–5
  Usambara Forest 77–8
  White-headed Dwarf 110–11
  Yellow-headed Dwarf 111–12
  Zebra Dwarf 108–9

Geocalamus acutus 317–18
  modestus 318

Gerrhosaurus flavigularis 216
  nigrolineatus 217

Gonionotophis brussauxi 406–7
  capensis 407–8
  chanleri 408–9
  nyassae 409–10

poensis 410-11
savorgnani 409
stenophthalmus 411-12

Grayia smythii 536-7
tholloni 537-8

H

Hapsidophrys lineatus 498-9
smaragdinus 499-500

Heliobolus neumanni 197-8
nitidus 198
spekii 199

Hemidactylus angulatus 84-5
barbierii 86-7
barbouri 85-6
bavazzanoi 87
frenatus 88
funaiolli 88-9
isolepsis 89-90
mabouia 90-1
macropholis 91-2
modestus 92
mrimaensis 92-3
platycephalus 93-4
richardsoni 94-5
robustus 95
ruspolii 96
squamulatus 96-7
tanganicus 97-8
tropidolepsis 98

Hemirhagerrhis hildebrandtii 428
kelleri 427
nototaenia 428-9

Holaspis guentheri 188-9
laevis 189-91

Holodactylus africanus 70

Homopholis fasciata 121

Hormonotus modestus 404-5

Hydrophis platurus 566

I

Ichnotropis bivittata 202-3
tanganicana 203

Indotyphlops braminus 354-5

K

Kinixys belliana 32
erosa 30-1
spekii 33-4
zombensis 34

Kinyongia adolfifriderici 264
asheorum 265
boehmei 265-6
carpenteri 267
excubitor 267-8
fischeri 268-9
magomberae 269-70
matschiei 270-1
msuyae 271
multituberculata 272
oxyrhina 272-3
tavetana 273-4
tenuis 274-5
uluguruensis 276
uthmoelleri 276-7
vanheygeni 277-8
vosseleri 278
xenorhina 279-80

L

Latastia caeruleopunctata 194-5
johnstoni 195-6
longicaudata 196-7

Lepidochelys olivacea 40-1

Lepidothyris hinkeli 155-6

Leptosiaphos aloysiisabaudiae 158
blochmanni 158-9
graueri 159-60
hackarsi 160
kilimensis 160-1
meleagris 161-2
rhomboidalis 162-3

Leptotyphlops aethiopicus 366
  emini 366-7
  howelli 367
  keniensis 367-8
  latirostris 368
  macrops 368-9
  mbanjensis 369
  merkeri 369-70
  nigroterminus 370
  parkeri 370-1
  pembae 371-2
  pitmani 372-3
  tanae 373

Letheobia gracilis 356
  graueri 356-7
  lumbriciformis 357
  mbeerensis 358
  pallida 358
  pembana 359
  swahilica 359-60
  uluguruensis 360

Lizard, Angolan Rough-scaled 202-3
  Barker's Sharp-snouted Worm 323-4
  Black-lined Plated 217
  Blue-spotted Long-tailed 194-5
  Boulenger's Scrub 192-3
  East African Highland Grass 207-8
  Eastern Blue-tailed Gliding 189-91
  Ewerbeck's Round-headed Worm 320-1
  Glittering Sand 198
  Great Plated 214-15
  Green Keel-bellied 186
  Indian Garden 241
  Ionides' Sharp-snouted Worm 324
  Jackson's Forest 182-4
  Johnston's Long-tailed 195-6
  Liwale Round-snouted Worm 319
  Maasai Girdled 210-11
  Malawi Long-tailed 195-6
  Mozambique Rough-scaled 203-4
  Mpwapwa Round-headed Worm 321
  Mpwapwa Wedge-snouted Worm 318
  Mt Kenya Alpine-Meadow 182
  Multi-scaled Forest 181
  Neumann's Sand 197-8
  Ornate Scrub 193-4
  Parker's Long-tailed 194-5
  Peters' Sand 200
  Rondo Round-headed Worm 322
  Smith's Sand 201
  Southern Long-tailed 196-7
  Sparse-scaled Forest 185
  Speke's Sand 199
  Spotted Flat 208-9
  Striped Keel-bellied 187-8
  Swynnerton's Round-headed Worm 322-3
  Tanzanian Rough-scaled 203
  Tropical Girdled 211-12
  Turkana Shield-backed Ground 191-2
  Ujiji Round-snouted Worm 319-20
  Ukinga Girdled 212
  Voi Wedge-snouted Worm 317-18
  Western Alpine-Meadow 184
  Western Blue-tailed Gliding 188-9
  Yellow-throated Plated 216
  Zambian Snake/Grass 206-7

Loveridgea ionidesi 319
  phylofiniens 319-20

Lycodonomorphus bicolor 389-90
  whytii 390

Lycophidion acutirostris 396
  capense 396-7
  depressirostre 398
  laterale 398-9
  meleagre 399-400
  multimaculatum 400
  ornatum 401
  pembanum 401-2
  taylori 402-3
  uzungwense 403

Lygodactylus angolensis 100
  angularis 100-1
  broadleyi 101
  capensis 102
  conradti 103
  grandisonae 103-4
  gravis 104-5
  grotei 105
  gutturalis 105-6
  howelli 106-7
  inexpectatus 107
  keniensis 107-8
  kimhowelli 108-9
  laterimaculatus 109-10
  manni 110
  mombasicus 110-11
  picturatus 111-12
  scheffleri 113
  scorteccii 113-14
  somalicus 114-15
  tsavoensis 115
  williamsi 115-16
  wojnowskii 116-17

M

Malacochersus tornieri 35-6
Mamba, Black 564-5
  Green 561-2
  Jameson's 562-4

Mecistops cataphractus 327-9

Meizodon krameri 480-1
  plumbiceps 481
  regularis 481-2
  semiornatus 482-3

Melanoseps ater 171
  emmrichi 171-2
  longicauda 172
  loveridgei 173
  pygmaeus 173-4
  rondoensis 174
  uzungwensis 174-5

Meroles squamulosa 203-4

Micrelaps bicoloratus 458-9
  vaillanti 459-60

Mochlus afer 150-1
  mabuiiforme 152-3
  mafianum 153
  pembanum 153-4
  somalicum 154-5
  sundevalli 151-2
  tanae 155

Monitor, Nile 309-10
  Water 309-10
  Western Savanna 313-14
  White-throated Savanna 311-12

Montatheris hindii 594-6

Myriophilis braccianii 373
  ionidesi 374
  macrorhyncha 374

N

Naja annulata 549-51
  ashei 557-8
  haje 551-2
  melanoleuca 552-4
  mossambica 554-5
  nigricollis 555-7
  pallida 558-9

Natriciteres olivacea 539-40
  pembana 540-1
  sylvatica 540

Nucras boulengeri 192-3
  ornata 193-4

O

Osteolaemus tetraspis 333-4

P

Panaspis megalurus 164-5
  wahlbergi 165

Pelomedusa kobe 51-2
  neumanni 51-2
  subrufa 51-2
Pelusios adansoni 53-4
  broadleyi 54-5
  castanoides 55-6
  chapini 56
  gabonensis 57
  nanus 58-9
  rhodesianus 59-60
  sinuatus 60-1
  subniger 61-3
  williamsi 63-4

Phelsuma dubia 123-4
  parkeri 122-3

Philochortus rudolfensis 191-2

Philothamnus angolensis 484-5
  battersbyi 485-6
  bequaerti 487
  carinatus 487-8
  heterodermus 488
  heterolepidotus 489
  hoplogaster 489-90
  hughesi 491
  macrops 491-2
  nitidus 492-3
  ornatus 493-4
  punctatus 494-5

ruandae 495–6
semivariegatus 496–7

Platyceps brevis (smithi) 479–80
florulentus 478–9

Platysaurus maculatus 208–9

Polemon christyi 461
collaris 461–2
fulvicollis 463–4
gabonensis 464
graueri 462–3

Pristurus crucifer 71–2

Proatheris superciliaris 596–7

Proscelotes eggeli 175

Prosymna ambigua 420–1
greigerti 421–2
ornatissima 422
pitmani 423
ruspolii 423–4
semifasciata 424–5
stuhlmanni 425–6

Psammophis angolensis 431–2
biseriatus 439–40
langanicus 440–1
lineatus 433–4
mossambicus 434–5
orientalis 435–6
pulcher 432–3
punctulatus 438–9
rukwae 436–7
sudanensis 437–8

Psammophylax acutus 445
multisquamis 442–3
togoensis 445–6
tritaeniatus 443–4
variabilis 444

Pseudaspis cana 412–14

Pseuderemias smithii 201
striata 200

Pseudohaje goldii 559–61

Python, Central African Rock 380–1
Royal 382
Southern African Rock 379–80

Python natalensis 379–80
regius 382
sebae 380–1

R

Racer, Flowered 478–9
Smith's 479–80

Rhamnophis aethiopissa 502–3

Rhamphiophis oxyrhynchus 447–8
rostratus 448–9
rubropunctatus 449–50

Rhampholeon acuminatus 249–50
beraduccii 250
boulengeri 250–1
moyeri 252
nchisiensis 253
spinosus 253–4
temporalis 255
uluguruensis 255–6
viridis 256–7

Rhinotyphlops ataeniatus 361
unitaeniatus 362

Rieppeleon brachyurus 245–6
brevicaudatus 246–7
kerstenii 247–8

S

Scaphiophis albopunctatus 534–5
raffreyi 535–6
Scelotes uluguruensis 176–7
Scolecoseps sp. 178
acontias 177
litipoensis 178

Seps, Ellenberger's Long-tailed 218–19
Udzungwa Long-tailed 219

Sepsina tetradactyla 167–8

Shovel-snout, Angolan 420–1
Banded 424–5
East African 425–6
Miombo 420–1
Ornate 422

Pitman's 423
Prince Ruspoli's 423–4
Southern Sahel Speckled 421–2

Skaapsteker, Grey-bellied 444
    Kenyan Striped 442–3
    Northern Sharp-nosed 445–6
    Southern Sharp-nosed 445
    Southern Striped 443–4

Skink, Alpine Meadow 139
    Bayon's 134–5
    Bi-coloured 138–9
    Black Limbless 171
    Blue-tailed Snake-eyed 164–5
    Boulenger's 135–6
    Coral Rag 163–4
    Five-lined 146–7
    Four-toed Fossorial 167–8
    Gede Sand 178
    Grass Top 143–4
    Hinkel's Red-flanked 155–6
    Katavi Blind Dart 168–9
    Kilimanjaro Five-toed 160–1
    Kivu Three-toed 158–9
    Legless Sand 177
    Litipo Sand 178
    Lolui Island 140
    Long-tailed 143–4
    Long-tailed Limbless 172
    Loveridge's Limbless 173
    Mabuya-like Writhing 152–3
    Mafia Writhing 153
    Ocellated 166–7
    Orange-flanked 144–5
    Pemba Island Writhing 153–4
    Pemba Speckle-lipped 133–4
    Percival's Legless 129
    Peter's Writhing 150–1
    Pigmy Limbless 173–4
    Rainbow 142–3
    Rondo Limbless 174
    Rwanda Five-toed/Four-toed 159–60
    Rwenzori Four-toed 161–2
    Short-necked 137–8
    Somali Writhing 154–5
    Speckle-lipped 141–2
    Striped 147–8
    Sundevall's Writhing 151–2
    Tana River Writhing 155
    Tree 145–6
    Udzungwa Five-toed 162–3
    Udzungwa Mountain Limbless 174–5
    Uganda Five-toed 158
    Ukinga Mountain 136–7

    Uluguru Fossorial 176–7
    Uluguru Limbless 171–2
    Usambara Five-toed Fossorial 175
    Variable 148–9
    Virunga Four-toed 160
    Wahlberg's Snake-eyed 165
    Western Forest Limbless 169–70
    Western Serpentiform 130–1
    White-lipped Forest 133
    Zebra 138–9

Slug Eater 418–19

Snake, Angle-snouted Blind 344–5
    Angola Blind 343–4
    Angola Green 484–5
    Battersby's Green 485–6
    Beautiful Sand 432–3
    Bicoloured Blind 350–1
    Black File 409–10
    Black-headed 459–60
    Black-headed Smooth 481
    Black-lined Green 498–9
    Black-tipped Worm 370
    Blanding's Tree 511–12
    Blotched Blind 345–6
    Blotched Wolf 400
    Boulenger's Garter 544–5
    Brahminy Blind 354–5
    Brown House 391–3
    Cape File 407–8
    Cape Wolf 396–7
    Central African Garter 545–6
    Chanler's File 408–9
    Common Purple-glossed 468–9
    Congo/De Brazza's File 409
    Dagger-tooth Vine 509–10
    Degen's Water 524–5
    Drewes' Worm 365–6
    Dwarf File 409–10
    Dwarf Sand 431–2
    East African Garter 546–7
    Eastern-crowned 481–2
    Eastern Stripe-bellied Sand 435–6
    Eastern Vine 507–8
    Emerald 499–500
    Emin Pasha's Worm 366–7
    Ethiopian Hook-nosed 535–6
    Ethiopian Worm 366
    Flat-snouted Wolf 398
    Flower Pot Blind 354–5
    Forest File 410–11
    Forest Green 488
    Forest Marsh 540
    Forest Vine/Twig 505–7

Forest Wolf 401
Gierra's Blind 346–7
Grauer's Gracile Blind 356–7
Guinea-fowl 459–60
Hissing Sand 434–5
Hook-nosed 534–5
Hook-snouted Worm 374
Hughes' Green 491
Ionides' Purple-glossed 468
Ionides' Worm 374
Ituri Banded 404
Jackson's Tree 500–1
Kakamega Blind 347
Kenyan Bark 428
Kenyan Dwarf Blind 349–50
Kim Howell's Worm 367
Lake Rukwa Sand 436–7
Lake Tanganyika Water 389–90
Lamu Worm 365
Large-eyed 520–1
Large-eyed Green Tree 502–3
Large-eyed Worm 368–9
Lineolate Blind 348
Link-marked Sand 439–40
Liwale Blind 352–3
Loveridge's Green 492–3
Mbanja Worm 369
Mbeere Gracile Blind 358
Merker's Worm 369–70
Meru Tree 501–2
Mocquard's File 406–7
Mole 412–14
Mozambique Wolf 396
Mpwapwa Purple-glossed 467–8
Mt Kenya Worm 367–8
Northern Stripe-bellied Sand 437–8
Olive House 393–4
Olive Marsh 539–40
Olive Sand 434–5
Pale-headed Forest 414–15
Pallid Gracile Blind 358
Parker's Worm 370–1
Pemba Gracile Blind 359
Pemba Marsh 540–1
Pemba Wolf 401–2
Pemba Worm 371–2
Pitman's Worm 372–3
Powdered Tree 512–13
Radford's House 394
Red and Black Striped 388–9
Red-snouted Wolf 403
Red-spotted Beaked 449–50
Rondo Blind 352
Rufous Beaked 448–9
Rwandan Green 495–6

Scortecci's Worm 373
Semi-ornate 482–3
Slender Gracile Blind 356
Slender Green 489
Small-eyed File 411–12
Smyth's Water 536–7
Somali Blind 361
South-eastern Bark 428–9
South-eastern Green 489–90
Southern Wedge-snouted Blind 345
Speckled Green 494–5
Speckled Sand 438–9
Speckled Wolf 399–400
Spotted Blind 351–2
Spotted Bush/Wood 496–7
Stripe-backed Green 493–4
Striped Bark 427
Striped Olympic Sand 433–4
Swahili Gracile Blind 359–60
Taita Hills Purple-glossed 470
Tana Delta Smooth 480–1
Tana Delta Worm 373
Tana Herald 523–4
Tanga Blind 351
Tanganyika Blind 352–3
Tanganyika Gracile Blind 356–7
Tanganyika Sand 440–1
Taylor's Wolf 402–3
Thirteen-scaled Green 487–8
Thollon's Water 537–8
Tiger 522
Tornier's Cat 526
Two-coloured 458–9
Uganda Green 487
Uluguru Forest 415–16
Uluguru Gracile Blind 360
Usambara Blotched Blind 353–4
Usambara Forest 416–17
Usambara Garter 547–8
Usambara Green 491–2
Usambara Vine 508–9
Uvira Worm 368
Western Beaked 447–8
Western Forest Wolf 398–9
Western Purple-glossed 470–1
White-lipped 525–6
Whyte's Water 390
Worm-like/Zanzibar Gracile Blind 357
Yellow Forest 404–5
Yellow-bellied Sea 566
Yellow-flanked 524–5
Yellow-striped Blind 362
Zambezi Giant Blind 348–9

Snake-eater, Christy's 461

Fawn-headed 461–2
Gabon 464
Pale-collared 462–3
Yellow-necked 463–4

Stenodactylus sthenodactylus 76

Stigmochelys pardalis 28–30

T

Telescopus dhara 520–1
    semiannulatus 522

Terrapin, Adanson's Hinged 53–4
    Black-bellied Hinged 61–3
    Congo Hinged 56
    Dwarf Hinged 58–9
    Forest Hinged 57
    Helmeted 51–2
    Lake Turkana Hinged 54–5
    Neumann's Marsh 51–2
    Pan Hinged 61–3
    Serrated Hinged 60–1
    Tanzanian Marsh 51–2
    Williams' Hinged 63–4
    Yellow-bellied Hinged 55–6
    Zambian Hinged 59–60

Tetradactylus ellenbergeri 218–19
    udzungwensis 219

Thelotornis kirtlandii 505–7
    mossambicanus 507–8
    usambaricus 508–9

Thrasops jacksoni 500–1
    schmidti 501–2

Tortoise, Aldabran 36–7
    Bell's Hinged 32
    Eastern Hinged 34
    Flat 35–6
    Forest Hinged 30–1
    Leopard 28–30
    Pancake 35–6
    Speke's Hinged 33–4

Toxicodryas blandingii 511–12
    pulverulenta 512–13

Trachylepis albilabris 133
    albotaeniata 133–4

bayoni 134–5
boulengeri 135–6
brauni 136–7
brevicollis 137–8
dichroma 138–9
irregularis 139
loluiensis 140
maculilabris 141–2
margaritifer 142–3
megalura 143–4
perroteti 144–5
planifrons 145–6
quinquetaeniata 146–7
striata 147–8
varia 148–9

Tree-snake, Black-tailed 531–2
    Cross-barred 528–9
    Gunther's Green 530
    Laurent's Green 530–1
    Marbled 527–8
    Shreve's 529–30
    Werner's 532

Trioceros bitaeniatus 282–3
    conirostratus 283–4
    deremensis 285
    ellioti 286
    fuelleborni 287
    goetzei 287–8
    hanangensis 288–9
    hoehnelii 289–90
    incornutus 290–1
    ituriensis 291–2
    jacksoni 292–3
    johnstoni 293–4
    kinangopensis 295
    laterispinis 296–7
    marsabitensis 297
    melleri 298–9
    narraioca 299–300
    ntunte 300
    nyirit 301
    rudis 301–2
    schoutedeni 302–3
    schubotzi 303–4
    sternfeldi 304–5
    tempeli 305–6
    werneri 306–7

Trionyx triunguis 47–8

Turtle, Green 43–4
    Hawksbill 42–3
    Leatherback 45–6

Loggerhead 41–2
Nile Soft-shelled 47–8
Olive Ridley 40–1
Zambezi Flap-shelled/Soft-shelled 48–9

Typhlacontias kataviensis 168–9

U

Urocotyledon rasmusseni 126
wolterstorffi 124–5

V

Varanus albigularis 311–12
exanthematicus 313–14
niloticus 309–10

Viper, Acuminate 586
Barbour's Short-headed 585–6
Floodplain 596–7
Gaboon 577–9
Great Lakes Bush 591–2
Green Bush 593–4
Horned Bush 586–7
Kenya Horned 581–2
Kenya Montane 594–6
Matilda's Bush 588–9
Mt Kenya Bush 589–90
Mt Rungwe Bush 592–3
North-east African Carpet 582–3
Rhinoceros/Rhino-horned 579–81
Rough-scaled Bush 590–1
Saw-scaled 582–4
Udzungwa 585–6
Usambara Bush 586–7

W

White-lip 525–6

X

Xyelodontophis uluguruensis 509–10